Student Solutions Manual

FOR

Kaufmann and Schwitters'

ELEMENTARY
ALGEBRA

SEVENTH EDITION

Student Solutions Manual

FOR

Kaufmann and Schwitters'

ELEMENTARY ALGEBRA

Seventh Edition

Karen Schwitters
Seminole Community College

Laurel Fischer

Jesse Turner
Seminole Community College

THOMSON

BROOKS/COLE

Australia • Canada • Mexico • Singapore • Spain • United Kingdom • United States

ISBN: 0-534-40043-4

Printer: Globus Printing

For more information about our products,
contact us at:
Thomson Learning Academic Resource Center
1-800-423-0563

For permission to use material from this text,
contact us by:
Phone: 1-800-730-2214
Fax: 1-800-731-2215
Web: http://www.thomsonrights.com

Cover Designer: Roger Knox
Cover Image: Rowan Moore

For more information contact:
Brooks/Cole-Thomson Learning, Inc.
10 Davis Drive
Belmont, CA 94002-3098
USA

Asia
Thomson Learning
5 Shenton Way #01-01
UIC Building
Singapore 068808

Australia/ New Zealand
Thomson Learning
102 Dodds Street
Southbank, Victoria 3006
Australia

Canada
Nelson
1120 Birchmount Road
Toronto, Ontario M1K 5G4
Canada

Europe/Middle East/South Africa
Thomson Learning
High Holborn House
50/51 Bedford Row
London WC1R 4LR
United Kingdom

Latin America
Thomson Learning
Seneca, 53
Colonia Polanco
11560 Mexico D.F.
Mexico

Spain/ Portugal
Paraninfo
Calle/Magallanes, 25
28015 Madrid, Spain

Table of Contents

Chapter 11: Additional Topics

Chapter 1 Some Basic Concepts of Arithmetic and Algebra

PROBLEM SET **1.1** Numerical and Algebraic Expressions

1. $9 + 14 - 7$
$23 - 7$
16

3. $7(14 - 9)$
$7(5)$
35

5. $16 + 5 \cdot 7$
$16 + 35$
51

7. $4(12 + 9) - 3(8 - 4)$
$4(21) - 3(4)$
$84 - 12$
72

9. $4(7) + 6(9)$
$28 + 54$
82

11. $6 \cdot 7 + 5 \cdot 8 - 3 \cdot 9$
$42 + 40 - 27$
$82 - 27$
55

13. $(6 + 9)(8 - 4)$
$(15)(4)$
60

15. $6 + 4[3(9 - 4)]$
$6 + 4[3(5)]$
$6 + 4(15)$
$6 + 60$
66

17. $16 \div 8 \cdot 4 + 36 \div 4 \cdot 2$
$2 \cdot 4 + 36 \div 4 \cdot 2$
$8 + 36 \div 4 \cdot 2$
$8 + 9 \cdot 2$
$8 + 18$
26

19. $\dfrac{8 + 12}{4} - \dfrac{9 + 15}{8}$
$\dfrac{20}{4} - \dfrac{24}{8}$
$5 - 3$
2

21. $56 - [3(9 - 6)]$
$56 - [3(3)]$
$56 - (9)$
$56 - 9$
47

23. $7 \cdot 4 \cdot 2 \div 8 + 14$
$28 \cdot 2 \div 8 + 14$
$56 \div 8 + 14$
$7 + 14$
21

Problem Set 1.1

25. $32 \div 8 \cdot 2 + 24 \div 6 - 1$
$4 \cdot 2 + 4 - 1$
$8 + 4 - 1$
$12 - 1$
11

27. $4 \cdot 9 \div 12 + 18 \div 2 + 3$
$36 \div 12 + 18 \div 2 + 3$
$3 + 18 \div 2 + 3$
$3 + 9 + 3$
$12 + 3$
15

29. $\dfrac{6(8-3)}{3} + \dfrac{12(7-4)}{9}$
$\dfrac{6(5)}{3} + \dfrac{12(3)}{9}$
$\dfrac{30}{3} + \dfrac{36}{9}$
$10 + 4$
14

31. $83 - \dfrac{4(12-7)}{5}$
$83 - \dfrac{4(5)}{5}$
$83 - \dfrac{20}{5}$
$83 - 4$
79

33. $\dfrac{4 \cdot 6 + 5 \cdot 3}{7 + 2 \cdot 3} + \dfrac{7 \cdot 9 + 6 \cdot 5}{3 \cdot 5 + 8 \cdot 2}$
$\dfrac{24 + 15}{7 + 6} + \dfrac{63 + 30}{15 + 16}$
$\dfrac{39}{13} + \dfrac{93}{31}$
$3 + 3$
6

35. $7x + 4y$ for $x = 6$, $y = 8$
$7(6) + 4(8)$
$42 + 32$
74

37. $16a - 9b$ for $a = 3$, $b = 4$
$16(3) - 9(4)$
$48 - 36$
12

39. $4x + 7y + 3xy$ for $x = 4$, $y = 9$
$4(4) + 7(9) + 3(4)(9)$
$16 + 63 + 108$
$79 + 108$
187

41. $14xz + 6xy - 4yz$ for $x = 8$, $y = 5$, $z = 7$
$14(8)(7) + 6(8)(5) - 4(5)(7)$
$784 + 240 - 140$
$1024 - 140$
884

43. $\dfrac{54}{n} + \dfrac{n}{3}$ for $n = 9$
$\dfrac{54}{9} + \dfrac{9}{3}$
$6 + 3$
9

45. $\dfrac{y + 16}{6} + \dfrac{50 - y}{3}$ for $y = 8$
$\dfrac{8 + 16}{6} + \dfrac{50 - 8}{3}$
$\dfrac{24}{6} + \dfrac{42}{3}$
$4 + 14$
18

2

47. $(x + y)(x - y)$ for $x = 8$, $y = 3$
$(8 + 3)(8 - 3)$
$(11)(5)$
55

49. $(5x - 2y)(3x + 4y)$ for $x = 3$, $y = 6$
$[5(3) - 2(6)][3(3) + 4(6)]$
$[15 - 12][9 + 24]$
$(3)(33)$
99

51. $6 + 3[2(x + 4)]$ for $x = 7$
$6 + 3[2(7 + 4)]$
$6 + 3[2(11)]$
$6 + 3(22)$
$6 + 66$
72

53. $81 - 2[5(n + 4)]$ for $n = 3$
$81 - 2[5(3 + 4)]$
$81 - 2[5(7)]$
$81 - 2[35]$
$81 - 70$
11

55. $\dfrac{bh}{2}$, for $b = 8$, $h = 12$
$\dfrac{8(12)}{2}$
$\dfrac{96}{2}$
48

57. $\dfrac{bh}{2}$, for $b = 7$, $h = 6$
$\dfrac{7(6)}{2}$
$\dfrac{42}{2}$
21

59. $\dfrac{bh}{2}$, for $b = 16$, $h = 5$
$\dfrac{16(5)}{2}$
$\dfrac{80}{2}$
40

61. $\dfrac{Bh}{3}$, for $B = 27$, $h = 9$
$\dfrac{27(9)}{3}$
$\dfrac{243}{3}$
81

63. $\dfrac{Bh}{3}$, for $B = 25$, $h = 12$
$\dfrac{25(12)}{3}$
$\dfrac{300}{3}$
100

65. $\dfrac{Bh}{3}$, for $B = 36$, $h = 7$
$\dfrac{36(7)}{3}$
$\dfrac{252}{3}$
84

67. $\dfrac{h(b_1 + b_2)}{2}$, for $h = 17$, $b_1 = 14$, $b_2 = 6$
$\dfrac{17(14 + 6)}{2}$
$\dfrac{17(20)}{2}$
$\dfrac{340}{2}$
170

Problem Set 1.1

69. $\dfrac{h(b_1 + b_2)}{2}$, for $h = 8$, $b_1 = 17$, $b_2 = 24$

$\dfrac{8(17 + 24)}{2}$

$\dfrac{8(41)}{2}$

$\dfrac{328}{2}$

164

71. $\dfrac{h(b_1 + b_2)}{2}$, $h = 18$, $b_1 = 6$, $b_2 = 11$

$\dfrac{18(6 + 11)}{2}$

$\dfrac{18(17)}{2}$

$\dfrac{306}{2}$

153

PROBLEM SET 1.2 Prime and Composite Numbers

1. Since $8(7) = 56$, "8 divides 56" is a true statement.

3. Since $6(9) = 54$, "6 does not divide 54" is a false statement.

5. Since $8(12) = 96$, "96 is a multiple of 8" is a true statement.

7. Since there is no whole number, k, such that $4(k) = 54$, "54 is not a multiple of 4" is a true statement.

9. Since $4(36) = 144$, "144 is divisible by 4" is a true statement.

11. Since there is no whole number, k, such that $3(k) = 173$, "173 is divisible by 3" is a false statement.

13. Since $11(13) = 143$, "11 is a factor of 143" is a true statement.

15. Since there is no whole number, k, such that $9(k) = 119$, "9 is a factor of 119" is a false statement.

17. Since $3(19) = 57$ and 3 is a prime number, "3 is a prime factor of 57" is a true statement.

19. Since $2(2) = 4$, 4 is not a prime number, therefore "4 is a prime factor of 48" is a false statement.

21. Since 53 is only divisible by itself and 1, it is a prime number.

23. Since 59 is only divisible by itself and 1, it is a prime number.

25. Since $91 = 7(13)$, it is a composite number.

27. Since 89 is only divisible by itself and 1, it is a prime number.

29. Since $111 = 3(37)$, it is a composite number.

31. $26 = 2 \cdot 13$

33. $36 = 4 \cdot 9 = 2 \cdot 2 \cdot 3 \cdot 3$

35. $49 = 7 \cdot 7$

37. $56 = 8 \cdot 7 = 2 \cdot 4 \cdot 7 = 2 \cdot 2 \cdot 2 \cdot 7$

39. $120 = 3 \cdot 40 = 3 \cdot 4 \cdot 10 = 3 \cdot 2 \cdot 2 \cdot 2 \cdot 5 = 2 \cdot 2 \cdot 2 \cdot 3 \cdot 5$

41. $135 = 5 \cdot 27 = 5 \cdot 3 \cdot 9 = 5 \cdot 3 \cdot 3 \cdot 3 = 3 \cdot 3 \cdot 3 \cdot 5$

43. $12 = 2 \cdot 2 \cdot 3$ The greatest common
$16 = 2 \cdot 2 \cdot 2 \cdot 2$ factor is $2 \cdot 2 = 4$.

45. $56 = 2 \cdot 2 \cdot 2 \cdot 7$ The greatest
$64 = 2 \cdot 2 \cdot 2 \cdot 2 \cdot 2 \cdot 2$ common factor is
$2 \cdot 2 \cdot 2 = 8$.

47. $63 = 3 \cdot 3 \cdot 7$ The greatest common
$81 = 3 \cdot 3 \cdot 3 \cdot 3$ factor is $3 \cdot 3 = 9$.

49. $84 = 2 \cdot 2 \cdot 3 \cdot 7$ The greatest
$96 = 2 \cdot 2 \cdot 2 \cdot 2 \cdot 2 \cdot 3$ common factor is
     $2 \cdot 2 \cdot 3 = 12$.

51. $36 = 2 \cdot 2 \cdot 3 \cdot 3$ The greatest
$72 = 2 \cdot 2 \cdot 2 \cdot 3 \cdot 3$ common factor is
$90 = 2 \cdot 3 \cdot 3 \cdot 5$ $2 \cdot 3 \cdot 3 = 18$.

53. $48 = 2 \cdot 2 \cdot 2 \cdot 2 \cdot 3$ The greatest
$60 = 2 \cdot 2 \cdot 3 \cdot 5$ common factor is
$84 = 2 \cdot 2 \cdot 3 \cdot 7$ $2 \cdot 2 \cdot 3 = 12$.

55. $6 = 2 \cdot 3$ The least common
$8 = 2 \cdot 2 \cdot 2$ multiple is
     $2 \cdot 2 \cdot 2 \cdot 3 = 24$.

57. $12 = 2 \cdot 2 \cdot 3$ The least common
$16 = 2 \cdot 2 \cdot 2 \cdot 2$ multiple is
     $2 \cdot 2 \cdot 2 \cdot 2 \cdot 3 = 48$.

59. $28 = 2 \cdot 2 \cdot 7$ The least common
$35 = 5 \cdot 7$ multiple is
     $2 \cdot 2 \cdot 5 \cdot 7 = 140$.

61. $49 = 7 \cdot 7$ The least common
$56 = 2 \cdot 2 \cdot 2 \cdot 7$ multiple is
     $2 \cdot 2 \cdot 2 \cdot 7 \cdot 7 = 392$.

63. $8 = 2 \cdot 2 \cdot 2$ The least common
$12 = 2 \cdot 2 \cdot 3$ multiple is
$28 = 2 \cdot 2 \cdot 7$ $2 \cdot 2 \cdot 2 \cdot 3 \cdot 7 = 168$.

65. $9 = 3 \cdot 3$ The least common
$15 = 3 \cdot 5$ multiple is
$18 = 2 \cdot 3 \cdot 3$ $2 \cdot 3 \cdot 3 \cdot 5 = 90$.

69. Since all even numbers are divisible by 2, 2 is the only even number that is divisible only by itself and 1.

71. The smallest whole number, greater than 1, that would produce a remainder of 1 when divided by 2,3,4,5,6 would be the least common multiple of those numbers plus 1. The prime factors are: $2 = 2$, $3 = 3, 4 = 2 \cdot 2, 5 = 5, 6 = 2 \cdot 3$. The least common multiple is $2 \cdot 2 \cdot 3 \cdot 5 = 60$. The least common multiple plus 1 is 61.

73. The greatest common factor of x and y if x and y are nonzero whole numbers and y is a multiple of x is x, because x is a factor of x and y and x would be the largest whole number that would divide x.

75. The least common multiple of x and y if the greatest common factor of x and y is 1 would be xy, because all prime factors of x and y would need to be included in the least common multiple.

77. $76 = 2 \cdot 38 = 2 \cdot 2 \cdot 19$
Rule for 2 used twice.

79. $123 = 3 \cdot 41$ Rule for 3

81. $115 = 5 \cdot 23$ Rule for 5

83. $441 = 9 \cdot 49 = 3 \cdot 3 \cdot 7 \cdot 7$
Rule for 9
Rule for 3

85. $153 = 9 \cdot 17 = 3 \cdot 3 \cdot 17$
Rule for 9
Rule for 3

Problem Set 1.3

1. $5 + (-3) = 2$

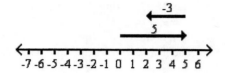

3. $-6 + 2 = -4$

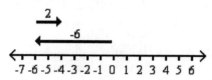

5. $-3 + (-4) = -7$

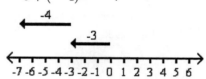

7. $8 + (-2) = 6$

9. $5 + (-11) = -6$

11. $17 + (-9) = |17| - |-9| = 17 - 9 = 8$

13. $8 + (-19) = -(|-19| - |8|) = -(19 - 8) = -11$

15. $-7 + (-8) = -(|-7| + |-8|) = -(7 + 8) = -15$

17. $-15 + 8 = -(|-15| - |8|) = -(15 - 8) = -7$

19. $-13 + (-18) = -(|-13| + |-18|) = -(13 + 18) = -31$

21. $-27 + 8 = -(|-27| - |8|) = -(27 - 8) = -19$

23. $32 + (-23) = +(|32| - |-23|) = +(32 - 23) = 9$

25. $-25 + (-36) = -(|-25| + |-36|) = -(25 + 36) = -61$

27. $54 + (-72) = -(|-72| - |54|) = -(72 - 54) = -18$

29. $-34 + (-58) = -(|-34| + |-58|) = -(34 + 58) = -92$

31. $3 - 8 = 3 + (-8) = -5$

33. $-4 - 9 = -4 + (-9) = -13$

35. $5 - (-7) = 5 + 7 = 12$

37. $-6 - (-12) = -6 + 12 = 6$

39. $-11 - (-10) = -11 + 10 = -1$

41. $-18 - 27 = -18 + (-27) = -45$

43. $34 - 63 = 34 + (-63) = -29$

45. $45 - 18 = 45 + (-18) = 27$

47. $-21 - 44 = -21 + (-44) = -65$

49. $-53-(-24)=-53+24=-29$

51. $6-8-9=6+(-8)+(-9)=-11$

53. $-4-(-6)+5-8=$
$-4+6+5+(-8)=-1$

55. $5+7-8-12=$
$5+7+(-8)+(-12)=-8$

57. $-6-4-(-2)+(-5)=$
$-6+(-4)+2+(-5)=-13$

59. $-6-5-9-8-7=$
$-6+(-5)+(-9)+(-8)+(-7)=-35$

61. $7-12+14-15-9=$
$7+(-12)+14+(-15)+(-9)=-15$

63. $-11-(-14)+(-17)-18=$
$-11+14+(-17)+(-18)=-32$

65. $16-21+(-15)-(-22)=$
$16+(-21)+(-15)+22=2$

67.
$$\begin{array}{r} 5 \\ \underline{-\ 9} \\ -4 \end{array}$$
This problem in horizontal format is $5+(-9)=-4$

69.
$$\begin{array}{r} -13 \\ \underline{-18} \\ -31 \end{array}$$
This problem in horizontal format is $(-13)+(-18)=-31$

71.
$$\begin{array}{r} -18 \\ \underline{\ \ \ 9} \\ -9 \end{array}$$
This problem in horizontal format is $(-18)+9=-9$

73.
$$\begin{array}{r} -21 \\ \underline{\ \ 39} \\ 18 \end{array}$$
This problem in horizontal format is $(-21)+39=18$

75.
$$\begin{array}{r} 27 \\ \underline{-19} \\ 8 \end{array}$$
This problem in horizontal format is $27+(-19)=8$

77.
$$\begin{array}{r} -53 \\ \underline{\ \ 24} \\ -29 \end{array}$$
This problem in horizontal format is $(-53)+24=-29$

79.
$$\begin{array}{r} 5 \\ \underline{\ 12} \\ -7 \end{array}$$
Change the sign of the bottom number and add
$$\begin{array}{r} 5 \\ \underline{-12} \\ -7 \end{array}$$

81.
$$\begin{array}{r} 6 \\ \underline{-\ 9} \\ 15 \end{array}$$
Change the sign of the bottom number and add
$$\begin{array}{r} 6 \\ \underline{\ \ 9} \\ 15 \end{array}$$

83.
$$\begin{array}{r} -7 \\ \underline{-8} \\ 1 \end{array}$$
Change the sign of the bottom number and add
$$\begin{array}{r} -7 \\ \underline{\ 8} \\ 1 \end{array}$$

85.
$$\begin{array}{r} 17 \\ \underline{-19} \\ 36 \end{array}$$
Change the sign of the bottom number and add
$$\begin{array}{r} 17 \\ \underline{\ 19} \\ 36 \end{array}$$

87.
$$\begin{array}{r} -23 \\ \underline{\ \ 16} \\ -39 \end{array}$$
Change the sign of the bottom number and add
$$\begin{array}{r} -23 \\ \underline{-16} \\ -39 \end{array}$$

89.
$$\begin{array}{r} -12 \\ \underline{\ \ 12} \\ -24 \end{array}$$
Change the sign of the bottom number and add
$$\begin{array}{r} -12 \\ \underline{-12} \\ -24 \end{array}$$

91. $x-y$ for $x=-6, y=-13$
$(-6)-(-13)$
$(-6)+13$
7

93. $-x+y-z$ for $x=3, y=-4, z=-6$
$-(3)+(-4)-(-6)$
$-3+(-4)+6$
-1

95. $-x-y-z$ for $x=-2, y=3, z=-11$
$-(-2)-(3)-(-11)=$
$2+(-3)+11=10$

97. $-x+y+z$ for $x=-11, y=7, z=-9$
$-(-11)+(7)+(-9)=$
$11+7+(-9)=9$

99. $x-y-z$ for $x=-15, y=12, z=-10$
$(-15)-(12)-(-10)=$
$(-15)+(-12)+10=-17$

101. $4+(-7)=-3$ (3 yards behind original scrimmage line)

Problem Set 1.3

103. $-4 + (-6) = -10$ (10 yards behind original scrimmage line)

105. $-5 + 2 = -3$ (3 yards behind original scrimmage line)

107. $-4 + 15 = 11$ (11 yards ahead of original scrimmage line)

109. $-12 + 17 = 5$ (5 yards ahead of original scrimmage line)

111. $60 + (-125) = -65$ (overall expense of $65)

113. $-55 + (-45) = -100$ (overall expense of $100)

115. $-70 + 45 = -25$ (overall expense of $25)

117. $-120 + 250 = 130$ (overall income of $130)

119. $145 + (-65) = 80$ (overall income of $80)

121. $-17 + 14 = -3$ (temperature at noon is $-3°F$.)

123. $(+3) + (-2) + (-3) + (-5) = -7$ (7 under par for tournament)

125. $(-2) + (+1) + (+3) + (+1) + (-2) = +1$ ($1 gain for the five-day period)

PROBLEM SET **1.4** **Integers: Multiplication and Division**

1. $5(-6) = -(|5| \cdot |-6|) = -(5 \cdot 6) = -30$

3. $\dfrac{-27}{3} = -\left(\dfrac{|-27|}{|3|}\right) = -\left(\dfrac{27}{3}\right) = -9$

5. $\dfrac{-42}{-6} = \dfrac{|-42|}{|-6|} = \dfrac{42}{6} = 7$

7. $(-7)(8) = -(|-7| \cdot |8|) = -(7 \cdot 8) = -(56) = -56$

9. $(-5)(-12) = |-5| \cdot |-12|$ $5 \cdot 12 = 60$

11. $\dfrac{96}{-8} = -\left(\dfrac{|96|}{|-8|}\right) = -\left(\dfrac{96}{8}\right) = -12$

13. $14(-9) = -(|14| \cdot |-9|) = -(14 \cdot 9) = -126$

15. $(-11)(-14) = |-11| \cdot |-14| = 11 \cdot 14 = 154$

17. $\dfrac{135}{-15} = -\left(\dfrac{|135|}{|-15|}\right) = -\left(\dfrac{135}{15}\right) = -9$

19. $\dfrac{-121}{-11} = \dfrac{|-121|}{|-11|} = \dfrac{121}{11} = 11$

21. $(-15)(-15) = |-15| \cdot |-15| = 15 \cdot 15 = 225$

23. $\dfrac{112}{-8} = -\left(\dfrac{|112|}{|-8|}\right) = -\left(\dfrac{112}{8}\right) = -14$

25. $\dfrac{0}{-8} = 0$ because $(-8)(0) = 0$

27. $\dfrac{-138}{-6} = \dfrac{|-138|}{|-6|} = \dfrac{138}{6} = 23$

8

29. $\dfrac{76}{-4} = -\left(\dfrac{|76|}{|-4|}\right) =$
$-\left(\dfrac{76}{4}\right) = -(19) = -19$

31. $(-6)(-15) = |-6| \cdot |-15| =$
$(6)(15) = 90$

33. $(-56) \div (-4) = |-56| \div |-4| =$
$56 \div 4 = 14$

35. $(-19) \div 0$ is undefined because no number times zero produces -19.

37. $(-72) \div 18 = -(|-72| \div |18|) =$
$-(72 \div 18) = -4$

39. $(-36)(27) = -(|-36| \cdot |27|) =$
$-(36 \cdot 27) = -972$

41. $3(-4) + 5(-7) =$
$-12 + (-35) = -47$

43. $7(-2) - 4(-8) = -14 - (-32) =$
$-14 + 32 = 18$

45. $(-3)(-8) + (-9)(-5) =$
$24 + 45 = 69$

47. $5(-6) - 4(-7) + 3(2) =$
$-30 - (-28) + 6 =$
$-30 + 28 + 6 = 4$

49. $\dfrac{13 + (-25)}{-3} = \dfrac{-12}{-3} = 4$

51. $\dfrac{12 - 48}{6} = \dfrac{-36}{6} = -6$

53. $\dfrac{-7(10) + 6(-9)}{-4} = \dfrac{-70 + (-54)}{-4} =$
$\dfrac{-124}{-4} = 31$

55. $\dfrac{4(-7) - 8(-9)}{11} = \dfrac{-28 - (-72)}{11} =$
$\dfrac{-28 + 72}{11} = \dfrac{44}{11} = 4$

57. $-2(3) - 3(-4) + 4(-5) - 6(-7) =$
$-6 - (-12) + (-20) - (-42) =$
$-6 + 12 + (-20) + 42 = 28$

59. $-1(-6) - 4 + 6(-2) - 7(-3) - 18 =$
$6 - 4 + (-12) - (-21) - 18 =$
$6 - 4 + (-12) + 21 - 18 = -7$

61. $7x + 5y$ for $x = -5, y = 9$
$7(-5) + 5(9) =$
$-35 + 45 = 10$

63. $9a - 2b$ for $a = -5, b = 7$
$9(-5) - 2(7) = -45 - 14 =$
$-45 + (-14) = -59$

65. $-6x - 7y$, for $x = -4, y = -6$
$-6(-4) - 7(-6) =$
$24 + 42 = 66$

67. $\dfrac{5x - 3y}{-6}$ for $x = -6, y = 4$
$\dfrac{5(-6) - 3(4)}{-6} = \dfrac{-30 - 12}{-6} =$
$\dfrac{-30 + (-12)}{-6} = \dfrac{-42}{-6} = 7$

69. $3(2a - 5b)$ for $a = -1, b = -5$
$3[2(-1) - 5(-5)] =$
$3[-2 - (-25)] =$
$3[-2 + 25] = 3(23) = 69$

71. $-2x + 6y - xy$ for $x = 7, y = -7$
$-2(7) + 6(-7) - (7)(-7) =$
$-14 + (-42) - (-49) =$
$-56 + 49 = -7$

73. $-4ab - b$ for $a = 2, b = -14$
$-4(2)(-14) - (-14) =$
$-8(-14) + 14 =$
$112 + 14 = 126$

Problem Set 1.4

75. $(ab + c)(b - c)$ for $a = -2$, $b = -3$, $c = 4$
$[(-2)(-3) + (4)][(-3) - (4)] =$
$(6 + 4)(-7) = 10(-7) = -70$

77. $F = 59$: $\dfrac{5(F - 32)}{9} = \dfrac{5[(59) - 32]}{9} =$
$\dfrac{5(27)}{9} = \dfrac{135}{9} = 15$

79. $F = 14$: $\dfrac{5(F - 32)}{9} = \dfrac{5[(14) - 32]}{9} =$
$\dfrac{5(-18)}{9} = \dfrac{-90}{9} = -10$

81. $F = -13$: $\dfrac{5(F - 32)}{9} =$
$\dfrac{5[(-13) - 32]}{9} =$
$\dfrac{5(-45)}{9} = \dfrac{-225}{9} = -25$

83. $C = 25$: $\dfrac{9C}{5} + 32 =$
$\dfrac{9(25)}{5} + 32 = \dfrac{225}{5} + 32 = 45 + 32 = 77$

85. $C = 40$: $\dfrac{9C}{5} + 32 =$
$\dfrac{9(40)}{5} + 32 = \dfrac{360}{5} + 32 = 72 + 32 = 104$

87. $C = -10$: $\dfrac{9C}{5} + 32 =$
$\dfrac{9(-10)}{5} + 32 = \dfrac{-90}{5} + 32 =$
$-18 + 32 = 14$

89. $\begin{pmatrix} \text{Value of} \\ \text{800 shares} \end{pmatrix} = \begin{pmatrix} \text{800 times the price} \\ \text{of shares at closing} \end{pmatrix}$:

$\text{Value} = 800\left[\text{price} + \begin{pmatrix} \text{1 day's} \\ \text{increase} \end{pmatrix} + \begin{pmatrix} \text{4 day's} \\ \text{decrease} \end{pmatrix}\right]$
$= 800[19 + 2 + 4(-1)]$
$= 800[19 + 2 + (-4)]$
$= 800(17)$
$= \$13,600$

91. $\begin{pmatrix} \text{Temperature} \\ \text{at 10 pm} \end{pmatrix} =$
$\begin{pmatrix} \text{Starting} \\ \text{temp} \end{pmatrix} + \begin{pmatrix} \text{Hours temp} \\ \text{dropped} \end{pmatrix} \begin{pmatrix} \text{Degree change} \\ \text{per hour} \end{pmatrix} =$
$5 + 4(-3) = 5 + (-12) = -7$ degrees

PROBLEM SET **1.5** **Use of Properties**

1. $3(7 + 8) = 3(7) + 3(8)$:
Distributive property

3. $-2 + (5 + 7) = (-2 + 5) + 7$:
Associative property of addition

5. $143(-7) = -7(143)$:
Commutative property of multiplication

7. $-119 + 119 = 0$:
Additive inverse property

9. $-56 + 0 = -56$:
Identity property of addition

11. $[5(-8)]4 = 5[-8(4)]$:
Associative property of multiplication

13. $(-18 + 56) + 18 =$
$[56 + (-18)] + 18 =$
$56 + (-18 + 18) =$
$56 + 0 = 56$

15. $36 - 48 - 22 + 41 =$
$-12 - 22 + 41 =$
$-34 + 41 = 7$

17. $(25)(-18)(-4) =$
$(25)(-4)(-18) =$
$(-100)(-18) = 1,800$

19. $(4)(-16)(-9)(-25) =$
$(4)(-25)(-16)(-9) =$
$(-100)(144) = -14,400$

21. $37(-42-58) = 37(-100) = -3,700$

23. $59(36) + 59(64) = 59(36+64) =$
$59(100) = 5,900$

25. $15(-14) + 16(-8) =$
$-210 - 128 = -338$

27. $17 + (-18) - 19 - 14 + 13 - 17 =$
$[17 + (-17)] + (-18) - 19 - 14 + 13 =$
$0 + (-37) + (-14 + 13) =$
$-37 - 1 = -38$

29. $-21 + 22 - 23 + 27 + 21 - 19 =$
$(-21 + 22) + (-23 + 27) + (21 - 19) =$
$1 + 4 + 2 = 7$

31. $9x - 14x = (9 - 14)x = 5x$

33. $4m + m - 8m = (4 + 1 - 8)m = -3m$

35. $-9y + 5y - 7y =$
$(-9 + 5 - 7)y = -11y$

37. $4x - 3y - 7x + y =$
$4x - 7x - 3y + y =$
$(4 - 7)x + (-3 + 1)y =$
$-3x - 2y$

39. $-7a - 7b - 9a + 3b =$
$-7a - 9a - 7b + 3b =$
$(-7 - 9)a + (-7 + 3)b =$
$-16a - 4b$

41. $6xy - x - 13xy + 4x =$
$6xy - 13xy - x + 4x =$
$-7xy + 3x$

43. $5x - 4 + 7x - 2x + 9 =$
$5x + 7x - 2x - 4 + 9 =$
$10x + 5$

45. $-2xy + 12 + 8xy - 16 =$
$-2xy + 8xy + 12 - 16 =$
$6xy - 4$

47. $-2a + 3b - 7b - b + 5a - 9a =$
$-2a + 5a - 9a + 3b - 7b - b =$
$-6a - 5b$

49. $13ab + 2a - 7a - 9ab + ab - 6a =$
$13ab - 9ab + ab + 2a - 7a - 6a =$
$5ab - 11a$

51. $3(x + 2) + 5(x + 6) =$
$3(x) + 3(2) + 5(x) + 5(6) =$
$3x + 6 + 5x + 30 =$
$3x + 5x + 6 + 30 =$
$8x + 36$

53. $5(x - 4) + 6(x + 8) =$
$5x - 20 + 6x + 48 =$
$11x + 28$

55. $9(x + 4) - (x - 8) =$
$9(x + 4) - 1(x - 8) =$
$9x + 36 - x + 8 =$
$8x + 44$

57. $3(a - 1) - 2(a - 6) + 4(a + 5) =$
$3a - 3 - 2a + 12 + 4a + 20 =$
$5a + 29$

59. $-2(m + 3) - 3(m - 1) + 8(m + 4) =$
$-2m - 6 - 3m + 3 + 8m + 32 =$
$3m + 29$

61. $(y + 3) - 1(y - 2) - 1(y + 6) - 7(y - 1) =$
$y + 3 - y + 2 - y - 6 - 7y + 7 =$
$-8y + 6$

63. $3x + 5y + 4x - 2y = 7x + 3y =$
$7(-2) + 3(3)$ for $x = -2$ and $y = 3$
$-14 + 9 = -5$

65. $5(x - 2) + 8(x + 6) =$
$5x - 10 + 8x + 48 =$
$13x + 38$ for $x = -6$
$13(-6) + 38 = -78 + 38 = -40$

Problem Set 1.5

67. $8(x + 4) - 10(x - 3) =$
  $8x + 32 - 10x + 30 =$
  $-2x + 62$ for $x = -5$
  $-2(-5) + 62 =$
  $10 + 62 = 72$

69. $(x - 6) - (x + 12) =$
  $x - 6 - x - 12 = -18$
  -18 for $x = -3$

71. $2(x + y) - 3(x - y) =$
  $2x + 2y - 3x + 3y =$
  $-x + 5y$ for $x = -2,\ y = 7$
  $-(-2) + 5(7) = 2 + 35 = 37$

73. $2xy + 6 + 7xy - 8 =$
  $9xy - 2$ for $x = 2,\ y = -4$
  $9(2)(-4) - 2 = -72 - 2 = -74$

75. $5x - 9xy + 3x + 2xy =$
  $8x - 7xy$ for $x = 12$ and $y = -1$
  $8(12) - 7(12)(-1) = 96 + 84 = 180$

77. $(a - b) - (a + b) =$
  $a - b - a - b =$
  $-2b$ for $a = 19,\ b = -1$
  $-2(-17) = 34$

79. $-3x + 7x + 4x - 2x - x =$
  $5x$ for $x = -13$
  $5(-13) = -65$

CHAPTER 1 **Review Problem Set**

1. $7 + (-10) = -(|-10| - |7|) =$
  $-(10 - 7) = -(3) = -3$

2. $(-12) + (-13) = -(|-12| + |-13|) =$
  $-(12 + 13) = -(25) = -25$

3. $8 - 13 = 8 + (-13) = -5$

4. $-6 - 9 = -6 + (-9) = -15$

5. $-12 - (-11) = -12 + 11 = -1$

6. $-17 - (-19) = -17 + 19 = 2$

7. $(13)(-12) = -(|13| \cdot |-12|) =$
  $-(13 \cdot 12) = -(156) = -156$

8. $(-14)(-18) = |-14| \cdot |-18| =$
  $14 \cdot 18 = 252$

9. $(-72) \div (-12) = |-72| \div |-12| =$
  $72 \div 12 = 6$

10. $(117) \div (-9) = -(|117| \div |-9|) =$
  $-(117 \div 9) = -(13) = -13$

11. 73, prime number

12. Because $87 = 3(29)$, it is a composite number.

13. Because $63 = 9(7)$, it is a composite number.

14. Because $81 = 9(9)$, it is a composite number.

15. Because $91 = 7(13)$, it is a composite number.

16. $24 = 4 \cdot 6 = 2 \cdot 2 \cdot 2 \cdot 3$

17. $63 = 9 \cdot 7 = 3 \cdot 3 \cdot 7$

18. $57 = 3 \cdot 19$

19. $64 = 16 \cdot 4 = 4 \cdot 4 \cdot 4 = 2 \cdot 2 \cdot 2 \cdot 2 \cdot 2 \cdot 2$

20. $84 = 4 \cdot 21 = 2 \cdot 2 \cdot 3 \cdot 7$

21. $36 = 2 \cdot 2 \cdot 3 \cdot 3$ The greatest common
  $54 = 2 \cdot 3 \cdot 3 \cdot 3$ factor is $2 \cdot 3 \cdot 3 = 18$.

22. $48 = 2 \cdot 2 \cdot 2 \cdot 2 \cdot 3$ The greatest
  $60 = 2 \cdot 2 \cdot 3 \cdot 5$ common factor
  $84 = 2 \cdot 2 \cdot 3 \cdot 7$ is $2 \cdot 2 \cdot 3 = 12$.

23. $18 = 2 \cdot 3 \cdot 3$ The least common multiple
$20 = 2 \cdot 2 \cdot 5$ is $2 \cdot 2 \cdot 3 \cdot 3 \cdot 5 = 180.$

24. $15 = 3 \cdot 5$ The least common multiple
$27 = 3 \cdot 3 \cdot 3$ is $3 \cdot 3 \cdot 3 \cdot 5 \cdot 7 = 945.$
$35 = 5 \cdot 7$

25. $(19 + 56) + (-9) = 75 + (-9) = 66$

26. $43 - 62 + 12 = 43 + (-62) + 12 =$
$-19 + 12 = -7$

27. $8 + (-9) + (-16) + (-14) + 17 + 12 =$
$37 + (-39) = -2$

28. $19 - 23 - 14 + 21 + 14 - 13 = 4$

29. $3(-4) - 6 = -12 - 6 = -18$

30. $(-5)(-4) - 8 = 20 - 8 = 12$

31. $5(-2) + 6(-4) = -10 - 24 = -34$

32. $(-6)(8) + (-7)(-3) =$
$-48 + 21 = -27$

33. $(-6)(3) - (-4)(-5) =$
$-18 - 20 = -38$

34. $(-7)(9) - (6)(5) = -63 - 30 = -93$

35. $\dfrac{4(-7) - 3(-2)}{-11} = \dfrac{-28 - (-6)}{-11} =$
$\dfrac{-28 + 6}{-11} = \dfrac{-22}{-11} = 2$

36. $\dfrac{(-4)(9) + (5)(-3)}{1 - 18} = \dfrac{-36 + (-15)}{-17} =$
$\dfrac{-51}{-17} = 3$

37. $3 - 2[4(-3 - 1)] =$
$3 - 2[4(-3 + (-1))] =$
$3 - 2[4(-4)] = 3 - 2(-16) =$
$3 - (-32) = 3 + 32 = 35$

38. $-6 - [3(-4 - 7)] =$
$-6 - [3(-4 + (-7))] =$
$-6 - [3(-11)] = -6 - (-33) =$
$-6 + 33 = 27$

39. $12x + 3x - 7x = (12 + 3 - 7)x = 8x$

40. $9y + 3 - 14y - 12 = 9y - 14y + 3 - 12 =$
$(9 - 14)y + (3 - 12) = -5y + (-9) =$
$-5y - 9$

41. $8x + 5y - 13x - y = 8x - 13x + 5y - y =$
$(8 - 13)x + (5 - 1)y = -5x + 4y$

42. $9a + 11b + 4a - 17b = 9a + 4a + 11b - 17b =$
$(9 + 4)a + (11 - 17)b = 13a + (-6b) =$
$13a - 6b$

43. $3ab - 4ab - 2a = -ab - 2a$

44. $5xy - 9xy + xy - y = -3xy - y$

45. $3(x + 6) + 7(x + 8) =$
$3x + 18 + 7x + 56 = 10x + 74$

46. $5(x - 4) - 3(x - 9) =$
$5x - 20 - 3x + 27 = 2x + 7$

47. $-3(x - 2) - 4(x + 6) =$
$-3x + 6 - 4x - 24 = -7x - 18$

48. $-2x - 3(x - 4) + 2x =$
$-2x - 3x + 12 + 2x = -3x + 12$

49. $2(a - 1) - a - 3(a - 2) =$
$2a - 2 - a - 3a + 6 = -2a + 4$

50. $-(a - 1) + 3(a - 2) - 4a + 1 =$
$-a + 1 + 3a - 6 - 4a + 1 = -2a - 4$

51. $5x + 8y$ for $x = -7$ and $y = -3$
$5(-7) + 8(-3) = -35 - 24 = -59$

52. $7x - 9y$ for $x = -3$ and $y = 4$
$7(-3) - 9(4) = -21 - 36 = -57$

Chapter 1 Review Problem Set

53. $\dfrac{-5x - 2y}{-2x - 7}$ for $x = 6$ and $y = 4$

$\dfrac{-5(6) - 2(4)}{-2(6) - 7} = \dfrac{-30 - 8}{-12 - 7} = \dfrac{-38}{-19} = 2$

54. $\dfrac{-3x + 4y}{3x}$ for $x = -4$, $y = -6$

$\dfrac{-3(-4) + 4(-6)}{3(-4)} =$

$\dfrac{12 + (-24)}{-12} = \dfrac{-12}{-12} = 1$

55. $-2a + \dfrac{a - b}{a - 2}$ for $a = -5$, $b = 9$

$-2(-5) + \dfrac{(-5) - (9)}{(-5) - 2} =$

$10 + \dfrac{(-5) + (-9)}{(-5) + (-2)} =$

$10 + \dfrac{-14}{-7} = 10 + 2 = 12$

56. $\dfrac{2a + b}{b + 6} - 3b$ for a $= 3$, $b = -4$

$\dfrac{2(3) + (-4)}{(-4) + 6} - 3(-4) =$

$\dfrac{6 + (-4)}{2} - (-12) =$

$\dfrac{2}{2} + 12 = 1 + 12 = 13$

57. $5a + 6b - 7a - 2b =$
$-2a + 4b$ for $a = -1$, $b = 5$
$-2(-1) + 4(5) = 2 + 20 = 22$

58. $3x + 7y - 5x + y =$
$-2x + 8y$ for $x = -4$ and $y = 3$
$-2(-4) + 8(3) = 8 + 24 = 32$

59. $2xy + 6 + 5xy - 8 =$
$7xy - 2$ for $x = -1$, $y = 1$
$7(-1)(1) - 2 = (-7)(1) - 2 =$
$-7 - 2 = -9$

60. $7(x + 6) - 9(x + 1) =$
$7x + 42 - 9x - 9 =$
$-2x + 33$ for $x = -2$
$-2(-2) + 33 = 4 + 33 = 37$

61. $-3(x - 4) - 2(x + 8) =$
$-3x + 12 - 2x - 16 =$
$-5x - 4$ for $x = 7$
$-5(7) - 4 = -35 - 4 = -39$

62. $2(x - 1) - (x + 2) + 3(x - 4) =$
$2x - 2 - x - 2 + 3x - 12 =$
$4x - 16$ for $x = -4$
$4(-4) - 16 = -16 - 16 = -32$

63. $(a - b) - (a + b) - b =$
$a - b - a - b - b =$
$-3b$ for $a = -1$, $b = -3$
$-3(-3) = 9$

64. $2ab - 3(a - b) + b + a =$
$2ab - 3a + 3b + b + a =$
$2ab - 2a + 4b$ for $a = 2$, $b = -5$
$2(2)(-5) - 2(2) + 4(-5) =$
$4(-5) - 4 - 20 = -20 - 4 - 20 = -44$

CHAPTER 1 Test

1. $6 + (-7) - 4 + 12$
$-1 - 4 + 12 =$
$-5 + 12 = 7$

2. $7 + 4(9) + 2 =$
$7 + 36 + 2 = 45$

3. $-4(2-8)+14=$
$-4(-6)+14=$
$24+14=38$

4. $5(-7)-(-3)(8)=$
$-35-(-24)=$
$-35+24=-11$

5. $8 \div (-4)+(-6)(9)-2=$
$-2+(-6)(9)-2=$
$-2+(-54)-2=-58$

6. $(-8)(-7)+(-6)-(9)(12)=$
$56+(-6)-108=-58$

7. $\dfrac{6(-4)-(-8)(-5)}{-16}=\dfrac{-24-40}{-16}=$
$\dfrac{-64}{-16}=4$

8. $-14+23-17-19+26=$
$-14-17-19+23+26=$
$-50+49=-1$

9. $(-14)(4) \div 4+(-6)=$
$-56 \div 4+(-6)=$
$-14+(-6)=-20$

10. $6(-9)-(-8)-(-7)(4)=$
$-54+8-(-28)+11=$
$-54+8+28+11=-7$

11. $7x-9y$ for $x=-4, y=-6$
$7(-4)-9(-6)=$
$-28-(-54)=$
$-28+54=26$

12. $-4a-6b$ for $a=-9, b=12$
$-4(-9)-6(12)=$
$36-72=-36$

13. $3xy-8y+5x$ for $x=7, y=-2$
$3(7)(-2)-8(-2)+5(7)=$
$21(-2)-(-16)+35=$
$-42+16+35=9$

14. $5(x-4)-6(x+7)=$
$5x-20-6x-42=$

$-x-62$ for $x=-5$
$-(-5)-62=5-62=-57$

15. $3x-2y-4x-x+7y=$
$-2x+5y$ for $x=6, y=-7$
$-2(6)+5(-7)=-12+(-35)=-47$

16. $3(x-2)-5(x-4)+6(x-1)=$
$3x-6-5x+20+6x-6=$
$4x+8$ for $x=-3$
$4(-3)+8=-12+8=-4$

17. $\dfrac{-x-y}{y-x}$ for $x=-9, y=-6$

$\dfrac{-(-9)-(-6)}{(-6)-(-9)}=\dfrac{9+6}{-6+9}=$
$\dfrac{15}{3}=5$

18. 79, prime number

19. $360=8 \cdot 45=8 \cdot 9 \cdot 5$
$=2 \cdot 2 \cdot 2 \cdot 3 \cdot 3 \cdot 5$

20. $36=2 \cdot 2 \cdot 3 \cdot 3$ The greatest
$60=2 \cdot 2 \cdot 3 \cdot 5$ common factor
$84=2 \cdot 2 \cdot 3 \cdot 7$ is $2 \cdot 2 \cdot 3=12$.

21. $9=3 \cdot 3$ The least common
$24=2 \cdot 2 \cdot 2 \cdot 3$ multiple is
                     $2 \cdot 2 \cdot 2 \cdot 3 \cdot 3=72$.

22. $[-3+(-4)]+(-6)=-3+[(-4)+(-6)]$
Associative Property of Addition

23. $8(25+37)=8(25)+8(37)$:
Distributive Property

24. $-7x+9y-y+x-2y-7x=$
$-7x+x-7x+9y-y-2y=$
$-13x+6y$

25. $-2(x-4)-5(x+7)-6(x-1)=$
$-2x+8-5x-35-6x+6=$
$-2x-5x-6x+8-35+6=$
$-13x-21$

Chapter 2 Real Numbers

PROBLEM SET **2.1**   **Rational Numbers: Multiplication and Division**

1. $\dfrac{8}{12} = \dfrac{4 \cdot 2}{4 \cdot 3} = \dfrac{2}{3}$

3. $\dfrac{16}{24} = \dfrac{8 \cdot 2}{8 \cdot 3} = \dfrac{2}{3}$

5. $\dfrac{15}{9} = \dfrac{3 \cdot 5}{3 \cdot 3} = \dfrac{5}{3}$

7. $\dfrac{-8}{48} = -\dfrac{2 \cdot 2 \cdot 2}{2 \cdot 2 \cdot 2 \cdot 3} = -\dfrac{1}{6}$

9. $\dfrac{27}{-36} = -\dfrac{9 \cdot 3}{9 \cdot 4} = -\dfrac{3}{4}$

11. $\dfrac{-54}{-56} = \dfrac{2 \cdot 27}{2 \cdot 28} = \dfrac{27}{28}$

13. $\dfrac{24x}{44x} = \dfrac{4 \cdot 6 \cdot x}{4 \cdot 11 \cdot x} = \dfrac{6}{11}$

15. $\dfrac{9x}{21y} = \dfrac{3 \cdot 3 \cdot x}{3 \cdot 7 \cdot y} = \dfrac{3x}{7y}$

17. $\dfrac{14xy}{35y} = \dfrac{2 \cdot 7 \cdot x \cdot y}{5 \cdot 7 \cdot y} = \dfrac{2x}{5}$

19. $\dfrac{-20ab}{52bc} = -\dfrac{4 \cdot 5 \cdot a \cdot b}{4 \cdot 13 \cdot b \cdot c} = -\dfrac{5a}{13c}$

21. $\dfrac{-56yz}{-49xy} = \dfrac{7 \cdot 8 \cdot y \cdot z}{7 \cdot 7 \cdot x \cdot y} = \dfrac{8z}{7x}$

23. $\dfrac{65abc}{91ac} = \dfrac{5 \cdot 13 \cdot a \cdot b \cdot c}{7 \cdot 13 \cdot a \cdot c} = \dfrac{5b}{7}$

25. $\dfrac{3}{4} \cdot \dfrac{5}{7} = \dfrac{3 \cdot 5}{4 \cdot 7} = \dfrac{15}{28}$

27. $\dfrac{2}{7} \div \dfrac{3}{5} = \dfrac{2}{7} \cdot \dfrac{5}{3} = \dfrac{10}{21}$

29. $\dfrac{3}{8} \cdot \dfrac{12}{15} = \dfrac{3 \cdot \overset{1}{\cancel{3}} \cdot \overset{3}{\cancel{12}}}{\underset{2}{\cancel{8}} \cdot \underset{5}{\cancel{15}}} = \dfrac{3}{10}$

31. $\dfrac{-6}{13} \cdot \dfrac{26}{9} = -\dfrac{\overset{2}{\cancel{6}} \cdot \overset{2}{\cancel{26}}}{\underset{1}{\cancel{13}} \cdot \underset{3}{\cancel{9}}} = -\dfrac{4}{3}$

33. $\dfrac{7}{9} \div \dfrac{5}{9} = \dfrac{7}{9} \cdot \dfrac{9}{5} = \dfrac{7 \cdot \overset{1}{\cancel{9}}}{\underset{1}{\cancel{9}} \cdot 5} = \dfrac{7}{5}$

35. $\dfrac{1}{4} \div \dfrac{-5}{6} = -\dfrac{1}{4} \cdot \dfrac{6}{5} = -\dfrac{1 \cdot \overset{3}{\cancel{6}}}{\underset{2}{\cancel{4}} \cdot 5} = -\dfrac{3}{10}$

37. $\left(-\dfrac{8}{10}\right)\left(-\dfrac{10}{32}\right) = \dfrac{\overset{1}{\cancel{8}} \cdot \overset{1}{\cancel{10}}}{\underset{1}{\cancel{10}} \cdot \underset{4}{\cancel{32}}} = \dfrac{1}{4}$

39. $-9 \div \dfrac{1}{3} = -\dfrac{9}{1} \cdot \dfrac{3}{1} = -27$

41. $\dfrac{5x}{9y} \cdot \dfrac{7y}{3x} = \dfrac{5 \cdot 7 \cdot \cancel{x} \cdot \cancel{y}}{9 \cdot 3 \cdot \cancel{x} \cdot \cancel{y}} = \dfrac{35}{27}$

43. $\dfrac{6a}{14b} \cdot \dfrac{16b}{18a} = \dfrac{\overset{1}{\cancel{6}} \cdot \overset{8}{\cancel{16}} \cdot \cancel{a} \cdot \cancel{b}}{\underset{7}{\cancel{14}} \cdot \underset{3}{\cancel{18}} \cdot \cancel{a} \cdot \cancel{b}} = \dfrac{8}{21}$

45. $\dfrac{10x}{-9y} \cdot \dfrac{15}{20x} = -\dfrac{\overset{1}{\cancel{10}} \cdot \overset{5}{\cancel{15}} \cdot \overset{1}{\cancel{x}}}{\underset{3}{\cancel{9}} \cdot \underset{2}{\cancel{20}} \cdot \underset{1}{\cancel{x}} \cdot y} = -\dfrac{5}{6y}$

47. $ab \cdot \dfrac{2}{b} = \dfrac{a \cdot \overset{1}{\cancel{b}} \cdot 2}{\underset{1}{\cancel{b}}} = 2a$

49. $\left(-\dfrac{7x}{12y}\right)\left(-\dfrac{24y}{35x}\right) = \dfrac{\overset{1}{\cancel{7}} \cdot \overset{2}{\cancel{24}} \cdot \overset{1}{\cancel{x}} \cdot \overset{1}{\cancel{y}}}{\underset{1}{\cancel{12}} \cdot \underset{5}{\cancel{35}} \cdot \underset{1}{\cancel{x}} \cdot \underset{1}{\cancel{y}}} = \dfrac{2}{5}$

51. $\dfrac{3}{x} \div \dfrac{6}{y} = \dfrac{3}{x} \cdot \dfrac{y}{\overset{2}{\cancel{6}}} = \dfrac{y}{2x}$

$-\dfrac{1}{2} \div \dfrac{1}{8} = -\dfrac{1}{2} \cdot 8 = -4$

67. $\dfrac{5}{7} \div \left(-\dfrac{5}{6}\right)\left(-\dfrac{6}{7}\right) =$

$\dfrac{5}{7} \cdot \left(-\dfrac{6}{5}\right)\left(-\dfrac{6}{7}\right) =$

$\dfrac{\overset{1}{\cancel{5}} \cdot 6 \cdot 6}{7 \cdot \underset{1}{\cancel{5}} \cdot 7} = \dfrac{36}{49}$

53. $\dfrac{5x}{9y} \div \dfrac{13x}{36y} = \dfrac{5x}{9y} \cdot \dfrac{36y}{13x} =$

$\dfrac{5 \cdot \overset{4}{\cancel{36}} \cdot \cancel{x} \cdot \cancel{y}}{\underset{1}{\cancel{9}} \cdot 13 \cdot \cancel{x} \cdot \cancel{y}} = \dfrac{20}{13}$

55. $\dfrac{-7}{x} \div \dfrac{9}{x} = \dfrac{-7}{\cancel{x}} \cdot \dfrac{\cancel{x}}{9} = -\dfrac{7}{9}$

69. $\left(-\dfrac{6}{7}\right) \div \left(\dfrac{5}{7}\right)\left(-\dfrac{5}{6}\right) =$

$\left(-\dfrac{6}{7}\right) \cdot \left(\dfrac{7}{5}\right)\left(-\dfrac{5}{6}\right) =$

$\dfrac{\overset{1}{\cancel{6}} \cdot \overset{1}{\cancel{7}} \cdot \overset{1}{\cancel{5}}}{\underset{1}{\cancel{7}} \cdot \underset{1}{\cancel{5}} \cdot \underset{1}{\cancel{6}}} = \dfrac{1}{1} = 1$

57. $\dfrac{-4}{n} \div \dfrac{-18}{n} = \dfrac{-\overset{2}{\cancel{4}}}{\cancel{n}} \cdot \dfrac{\cancel{n}}{\underset{9}{\cancel{-18}}} = \dfrac{2}{9}$

59. $\dfrac{3}{4} \cdot \dfrac{8}{9} \cdot \dfrac{12}{20} = \dfrac{\overset{1}{\cancel{3}} \cdot \overset{2}{\cancel{8}} \cdot \overset{3}{\cancel{12}}}{\underset{1}{\cancel{4}} \cdot \underset{3}{\cancel{9}} \cdot \underset{5}{\cancel{20}}} = \dfrac{6}{15} = \dfrac{2}{5}$

71. $\left(\dfrac{4}{9}\right)\left(-\dfrac{9}{8}\right) \div \left(-\dfrac{3}{4}\right) =$

$\left(\dfrac{4}{9}\right)\left(-\dfrac{9}{8}\right) \cdot \left(-\dfrac{4}{3}\right) =$

$\dfrac{\overset{1}{\cancel{4}} \cdot \overset{1}{\cancel{9}} \cdot 4}{\underset{1}{\cancel{9}} \cdot \underset{2}{\cancel{8}} \cdot 3} = \dfrac{\overset{2}{\cancel{4}}}{\underset{1}{\cancel{2}} \cdot 3} = \dfrac{2}{3}$

61. $\left(-\dfrac{3}{8}\right)\left(\dfrac{13}{14}\right)\left(-\dfrac{12}{9}\right) =$

$\dfrac{\overset{1}{\cancel{3}} \cdot 13 \cdot \overset{3}{\cancel{12}}}{\underset{2}{\cancel{8}} \cdot 14 \cdot \underset{3}{\cancel{9}}} = \dfrac{13 \cdot \overset{1}{\cancel{3}}}{14 \cdot \underset{2}{\cancel{6}}} = \dfrac{13}{28}$

63. $\left(\dfrac{3x}{4y}\right)\left(\dfrac{8}{9x}\right)\left(\dfrac{12y}{5}\right) =$

$\dfrac{\overset{1}{\cancel{3}} \cdot \overset{2}{\cancel{8}} \cdot 12 \cdot \cancel{x} \cdot \cancel{y}}{\underset{1}{\cancel{4}} \cdot \underset{3}{\cancel{9}} \cdot 5 \cdot \cancel{x} \cdot \cancel{y}} = \dfrac{2 \cdot \overset{4}{\cancel{12}}}{\underset{1}{\cancel{3}} \cdot 5} = \dfrac{8}{5}$

73. $\left(\dfrac{5}{2}\right)\left(\dfrac{2}{3}\right) \div \left(-\dfrac{1}{4}\right) \div (-3) =$

$\left(\dfrac{5}{2}\right)\left(\dfrac{2}{3}\right) \cdot \left(-\dfrac{4}{1}\right) \cdot \left(-\dfrac{1}{3}\right) =$

$\dfrac{5 \cdot \overset{1}{\cancel{2}} \cdot 4 \cdot 1}{\underset{1}{\cancel{2}} \cdot 3 \cdot 1 \cdot 3} = \dfrac{5 \cdot 4}{3 \cdot 3} = \dfrac{20}{9}$

65. $\left(-\dfrac{2}{3}\right)\left(\dfrac{3}{4}\right) \div \dfrac{1}{8} = -\dfrac{\overset{1}{\cancel{2}} \cdot \overset{1}{\cancel{3}}}{\underset{1}{\cancel{3}} \cdot \underset{2}{\cancel{4}}} \div \dfrac{1}{8} =$

Problem Set 2.1

75. $\left(\begin{array}{c}\text{portions of accounts}\\\text{in Maria's dept.}\end{array}\right) \cdot \left(\begin{array}{c}\text{portions of accounts}\\\text{Maria has}\end{array}\right) = \left(\begin{array}{c}\text{Maria's portion of all}\\\text{the agency's accounts}\end{array}\right)$

$$\left(\frac{3}{4}\right) \cdot \left(\frac{1}{3}\right) = \left(\begin{array}{c}\text{Maria's portions of all}\\\text{the agency's accounts}\end{array}\right)$$

$$\frac{3}{12} = \frac{1}{4}$$

Maria is responsible for $\frac{1}{4}$ of all the agency's accounts.

77. $\left(\begin{array}{c}\text{cups of}\\\text{sugar}\end{array}\right) \cdot \left(\begin{array}{c}\text{portion of}\\\text{recipe}\end{array}\right) = \left(\begin{array}{c}\text{amount of}\\\text{sugar}\end{array}\right)$

$$\left(\frac{3}{4}\right) \cdot \left(\frac{1}{2}\right) = \left(\begin{array}{c}\text{amount of}\\\text{sugar}\end{array}\right)$$

$$\frac{3 \cdot 1}{4 \cdot 2} = \frac{3}{8}$$

The amount of sugar needed is $\frac{3}{8}$ cups of sugar.

79. $\left(\begin{array}{c}\text{amount of milk in}\\\text{original recipe}\end{array}\right) \cdot \left(\frac{1}{2}\right) = \left(\begin{array}{c}\text{amount of milk in}\\\text{half of recipe}\end{array}\right)$

$$\left(3\frac{1}{2}\right) \cdot \left(\frac{1}{2}\right) = \left(\begin{array}{c}\text{amount of milk in}\\\text{half of recipe}\end{array}\right)$$

$$\left(\frac{7}{2}\right) \cdot \left(\frac{1}{2}\right) = \frac{7 \cdot 1}{2 \cdot 2} = \frac{7}{4} = 1\frac{3}{4}$$

She should use $1\frac{3}{4}$ cups of sugar.

81. $\left(\begin{array}{c}\text{material to make}\\\text{one dress}\end{array}\right) \cdot \left(\begin{array}{c}\text{number of}\\\text{dresses}\end{array}\right) = \left(\begin{array}{c}\text{material to make}\\\text{20 dresses}\end{array}\right)$

$$\left(3\frac{1}{4}\right) \cdot \left(20\right) = \left(\begin{array}{c}\textbf{material to make}\\\text{20 dresses}\end{array}\right)$$

$$\left(\frac{13}{4}\right) \cdot \left(\frac{20}{1}\right) = \frac{13 \cdot \overset{5}{\cancel{20}}}{\underset{1}{\cancel{4}} \cdot 1} = \frac{65}{1} = 65$$

The amount of material needed to make
20 dresses is 65 yards.

87   **a.** larger   **b.** larger   **c.** smaller
      d. larger   **e.** larger   **f.** smaller

PROBLEM SET   **2.2**   **Addition and Subtraction of Rational Numbers**

1. $\dfrac{2}{7} + \dfrac{3}{7} = \dfrac{2+3}{7} = \dfrac{5}{7}$

$\dfrac{21}{30} + \dfrac{16}{30} = \dfrac{37}{30}$

3. $\dfrac{7}{9} - \dfrac{2}{9} = \dfrac{7-2}{9} = \dfrac{5}{9}$

23. $\dfrac{11}{24} + \dfrac{5}{32} = \left(\dfrac{11 \cdot 4}{24 \cdot 4}\right) + \left(\dfrac{5 \cdot 3}{32 \cdot 3}\right) =$

$\dfrac{44}{96} + \dfrac{15}{96} = \dfrac{59}{96}$

5. $\dfrac{3}{4} + \dfrac{9}{4} = \dfrac{3+9}{4} = \dfrac{12}{4} = 3$

7. $\dfrac{11}{12} - \dfrac{3}{12} = \dfrac{11-3}{12} = \dfrac{8}{12} = \dfrac{2}{3}$

25. $\dfrac{5}{18} - \dfrac{13}{24} = \left(\dfrac{5 \cdot 4}{18 \cdot 4}\right) - \left(\dfrac{13 \cdot 3}{24 \cdot 3}\right) =$

$\dfrac{20}{72} - \dfrac{39}{72} = -\dfrac{19}{72}$

9. $\dfrac{1}{8} - \dfrac{5}{8} = \dfrac{1-5}{8} = \dfrac{-4}{8} = -\dfrac{1}{2}$

11. $\dfrac{5}{24} + \dfrac{11}{24} = \dfrac{5+11}{24} = \dfrac{16}{24} = \dfrac{2}{3}$

27. $\dfrac{5}{8} - \dfrac{2}{3} = \left(\dfrac{5 \cdot 3}{8 \cdot 3}\right) - \left(\dfrac{2 \cdot 8}{3 \cdot 8}\right) =$

$\dfrac{15}{24} - \dfrac{16}{24} = -\dfrac{1}{24}$

13. $\dfrac{8}{x} + \dfrac{7}{x} = \dfrac{8+7}{x} = \dfrac{15}{x}$

15. $\dfrac{5}{3y} + \dfrac{1}{3y} = \dfrac{5+1}{3y} = \dfrac{6}{3y} = \dfrac{2}{y}$

29. $-\dfrac{2}{13} - \dfrac{7}{39} = -\left(\dfrac{2 \cdot 3}{13 \cdot 3}\right) - \dfrac{7}{39} =$

$-\dfrac{6}{39} - \dfrac{7}{39} = -\dfrac{13}{39} = -\dfrac{1}{3}$

17. $\dfrac{1}{3} + \dfrac{1}{5} = \left(\dfrac{1 \cdot 5}{3 \cdot 5}\right) + \left(\dfrac{1 \cdot 3}{5 \cdot 3}\right) =$

$\dfrac{5}{15} + \dfrac{3}{15} = \dfrac{8}{15}$

31. $-\dfrac{3}{14} + \dfrac{1}{21} = -\left(\dfrac{3 \cdot 3}{14 \cdot 3}\right) + \left(\dfrac{1 \cdot 2}{21 \cdot 2}\right) =$

$-\dfrac{9}{42} + \dfrac{2}{42} = -\dfrac{7}{42} = -\dfrac{1}{6}$

19. $\dfrac{15}{16} - \dfrac{3}{8} = \dfrac{15}{16} - \left(\dfrac{3 \cdot 2}{8 \cdot 2}\right) =$

$\dfrac{15}{16} - \dfrac{6}{16} = \dfrac{9}{16}$

33. $-4 - \dfrac{3}{7} = -\left(\dfrac{4}{1} \cdot \dfrac{7}{7}\right) - \dfrac{3}{7} =$

$-\dfrac{28}{7} - \dfrac{3}{7} = -\dfrac{31}{7}$

21. $\dfrac{7}{10} + \dfrac{8}{15} = \left(\dfrac{7 \cdot 3}{10 \cdot 3}\right) + \left(\dfrac{8 \cdot 2}{15 \cdot 2}\right) =$

Problem Set 2.2

35. $\dfrac{3}{4} - 6 = \dfrac{3}{4} - \left(\dfrac{6}{1} \cdot \dfrac{4}{4}\right) =$

$\dfrac{3}{4} - \dfrac{24}{4} = -\dfrac{21}{4}$

37. $\dfrac{3}{x} + \dfrac{4}{y} = \left(\dfrac{3 \cdot y}{x \cdot y}\right) + \left(\dfrac{4 \cdot x}{y \cdot x}\right) =$

$\dfrac{3y}{xy} + \dfrac{4x}{xy} = \dfrac{3y + 4x}{xy}$

39. $\dfrac{7}{a} - \dfrac{2}{b} = \left(\dfrac{7 \cdot b}{a \cdot b}\right) - \left(\dfrac{2 \cdot a}{b \cdot a}\right) =$

$\dfrac{7b}{ab} - \dfrac{2a}{ab} = \dfrac{7b - 2a}{ab}$

41. $\dfrac{2}{x} + \dfrac{7}{2x} = \left(\dfrac{2 \cdot 2}{x \cdot 2}\right) + \dfrac{7}{2x} =$

$\dfrac{4}{2x} + \dfrac{7}{2x} = \dfrac{11}{2x}$

43. $\dfrac{10}{3x} - \dfrac{2}{x} = \dfrac{10}{3x} - \left(\dfrac{2 \cdot 3}{x \cdot 3}\right) =$

$\dfrac{10}{3x} - \dfrac{6}{3x} = \dfrac{4}{3x}$

45. $\dfrac{1}{x} - \dfrac{7}{5x} = \left(\dfrac{1 \cdot 5}{x \cdot 5}\right) - \dfrac{7}{5x} =$

$\dfrac{5}{5x} - \dfrac{7}{5x} = -\dfrac{2}{5x}$

47. $\dfrac{3}{2y} + \dfrac{5}{3y} = \left(\dfrac{3 \cdot 3}{2y \cdot 3}\right) + \left(\dfrac{5 \cdot 2}{3y \cdot 2}\right) =$

$\dfrac{9}{6y} + \dfrac{10}{6y} = \dfrac{19}{6y}$

49. $\dfrac{5}{12y} - \dfrac{3}{8y} = \left(\dfrac{5 \cdot 2}{12y \cdot 2}\right) - \left(\dfrac{3 \cdot 3}{8y \cdot 3}\right) =$

$\dfrac{10}{24y} - \dfrac{9}{24y} = \dfrac{1}{24y}$

51. $\dfrac{1}{6n} - \dfrac{7}{8n} = \left(\dfrac{1 \cdot 4}{6n \cdot 4}\right) - \left(\dfrac{7 \cdot 3}{8n \cdot 3}\right) =$

$\dfrac{4}{24n} - \dfrac{21}{24n} = -\dfrac{17}{24n}$

53. $\dfrac{5}{3x} + \dfrac{7}{3y} = \left(\dfrac{5 \cdot y}{3x \cdot y}\right) + \left(\dfrac{7 \cdot x}{3y \cdot x}\right) =$

$\dfrac{5y}{3xy} + \dfrac{7x}{3xy} = \dfrac{5y + 7x}{3xy}$

55. $\dfrac{8}{5x} + \dfrac{3}{4y} = \left(\dfrac{8 \cdot 4y}{5x \cdot 4y}\right) + \left(\dfrac{3 \cdot 5x}{4y \cdot 5x}\right) =$

$\dfrac{32y}{20xy} + \dfrac{15x}{20xy} = \dfrac{32y + 15x}{20xy}$

57. $\dfrac{7}{4x} - \dfrac{5}{9y} = \left(\dfrac{7 \cdot 9y}{4x \cdot 9y}\right) - \left(\dfrac{5 \cdot 4x}{9y \cdot 4x}\right) =$

$\dfrac{63y}{36xy} - \dfrac{20x}{36xy} = \dfrac{63y - 20x}{36xy}$

59. $-\dfrac{3}{2x} - \dfrac{5}{4y} = -\left(\dfrac{3 \cdot 2y}{2x \cdot 2y}\right) - \left(\dfrac{5 \cdot x}{4y \cdot x}\right) =$

$-\dfrac{6y}{4xy} - \dfrac{5x}{4xy} = \dfrac{-6y - 5x}{4xy}$

61. $3 + \dfrac{2}{x} = \left(\dfrac{3 \cdot x}{1 \cdot x}\right) + \dfrac{2}{x} =$

$\dfrac{3x}{x} + \dfrac{2}{x} = \dfrac{3x + 2}{x}$

63. $2 - \dfrac{3}{2x} = \left(\dfrac{2}{1} \cdot \dfrac{2x}{2x}\right) - \dfrac{3}{2x} =$

$\dfrac{4x}{2x} - \dfrac{3}{2x} = \dfrac{4x - 3}{2x}$

65. $\dfrac{1}{4} - \dfrac{3}{8} + \dfrac{5}{12} - \dfrac{1}{24} =$

$\dfrac{6}{24} - \dfrac{9}{24} + \dfrac{10}{24} - \dfrac{1}{24} = \dfrac{6}{24} = \dfrac{1}{4}$

67. $\dfrac{5}{6} + \dfrac{2}{3} \cdot \dfrac{3}{4} - \dfrac{1}{4} \cdot \dfrac{2}{5} =$

$\dfrac{5}{6} + \dfrac{1}{2} - \dfrac{1}{10} =$

$\dfrac{25}{30} + \dfrac{15}{30} - \dfrac{3}{30} = \dfrac{37}{30}$

69. $\dfrac{3}{4} \cdot \dfrac{6}{9} - \dfrac{5}{6} \cdot \dfrac{8}{10} + \dfrac{2}{3} \cdot \dfrac{6}{8} =$

$\dfrac{1}{2} - \dfrac{2}{3} + \dfrac{1}{2} =$

$\dfrac{3}{6} - \dfrac{4}{6} + \dfrac{3}{6} = \dfrac{2}{6} = \dfrac{1}{3}$

71. $4 - \dfrac{2}{3} \cdot \dfrac{3}{5} - 6 = 4 - \dfrac{2}{5} - 6$

$= -2 - \dfrac{2}{5} = -\left(\dfrac{2}{1} \cdot \dfrac{5}{5}\right) - \dfrac{2}{5}$

$= -\dfrac{10}{5} - \dfrac{2}{5} = -\dfrac{12}{5}$

73. $\dfrac{4}{5} - \dfrac{10}{12} - \dfrac{5}{6} \div \dfrac{14}{8} + \dfrac{10}{21} =$

$\dfrac{4}{5} - \dfrac{10}{12} - \dfrac{5}{6} \cdot \dfrac{8}{14} + \dfrac{10}{21} =$

$\dfrac{4}{5} - \dfrac{10}{12} - \dfrac{10}{21} + \dfrac{10}{21} =$

$\dfrac{4}{5} - \dfrac{10}{12} = \left(\dfrac{4}{5} \cdot \dfrac{12}{12}\right) - \left(\dfrac{10}{12} \cdot \dfrac{5}{5}\right) =$

$\dfrac{48}{60} - \dfrac{50}{60} = -\dfrac{2}{60} = -\dfrac{1}{30}$

75. $24\left(\dfrac{3}{4} - \dfrac{1}{6}\right) =$

$24\left(\dfrac{3}{4}\right) - 24\left(\dfrac{1}{6}\right) =$

$18 - 4 = 14$

77. $64\left(\dfrac{3}{16} + \dfrac{5}{8} - \dfrac{1}{4} + \dfrac{1}{2}\right) =$

$64\left(\dfrac{3}{16}\right) + 64\left(\dfrac{5}{8}\right) - 64\left(\dfrac{1}{4}\right) + 64\left(\dfrac{1}{2}\right) =$

$12 + 40 - 16 + 32 = 68$

79. $\dfrac{7}{13}\left(\dfrac{2}{3} - \dfrac{1}{6}\right) = \dfrac{7}{13}\left(\dfrac{2}{3}\right) - \dfrac{7}{13}\left(\dfrac{1}{6}\right) =$

$\dfrac{14}{39} - \dfrac{7}{78} = \left(\dfrac{14 \cdot 2}{39 \cdot 2}\right) - \dfrac{7}{78} =$

$\dfrac{28}{78} - \dfrac{7}{78} = \dfrac{21}{78} = \dfrac{7}{26}$

81. $\dfrac{1}{3}x + \dfrac{2}{5}x = \left(\dfrac{1}{3} + \dfrac{2}{5}\right)x =$

$\left(\dfrac{5}{15} + \dfrac{6}{15}\right)x = \dfrac{11}{15}x$

83. $\dfrac{1}{3}a - \dfrac{1}{8}a = \left(\dfrac{1}{3} - \dfrac{1}{8}\right)a =$

$\left(\dfrac{8}{24} - \dfrac{3}{24}\right)a = \dfrac{5}{24}a$

85. $\dfrac{1}{2}x + \dfrac{2}{3}x + \dfrac{1}{6}x = \left(\dfrac{1}{2} + \dfrac{2}{3} + \dfrac{1}{6}\right)x =$

$\left(\dfrac{3}{6} + \dfrac{4}{6} + \dfrac{1}{6}\right)x = \dfrac{8}{6}x = \dfrac{4}{3}x$

87. $\dfrac{3}{5}n - \dfrac{1}{4}n + \dfrac{3}{10}n = \left(\dfrac{3}{5} - \dfrac{1}{4} + \dfrac{3}{10}\right)n =$

$\left(\dfrac{12}{20} - \dfrac{5}{20} + \dfrac{6}{20}\right)n = \dfrac{13}{20}n$

89. $n + \dfrac{4}{3}n - \dfrac{1}{9}n = \left(1 + \dfrac{4}{3} - \dfrac{1}{9}\right)n =$

$\left(\dfrac{9}{9} + \dfrac{12}{9} - \dfrac{1}{9}\right)n = \dfrac{20}{9}n$

Problem Set 2.2

91. $-n - \dfrac{7}{9}n - \dfrac{5}{12}n = \left(-1 - \dfrac{7}{9} - \dfrac{5}{12}\right)n =$

$\left(-\dfrac{36}{36} - \dfrac{28}{36} - \dfrac{15}{36}\right)n = -\dfrac{79}{36}n$

93. $\dfrac{3}{7}x + \dfrac{1}{4}y + \dfrac{1}{2}x + \dfrac{7}{8}y =$

$\dfrac{3}{7}x + \dfrac{1}{2}x + \dfrac{1}{4}y + \dfrac{7}{8}y =$

$\left(\dfrac{3}{7} + \dfrac{1}{2}\right)x + \left(\dfrac{1}{4} + \dfrac{7}{8}\right)y =$

$\left(\dfrac{6}{14} + \dfrac{7}{14}\right)x + \left(\dfrac{2}{8} + \dfrac{7}{8}\right)y =$

$\dfrac{13}{14}x + \dfrac{9}{8}y$

95. $\dfrac{2}{9}x + \dfrac{5}{12}y - \dfrac{7}{15}x - \dfrac{13}{15}y =$

$\dfrac{2}{9}x - \dfrac{7}{15}x + \dfrac{5}{12}y - \dfrac{13}{15}y =$

$\left(\dfrac{2}{9} - \dfrac{7}{15}\right)x + \left(\dfrac{5}{12} - \dfrac{13}{15}\right)y =$

$\left(\dfrac{10}{45} - \dfrac{21}{45}\right)x + \left(\dfrac{25}{60} - \dfrac{52}{60}\right)y =$

$-\dfrac{11}{45}x - \dfrac{27}{60}y = -\dfrac{11}{45}x - \dfrac{9}{20}y$

97. $\left(\begin{smallmatrix}\text{Remaining}\\\text{piece}\end{smallmatrix}\right) = \left(\begin{smallmatrix}\text{Original}\\\text{piece}\end{smallmatrix}\right) - \left(\begin{smallmatrix}\text{Piece}\\\text{cut off}\end{smallmatrix}\right)$

$= \left(12\dfrac{1}{2}\right) - \left(1\dfrac{3}{4}\right)$

$= \left(12 + \dfrac{1}{2}\right) - \left(1 + \dfrac{3}{4}\right)$

$= \dfrac{48}{4} + \dfrac{2}{4} - \dfrac{4}{4} - \dfrac{3}{4}$

$= \dfrac{43}{4} = 10\dfrac{3}{4} \text{ feet}$

99. $\left(\begin{smallmatrix}\text{Amount walk}\\\text{shortened}\end{smallmatrix}\right) = \left(\begin{smallmatrix}\text{Daily}\\\text{walk}\end{smallmatrix}\right) - \left(\begin{smallmatrix}\text{Amount}\\\text{walked}\end{smallmatrix}\right)$

$2\dfrac{1}{2} - \dfrac{3}{4} = \left(2 + \dfrac{1}{2}\right) - \dfrac{3}{4} =$

$\dfrac{8}{4} + \dfrac{2}{4} - \dfrac{3}{4} = \dfrac{7}{4} = 1\dfrac{3}{4} \text{ miles}$

101. Perimeter is the sum of the lengths of the three sides of the triangle.

$P = a + b + c$

$= 14\dfrac{1}{2} + 12\dfrac{1}{3} + 9\dfrac{5}{6}$

$= 14 + \dfrac{1}{2} + 12 + \dfrac{1}{3} + 9 + \dfrac{5}{6}$

$= 14 + 12 + 9 + \dfrac{1}{2} + \dfrac{1}{3} + \dfrac{5}{6}$

$= 35 + \dfrac{3}{6} + \dfrac{2}{6} + \dfrac{5}{6}$

$= 35 + \dfrac{10}{6} = 35 + 1\dfrac{4}{6} = 36\dfrac{2}{3}$

The plot of ground needs $36\frac{2}{3}$ yards of fencing.

PROBLEM SET **2.3** **Real Numbers and Algebraic Expressions**

1. $.37 + .25 = .62$

3. $2.93 - 1.48 = 1.45$

5. $(7.6) + (-3.8) = 3.8$

7. $(-4.7) + 1.4 = -3.3$

9. $-3.8 + 11.3 = 7.5$

11. $6.6 - (-1.2) = 6.6 + 1.2 = 7.8$

13. $-11.5 - (-10.6) = -11.5 + 10.6 = -.9$

15. $-17.2 - (-9.4) = -17.2 + 9.4 = -7.8$

17. $(.4)(2.9) = 1.16$

19. $(-.8)(.34) = -.272$

21. $(9)(-2.7) = -24.3$

23. $(-.7)(-64) = 44.8$

25. $(-.12)(-.13) = .0156$

27. $1.56 \div 1.3 = 1.2$

29. $5.92 \div (-.8) = -7.4$

31. $-.266 \div (-.7) = .38$

33. $16.5 - 18.7 + 9.4 =$
$16.5 + 9.4 - 18.7 =$
$25.9 - 18.7 = 7.2$

35. $.34 - .21 - .74 + .19 =$
$.34 + .19 - .21 - .74 =$
$.53 - .95 = -.42$

37. $.76(.2 + .8) = .76(1.0) = .76$

39. $.6(4.1) + .7(3.2) = 2.46 + 2.24 = 4.7$

41. $7(.6) + .9 - 3(.4) + .4 =$
$4.2 + .9 - 1.2 + .4 =$
$4.2 + .9 + .4 - 1.2 =$
$5.5 - 1.2 = 4.3$

43. $(.96) \div (-.8) + 6(-1.4) - 5.2 =$
$-1.2 + 6(-1.4) - 5.2 =$
$-1.2 + (-8.4) - 5.2 = -14.8$

45. $5(2.3) - 1.2 - 7.36 \div .8 + .2 =$
$11.5 - 1.2 - 7.36 \div .8 + .2 =$
$11.5 - 1.2 - 9.2 + .2 =$
$11.5 + .2 - 1.2 - 9.2 =$
$11.7 - 10.4 = 1.3$

47. $x - .4x - 1.8x =$
$(1 - .4 - 1.8)x = -1.2x$

49. $5.4n - .8n - 1.6n =$
$(5.4 - .8 - 1.6)n = 3n$

51. $-3t + 4.2t - .9t + .2t =$
$(-3 + 4.2 - .9 + .2)t = .5t$

53. $3.6x - 7.4y - 9.4x + 10.2y =$
$3.6x - 9.4x - 7.4y + 102.y =$
$(3.6 - 9.4)x + (-7.4 + 10.2)y =$
$-5.8x + 2.8y$

55. $.3(x - 4) + .4(x + 6) - .6x =$
$.3x - 1.2 + .4x + 2.4 - .6x =$
$.3x + .4x - .6x - 1.2 + 2.4 =$
$(.3 + .4 - .6)x + (-1.2 + 2.4) =$
$.1x + 1.2$

57. $6(x - 1.1) - 5(x - 2.3) - 4(x + 1.8) =$
$6x - 6.6 - 5x + 11.5 - 4x - 7.2 =$
$6x - 5x - 4x - 6.6 + 11.5 - 7.2 =$
$(6 - 5 - 4)x + (-6.6 + 11.5 - 7.2) =$
$-3x - 2.3$

Problem Set 2.3

59.
$5(x - .5) + .3(x - 2) - .7(x + 7) =$
$5x - 2.5 + .3x - .6 - .7x - 4.9 =$
$5x + .3x - .7x - 2.5 - .6 - 4.9 =$
$(5 + .3 - .7)x + (-2.5 - .6 - 4.9) =$
$4.6x - 8$

61. $x + 2y + 3z$ for $x = \dfrac{3}{4}$, $y = \dfrac{1}{3}$, $z = -\dfrac{1}{6}$

$\left(\dfrac{3}{4}\right) + 2\left(\dfrac{1}{3}\right) + 3\left(-\dfrac{1}{6}\right) =$

$\dfrac{3}{4} + \dfrac{2}{3} - \dfrac{1}{2} = \dfrac{11}{12}$

63. $\dfrac{3}{5}y - \dfrac{2}{3}y - \dfrac{7}{15}y =$

$\left(\dfrac{3}{5} - \dfrac{2}{3} - \dfrac{7}{15}\right)y =$

$-\dfrac{8}{15}y$ for $y = -\dfrac{5}{2}$

$-\dfrac{8}{15}\left(-\dfrac{5}{2}\right) = \dfrac{4}{3}$

65. $-x - 2y + 4z$ for $x = 1.7$, $y = -2.3$, $z = 3.6$
$-(1.7) - 2(-2.3) + 4(3.6) =$
$-1.7 + 4.6 + 14.4 = 17.3$

67. $5x - 7y$ for $x = -7.8$, $y = 8.4$
$5(-7.8) - 7(8.4) =$
$-39 - 58.8 = -97.8$

69. $.7x + .6y$ for $x = -2$, $y = 6$
$.7(-2) + .6(6) =$
$-1.4 + 3.6 = 2.2$

71. $1.2x + 2.3x - 1.4x - 7.6x =$
$(1.2 + 2.3 - 1.4 - 7.6)x =$
$-5.5x$ for $x = -2.5$
$-5.5(-2.5) = 13.75$

73. $-3a - 1 + 7a - 2 =$
$-3a + 7a - 1 - 2 =$
$4a - 3$ for $a = .9$
$4(.9) - 3 = 3.6 - 3 = .6$

75. $\begin{pmatrix} \text{value of} \\ \text{stock} \end{pmatrix} = \begin{pmatrix} \text{number of} \\ \text{shares} \end{pmatrix} \cdot \begin{pmatrix} \text{price per} \\ \text{share} \end{pmatrix}$

$\begin{pmatrix} \text{value of} \\ \text{400 shares} \end{pmatrix} = \left(400\right) \cdot \left(14.78\right) = 5912$

$\begin{pmatrix} \text{value of} \\ \text{250 shares} \end{pmatrix} = \left(250\right) \cdot \left(16.36\right) = 4090$

$\begin{matrix} \text{value of} \\ \text{650 shares} \end{matrix} = 5912 + 4090 = 10,002$

Total value of stock is \$10,002.

77. $\begin{pmatrix} \text{length of} \\ \text{each piece} \end{pmatrix} = \begin{pmatrix} \text{total} \\ \text{length} \end{pmatrix} \div \begin{pmatrix} \text{number of} \\ \text{pieces} \end{pmatrix}$

$= 76.4 \div 4 = 19.1$

The length of each piece is 19.1 cm.

79. Perimeter $= 4 \cdot \begin{pmatrix} \text{length of} \\ \text{each side} \end{pmatrix}$

$\dfrac{\text{Perimeter}}{4} = \text{length of each side}$

$\dfrac{18.8}{4} = 4.7$

The length of each side of the square is 4.7 cm.

81. $\begin{matrix} \text{cost of} \\ \text{apples} \end{matrix} = \begin{pmatrix} \text{number of} \\ \text{pounds} \end{pmatrix} \cdot \begin{pmatrix} \text{price per} \\ \text{pound} \end{pmatrix}$

$\begin{matrix} \text{price of} \\ \text{gala apples} \end{matrix} = (2)(1.79) = 3.58$

$\begin{matrix} \text{price of} \\ \text{fuji apples} \end{matrix} = (3)(.99) = 2.97$

Total cost of apples $= 3.58 + 2.97 = \$6.55$.

87. **(a)** $\dfrac{1}{7} = .\overline{142857}$

(b) $\dfrac{2}{7} = .\overline{285714}$

(c) $\dfrac{4}{9} = .\overline{4}$

(d) $\dfrac{5}{6} = .8\overline{3}$

(e) $\dfrac{3}{11} = .\overline{27}$

(f) $\dfrac{1}{12} = .08\overline{3}$

PROBLEM SET **2.4** **Exponents**

1. $2^6 = 2 \cdot 2 \cdot 2 \cdot 2 \cdot 2 \cdot 2 = 64$

3. $3^4 = 3 \cdot 3 \cdot 3 \cdot 3 = 81$

5. $(-2)^3 = (-2)(-2)(-2) = -8$

7. $-3^2 = -(3 \cdot 3) = -9$

9. $(-4)^2 = (-4)(-4) = 16$

11. $\left(\dfrac{2}{3}\right)^4 = \left(\dfrac{2}{3}\right)\left(\dfrac{2}{3}\right)\left(\dfrac{2}{3}\right)\left(\dfrac{2}{3}\right) = \dfrac{16}{81}$

13. $-\left(\dfrac{1}{2}\right)^3 = -\left(\dfrac{1}{2}\right)\left(\dfrac{1}{2}\right)\left(\dfrac{1}{2}\right) = -\dfrac{1}{8}$

15. $\left(-\dfrac{3}{2}\right)^2 = \left(-\dfrac{3}{2}\right)\left(-\dfrac{3}{2}\right) = \dfrac{9}{4}$

17. $(0.3)^3 = (0.3)(0.3)(0.3) = 0.027$

19. $-(1.2)^2 = -(1.2)(1.2) = -1.44$

21. $3^2 + 2^3 - 4^3 = 9 + 8 - 64 = -47$

23. $(-2)^3 - 2^4 - 3^2 =$
$-8 - 16 - 9 = -33$

25. $5(2)^2 - 4(2) - 1 = 5(4) - 8 - 1 =$
$20 - 8 - 1 = 11$

27. $-2(3)^3 - 3(3)^2 + 4(3) - 6 =$
$-2(27) - 3(9) + 12 - 6 =$
$-54 - 27 + 12 - 6 = -75$

29. $-7^2 - 6^2 + 5^2 =$
$-49 - 36 + 25 = -60$

31. $-3(-4)^2 - 2(-3)^3 + (-5)^2 =$
$-3(16) - 2(-27) + 25 =$
$-48 + 54 + 25 = 31$

33. $\dfrac{-3(2)^4}{12} + \dfrac{5(-3)^3}{15} =$
$\dfrac{-3(16)}{12} + \dfrac{5(-27)}{15} =$
$\dfrac{-48}{12} + \dfrac{-135}{15} =$
$-4 - 9 = -13$

35. $9 \cdot x \cdot x = 9x^2$

37. $3 \cdot 4 \cdot x \cdot y \cdot y = 12xy^2$

39. $-2 \cdot 9 \cdot x \cdot x \cdot x \cdot x \cdot y = -18x^4 y$

41. $(5x)(3y) = 5 \cdot 3 \cdot x \cdot y = 15xy$

43. $(6x^2)(2x^2) = 6 \cdot 2 \cdot x \cdot x \cdot x \cdot x = 12x^4$

45. $(-4a^2)(-2a^3) =$
$(-4)(-2) \cdot a \cdot a \cdot a \cdot a \cdot a =$
$8a^5$

47. $3x^2 - 7x^2 - 4x^2 =$
$(3 - 7 - 4)x^2 = -8x^2$

49. $-12y^3 + 17y^3 - y^3 =$
$(-12 + 17 - 1)y^3 = 4y^3$

Problem Set 2.4

51. $7x^2 - 2y^2 - 9x^2 + 8y^2 =$
$(7-9)x^2 + (-2+8)y^2 =$
$-2x^2 + 6y^2$

53. $\dfrac{2}{3}n^2 - \dfrac{1}{4}n^2 - \dfrac{3}{5}n^2 =$
$\left(\dfrac{2}{3} - \dfrac{1}{4} - \dfrac{3}{5}\right)n^2 =$
$\left(\dfrac{40}{60} - \dfrac{15}{60} - \dfrac{36}{60}\right)n^2 =$
$-\dfrac{11}{60}n^2$

55. $5x^2 - 8x - 7x^2 + 2x =$
$(5-7)x^2 + (-8+2)x =$
$-2x^2 - 6x$

57. $x^2 - 2x - 4 + 6x^2 - x + 12 =$
$x^2 + 6x^2 - 2x - x - 4 + 12 =$
$7x^2 - 3x + 8$

59. $\dfrac{9xy}{15x} = \dfrac{\overset{3}{\cancel{9}} \cdot \cancel{x} \cdot y}{\underset{5}{\cancel{15}} \cdot \cancel{x}} = \dfrac{3y}{5}$

61. $\dfrac{22xy^2}{6xy^3} = \dfrac{\overset{11}{\cancel{22}} \cdot \cancel{x} \cdot \cancel{y} \cdot \cancel{y}}{\underset{3}{\cancel{6}} \cdot \cancel{x} \cdot \cancel{y} \cdot \cancel{y} \cdot y} = \dfrac{11}{3y}$

63. $\dfrac{7a^2b^3}{17a^3b} = \dfrac{7 \cdot \cancel{a} \cdot \cancel{a} \cdot \cancel{b} \cdot b \cdot b}{17 \cdot \cancel{a} \cdot \cancel{a} \cdot a \cdot \cancel{b}} = \dfrac{7b^2}{17a}$

65. $\dfrac{-24abc^2}{32bc} = \dfrac{\overset{3}{\cancel{24}} \cdot a \cdot \cancel{b} \cdot \cancel{c} \cdot c}{\underset{4}{\cancel{32}} \cdot \cancel{b} \cdot \cancel{c}} = -\dfrac{3ac}{4}$

67. $\dfrac{-5x^4y^3}{-20x^2y} = \dfrac{\cancel{5} \cdot \cancel{x} \cdot \cancel{x} \cdot x \cdot x \cdot \cancel{y} \cdot y \cdot y}{4 \cdot \cancel{5} \cdot \cancel{x} \cdot \cancel{x} \cdot \cancel{y}} =$
$\dfrac{x^2y^2}{4}$

69. $\left(\dfrac{7x^2}{9y}\right)\left(\dfrac{12y}{21x}\right) = \dfrac{7 \cdot \overset{4}{\cancel{12}} \cdot \cancel{x} \cdot x \cdot \cancel{y}}{\underset{3}{\cancel{9}} \cdot \underset{3}{\cancel{21}} \cdot \cancel{x} \cdot \cancel{y}} = \dfrac{4x}{9}$

71. $\left(\dfrac{5c}{a^2b^2}\right) \div \left(\dfrac{12c}{ab}\right) = \dfrac{5c}{a^2b^2} \cdot \dfrac{ab}{12c} =$
$\dfrac{5 \cdot \cancel{a} \cdot \cancel{b} \cdot \cancel{c}}{12 \cdot \cancel{a} \cdot a \cdot \cancel{b} \cdot b \cdot \cancel{c}} = \dfrac{5}{12ab}$

73. $\dfrac{6}{x} + \dfrac{5}{y^2} = \dfrac{6 \cdot y^2}{x \cdot y^2} + \dfrac{5 \cdot x}{y^2 \cdot x} =$
$\dfrac{6y^2}{xy^2} + \dfrac{5x}{xy^2} = \dfrac{6y^2 + 5x}{xy^2}$

75. $\dfrac{5}{x^4} - \dfrac{7}{x^2} = \dfrac{5}{x^4} - \dfrac{7 \cdot x \cdot x}{x \cdot x \cdot x \cdot x} =$
$\dfrac{5}{x^4} - \dfrac{7x^2}{x^4} = \dfrac{5 - 7x^2}{x^4}$

77. $\dfrac{3}{2x^3} + \dfrac{6}{x} = \dfrac{3}{2x^3} + \dfrac{6 \cdot 2 \cdot x \cdot x}{x \cdot 2 \cdot x \cdot x} =$
$\dfrac{3}{2x^3} + \dfrac{12x^2}{2x^3} = \dfrac{3 + 12x^2}{2x^3}$

79. $\dfrac{-5}{4x^2} + \dfrac{7}{3x^2} = \dfrac{-5 \cdot 3}{4x^2 \cdot 3} + \dfrac{7 \cdot 4}{3x^2 \cdot 4} =$
$\dfrac{-15}{12x^2} + \dfrac{28}{12x^2} = \dfrac{13}{12x^2}$

81. $\dfrac{11}{a^2} - \dfrac{14}{b^2} = \dfrac{11 \cdot b^2}{a^2 \cdot b^2} - \dfrac{14 \cdot a^2}{b^2 \cdot a^2} =$
$\dfrac{11b^2}{a^2b^2} - \dfrac{14a^2}{a^2b^2} = \dfrac{11b^2 - 14a^2}{a^2b^2}$

83. $\dfrac{1}{2x^3} - \dfrac{4}{3x^2} = \dfrac{1 \cdot 3}{2x^3 \cdot 3} - \dfrac{4 \cdot 2x}{3x^2 \cdot 2x} =$
$\dfrac{3}{6x^3} - \dfrac{8x}{6x^3} = \dfrac{3 - 8x}{6x^3}$

85. $\dfrac{3}{x} - \dfrac{4}{y} - \dfrac{5}{xy} = \dfrac{3 \cdot y}{x \cdot y} - \dfrac{4 \cdot x}{y \cdot x} - \dfrac{5}{xy} =$
$\dfrac{3y}{xy} - \dfrac{4x}{xy} - \dfrac{5}{xy} = \dfrac{3y - 4x - 5}{xy}$

87. $4x^2 + 7y^2$ for $x = -2, y = -3$
$4(-2)^2 + 7(-3)^2 =$
$4(4) + 7(9) =$
$16 + 63 = 79$

26

89. $3x^2 - y^2$ for $x = \frac{1}{2}$, $y = -\frac{1}{3}$

$3\left(\frac{1}{2}\right)^2 - \left(-\frac{1}{3}\right)^2 =$

$3\left(\frac{1}{4}\right) - \left(\frac{1}{9}\right) =$

$\frac{3}{4} - \frac{1}{9} = \frac{3 \cdot 9}{4 \cdot 9} - \frac{1 \cdot 4}{9 \cdot 4} =$

$\frac{27}{36} - \frac{4}{36} = \frac{23}{36}$

91. $x^2 - 2xy + y^2$ for $x = -\frac{1}{2}$, $y = 2$

$\left(-\frac{1}{2}\right)^2 - 2\left(-\frac{1}{2}\right)(2) + (2)^2 =$

$\left(\frac{1}{4}\right) + 2 + 4 = \frac{1}{4} + \frac{2 \cdot 4}{1 \cdot 4} + \frac{4 \cdot 4}{1 \cdot 4} =$

$\frac{1}{4} + \frac{8}{4} + \frac{16}{4} = \frac{25}{4}$

93. $-x^2$ for $x = -8$
$-x^2 = -(-8)^2 =$
$-(-8)(-8) = -(64) = -64$

95. $-x^2 - y^2$ for $x = -3$, $y = -4$
$-(-3)^2 - (-4)^2 =$
$-9 - 16 = -25$

97. $-a^2 - 3b^3$ for $a = -6$, $b = -1$
$-(-6)^2 - 3(-1)^3 =$
$-36 - 3(-1) =$
$-36 + 3 = -33$

99. $y^2 - 3xy$ for $x = 0.4$, $y = -0.3$
$(-0.3)^2 - 3(0.4)(-0.3) =$
$0.09 - 3(-0.12) =$
$0.09 + 0.36 = 0.45$

PROBLEM SET **2.5** **Translating from English to Algebra**

For problems, 1-11, the answers will vary.

1. The difference of a and b

3. One-third of the product of B and h

5. Two times the quantity, l plus w

7. The quotient of A divided by w

9. The quantity, a plus b, divided by 2

11. Two more than three times y

13. $l + w$

15. ab

17. $\frac{d}{t}$

19. lwh

21. $y - x$

23. $xy + 2$

25. $7 - y^2$

27. $\frac{x - y}{4}$

29. $10 - x$

31. $10(n + 2)$

33. $xy - 7$

35. $xy - 12$

For problems 37-67, it may help to do a specific example before trying to formulate the general expression.

27

37. Suppose that the sum of two numbers is 35 and one of the numbers is 14. Then we subtract to find the other number, $35 - 14$. Thus, if one of the numbers is n, then the other number is $35 - n$.

39. Since the smaller number plus the difference must equal the larger number, the other number must be $n + 45$.

41. In ten years, Janet's age will be increased by 10, so it is represented as $y + 10$.

43. Twice Debra's age is $2x$, so Debra's mother is 3 years less than $2x$ or $2x - 3$.

45. Three dimes and five quarters is $10(3) + 25(2) = 80$ cents. Thus d dimes and q quarters is $10d + 25q$ cents.

47. If a car travels 200 miles in 5 hours, then it would be traveling at a rate of
$$\frac{200 \text{ miles}}{5 \text{ hours}} = 40 \text{ miles per hour.}$$
Therefore, the rate of the car is $\frac{d}{t}$.

49. If 5 pounds of candy cost $15, then to find the price per pound we divide 15 by 5.

Thus, if p pounds cost d dollars, $\frac{d}{p}$ represents the price per pound.

51. To find a monthly salary, the annual salary is divided by 12, so Larry's monthly salary is $\frac{d}{12}$.

53. If 6 is the whole number, then the next larger whole number would be $6 + 1 = 7$. Thus, if n is the whole number, the next larger whole number is $n + 1$.

55. Suppose 7 is the odd number, then the next larger odd number would be $7 + 2 = 9$. Thus, if n is the odd number, the next larger odd number would be $n + 2$.

57. If Willie is y years old, then twice Willie's age is $2y$. Since his father is 2 years less than twice Willie's age, the father's age is $2y - 2$. The sum of their ages is $y + (2y - 2) = 3y - 2$.

59. To convert yards to inches, the yards must be multiplied by 36. To convert feet to inches, the number of feet must be multiplied by 12. If the perimeter of a rectangle is 5 yards and 2 feet, then the perimeter in inches is $36(5) + 12(2) = 204$ inches. Thus, the perimeter of a rectangle that is y yards and f feet is $36y + 12f$ inches.

61. To change feet to yards, we divide the number of feet by 3. Therefore, f feet equals $\frac{f}{3}$ yards.

63. The width of a rectangle is w feet and the length is three times the width, or $3w$ feet. The perimeter of the rectangle is the sum of the lengths of the four sides. There are 2 widths and 2 lengths in a rectangle so the perimeter is 2 times the width plus 2 times the length or $2(w) + 2(3w) = 8w$ feet.

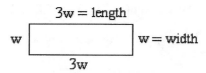

65. See problem 63. The length is l inches and the width is $\left(\frac{l}{2} - 2\right)$ inches. The perimeter is $2l + 2(\frac{l}{2} - 2) = 3l - 4$ inches.

l = length

$\frac{l}{2} - 2$ = width

28

67. The first side of a triangle is f feet long. The second side is 2 feet longer, or $f + 2$ feet long. The third side is twice as long as the second side, or $2(f + 2)$. The perimeter is the sum of the lengths of the sides, or $f + (f + 2) + 2(f + 2) = 4f + 6$ feet. To convert to inches, multiply the number of feet by 12. The perimeter of the triangle in inches is $12(4f + 6) = 48f + 72$ inches.

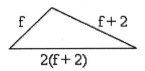

CHAPTER 2 **Review Problem Set**

1. $2^6 = 2 \cdot 2 \cdot 2 \cdot 2 \cdot 2 \cdot 2 = 64$

2. $(-3)^3 = (-3)(-3)(-3) = -27$

3. $-4^2 = -(4 \cdot 4) = -(16) = -16$

4. $\left(\dfrac{3}{4}\right)^2 = \left(\dfrac{3}{4}\right)\left(\dfrac{3}{4}\right) = \dfrac{9}{16}$

5. $\left(\dfrac{1}{2} + \dfrac{2}{3}\right)^2 = \left(\dfrac{3}{6} + \dfrac{4}{6}\right)^2 =$
$\left(\dfrac{7}{6}\right)^2 = \left(\dfrac{7}{6}\right)\left(\dfrac{7}{6}\right) = \dfrac{49}{36}$

6. $(0.6)^3 = (0.6)(0.6)(0.6) = 0.216$

7. $(0.12)^2 = (0.12)(0.12) = 0.0144$

8. $(0.06)^2 = (0.06)(0.06) = 0.0036$

9. $\left(-\dfrac{2}{3}\right)^3 =$
$\left(-\dfrac{2}{3}\right)\left(-\dfrac{2}{3}\right)\left(-\dfrac{2}{3}\right) = -\dfrac{8}{27}$

10. $\left(-\dfrac{1}{2}\right)^4 =$
$\left(-\dfrac{1}{2}\right)\left(-\dfrac{1}{2}\right)\left(-\dfrac{1}{2}\right)\left(-\dfrac{1}{2}\right) = \dfrac{1}{16}$

11. $\dfrac{3}{8} + \dfrac{5}{12} = \dfrac{3}{8} \cdot \dfrac{3}{3} + \dfrac{5}{12} \cdot \dfrac{2}{2} =$
$\dfrac{9}{24} + \dfrac{10}{24} = \dfrac{19}{24}$

12. $\dfrac{9}{14} - \dfrac{3}{35} = \dfrac{9}{14} \cdot \dfrac{5}{5} - \dfrac{3}{35} \cdot \dfrac{2}{2} =$
$\dfrac{45}{70} - \dfrac{6}{70} = \dfrac{39}{70}$

13. $\dfrac{2}{3} + \dfrac{-3}{5} = \dfrac{2}{3} \cdot \dfrac{5}{5} + \dfrac{-3}{5} \cdot \dfrac{3}{3} =$
$\dfrac{10}{15} + \dfrac{-9}{15} = \dfrac{1}{15}$

14. $\dfrac{7}{x} + \dfrac{9}{2y} = \dfrac{7 \cdot 2y}{x \cdot 2y} + \dfrac{9 \cdot x}{2y \cdot x} =$
$\dfrac{14y}{2xy} + \dfrac{9x}{2xy} = \dfrac{14y + 9x}{2xy}$

15. $\dfrac{5}{xy} - \dfrac{8}{x^2} = \dfrac{5 \cdot x}{xy \cdot x} - \dfrac{8 \cdot y}{x^2 \cdot y} =$
$\dfrac{5x}{x^2 y} - \dfrac{8y}{x^2 y} = \dfrac{5x - 8y}{x^2 y}$

16. $\left(\dfrac{7y}{8x}\right)\left(\dfrac{14x}{35}\right) = \dfrac{7y}{\cancel{8}\cancel{x}} \cdot \dfrac{\overset{7}{\cancel{14}}\cancel{x}}{\underset{5}{\cancel{35}}} = \dfrac{7y}{20}$

17. $\left(\dfrac{6xy}{9y^2}\right) \div \left(\dfrac{15y}{18x^2}\right) = \dfrac{6xy}{9y^2} \cdot \dfrac{18x^2}{15y} =$
$\dfrac{108x^3 y}{135y^3} = \dfrac{4x^3}{5y^2}$

Chapter 2 Review Problem Set

18. $\left(\dfrac{-3x}{12y}\right)\left(\dfrac{8y}{-7x}\right) = \dfrac{-3x}{12y} \cdot \dfrac{8y}{-7x} =$

$\dfrac{-24xy}{-84xy} = \dfrac{2}{7}$

19. $\left(-\dfrac{4y}{3x}\right)\left(-\dfrac{3x}{4y}\right) = -\dfrac{4y}{3x} \cdot -\dfrac{3x}{4y} =$

$\dfrac{12xy}{12xy} = 1$

20. $\left(\dfrac{6n}{7}\right)\left(\dfrac{9n}{8}\right) = \dfrac{6n}{7} \cdot \dfrac{9n}{8} =$

$\dfrac{54n^2}{56} = \dfrac{27n^2}{28}$

21. $\dfrac{1}{2} + \dfrac{2}{3} \cdot \dfrac{3}{4} - \dfrac{5}{6} \div \dfrac{8}{6} =$

$\dfrac{1}{6} + \dfrac{2}{4} - \dfrac{5}{6} \div \dfrac{8}{6} =$

$\dfrac{1}{6} + \dfrac{1}{2} - \dfrac{5}{6} \cdot \dfrac{6}{8} =$

$\dfrac{1}{6} + \dfrac{1}{2} - \dfrac{5}{8} =$

$\dfrac{4}{24} + \dfrac{12}{24} - \dfrac{15}{24} = \dfrac{1}{24}$

22. $\dfrac{3}{4} \cdot \dfrac{1}{2} - \dfrac{4}{3} \cdot \dfrac{3}{2} =$

$\dfrac{3}{8} - \dfrac{4}{2} = \dfrac{3}{8} - \dfrac{16}{8} = -\dfrac{13}{8}$

23. $\dfrac{7}{9} \cdot \dfrac{3}{5} + \dfrac{7}{9} \cdot \dfrac{2}{5} = \dfrac{7}{9}\left(\dfrac{3}{5} + \dfrac{2}{5}\right) =$

$\dfrac{7}{9}\left(\dfrac{3}{5} + \dfrac{2}{5}\right) = \dfrac{7}{9}\left(\dfrac{5}{5}\right) = \dfrac{7}{9} \cdot 1 = \dfrac{7}{9}$

24. $\dfrac{4}{5} \div \dfrac{1}{5} \cdot \dfrac{2}{3} - \dfrac{1}{4} = \left(\dfrac{4}{5} \cdot \dfrac{5}{1}\right) \cdot \dfrac{2}{3} - \dfrac{1}{4} =$

$\dfrac{4}{1} \cdot \dfrac{2}{3} - \dfrac{1}{4} = \dfrac{8}{3} - \dfrac{1}{4} = \dfrac{8 \cdot 4}{3 \cdot 4} - \dfrac{1 \cdot 3}{4 \cdot 3} =$

$\dfrac{32}{12} - \dfrac{3}{12} = \dfrac{29}{12}$

25. $\dfrac{2}{3} \cdot \dfrac{1}{4} \div \dfrac{1}{2} + \dfrac{2}{3} \cdot \dfrac{1}{4} =$

$\dfrac{2}{12} \div \dfrac{1}{2} + \dfrac{2}{12} =$

$\left(\dfrac{2}{12} \cdot \dfrac{2}{1}\right) + \dfrac{2}{12} =$

$\dfrac{4}{12} + \dfrac{2}{12} = \dfrac{6}{12} = \dfrac{1}{2}$

26. $0.48 + 0.72 - 0.35 - 0.18 =$
$1.20 - 0.53 = 0.67$

27. $0.81 + (0.6)(0.4) - (0.7)(0.8) =$
$0.81 + 0.24 - 0.56 =$
$1.05 - 0.56 = 0.49$

28. $1.28 \div .8 - .81 \div .9 + 1.7 =$
$1.6 - .81 \div .9 + 1.7 =$
$1.6 - .9 + 1.7 = 2.4$

29. $(.3)^2 + (.4)^2 - (.6)^2 =$
$.09 + .06 - .36 = -.11$

30. $(1.76)(.8) + (1.76)(.2) =$
$1.76(.8 + .2) = 1.76(1.0) = 1.76$

31. $\dfrac{3}{8}x^2 - \dfrac{2}{5}y^2 - \dfrac{2}{7}x^2 + \dfrac{3}{4}y^2 =$

$\dfrac{3}{8}x^2 - \dfrac{2}{7}x^2 - \dfrac{2}{5}y^2 + \dfrac{3}{4}y^2 =$

$\left(\dfrac{3}{8} - \dfrac{2}{7}\right)x^2 + \left(-\dfrac{2}{5} + \dfrac{3}{4}\right)y^2 =$

$\dfrac{5}{56}x^2 + \dfrac{7}{20}y^2$

32. $.24ab + .73bc - .82ab - .37bc =$
$.24ab - .82ab + .73bc - .37bc =$
$(.24 - .82)ab + (.73 - .37)bc =$
$-.58ab + .36bc$

33. $\dfrac{1}{2}x + \dfrac{3}{4}x - \dfrac{5}{6}x + \dfrac{1}{24}x =$

$\dfrac{12}{24}x + \dfrac{18}{24}x - \dfrac{20}{24}x + \dfrac{1}{24}x =$

$\dfrac{31}{24}x - \dfrac{20}{24}x = \dfrac{11}{24}x$

34. $1.4a - 1.9b + 0.8a + 3.6b = 2.2a + 1.7b$

35. $\dfrac{2}{5}n + \dfrac{1}{3}n - \dfrac{5}{6}n =$

$\dfrac{12}{30}n + \dfrac{10}{30}n - \dfrac{25}{30}n =$

$\dfrac{22}{30}n - \dfrac{25}{30}n = -\dfrac{3}{30}n = -\dfrac{1}{10}n$

36. $n - \dfrac{3}{4}n + 2n - \dfrac{1}{5}n =$

$\left(1 - \dfrac{3}{4} + 2 - \dfrac{1}{5}\right)n =$

$\left(\dfrac{20}{20} - \dfrac{15}{20} + \dfrac{40}{20} - \dfrac{4}{20}\right)n = \dfrac{41}{20}n$

37. $\dfrac{1}{4}x - \dfrac{2}{5}y$ for $x = \dfrac{2}{3}$, $y = -\dfrac{5}{7}$

$\dfrac{1}{4}\left(\dfrac{2}{3}\right) - \dfrac{2}{5}\left(-\dfrac{5}{7}\right) =$

$\dfrac{1}{6} + \dfrac{2}{7} = \dfrac{7}{42} + \dfrac{12}{42} = \dfrac{19}{42}$

38. $a^3 + b^2$ for $a = -\dfrac{1}{2}$, $b = \dfrac{1}{3}$

$\left(-\dfrac{1}{2}\right)\left(-\dfrac{1}{2}\right)\left(-\dfrac{1}{2}\right) + \left(\dfrac{1}{3}\right)\left(\dfrac{1}{3}\right) =$

$-\dfrac{1}{8} + \dfrac{1}{9} = -\dfrac{9}{72} + \dfrac{8}{72} = -\dfrac{1}{72}$

39. $2x^2 - 3y^2$ for $x = 0.6$, $y = 0.7$

$2(0.6)^2 - 3(0.7)^2 =$

$2(0.36) - 3(0.49) =$

$0.72 - 1.47 = -0.75$

40. $0.7w + 0.9z$ for $w = 0.4$, $z = -0.7$

$0.7(0.4) + 0.9(-0.7) =$

$0.28 - 0.63 = -0.35$

41. $\dfrac{3}{5}x - \dfrac{1}{3}x + \dfrac{7}{15}x - \dfrac{2}{3}x =$

$\left(\dfrac{3}{5} - \dfrac{1}{3} + \dfrac{7}{15} - \dfrac{2}{3}\right)x =$

$\left(\dfrac{9}{15} - \dfrac{5}{15} + \dfrac{7}{15} - \dfrac{10}{15}\right)x =$

$\dfrac{1}{15}x$ for $x = \dfrac{15}{17}$

$\dfrac{1}{15}\left(\dfrac{15}{17}\right) = \dfrac{1}{17}$

42. $\dfrac{1}{3}n + \dfrac{2}{7}n - n =$

$\left(\dfrac{1}{3} + \dfrac{2}{7} - 1\right)n =$

$\left(\dfrac{7}{21} + \dfrac{6}{21} - \dfrac{21}{21}\right)n =$

$-\dfrac{8}{21}n$ for $n = 21$

$-\dfrac{8}{21}\left(21\right) = -8$

43. If one number is n,
the other number is $72 - n$.

44. If Joan has 3 pennies and 4 dimes,
she has $3 + 10(4) = 43$ cents.
Thus, if she has p pennies and d dimes,
she has $p + 10d$ cents.

45. 1 hour = 60 minutes

$\dfrac{x \text{ words}}{1 \text{ hour}} \cdot \dfrac{1 \text{ hour}}{60 \text{ minutes}} = \dfrac{x}{60}$ words/minute

46. Twice Harry's age is $2y$. If his brother is
3 years less than twice Harry's age,
his brother is $2y - 3$ years old.

47. Larry chose n. Cindy chose 3 more than
5 times Larry's number or $5n + 3$.

Chapter 2 Review Problem Set

48. To convert yards to inches, the yards must be multiplied by 36. To convert feet to inches, the feet must be multiplied by 12. If the file cabinet is y yards and f feet tall, then it would be $36y + 12f$ inches tall.

49. $100\,\text{cm} = 1\text{m}$
m meters $\cdot\ 100\text{cm/m} = 100\text{m}$

50. If Corinne has 3 nickels, 4 dimes, and 2 quarters, she has
$5(3) + 10(4) + 25(2) = 105$ cents.
Thus, if she has n nickels, d dimes, and q quarters she has $5n + 10d + 25q$ cents.

51. $n - 5$

52. $5 - n$

53. $10(x - 2)$

54. $10x - 2$

55. $x - 3$

56. $\dfrac{d}{r}$

57. $x^2 + 9$

58. $(x + 9)^2$

59. $x^3 + y^3$

60. $xy - 4$

1. $(-3)^4 = (-3)(-3)(-3)(-3) = 81$
$-2^6 = -(2)(2)(2)(2)(2)(2) = -64$

2. $\dfrac{42}{54} = \dfrac{6 \cdot 7}{6 \cdot 9} = \dfrac{7}{9}$

3. $\dfrac{18xy^2}{32y} = \dfrac{2 \cdot 9 \cdot x \cdot y \cdot y}{2 \cdot 16 \cdot y} = \dfrac{9xy}{16}$

4. $5.7 - 3.8 + 4.6 - 9.1 =$
$5.7 + 4.6 - 3.8 - 9.1 =$
$10.3 - 12.9 = -2.6$

5. $.2(.4) - .6(.9) + .5(7) =$
$.08 - .54 + 3.5 = 3.04$

6. $-.4^2 + .3^2 - .7^2 =$
$-.16 + .09 - .49 = -.56$

7. $4(.21) - 3(.17) - 6(.04) =$
$.84 - .51 - .24 = .09$

8. $\dfrac{5}{12} \div \dfrac{15}{8} = \dfrac{\overset{1}{\cancel{5}}}{\underset{3}{\cancel{12}}} \cdot \dfrac{\overset{2}{\cancel{8}}}{\underset{3}{\cancel{15}}} = \dfrac{2}{9}$

9. $-\dfrac{2}{3} - \dfrac{1}{2}\left(\dfrac{3}{4}\right) + \dfrac{5}{6} =$
$-\dfrac{2}{3} - \dfrac{3}{8} + \dfrac{5}{6} =$
$-\dfrac{16}{24} - \dfrac{9}{24} + \dfrac{20}{24} = -\dfrac{5}{24}$

10. $3\left(\dfrac{2}{5}\right) - 4\left(\dfrac{5}{6}\right) + 6\left(\dfrac{7}{8}\right) =$
$\dfrac{6}{5} - \dfrac{10}{3} + \dfrac{21}{4} =$
$\dfrac{72}{60} - \dfrac{200}{60} + \dfrac{315}{60} = \dfrac{187}{60}$

11. $4\left(\dfrac{1}{2}\right)^3 - 3\left(\dfrac{2}{3}\right)^2 + 9\left(\dfrac{1}{4}\right)^2 =$
$4\left(\dfrac{1}{8}\right) - 3\left(\dfrac{4}{9}\right) + 9\left(\dfrac{1}{16}\right) =$

$$\frac{1}{2} - \frac{4}{3} + \frac{9}{16} =$$

$$\frac{24}{48} - \frac{64}{48} + \frac{27}{48} = -\frac{13}{48}$$

12. $\dfrac{8x}{15y} \cdot \dfrac{9y^2}{6x} = \dfrac{\cancel{72}^{\,4}\, xy^2}{\cancel{90}_{\,5}\, xy} = \dfrac{4y}{5}$

13. $\dfrac{6xy}{9} \div \dfrac{y}{3x} = \dfrac{6xy}{9} \cdot \dfrac{3x}{y} =$

$$\frac{18x^2 y}{9y} = 2x^2$$

14. $\dfrac{4}{x} - \dfrac{5}{y^2} = \left(\dfrac{4}{x} \cdot \dfrac{y^2}{y^2}\right) - \left(\dfrac{5}{y^2} \cdot \dfrac{x}{x}\right) =$

$$\frac{4y2}{xy^2} - \frac{5x}{xy^2} = \frac{4y^2 - 5x}{xy^2}$$

15. $\dfrac{3}{2x} + \dfrac{7}{6x} = \left(\dfrac{3}{2x} \cdot \dfrac{3}{3}\right) + \dfrac{7}{6x} =$

$$\frac{9}{6x} + \frac{7}{6x} = \frac{16}{6x} = \frac{8}{3x}$$

16. $\dfrac{5}{3y} + \dfrac{9}{7y^2} = \left(\dfrac{5}{3y} \cdot \dfrac{7y}{7y}\right) + \left(\dfrac{9}{7y^2} \cdot \dfrac{3}{3}\right) =$

$$\frac{35y}{21y^2} + \frac{27}{21y^2} = \frac{35y + 27}{21y^2}$$

17. $\left(\dfrac{15a^2 b}{12a}\right)\left(\dfrac{8ab}{9b}\right) = \dfrac{120a^3 b^2}{108ab} = \dfrac{10a^2 b}{9}$

18. $\dfrac{3}{x^2 y} - \dfrac{4}{x} + \dfrac{6}{y} =$

$$\frac{3}{x^2 y} - \left(\frac{4}{x} \cdot \frac{xy}{xy}\right) + \left(\frac{6}{y} \cdot \frac{x^2}{x^2}\right) =$$

$$\frac{3}{x^2 y} - \frac{4xy}{x^2 y} + \frac{6x^2}{x^2 y} =$$

$$\frac{3 - 4xy + 6x^2}{x^2 y}$$

19. $\dfrac{3}{5x} - \dfrac{5}{4} + \dfrac{7}{10x} =$

$$\left(\frac{3}{5x} \cdot \frac{4}{4}\right) - \left(\frac{5}{4} \cdot \frac{5x}{5x}\right) + \left(\frac{7}{10x} \cdot \frac{2}{2}\right) =$$

$$\frac{12}{20x} - \frac{25x}{20x} + \frac{14}{20x} =$$

$$\frac{12 - 25x + 14}{20x} = \frac{26 - 25x}{20x}$$

20. $x^2 - xy + y^2$ for $x = \dfrac{1}{2},\ y = -\dfrac{2}{3}$

$$\left(\frac{1}{2}\right)^2 - \left(\frac{1}{2}\right)\left(-\frac{2}{3}\right) + \left(-\frac{2}{3}\right)^2 =$$

$$\frac{1}{4} - \left(-\frac{1}{3}\right) + \frac{4}{9} =$$

$$\frac{9}{36} + \frac{12}{36} + \frac{16}{36} = \frac{37}{36}$$

21. $.2x - .3y - xy$ for $x = .4,\ y = .8$
$.2(.4) - .3(.8) - (.4)(.8) =$
$.08 - .24 - .32 = -.48$

22. $\dfrac{3}{4}x - \dfrac{2}{3}y$ for $x = -\dfrac{1}{2},\ y = \dfrac{3}{5}$

$$\frac{3}{4}\left(-\frac{1}{2}\right) - \frac{2}{3}\left(\frac{3}{5}\right) =$$

$$-\frac{3}{8} - \frac{2}{5} = -\frac{3}{8} \cdot \frac{5}{5} - \frac{2}{5} \cdot \frac{8}{8} =$$

$$-\frac{15}{40} - \frac{16}{40} = -\frac{31}{40}$$

23. $3x - 2y + xy$ for $x = .5,\ y = -.9$
$3(.5) - 2(-.9) + (.5)(-.9) =$
$1.5 - (-1.8) + (-.45) =$
$1.5 + 1.8 + (-.45) = 2.85$

24. If David has 3 nickels, 4 dimes, and 2 quarters, he has $5(3) + 10(4) + 25(2) = 105$ cents. So, if he has n nickels, d dimes, and q quarters, then he has $5n + 10d + 25q$ cents.

25. Hal chose n. Sheila chose 3 less than 4 times Hal's number or $4n - 3$.

Chapters 1-2 Cumulative Review

1. $16 - 18 - 14 + 21 - 14 + 19 =$
$56 - 46 = 10$

2. $7(-6) - 8(-6) + 4(-9) =$
$-42 - (-48) + (-36) =$
$-42 + 48 + (-36) = -30$

3. $6 - [3 - (10 - 12)] =$
$6 - [3 - (-2)] =$
$6 - [3 + 2] =$
$6 - (5) =$
$6 - 5 = 1$

4. $-9 - 2[4 - (-10 + 6)] - 1 =$
$-9 - 2[4 - (-4)] - 1 =$
$-9 - 2(8) - 1 =$
$-9 - 16 - 1 = -26$

5. $\dfrac{-7(-4) - 5(-6)}{-2} =$

$\dfrac{28 + 30}{-2} = \dfrac{58}{-2} = -29$

6. $\dfrac{5(-3) + (-4)(6) - 3(4)}{-3} =$

$\dfrac{-15 + (-24) - 12}{-3} =$

$\dfrac{-51}{-3} = 17$

7. $\dfrac{3}{4} + \dfrac{1}{3} \div \dfrac{4}{3} - \dfrac{1}{2} =$

$\dfrac{3}{4} + \dfrac{1}{3} \cdot \dfrac{3}{4} - \dfrac{1}{2} =$

$\dfrac{3}{4} + \dfrac{1}{4} - \dfrac{1}{2} =$

$\dfrac{4}{4} - \dfrac{1}{2} =$

$1 - \dfrac{1}{2} = \dfrac{1}{2}$

8. $\left(\dfrac{2}{3}\right)\left(-\dfrac{3}{4}\right) - \left(\dfrac{5}{6}\right)\left(\dfrac{4}{5}\right) =$

$-\dfrac{1}{2} - \dfrac{2}{3} = -\dfrac{3}{6} - \dfrac{4}{6} = -\dfrac{7}{6}$

9. $\left(\dfrac{1}{2} - \dfrac{2}{3}\right)^2 = \left(\dfrac{3}{6} - \dfrac{4}{6}\right)^2 =$

$\left(-\dfrac{1}{6}\right)^2 = \left(-\dfrac{1}{6}\right) \cdot \left(-\dfrac{1}{6}\right) = \dfrac{1}{36}$

10. $-4^3 = -(4^3) = -(4 \cdot 4 \cdot 4) =$
$-(64) = -64$

11. $\left(-\dfrac{1}{2}\right)^3 - \left(-\dfrac{3}{4}\right)^2 =$

$-\dfrac{1}{8} - \dfrac{9}{16} = -\dfrac{2}{16} - \dfrac{9}{16} = -\dfrac{11}{16}$

12. $(0.2)^2 - (0.3)^3 + (0.4)^2 =$
$0.04 - 0.027 + 0.16 =$
0.173

13. $3xy - 2x - 4y$ for $x = -6$ and $y = 7$
$3(-6)(7) - 2(-6) - 4(7) =$
$-126 + 12 - 28 =$
$-154 + 12 = -142$

14. $-4x^2y - 2xy^2 + xy$ for $x = -2$, $y = -4$
$-4(-2)^2(-4) - 2(-2)(-4)^2 + (-2)(-4) =$
$-4(4)(-4) - 2(-2)(16) + 8 =$
$64 + 64 + 8 = 136$

34

15. $\dfrac{5x - 2y}{3x}$ for $x = \dfrac{1}{2},\ y = -\dfrac{1}{3}$

$$\dfrac{5\left(\dfrac{1}{2}\right) - 2\left(-\dfrac{1}{3}\right)}{3\left(\dfrac{1}{2}\right)} =$$

$$\dfrac{\dfrac{5}{2} - \left(-\dfrac{2}{3}\right)}{\dfrac{3}{2}} = \dfrac{\dfrac{5}{2} + \dfrac{2}{3}}{\dfrac{3}{2}} =$$

$$\dfrac{\dfrac{15}{6} + \dfrac{4}{6}}{\dfrac{3}{2}} = \dfrac{\dfrac{19}{6}}{\dfrac{3}{2}} =$$

$$\dfrac{19}{6} \div \dfrac{3}{2} = \dfrac{19}{6} \cdot \dfrac{2}{3} = \dfrac{19}{9}$$

16. $.2x - .3y + 2xy$ for $x = .1,\ y = .3$
$.2(.1) - .3(.3) + 2(.1)(.3) =$
$.02 - .09 + .06 = -.01$

17. $-7x + 4y + 6x - 9y + x - y =$
$-7x + 6x + x + 4y - 9y - y =$
$-6y$ for $x = .2,\ y = .4$
$-6(.4) = -2.4$

18. $\dfrac{2}{3}x - \dfrac{3}{5}y + \dfrac{3}{4}x - \dfrac{1}{2}y =$
$\dfrac{2}{3}x + \dfrac{3}{4}x - \dfrac{3}{5}y - \dfrac{1}{2}y =$
$\left(\dfrac{2}{3} + \dfrac{3}{4}\right)x + \left(-\dfrac{3}{5} - \dfrac{1}{2}\right)y =$
$\dfrac{17}{12}x + \left(-\dfrac{11}{10}\right)y$ for $x = \dfrac{6}{5},\ y = -\dfrac{1}{4}$
$\dfrac{17}{12}\left(\dfrac{6}{5}\right) + \left(-\dfrac{11}{10}\right)\left(-\dfrac{1}{4}\right) =$
$\dfrac{17}{10} + \dfrac{11}{40} = \left(\dfrac{17}{10} \cdot \dfrac{4}{4}\right) + \dfrac{11}{40} =$
$\dfrac{68}{40} + \dfrac{11}{40} = \dfrac{79}{40}$

19. $\dfrac{1}{5}n - \dfrac{1}{3}n + n - \dfrac{1}{6}n =$
$\left(\dfrac{1}{5} - \dfrac{1}{3} + 1 - \dfrac{1}{6}\right)n =$
$\left(\dfrac{6}{30} - \dfrac{10}{30} + \dfrac{30}{30} - \dfrac{5}{30}\right)n =$
$\dfrac{21}{30}n = \dfrac{7}{10}n$ for $n = \dfrac{1}{5}$
$\dfrac{7}{10}\left(\dfrac{1}{5}\right) = \dfrac{7}{50}$

20. $-ab + \dfrac{1}{5}a - \dfrac{2}{3}b$ for $a = -2,\ b = \dfrac{3}{4}$
$-(-2)\left(\dfrac{3}{4}\right) + \dfrac{1}{5}(-2) - \left(\dfrac{2}{3}\right)\left(\dfrac{3}{4}\right) =$
$\dfrac{3}{2} + \left(-\dfrac{2}{5}\right) - \dfrac{1}{2} =$
$\dfrac{15}{10} + \left(-\dfrac{4}{10}\right) - \dfrac{5}{10} = \dfrac{6}{10} = \dfrac{3}{5}$

21. $54 = 2 \cdot 3 \cdot 3 \cdot 3$

22. $78 = 2 \cdot 3 \cdot 13$

23. $91 = 7 \cdot 13$

24. $153 = 3 \cdot 3 \cdot 17$

25. $\left.\begin{array}{l} 42 = 2 \cdot 3 \cdot 7 \\ 70 = 2 \cdot 5 \cdot 7 \end{array}\right\}$ The greatest common factor is $2 \cdot 7 = 14$.

26. $\left.\begin{array}{l} 63 = 3 \cdot 3 \cdot 7 \\ 81 = 3 \cdot 3 \cdot 3 \cdot 3 \end{array}\right\}$ The greatest common factor is $3 \cdot 3 = 9$.

27. $\left.\begin{array}{l} 28 = 2 \cdot 2 \cdot 7 \\ 36 = 2 \cdot 2 \cdot 3 \cdot 3 \\ 52 = 2 \cdot 2 \cdot 13 \end{array}\right\}$ The greatest common factor is $2 \cdot 2 = 4$.

28. $\left.\begin{array}{l} 48 = 2 \cdot 2 \cdot 2 \cdot 2 \cdot 3 \\ 66 = 2 \cdot 3 \cdot 11 \\ 78 = 2 \cdot 3 \cdot 13 \end{array}\right\}$ The greatest common factor is $2 \cdot 3 = 6$.

Chapters 1-2 Cumulative Review

29. $\left.\begin{array}{l} 20 = 2 \cdot 2 \cdot 5 \\ 28 = 2 \cdot 2 \cdot 7 \end{array}\right\}$ The least common multiple is $2 \cdot 2 \cdot 5 \cdot 7 = 140.$

30. $\left.\begin{array}{l} 40 = 2 \cdot 2 \cdot 2 \cdot 5 \\ 100 = 2 \cdot 2 \cdot 5 \cdot 5 \end{array}\right\}$ The least common multiple is $2 \cdot 2 \cdot 2 \cdot 5 \cdot 5 = 200.$

31. $\left.\begin{array}{l} 12 = 2 \cdot 2 \cdot 3 \\ 18 = 2 \cdot 3 \cdot 3 \\ 27 = 3 \cdot 3 \cdot 3 \end{array}\right\}$ The least common multiple is $2 \cdot 2 \cdot 3 \cdot 3 \cdot 3 = 108.$

32. $\left.\begin{array}{l} 16 = 2 \cdot 2 \cdot 2 \cdot 2 \\ 20 = 2 \cdot 2 \cdot 5 \\ 80 = 2 \cdot 2 \cdot 2 \cdot 2 \cdot 5 \end{array}\right\}$ The least common multiple is $2 \cdot 2 \cdot 2 \cdot 2 \cdot 5 = 80.$

33. $\dfrac{2}{3}x - \dfrac{1}{4}y - \dfrac{3}{4}x - \dfrac{2}{3}y =$

$\dfrac{2}{3}x - \dfrac{3}{4}x - \dfrac{1}{4}y - \dfrac{2}{3}y =$

$\dfrac{8}{12}x - \dfrac{9}{12}x - \dfrac{3}{12}y - \dfrac{8}{12}y =$

$-\dfrac{1}{12}x - \dfrac{11}{12}y$

34. $-n - \dfrac{1}{2}n + \dfrac{3}{5}n + \dfrac{5}{6}n =$

$\left(-1 - \dfrac{1}{2} + \dfrac{3}{5} + \dfrac{5}{6}\right)n =$

$\left(-\dfrac{30}{30} - \dfrac{15}{30} + \dfrac{18}{30} + \dfrac{25}{30}\right)n =$

$-\dfrac{2}{30}n = -\dfrac{1}{15}n$

35. $3.2a - 1.4b - 6.2a + 3.3b =$
$3.2a - 6.2a - 1.4b + 3.3b =$
$-3a + 1.9b$

36. $-(n-1) + 2(n-2) - 3(n-3) =$
$-n + 1 + 2n - 4 - 3n + 9 =$
$-2n + 6$

37. $-x + 4(x-1) - 3(x+2) - (x+5) =$
$-x + 4x - 4 - 3x - 6 - x - 5 =$
$-x + 4x - 3x - x - 4 - 6 - 5 =$
$-x - 15$

38. $2a - 5(a+3) - 2(a-1) - 4a =$
$2a - 5a - 15 - 2a + 2 - 4a =$
$2a - 5a - 2a - 4a - 15 + 2 =$
$-9a - 13$

39. $\dfrac{5}{12} - \dfrac{3}{16} = \dfrac{5}{12} \cdot \dfrac{4}{4} - \dfrac{3}{16} \cdot \dfrac{3}{3} =$

$\dfrac{20}{48} - \dfrac{9}{48} = \dfrac{11}{48}$

40. $\dfrac{3}{4} - \dfrac{5}{6} - \dfrac{7}{9} =$

$\dfrac{3}{4} \cdot \dfrac{9}{9} - \dfrac{5}{6} \cdot \dfrac{6}{6} - \dfrac{7}{9} \cdot \dfrac{4}{4} =$

$\dfrac{27}{36} - \dfrac{30}{36} - \dfrac{28}{36} = -\dfrac{31}{36}$

41. $\dfrac{5}{xy} - \dfrac{2}{x} + \dfrac{3}{y} =$

$\dfrac{5}{xy} - \dfrac{2}{x} \cdot \dfrac{y}{y} + \dfrac{3}{y} \cdot \dfrac{x}{x} =$

$\dfrac{5}{xy} - \dfrac{2y}{xy} + \dfrac{3x}{xy} =$

$\dfrac{5 - 2y + 3x}{xy}$

42. $-\dfrac{7}{x^2} + \dfrac{9}{xy} = \left(-\dfrac{7}{x^2} \cdot \dfrac{y}{y}\right) + \left(\dfrac{9}{xy} \cdot \dfrac{x}{x}\right) =$

$-\dfrac{7y}{x^2 y} + \dfrac{9x}{x^2 y} = \dfrac{-7y + 9x}{x^2 y}$

43. $\left(\dfrac{7x}{9y}\right)\left(\dfrac{12y}{14}\right) = \dfrac{\overset{1}{\cancel{7}}x}{\underset{3}{\cancel{9}}y} \cdot \dfrac{\overset{4}{\cancel{12}}\,y}{\underset{2}{\cancel{14}}} =$

$\dfrac{4xy}{6y} = \dfrac{2x}{3}$

44. $\left(-\dfrac{5a}{7b^2}\right)\left(-\dfrac{8ab}{15}\right) = -\dfrac{\overset{1}{\cancel{5}}a}{7b^2} \cdot -\dfrac{8ab}{\underset{3}{\cancel{15}}} =$

$\dfrac{8a^2b}{21b^2} = \dfrac{8a^2}{21b}$

45. $\left(\dfrac{6x^2y}{11}\right) \div \left(\dfrac{9y^2}{22}\right) =$

$\dfrac{\overset{2}{\cancel{6}}x^2y}{\underset{1}{\cancel{11}}} \cdot \dfrac{\overset{2}{\cancel{22}}}{\underset{3}{\cancel{9}}y^2} =$

$\dfrac{4x^2y}{3y^2} = \dfrac{4x^2}{3y}$

46. $\left(\dfrac{-9a}{8b}\right) \div \left(\dfrac{12a}{18b}\right) =$

$-\dfrac{\overset{3}{\cancel{9}}a}{\underset{4}{\cancel{8}}b} \cdot \dfrac{\overset{9}{\cancel{18}}\,b}{\underset{4}{\cancel{12}}\,a} = -\dfrac{27ab}{16ab} = -\dfrac{27}{16}$

47. p pennies and n nickels and d dimes is $p + 5n + 10d$ cents.

48. $4n - 5$

49. y yards and f feet and i inches is $36y + 12f + i$ inches.

50.

Perimeter of a Rectangle	=	the sum of the lengths of the sides

$\begin{aligned}\text{Perimeter in meters} &= x + x + y + y = \\ &= (2x + 2y) \text{ meters}\end{aligned}$

$1 \text{ meter} = 100 \text{ centimeters}$

To change to centimeters multiply meters by 100, so

$\begin{aligned}(2x + 2y)\text{meters} &= (2x + 2y) \cdot 100 \text{ cm} = \\ &= (200x + 200y)\text{cm}\end{aligned}$

Perimeter in centimeters is $(200x + 200y)\text{cm}$.

Problem Set 3.1

Chapter 3 First-Degree Equations and Inequalities of One Variable

PROBLEM SET **3.1** Solving First-Degree Equations

1.
$$x + 9 = 17$$
$$x + 9 - 9 = 17 - 9$$
Subtract 9 from both sides.
$$x = 8$$
The solution set is $\{8\}$.

3.
$$x + 11 = 5$$
$$x + 11 - 11 = 5 - 11$$
Subtract 11 from both sides.
$$x = -6$$
The solution set is $\{-6\}$.

5.
$$-7 = x + 2$$
$$-7 - 2 = x + 2 - 2$$
Subtract 2 from both sides.
$$-9 = x$$
The solution set is $\{-9\}$.

7.
$$8 = n + 14$$
$$8 - 14 = n + 14 - 14$$
Subtract 14 from both sides.
$$-6 = n$$
The solution set is $\{-6\}$.

9.
$$21 + y = 34$$
$$21 + y - 21 = 34 - 21$$
Subtract 21 from both sides.
$$y = 13$$
The solution set is $\{13\}$.

11.
$$x - 17 = 31$$
$$x - 17 + 17 = 31 + 17$$
Add 17 to both sides.
$$x = 48$$
The solution set is $\{48\}$.

13.
$$14 = x - 9$$
$$14 + 9 = x - 9 + 9$$
Add 9 to both sides.
$$23 = x$$
The solution set is $\{23\}$.

15.
$$-26 = n - 19$$
$$-26 + 19 = n - 19 + 19$$
Add 19 to both sides.
$$-7 = n$$
The solution set is $\{-7\}$.

17.
$$y - \frac{2}{3} = \frac{3}{4}$$
$$y - \frac{2}{3} + \frac{2}{3} = \frac{3}{4} + \frac{2}{3}$$
Add $\frac{2}{3}$ to both sides.
$$y = \frac{9}{12} + \frac{8}{12} = \frac{17}{12}$$
The solution set is $\left\{ \frac{17}{12} \right\}$.

19.
$$x + \frac{3}{5} = \frac{1}{3}$$
$$x + \frac{3}{5} - \frac{3}{5} = \frac{1}{3} - \frac{3}{5}$$
Subtract $\frac{3}{5}$ from both sides.
$$x = \frac{5}{15} - \frac{9}{15} = -\frac{4}{15}$$
The solution set is $\left\{ -\frac{4}{15} \right\}$.

21.
$$b + 0.19 = 0.46$$
$$b + 0.19 - 0.19 = 0.46 - 0.19$$
Subtract 0.19 from both sides.
$$b = 0.27$$
The solution set is $\{0.27\}$.

23.
$$n - 1.7 = -5.2$$
$$n - 1.7 + 1.7 = -5.2 + 1.7$$
Add 1.7 to both sides.
$$n = -3.5$$
The solution set is $\{-3.5\}$.

25.
$$15 - x = 32$$
$$15 - x - 15 = 32 - 15$$
Subtract 15 from both sides.
$$-x = 17$$
$$-1(-x) = -1(17)$$
Multiply both sides by -1.
$$x = -17$$
The solution set is $\{-17\}$.

27.
$$-14 - n = 21$$
$$-14 - n + 14 = 21 + 14$$
Add 14 to both sides.
$$-n = 35$$
$$-1(-n) = -1(35)$$
Multiply both sides by -1.
$$n = -35$$
The solution set is $\{-35\}$.

29.
$$7x = -56$$
$$\frac{7x}{7} = \frac{-56}{7}$$
Divide both sides by 7.
$$x = -8$$
The solution set is $\{-8\}$.

31.
$$-6x = 102$$
$$\frac{-6x}{-6} = \frac{102}{-6}$$
Divide both sides by -6.
$$x = -17$$
The solution set is $\{-17\}$.

33.
$$5x = 37$$
$$\frac{5x}{5} = \frac{37}{5}$$
Divide both sides by 5.
$$x = \frac{37}{5}$$
The solution set is $\left\{\frac{37}{5}\right\}$.

35.
$$-18 = 6n$$
$$\frac{-18}{6} = \frac{6n}{6}$$
Divide both sides by 6.
$$-3 = n$$
The solution set is $\{-3\}$.

37.
$$-26 = -4n$$
$$\frac{-26}{-4} = \frac{-4n}{-4}$$
Divide both sides by -4.
$$\frac{13}{2} = n$$
The solution set is $\left\{\frac{13}{2}\right\}$.

39.
$$\frac{t}{9} = 16$$
$$9\left(\frac{t}{9}\right) = 9(16)$$
Multiply both sides by 9.
$$t = 144$$
The solution set is $\{144\}$.

41.
$$\frac{n}{-8} = -3$$
$$-8\left(\frac{n}{-8}\right) = -8(-3)$$
Multiply both sides by -8.
$$n = 24$$
The solution set is $\{24\}$.

43.
$$-x = 15$$
$$-1(-x) = -1(15)$$
Multiply both sides by -1.
$$x = -15$$
The solution set is $\{-15\}$.

Problem Set 3.1

45.
$$\frac{3}{4}x = 18$$
$$\frac{4}{3}\left(\frac{3}{4}x\right) = \frac{4}{3}(18)$$
Multiply both sides by $\frac{4}{3}$.
$$x = 24$$
The solution set is $\{24\}$.

47.
$$-\frac{2}{5}n = 14$$
$$-\frac{5}{2}\left(-\frac{2}{5}n\right) = -\frac{5}{2}(14)$$
Multiply both sides by $-\frac{5}{2}$.
$$n = -35$$
The solution set is $\{-35\}$.

49.
$$\frac{2}{3}n = \frac{1}{5}$$
$$\frac{3}{2}\left(\frac{2}{3}n\right) = \frac{3}{2}\left(\frac{1}{5}\right)$$
Multiply both sides by $\frac{3}{2}$.
$$n = \frac{3}{10}$$
The solution set is $\left\{\frac{3}{10}\right\}$.

51.
$$\frac{5}{6}n = -\frac{3}{4}$$
$$\frac{6}{5}\left(\frac{5}{6}n\right) = \frac{6}{5}\left(-\frac{3}{4}\right)$$
Multiply both sides by $\frac{6}{5}$.
$$n = -\frac{9}{10}$$
The solution set is $\left\{-\frac{9}{10}\right\}$.

53.
$$\frac{3x}{10} = \frac{3}{20}$$
$$\frac{10}{3}\left(\frac{3x}{10}\right) = \frac{10}{3}\left(\frac{3}{20}\right)$$
Multiply both sides by $\frac{10}{3}$.
$$x = \frac{1}{2}$$
The solution set is $\left\{\frac{1}{2}\right\}$.

55.
$$\frac{-y}{2} = \frac{1}{6}$$
$$-2\left(\frac{-y}{2}\right) = -2\left(\frac{1}{6}\right)$$
Multiply both sides by -2.
$$y = -\frac{1}{3}$$
The solution set is $\left\{-\frac{1}{3}\right\}$.

57.
$$-\frac{4}{3}x = -\frac{9}{8}$$
$$-\frac{3}{4}\left(-\frac{4}{3}x\right) = -\frac{3}{4}\left(-\frac{9}{8}\right)$$
Multiply both sides by $-\frac{3}{4}$.
$$x = \frac{27}{32}$$
The solution set is $\left\{\frac{27}{32}\right\}$.

59.
$$-\frac{5}{12} = \frac{7}{6}x$$
$$\frac{6}{7}\left(-\frac{5}{12}\right) = \frac{6}{7}\left(\frac{7}{6}x\right)$$
Multiply both sides by $\frac{6}{7}$.
$$-\frac{5}{14} = x$$
The solution set is $\left\{-\frac{5}{14}\right\}$.

61.
$$-\frac{5}{7}x = 1$$

$$-\frac{7}{5}\left(-\frac{5}{7}x\right) = -\frac{7}{5}(1)$$

Multiply both sides by $-\frac{7}{5}$.

$$x = -\frac{7}{5}$$

The solution set is $\left\{-\frac{7}{5}\right\}$.

63.
$$-4n = \frac{1}{3}$$

$$-\frac{1}{4}(-4n) = -\frac{1}{4}\left(\frac{1}{3}\right)$$

Multiply both sides by $-\frac{1}{4}$.

$$n = -\frac{1}{12}$$

The solution set is $\left\{-\frac{1}{12}\right\}$.

65.
$$-8n = \frac{6}{5}$$

$$-\frac{1}{8}(-8n) = -\frac{1}{8}\left(\frac{6}{5}\right)$$

Multiply both sides by $-\frac{1}{8}$.

$$n = -\frac{3}{20}$$

The solution set is $\left\{-\frac{3}{20}\right\}$.

67.
$$1.2x = 0.36$$

$$\frac{1.2x}{1.2} = \frac{0.36}{1.2}$$

Divide both sides by 1.2.

$$x = 0.3$$

The solution set is $\{0.3\}$.

69.
$$30.6 = 3.4n$$

$$\frac{30.6}{3.4} = \frac{3.4n}{3.4}$$

Divide both sides by 3.4.

$$9 = n$$

The solution set is $\{9\}$.

71.
$$-3.4x = 17$$

$$\frac{-3.4x}{-3.4} = \frac{17}{-3.4}$$

Divide both sides by -3.4.

$$x = -5$$

The solution set is $\{-5\}$.

PROBLEM SET **3.2** **Equations and Problem Solving**

1.
$$2x + 5 = 13$$
$$2x + 5 - 5 = 13 - 5$$
Subtract 5 from both sides.
$$2x = 8$$
$$\frac{2x}{2} = \frac{8}{2}$$
Divide both sides by 2.
$$x = 4$$
The solution set is $\{4\}$.

3.
$$5x + 2 = 32$$
$$5x + 2 - 2 = 32 - 2$$
Subtract 2 from both sides.
$$5x = 30$$
$$\frac{5x}{5} = \frac{30}{5}$$
Divide both sides by 5.
$$x = 6$$
The solution set is $\{6\}$.

Problem Set 3.2

5.
$$3x - 1 = 23$$
$$3x - 1 + 1 = 23 + 1$$
Add 1 to both sides.
$$3x = 24$$
$$\frac{3x}{3} = \frac{24}{3}$$
Divide both sides by 3.
$$x = 8$$
The solution set is $\{8\}$.

7.
$$4n - 3 = 41$$
$$4n - 3 + 3 = 41 + 3$$
Add 3 to both sides.
$$4n = 44$$
$$\frac{4n}{4} = \frac{44}{4}$$
Divide both sides by 4.
$$x = 11$$
The solution set is $\{11\}$.

9.
$$6y - 1 = 16$$
$$6y - 1 + 1 = 16 + 1$$
Add 1 to both sides.
$$6y = 17$$
$$\frac{6y}{6} = \frac{17}{6}$$
Divide both sides by 6.
$$x = \frac{17}{6}$$
The solution set is $\left\{\dfrac{17}{6}\right\}$.

11.
$$2x + 3 = 22$$
$$2x + 3 - 3 = 22 - 3$$
Subtract 3 from both sides.
$$2x = 19$$
$$\frac{2x}{2} = \frac{19}{2}$$
Divide both sides by 2.
$$x = \frac{19}{2}$$
The solution set is $\left\{\dfrac{19}{2}\right\}$.

13.
$$10 = 3t - 8$$
$$10 + 8 = 3t - 8 + 8$$
Add 8 to both sides.
$$18 = 3t$$
$$\frac{18}{3} = \frac{3t}{3}$$
Divide both sides by 3.
$$6 = t$$
The solution set is $\{6\}$.

15.
$$5x + 14 = 9$$
$$5x + 14 - 14 = 9 - 14$$
Subtract 14 from both sides.
$$5x = -5$$
$$\frac{5x}{5} = \frac{-5}{5}$$
Divide both sides by 5.
$$x = -1$$
The solution set is $\{-1\}$.

17.
$$18 - n = 23$$
$$18 - n - 18 = 23 - 18$$
Subtract 18 from both sides.
$$-n = 5$$
$$-1(-n) = -1(5)$$
Multiply both sides by -1.
$$n = -5$$
The solution set is $\{-5\}$.

19.
$$-3x + 2 = 20$$
$$-3x + 2 - 2 = 20 - 2$$
Subtract 2 from both sides.
$$-3x = 18$$
$$\frac{-3x}{-3} = \frac{18}{-3}$$
Divide both sides by -3.
$$x = -6$$
The solution set is $\{-6\}$.

21.
$$7 + 4x = 29$$
$$7 + 4x - 7 = 29 - 7$$
Subtract 7 from both sides.
$$4x = 22$$
$$\frac{4x}{4} = \frac{22}{4}$$
Divide both sides by 4.
$$x = \frac{11}{2}$$
The solution set is $\left\{ \frac{11}{2} \right\}$.

23.
$$16 = -2 - 9a$$
$$16 + 2 = -2 - 9a + 2$$
Add 2 to both sides.
$$18 = -9a$$
$$\frac{18}{-9} = \frac{-9a}{-9}$$
Divide both sides by -9.
$$-2 = a$$
The solution set is $\{ -2 \}$.

25.
$$-7x + 3 = -7$$
$$-7x + 3 - 3 = -7 - 3$$
Subtract 3 from both sides.
$$-7x = -10$$
$$\frac{-7x}{-7} = \frac{-10}{-7}$$
Divide both sides by -7.
$$x = \frac{10}{7}$$
The solution set is $\left\{ \frac{10}{7} \right\}$.

27.
$$17 - 2x = -19$$
$$17 - 2x - 17 = -19 - 17$$
Subtract 17 from both sides.
$$-2x = -36$$
$$\frac{-2x}{-2} = \frac{-36}{-2}$$
Divide both sides by -2.
$$x = 18$$
The solution set is $\{18\}$.

29.
$$-16 - 4x = 9$$
$$-16 - 4x + 16 = 9 + 16$$
Add 16 to both sides.
$$-4x = 25$$
$$\frac{-4x}{-4} = \frac{25}{-4}$$
Divide both sides by -4.
$$x = -\frac{25}{4}$$
The solution set is $\left\{ -\frac{25}{4} \right\}$.

31.
$$-12t + 4 = 88$$
$$-12t + 4 - 4 = 88 - 4$$
Subtract 4 from both sides.
$$-12t = 84$$
$$\frac{-12t}{-12} = \frac{84}{-12}$$
Divide both sides by -12.
$$t = -7$$
The solution set is $\{ -7 \}$.

33.
$$14y + 15 = -33$$
$$14y + 15 - 15 = -33 - 15$$
Subtract 15 from both sides.
$$14y = -48$$
$$\frac{14y}{14} = \frac{-48}{14}$$
Divide both sides by 14.
$$y = -\frac{24}{7}$$
The solution set is $\left\{ -\frac{24}{7} \right\}$.

35.
$$32 - 16n = -8$$
$$32 - 16n - 32 = -8 - 32$$
Subtract 32 from both sides.
$$-16n = -40$$
$$\frac{-16n}{-16} = \frac{-40}{-16}$$
Divide both sides by -16.
$$n = \frac{5}{2}$$
The solution set is $\left\{ \frac{5}{2} \right\}$.

Problem Set 3.2

37.
$$17x - 41 = -37$$
$$17x - 41 + 41 = -37 + 41$$
Add 41 to both sides.
$$17x = 4$$
$$\frac{17x}{17} = \frac{4}{17}$$
Divide both sides by 17.
$$x = \frac{4}{17}$$
The solution set is $\left\{ \dfrac{4}{17} \right\}$.

39.
$$29 = -7 - 15x$$
$$29 + 7 = -7 - 15x + 7$$
Add 7 to both sides.
$$36 = -15x$$
$$\frac{36}{-15} = \frac{-15x}{-15}$$
Divide both sides by -15.
$$-\frac{12}{5} = x$$
The solution set is $\left\{ -\dfrac{12}{5} \right\}$.

41. Let n represent the number.
$$n + 12 = 21$$
$$n = 9$$
The number is 9.

43. Let n represent the number.
$$n - 9 = 13$$
$$n = 22$$
The number is 22.

45. Let c represent the cost of the other item.
$$c + 25 = 43$$
$$c = 18$$
The cost of the other item is $18.

47. Let x represent Nora's age now; therefore, $x + 6$ represents her age 6 years from now.
$$x + 6 = 41$$
$$x = 35$$
Nora is presently 35 years old.

49. Let h represent his hourly rate. Then the product of the number of hours times the hourly rate equals total amount earned.
$$6h = 39$$
$$h = \frac{39}{6} = 6.5$$
His hourly rate was $6.50.

51. Let x represent the number. Then 3 times the number is $3x$.
$$3x + 6 = 24$$
$$3x = 18$$
$$x = 6$$
The number is 6.

53. Let n represent the number. Then 3 times the number is $3n$.
$$19 = 3n + 4$$
$$15 = 3n$$
$$5 = n$$
The number is 5.

55. Let x represent the price of athletic socks, then $2.50 + x$ represents the price of dress socks.
$$2.50 + x + 6x = 21.75$$
$$2.50 + 7x = 21.75$$
$$7x = 19.25$$
$$x = 2.75$$
A pair of dress socks costs
$$2.50 + 2.75 = \$5.25.$$

57. Let x represent July's rainfall amount, then $2x - 1$ represents June's rainfall amount.
$$2x - 1 = 11.2$$
$$2x = 12.2$$
$$x = 6.1$$
The rainfall amount for July is 6.1 inches.

59. Let x represent the number.
$$27 - 8x = 3$$
$$-8x = -24$$
$$x = 3$$
The number is 3.

61. Let c represent the cost of the ring.
$550 = 2c - 50$
$600 = 2c$
$300 = c$
The cost of the ring was \$300.

63. Let w represent the width of the floor.
$18 = 5w - 2$
$20 = 5w$
$4 = w$
The width of the floor is 4 meters.

65. Let n represent the number of cars sold during December of 1998.
$32 = 2n + 4$
$28 = 2n$
$14 = n$
They sold 14 cars during December of 1998.

67. Let h represent the hours of labor.
$156 = 36 + 24h$
$120 = 24h$
$5 = h$
There were 5 hours of labor in the repair bill.

PROBLEM SET **3.3** **More on Solving Equations and Problem Solving**

1. $2x + 7 + 3x = 32$
$5x + 7 = 32$
$5x + 7 - 7 = 32 - 7$
$5x = 25$
$\dfrac{5x}{5} = \dfrac{25}{5}$
$x = 5$
The solution set is $\{5\}$.

3. $7x - 4 - 3x = -36$
$4x - 4 = -36$
$4x - 4 + 4 = -36 + 4$
$4x = -32$
$\dfrac{4x}{4} = \dfrac{-32}{4}$
$x = -8$
The solution set is $\{-8\}$.

5. $3y - 1 + 2y - 3 = 4$
$5y - 4 = 4$
$5y - 4 + 4 = 4 + 4$
$5y = 8$
$\dfrac{5y}{5} = \dfrac{8}{5}$
$y = \dfrac{8}{5}$
The solution set is $\left\{\dfrac{8}{5}\right\}$.

7. $5n - 2 - 8n = 31$
$-3n - 2 = 31$
$-3n - 2 + 2 = 31 + 2$
$-3n = 33$
$\dfrac{-3n}{-3} = \dfrac{33}{-3}$
$n = -11$
The solution set is $\{-11\}$.

9. $-2n + 1 - 3n + n - 4 = 7$
$-4n - 3 = 7$
$-4n - 3 + 3 = 7 + 3$
$-4n = 10$
$\dfrac{-4n}{-4} = \dfrac{10}{-4}$
$n = -\dfrac{5}{2}$
The solution set is $\left\{-\dfrac{5}{2}\right\}$.

11. $3x + 4 = 2x - 5$
$3x + 4 - 2x = 2x - 5 - 2x$
$x + 4 - 4 = -5 - 4$
$x = -9$
The solution set is $\{-9\}$.

Problem Set 3.3

13.
$$5x - 7 = 6x - 9$$
$$5x - 7 - 6x = 6x - 9 - 6x$$
$$-x - 7 + 7 = -9 + 7$$
$$-x = -2$$
$$x = 2$$
The solution set is $\{2\}$.

15.
$$6x + 1 = 3x - 8$$
$$6x + 1 - 3x = 3x - 8 - 3x$$
$$3x + 1 - 1 = -8 - 1$$
$$3x = -9$$
$$\frac{3x}{3} = \frac{-9}{3}$$
$$x = -3$$
The solution set is $\{-3\}$.

17.
$$7y - 3 = 5y + 10$$
$$7y - 3 - 5y = 5y + 10 - 5y$$
$$2y - 3 + 3 = 10 + 3$$
$$2y = 13$$
$$\frac{2y}{2} = \frac{13}{2}$$
$$y = \frac{13}{2}$$
The solution set is $\left\{\dfrac{13}{2}\right\}$.

19.
$$8n - 2 = 8n - 7$$
$$8n - 2 - 8n = 8n - 7 - 8n$$
$$-2 = -7$$
Since $-2 \neq -7$, this is a contradiction.
The solution set is $\emptyset$.

21.
$$-2x - 7 = -3x + 10$$
$$-2x - 7 + 3x = -3x + 10 + 3x$$
$$x - 7 + 7 = 10 + 7$$
$$x = 17$$
The solution set is $\{17\}$.

23.
$$-3x + 5 = -5x - 8$$
$$-3x + 5 + 5x = -5x - 8 + 5x$$
$$2x + 5 - 5 = -8 - 5$$
$$2x = -13$$
$$\frac{2x}{2} = \frac{-13}{2}$$
$$x = -\frac{13}{2}$$
The solution set is $\left\{-\dfrac{13}{2}\right\}$.

25.
$$-7 - 6x = 9 - 9x$$
$$-7 - 6x + 9x = 9 - 9x + 9x$$
$$-7 + 3x = 9$$
$$-7 + 3x + 7 = 9 + 7$$
$$3x = 16$$
$$\frac{3x}{3} = \frac{16}{3}$$
$$x = \frac{16}{3}$$
The solution set is $\left\{\dfrac{16}{3}\right\}$.

27.
$$2x - 1 - x = x - 1$$
$$x - 1 = x - 1$$
$$x - 1 - x = x - 1 - x$$
$$-1 = -1$$
Since $-1 = -1$, this is an identity.
The solution set is {all real numbers}.

29.
$$5n - 4 - n = -3n - 6 + n$$
$$4n - 4 = -2n - 6$$
$$4n - 4 + 2n = -2n - 6 + 2n$$
$$6n - 4 + 4 = -6 + 4$$
$$6n = -2$$
$$\frac{6n}{6} = \frac{-2}{6} = -\frac{1}{3}$$
The solution set is $\left\{-\dfrac{1}{3}\right\}$.

31.
$$-7 - 2n - 6n = 7n - 5n + 12$$
$$-7 - 8n = 2n + 12$$
$$-7 - 8n - 2n = 2n + 12 - 2n$$
$$-7 - 10n + 7 = 12 + 7$$
$$-10n = 19$$
$$\frac{-10n}{-10} = \frac{19}{-10} = -\frac{19}{10}$$

The solution set is $\left\{ -\dfrac{19}{10} \right\}$.

33. Let n represent the number. Then $4n$ represents four times the number.
$$n + 4n = 85$$
$$5n = 85$$
$$n = 17$$
The number is 17.

35. Let n represent the first odd number, then $n + 2$ represents the next odd number..
$$n + (n + 2) = 72$$
$$2n + 2 = 72$$
$$2n = 70$$
$$n = 35$$
The two consecutive odd numbers are 35 and 37.

37. Let n, $n + 2$, and $n + 4$ represent the three consecutive even numbers.
$$n + (n + 2) + (n + 4) = 114$$
$$3n + 6 = 114$$
$$3n = 108$$
$$n = 36$$
The numbers are 36, 38, and 40.

39. Let n represent the number.
$$3n + 2 = 7n - 4$$
$$-4n + 2 = -4$$
$$-4n = -6$$
$$n = \frac{-6}{-4} = \frac{3}{2}$$
The number is $\dfrac{3}{2}$.

41. Let n represent the number.
$$n + 5n = 3n - 18$$
$$6n = 3n - 18$$
$$3n = -18$$
$$n = -6$$
The number is -6.

43. Let a represent the first angle. Then the other angle is represented by $2a - 6$. Since the angles are complementary, the sum of their measures is $90°$.
$$a + (2a - 6) = 90$$
$$3a - 6 = 90$$
$$3a = 96$$
$$a = 32$$
The measures of the angles are $32°$ and $2(32) - 6 = 58°$.

45. Let a represent the smaller angle. Then $3a - 20$ represents the larger angle. Since they are supplementary angles, the sum of their measures is $180°$.
$$a + (3a - 20) = 180$$
$$4a - 20 = 180$$
$$4a = 200$$
$$a = 50$$
The measures of the angles are $50°$, and $3(50) - 20 = 130°$.

47. Let a represent the smaller of the other two angles; then $a + 10$ represents the larger angle. The sum of the measures of the three angles of a triangle is $180°$.
$$40 + a + a + 10 = 180$$
$$50 + 2a = 180$$
$$2a = 130$$
$$a = 65$$
The other two angles of the triangle would be $65°$ and $75°$.

Problem Set 3.3

49. Let x represent the price per share of the stock.
$$2x - 17 = 35$$
$$2x = 52$$
$$x = 26$$
He paid $26 per share for the stock.

51. Let h represent Bob's normal hourly rate. Bob worked 8 hours at the higher rate.
$$40h + 8(2h) = 504$$
$$40h + 16h = 504$$
$$56h = 504$$
$$h = 9$$
Bob's normal hourly rate is $9.00 per hour.

53. Let m represent the number of males; then $3m$ represents the number of females.
$$m + 3m = 600$$
$$4m = 600$$
$$m = 150$$
Therefore, 150 males and $3(150) = 450$ females attended the concert.

55. Let x represent the number of votes that Melton received; then Sanchez received $2x + 10$.
$$x + (2x + 10) = 1030$$
$$3x + 10 = 1030$$
$$3x = 1020$$
$$x = 340$$
Sanchez received $2(340) + 10 = 690$ votes.

57. Let x represent the length of the shorter piece; then $x + 8$ represents the length of the other piece.
$$x + (x + 8) = 20$$
$$2x + 8 = 20$$
$$2x = 12$$
$$x = 6$$
The shorter piece would be 6 feet long.

PROBLEM SET | **3.4** **Equations Involving Parentheses and Fractional Forms**

1. $7(x + 2) = 21$
$7x + 14 = 21$ Apply distributive property.
$\quad 7x = 7$ Subtract 14 from both sides.
$\quad\quad x = 1$ Divide both sides by 7.
The solution set is $\{1\}$.

3. $5(x - 3) = 35$
$5x - 15 = 35$ Apply distributive property.
$\quad 5x = 50$ Add 15 to both sides.
$\quad\quad x = 10$ Divide both sides by 5.
The solution set is $\{10\}$.

5. $-3(x + 5) = 12$
$-3x - 15 = 12$ Apply distributive property.
$\quad -3x = 27$ Add 15 to both sides.
$\quad\quad x = -9$ Divide both sides by -3.
The solution set is $\{-9\}$.

7. $4(n - 6) = 5$
$4n - 24 = 5$ Apply distributive property.
$\quad 4n = 29$ Add 24 to both sides.
$\quad n = \dfrac{29}{4}$ Divide both sides by 4.
The solution set is $\left\{ \dfrac{29}{4} \right\}$.

9. $6(n + 7) = 8$
$6n + 42 = 8$ Apply distributive property.
$\quad 6n = -34$ Subtract 42 from both sides.
$\quad n = \dfrac{-34}{6}$ Divide both sides by 6.
$\quad n = -\dfrac{17}{3}$ Reduce
The solution set is $\left\{ -\dfrac{17}{3} \right\}$.

48

11. $-10 = -5(t-8)$
  $-10 = -5t + 40$ Apply distributive property.
  $-50 = -5t$ Subtract 40 from both sides.
  $10 = t$ Divide both sides by -5.
  The solution set is $\{10\}$.

13. $5(x-4) = 4(x+6)$
  $5x - 20 = 4x + 24$ Apply distributive property.
  $x - 20 = 24$ Subtract $4x$ from both sides.
  $x = 44$ Add 20 to both sides.
  The solution set is $\{44\}$.

We will dicontinue giving reasons for each step but will continue to show enough of the work so that you can follow the steps. If a new technique is introduced, then we will indicate some of the reasons again.

15. $8(x+1) = 9(x-2)$
  $8x + 8 = 9x - 18$
  $-x + 8 = -18$
  $-x = -26$
  $x = 26$
  The solution set is $\{26\}$.

17. $8(t+5) = 6(t-6)$
  $8t + 40 = 6t - 36$
  $2t + 40 = -36$
  $2t = -76$
  $t = -38$
  The solution set is $\{-38\}$.

19. $3(2t+1) = 4(3t-2)$
  $6t + 3 = 12t - 8$
  $-6t + 3 = -8$
  $-6t = -11$
  $t = \dfrac{-11}{-6} = \dfrac{11}{6}$
  The solution set is $\left\{\dfrac{11}{6}\right\}$.

21. $-2(x-6) = -(x-9)$
  $-2x + 12 = -x + 9$
  $-x + 12 = 9$
  $-x = -3$
  $x = 3$
  The solution set is $\{3\}$.

23. $-3(t-4) - 2(t+4) = 9$
  $-3t + 12 - 2t - 8 = 9$
  $-5t + 4 = 9$
  $-5t = 5$
  $t = -1$
  The solution set is $\{-1\}$.

25. $3(n-10) - 5(n+12) = -86$
  $3n - 30 - 5n - 60 = -86$
  $-2n - 90 = -86$
  $-2n = 4$
  $n = -2$
  The solution set is $\{-2\}$.

27. $3(x+1) + 4(2x-1) = 5(2x+3)$
  $3x + 3 + 8x - 4 = 10x + 15$
  $11x - 1 = 10x + 15$
  $x - 1 = 15$
  $x = 16$
  The solution set is $\{16\}$.

29. $-(x+2) + 2(x-3) = -2(x-7)$
  $-x - 2 + 2x - 6 = -2x + 14$
  $x - 8 = -2x + 14$
  $3x - 8 = 14$
  $3x = 22$
  $x = \dfrac{22}{3}$
  The solution set is $\left\{\dfrac{22}{3}\right\}$.

31. $5(2x-1) - (3x+4) = 4(x+3) - 27$
  $10x - 5 - 3x - 4 = 4x + 12 - 27$
  $7x - 9 = 4x - 15$
  $3x - 9 = -15$
  $3x = -6$
  $x = -2$
  The solution set is $\{-2\}$.

Problem Set 3.4

33. $-(a-1)-(3a-2)=6+2(a-1)$
$-a+1-3a+2=6+2a-2$
$-4a+3=2a+4$
$-6a+3=4$
$-6a=1$
$$a=-\frac{1}{6}$$
The solution set is $\left\{-\frac{1}{6}\right\}$.

35. $3(x-1)+2(x-3)=-4(x-2)+10(x+4)$
$3x-3+2x-6=-4x+8+10x+40$
$5x-9=6x+48$
$-x-9=48$
$-x=57$
$x=-57$
The solution set is $\{-57\}$.

37. $3-7(x-1)=9-6(2x+1)$
$3-7x+7=9-12x-6$
$-7x+10=-12x+3$
$5x+10=3$
$5x=-7$
$$x=-\frac{7}{5}$$
The solution set is $\left\{-\frac{7}{5}\right\}$.

For Problems 39 − 59, we begin each solution by multiplying both sides of the given equation by the least common denominator of all of the denominators in the equation. This has the effect of "clearing the equation of all fractions."

39. $$\frac{3}{4}x-\frac{2}{3}=\frac{5}{6}$$
$$12\left(\frac{3}{4}x-\frac{2}{3}\right)=12\left(\frac{5}{6}\right)$$
$9x-8=10$
$9x=18$
$x=2$
The solution set is $\{2\}$.

41. $$\frac{5}{6}x+\frac{1}{4}=-\frac{9}{4}$$
$$12\left(\frac{5}{6}x+\frac{1}{4}\right)=12\left(-\frac{9}{4}\right)$$
$10x+3=-27$
$10x=-30$
$x=-3$
The solution set is $\{-3\}$.

43. $$\frac{1}{2}x-\frac{3}{5}=\frac{3}{4}$$
$$20\left(\frac{1}{2}x-\frac{3}{5}\right)=20\left(\frac{3}{4}\right)$$
$10x-12=15$
$10x=27$
$$x=\frac{27}{10}$$
The solution set is $\left\{\frac{27}{10}\right\}$.

45. $$\frac{n}{3}+\frac{5n}{6}=\frac{1}{8}$$
$$24\left(\frac{n}{3}+\frac{5n}{6}\right)=24\left(\frac{1}{8}\right)$$
$8n+20n=3$
$28n=3$
$$n=\frac{3}{28}$$
The solution set is $\left\{\frac{3}{28}\right\}$.

47. $$\frac{5y}{6}-\frac{3}{5}=\frac{2y}{3}$$
$$30\left(\frac{5y}{6}-\frac{3}{5}\right)=30\left(\frac{2y}{3}\right)$$
$25y-18=20y$
$5y-18=0$
$5y=18$
$$y=\frac{18}{5}$$
The solution set is $\left\{\frac{18}{5}\right\}$.

49.
$$\frac{h}{6} + \frac{h}{8} = 1$$
$$24\left(\frac{h}{6} + \frac{h}{8}\right) = 24(1)$$
$$4h + 3h = 24$$
$$7h = 24$$
$$h = \frac{24}{7}$$
The solution set is $\left\{\frac{24}{7}\right\}$.

51.
$$\frac{x+2}{3} + \frac{x+3}{4} = \frac{13}{3}$$
$$12\left(\frac{x+2}{3} + \frac{x+3}{4}\right) = 12\left(\frac{13}{3}\right)$$
$$4(x+2) + 3(x+3) = 52$$
$$4x + 8 + 3x + 9 = 52$$
$$7x + 17 = 52$$
$$7x = 35$$
$$x = 5$$
The solution set is $\{5\}$.

53.
$$\frac{x-1}{5} - \frac{x+4}{6} = -\frac{13}{15}$$
$$30\left(\frac{x-1}{5} - \frac{x+4}{6}\right) = 30\left(-\frac{13}{15}\right)$$
$$6(x-1) - 5(x+4) = -26$$
$$6x - 6 - 5x - 20 = -26$$
$$x - 26 = -26$$
$$x = 0$$
The solution set is $\{0\}$.

55.
$$\frac{x+8}{2} - \frac{x+10}{7} = \frac{3}{4}$$
$$28\left(\frac{x+8}{2} - \frac{x+10}{7}\right) = 28\left(\frac{3}{4}\right)$$
$$14(x+8) - 4(x+10) = 21$$
$$14x + 112 - 4x - 40 = 21$$
$$10x + 72 = 21$$
$$10x = -51$$
$$x = -\frac{51}{10}$$
The solution set is $\left\{-\frac{51}{10}\right\}$.

57.
$$\frac{x-2}{8} - 1 = \frac{x+1}{4}$$
$$8\left(\frac{x-2}{8} - 1\right) = 8\left(\frac{x+1}{4}\right)$$
$$1(x-2) - 8 = 2(x+1)$$
$$x - 2 - 8 = 2x + 2$$
$$x - 10 = 2x + 2$$
$$-10 = x + 2$$
$$-12 = x$$
The solution set is $\{-12\}$.

59.
$$\frac{x+1}{4} = \frac{x-3}{6} + 2$$
$$12\left(\frac{x+1}{4}\right) = 12\left(\frac{x-3}{6} + 2\right)$$
$$3(x+1) = 2(x-3) + 24$$
$$3x + 3 = 2x - 6 + 24$$
$$3x + 3 = 2x + 18$$
$$x + 3 = 18$$
$$x = 15$$
The solution set is $\{15\}$.

61. Let n and $n+1$ represent the consecutive whole numbers.
$$n + 4(n+1) = 39$$
$$n + 4n + 4 = 39$$
$$5n = 35$$
$$n = 7$$
The numbers are 7 and 8.

63. Let n, $n+1$, and $n+2$ represent the consecutive whole numbers.
$$2(n + n + 1) = 3(n+2) + 10$$
$$2n + 2n + 2 = 3n + 6 + 10$$
$$4n + 2 = 3n + 16$$
$$n = 14$$
The numbers are 14, 15, and 16.

Problem Set 3.4

65. Let n represent the smaller number; then, $17 - n$ represents the larger number.

$$2n = (17 - n) + 1$$
$$2n = 17 - n + 1$$
$$2n = 18 - n$$
$$3n = 18$$
$$n = 6$$

The numbers are 6 and 11.

67. Let n represent the number.

$$\frac{1}{3}n + 20 = \frac{3}{4}n$$
$$12\left(\frac{1}{3}n + 20\right) = 12\left(\frac{3}{4}n\right)$$
$$4n + 240 = 9n$$
$$240 = 5n$$
$$48 = n$$

The number is 48.

69. Let x represent the time waiting in line at the grocery store.

$$4 = \frac{1}{2}x - 3$$
$$2(4) = 2\left(\frac{1}{2}x - 3\right)$$
$$8 = 2\left(\frac{1}{2}x\right) - 2(3)$$
$$8 = x - 6$$
$$14 = x$$

Mrs. Nelson waited 14 minutes in line at the grocery store.

71. Let x represent the shorter piece of board; then $20 - x$ represents the longer piece.

$$4x = 3(20 - x) - 4$$
$$4x = 60 - 3x - 4$$
$$4x = 56 - 3x$$
$$7x = 56$$
$$x = 8$$

The length of the pieces are 8 feet and 12 feet.

73. Let x represent the number of nickels; then $35 - x$ is the number of quarters.

$$.05x + .25(35 - x) = 5.75$$
$$.05x + 8.75 - .25x = 5.75$$
$$-.20x + 8.75 = 5.75$$
$$-.20x = -3$$
$$x = \frac{-3}{-.20}$$
$$x = 15$$

There are 15 nickels and $35 - 15 = 20$ quarters.

75. Let n represent the number of nickels, $2n$ the number of dimes, and $2n + 10$ the number of quarters.

$$n + 2n + 2n + 10 = 210$$
$$5n + 10 = 210$$
$$5n = 200$$
$$n = 40$$

Max has 40 nickels, $2(40) = 80$ dimes, and $2(40) + 10 = 90$ quarters.

77. Let d represent the number of dimes and $18 - d$ the number of quarters. The value in cents of the dimes is $10d$ and $25(18 - d)$ is the value in cents of the quarters.

$$10d + 25(18 - d) = 330$$
$$10d + 450 - 25d = 330$$
$$-15d + 450 = 330$$
$$-15d = -120$$
$$d = 8$$

Maida has 8 dimes and $18 - 8 = 10$ quarters.

79. Let x represent the number of crabs; then $3x$ represents the number of fish; then $x + 2$ represents the number of plants.

$$x + 3x + x + 2 = 22$$
$$5x + 2 = 22$$
$$5x = 20$$
$$x = 4$$

There are 4 crabs, $3(4) = 12$ fish, and $4 + 2 = 6$ plants.

81. Let a represent the measure of the angle;
then $180 - a$ represents its supplement
and $90 - a$ represents its complement.
$$180 - a = 2(90 - a) + 30$$
$$180 - a = 180 - 2a + 30$$
$$180 - a = -2a + 210$$
$$a = 30$$
The angle has a measure of $30°$.

83. Let c represent the measure of angle C.
Then $\frac{1}{5}c - 2$ represents angle A and
$\frac{1}{2}c - 5$ represents angle B. The sum of the
measures of all angles in a triangle is $180°$.
$$c + \frac{1}{5}c - 2 + \frac{1}{2}c - 5 = 180$$
$$10\left[c + \frac{1}{5}c - 2 + \frac{1}{2}c - 5\right] = 10(180)$$
$$10c + 2c - 20 + 5c - 50 = 1800$$
$$17c - 70 = 1800$$
$$17c = 1870$$
$$c = 110$$
Angle A measures $\frac{1}{5}(110) - 2 = 20°$,
angle B measures $\frac{1}{2}(110) - 5 = 50°$, and
angle C measures $110°$.

85. Let a represent the measure of the angle;
then $180 - a$ represents its supplement
and $90 - a$ represents its complement.
$$180 - a = 3(90 - a) - 10$$
$$180 - a = 270 - 3a - 10$$
$$180 - a = -3a + 260$$
$$2a = 80$$
$$a = 40$$
The angle has a measure of $40°$.

91. Let n represent the first integer,
$n + 1$ the second integer, and
$n + 2$ the third integer.
$$n + n + 2 = 2(n + 1)$$
$$2n + 2 = 2n + 2$$
$$2 = 2 \text{ (identity)}$$
Any three consecutive integers
would satisfy the conditions given.

PROBLEM SET **3.5** **Inequalities**

1. The left side simplifies to
$2(3) - 4(5) = 6 - 20 = -14$
and the right side to
$5(3) - 2(-1) + 4 = 15 + 2 + 4 = 21$.
Since $-14 < 21$, the given inequality
is true.

3. The left side simplifies to
$\frac{2}{3} - \frac{3}{4} + \frac{1}{6} = \frac{8}{12} - \frac{9}{12} + \frac{2}{12} = \frac{1}{12}$
and the right side to
$\frac{1}{5} + \frac{3}{4} - \frac{7}{10} = \frac{4}{20} + \frac{15}{20} - \frac{14}{20} = \frac{5}{20} = \frac{1}{4}$.
Since $\frac{1}{12} < \frac{1}{4}$, the given inequality is false.

5. The left side simplifies to
$\left(-\frac{1}{2}\right)\left(\frac{4}{9}\right) = -\frac{2}{9}$ and the right side
to $\left(\frac{3}{5}\right)\left(-\frac{1}{3}\right) = -\frac{1}{5}$. Since
$-\frac{2}{9} < -\frac{1}{5}$, the given inequality is false.

7. The left side simplifies to
$\frac{3}{4} + \frac{2}{3} \div \frac{1}{5} = \frac{3}{4} + \frac{2}{3} \cdot \frac{5}{1} = \frac{3}{4} + \frac{10}{3}$
$= \frac{9}{12} + \frac{40}{12} = \frac{49}{12}$ and the right side to

Problem Set 3.5

$$\frac{2}{3} + \frac{1}{2} \div \frac{3}{4} = \frac{2}{3} + \frac{1}{2} \cdot \frac{4}{3} = \frac{2}{3} + \frac{2}{3} = \frac{4}{3}.$$

Since $\frac{49}{12} > \frac{4}{3}$, the given inequality is true.

9. The left side simplifies to
$0.16 + 0.34 = 0.50$
and the right side to
$0.23 + 0.17 = 0.40$.
Since $0.50 > 0.40$, the
given inequality is true.

11. $\{x | x > -2\}$ or $(-2, \infty)$

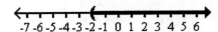

13. $\{x | x \le 3\}$ or $(-\infty, 3]$

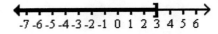

15. $\{x | x > 2\}$ or $(2, \infty)$

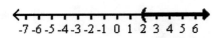

17. $\{x | x \le -2\}$ or $(-\infty, -2]$

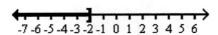

19. $\{x | x < -1\}$ or $(-\infty, -1)$

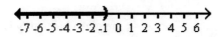

21. $\{x | x < 2\}$ or $(-\infty, 2)$

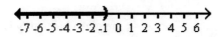

23. $$x + 6 < -14$$
Subtract 6 from both sides.
$$x + 6 - 6 < -14 - 6$$
$$x < -20$$
The solution set is
$\{x | x < -20\}$ or $(-\infty, -20)$.

25. $$x - 4 \ge -13$$
Add 4 to both sides.
$$x - 4 + 4 \ge -13 + 4$$
$$x \ge -9$$
The solution set is
$\{x | x \ge -9\}$ or $[-9, \infty)$.

27. $4x > 36$
Divide both sides by 4.
$$\frac{4x}{4} > \frac{36}{4}$$
$$x > 9$$
The solution set is
$\{x | x > 9\}$ or $(9, \infty)$.

29. $6x < 20$
Divide both sides by 6.
$$\frac{6x}{6} < \frac{20}{6}$$
$$x < \frac{10}{3}$$
The solution set is
$\left\{ x | x < \frac{10}{3} \right\}$ or $\left(-\infty, \frac{10}{3} \right)$.

31. $-5x > 40$
Divide both sides by -5,
which reverses the inequality.
$$\frac{-5x}{-5} < \frac{40}{-5}$$
$$x < -8$$
The solution set is
$\{x | x < -8\}$ or $(-\infty, -8)$.

33. $-7n \le -56$

Divide both sides by -7, which reverses the inequality.

$$\frac{-7n}{-7} \ge \frac{-56}{-7}$$
$$n \ge 8$$

The solution set is
$\{n|n \ge 8\}$ or $[8, \infty)$.

35. $48 > -14n$

Divide both sides by -14, which reverses the inequality.

$$\frac{48}{-14} < \frac{-14n}{-14}$$
$$-\frac{24}{7} < n$$

$-\frac{24}{7} < n$ means $n > -\frac{24}{7}$.

$$n > -\frac{24}{7}$$

The solution set is

$$\left\{n \middle| n > -\frac{24}{7}\right\} \text{ or } \left(-\frac{24}{7}, \infty\right).$$

37. $16 < 9 + n$

Subtract 9 from both sides.

$$16 - 9 < 9 + n - 9$$
$$7 < n$$

$7 < n$ means $n > 7$.

$$n > 7$$

The solution set is
$\{n|n > 7\}$ or $(7, \infty)$.

39. $3x + 2 > 17$

Subtract 2 from both sides.

$$3x + 2 - 2 > 17 - 2$$
$$3x > 15$$

Divide both sides by 3.

$$\frac{3x}{3} > \frac{15}{3}$$
$$x > 5$$

The solution set is
$\{x|x > 5\}$ or $(5, \infty)$.

41. $4x - 3 \le 21$

Add 3 to both sides.

$$4x - 3 + 3 \le 21 + 3$$
$$4x \le 24$$

Divide both sides by 4.

$$\frac{4x}{4} \le \frac{24}{4}$$
$$x \le 6$$

The solution set is
$\{x|x \le 6\}$ or $(-\infty, 6]$.

43. $-2x - 1 \ge 41$

Add 1 to both sides.

$$-2x - 1 + 1 \ge 41 + 1$$
$$-2x \ge 42$$

Divide both sides by -2, which reverses the inequality.

$$\frac{-2x}{-2} \le \frac{42}{-2}$$
$$x \le -21$$

The solution set is
$\{x|x \le -21\}$ or $(-\infty, -21]$.

45. $6x + 2 < 18$

Subtract 2 from both sides.

$$6x + 2 - 2 < 18 - 2$$
$$6x < 16$$

Divide both sides by 6.

$$\frac{6x}{6} < \frac{16}{6}$$
$$x < \frac{8}{3}$$

The solution set is

$$\left\{x \middle| x < \frac{8}{3}\right\} \text{ or } \left(-\infty, \frac{8}{3}\right).$$

Problem Set 3.5

47.
$$3 > 4x - 2$$
Add 2 to both sides.
$$3 + 2 > 4x - 2 + 2$$
$$5 > 4x$$
Divide both sides by 4.
$$\frac{5}{4} > \frac{4x}{4}$$
$$\frac{5}{4} > x$$
$$\frac{5}{4} > x \text{ means } x < \frac{5}{4}.$$
$$x < \frac{5}{4}$$
The solution set is
$$\left\{x \middle| x < \frac{5}{4}\right\} \text{ or } \left(-\infty, \frac{5}{4}\right).$$

49.
$$-2 < -3x + 1$$
Subtract 1 from both sides.
$$-2 - 1 < -3x + 1 - 1$$
$$-3 < -3x$$
Divide both sides by -3,
which reverses the inequality.
$$\frac{-3}{-3} > \frac{-3x}{-3}$$
$$1 > x$$
$1 > x$ means $x < 1$.
$$x < 1$$
The solution set is
$\{x | x < 1\}$ or $(-\infty, 1)$.

51.
$$-38 \geq -9t - 2$$
Add 2 to both sides.
$$-38 + 2 \geq -9t - 2 + 2$$
$$-36 \geq -9t$$
Divide both sides by -9,
which reverses the inequality.
$$\frac{-36}{-9} \leq \frac{-9t}{-9}$$
$$4 \leq t$$
$4 \leq t$ means $t \geq 4$.
$$t \geq 4$$
The solution set is
$\{t | t \geq 4\}$ or $[4, \infty)$.

53.
$$5x - 4 - 3x > 24$$
Combine similar terms on the left side.
$$2x - 4 > 24$$
Add 4 to both sides.
$$2x - 4 + 4 > 24 + 4$$
$$2x > 28$$
Divide both sides by 2.
$$\frac{2x}{2} > \frac{28}{2}$$
$$x > 14$$
The solution set is
$\{x | x > 14\}$ or $(14, \infty)$.

55.
$$4x + 2 - 6x < -1$$
Combine similar terms.
$$-2x + 2 < -1$$
Subtract 2 from both sides.
$$-2x + 2 - 2 < -1 - 2$$
$$-2x < -3$$
Divide both sides by -2,
which reverses the inequality.
$$\frac{-2x}{-2} > \frac{-3}{-2}$$
$$x > \frac{3}{2}$$
The solution set is
$$\left\{x \middle| x > \frac{3}{2}\right\} \text{ or } \left(\frac{3}{2}, \infty\right).$$

57.
$$-5 \geq 3t - 4 - 7t$$
Combine similar terms.
$$-5 \geq -4t - 4$$
Add 4 to both sides.
$$-5 + 4 \geq -4t - 4 + 4$$
$$-1 \geq -4t$$
Divide both sides by -4,
which reverses the inequality.
$$\frac{-1}{-4} \leq \frac{-4t}{-4}$$
$$\frac{1}{4} \leq t$$
$\frac{1}{4} \leq t$ means $t \geq \frac{1}{4}$
$$t \geq \frac{1}{4}$$
The solution set is
$$\left\{ t \mid t \geq \frac{1}{4} \right\} \text{ or } \left[\frac{1}{4}, \infty \right).$$

59.
$$-x - 4 - 3x > 5$$
Combine similar terms.
$$-4x - 4 > 5$$
Add 4 to both sides.
$$-4x - 4 + 4 > 5 + 4$$
$$-4x > 9$$
Divide both sides by -4,
which reverses the inequality.
$$\frac{-4x}{-4} < \frac{9}{-4}$$
$$x < -\frac{9}{4}$$
The solution set is
$$\left\{ x \mid x < -\frac{9}{4} \right\} \text{ or } \left(-\infty, -\frac{9}{4} \right).$$

65.
$$x - 4 < x + 6$$
$$-4 < 6 \text{ (identity)}$$
The solution set is $\{x \mid x \text{ is any real number}\}$.

67.
$$5x + 2 > 5x + 7$$
$$2 > 7 \text{ (contradiction)}$$
The solution set is $\emptyset$.

69.
$$-2x + 7 + 2x > 1$$
$$7 > 1 \text{ (identity)}$$
The solution set is $\{x \mid x \text{ is any real number}\}$.

71.
$$-7 \geq 5x - 2 - 5x$$
$$-7 \geq -2 \text{ (contradiction)}$$
The solution set is $\emptyset$.

Problem Set 3.6

1.
$$3x + 4 > x + 8$$
Subtract x from both sides.
$$3x + 4 - x > x + 8 - x$$
$$2x + 4 > 8$$
Subtract 4 from both sides.
$$2x + 4 - 4 > 8 - 4$$
$$2x > 4$$
Divide both sides by 2.
$$\frac{2x}{2} > \frac{4}{2}$$
$$x > 2$$
The solution set is
$\{x|x > 2\}$ or $(2, \infty)$.

3.
$$7x - 2 < 3x - 6$$
Subtract $3x$ from both sides.
$$7x - 2 - 3x < 3x - 6 - 3x$$
$$4x - 2 < -6$$
Add 2 to both sides.
$$4x - 2 + 2 < -6 + 2$$
$$4x < -4$$
Divide both sides by 4.
$$\frac{4x}{4} < \frac{-4}{4}$$
$$x < -1$$
The solution set is
$\{x|x < -1\}$ or $(-\infty, -1)$.

5.
$$6x + 7 > 3x - 3$$
Subtract $3x$ from both sides.
$$6x + 7 - 3x > 3x - 3 - 3x$$
$$3x + 7 > -3$$
Subtract 7 from both sides.
$$3x + 7 - 7 > -3 - 7$$
$$3x > -10$$
Divide both sides by 3.
$$\frac{3x}{3} > \frac{-10}{3}$$
$$x > -\frac{10}{3}$$
The solution set is
$$\left\{x\Big|x > -\frac{10}{3}\right\} \text{ or } \left(-\frac{10}{3}, \infty\right).$$

7.
$$5n - 2 \le 6n + 9$$
Subtract $6n$ from both sides.
$$5n - 2 - 6n \le 6n + 9 - 6n$$
$$-n - 2 \le 9$$
Add 2 to both sides.
$$-n - 2 + 2 \le 9 + 2$$
$$-n \le 11$$
Multiply by -1, which
reverses the inequality.
$$n \ge -11$$
The solution set is
$\{n|n \ge -11\}$ or $[-11, \infty)$.

9.
$$2t + 9 \geq 4t - 13$$
Subtract $4t$ from both sides.
$$2t + 9 - 4t \geq 4t - 13 - 4t$$
$$-2t + 9 \geq -13$$
Subtract 9 from both sides.
$$-2t + 9 - 9 \geq -13 - 9$$
$$-2t \geq -22$$
Divide both sides by -2,
which reverses the inequality.
$$\frac{-2t}{-2} \leq \frac{-22}{-2}$$
$$t \leq 11$$
The solution set is
$\{t | t \leq 11\}$ or $(-\infty, 11]$.

11.
$$-3x - 4 < 2x + 7$$
Subtract $2x$ from both sides.
$$-3x - 4 - 2x < 2x + 7 - 2x$$
$$-5x - 4 < 7$$
Add 4 to both sides.
$$-5x - 4 + 4 < 7 + 4$$
$$-5x < 11$$
Divide both sides by -5,
which reverses the inequality.
$$\frac{-5x}{-5} > \frac{11}{-5}$$
$$x > -\frac{11}{5}$$
The solution set is
$\left\{ x | x > -\frac{11}{5} \right\}$ or $\left(-\frac{11}{5}, \infty \right)$.

13.
$$-4x + 6 > -2x + 1$$
Add $2x$ to both sides.
$$-4x + 6 + 2x > -2x + 1 + 2x$$
$$-2x + 6 > 1$$
Subtract 6 from both sides.
$$-2x + 6 - 6 > 1 - 6$$
$$-2x > -5$$
Divide both sides by -2,
which reverses the inequality.
$$\frac{-2x}{-2} < \frac{-5}{-2}$$
$$x < \frac{5}{2}$$
The solution set is
$\left\{ x | x < \frac{5}{2} \right\}$ or $\left(-\infty, \frac{5}{2} \right)$.

15.
$$5(x - 2) \leq 30$$
Apply distributive property.
$$5x - 10 \leq 30$$
Add 10 to both sides.
$$5x - 10 + 10 \leq 30 + 10$$
$$5x \leq 40$$
Divide both sides by 5.
$$\frac{5x}{5} \leq \frac{40}{5}$$
$$x \leq 8$$
The solution set is
$\{x | x \leq 8\}$ or $(-\infty, 8]$.

17.
$$2(n + 3) > 9$$
Apply distributive property.
$$2n + 6 > 9$$
Subtract 6 from both sides.
$$2n + 6 - 6 > 9 - 6$$
$$2n > 3$$
Divide both sides by 2.
$$\frac{2n}{2} > \frac{3}{2}$$
$$n > \frac{3}{2}$$
The solution set is
$\left\{ n | n > \frac{3}{2} \right\}$ or $\left(\frac{3}{2}, \infty \right)$.

Problem Set 3.6

19. $-3(y-1) < 12$
Apply distributive property.
     $-3y + 3 < 12$
Subtract 3 from to both sides.
     $-3y + 3 - 3 < 12 - 3$
          $-3y < 9$
Divide both sides by -3,
which reverses the inequality.
          $\dfrac{-3y}{-3} > \dfrac{9}{-3}$
             $y > -3$
The solution set is
$\{y | y > -3\}$ or $(-3, \infty)$.

21. $-2(x+6) > -17$
Apply distributive property.
     $-2x - 12 > -17$
Add 12 to both sides.
     $-2x - 12 + 12 > -17 + 12$
          $-2x > -5$
Divide both sides by -2,
which reverses the inequality.
          $\dfrac{-2x}{-2} < \dfrac{-5}{-2}$
             $x < \dfrac{5}{2}$
The solution set is
$\left\{x | x < \dfrac{5}{2}\right\}$ or $\left(-\infty, \dfrac{5}{2}\right)$.

23. $3(x-2) < 2(x+1)$
Apply distributive property.
     $3x - 6 < 2x + 2$
Subtract $2x$ from both sides.
$3x - 6 - 2x < 2x + 2 - 2x$
     $x - 6 < 2$
Add 6 to both sides.
     $x - 6 + 6 < 2 + 6$
          $x < 8$
The solution set is
$\{x | x < 8\}$ or $(-\infty, 8)$.

25. $4(x+3) > 6(x-5)$
Apply distributive property.
          $4x + 12 > 6x - 30$
Subtract $6x$ from both sides.
     $4x + 12 - 6x > 6x - 30 - 6x$
          $-2x + 12 > -30$
Subtract 12 from both sides.
     $-2x + 12 - 12 > -30 - 12$
             $-2x > -42$
Divide both sides by -2,
which reverses the inequality.
          $\dfrac{-2x}{-2} < \dfrac{-42}{-2}$
             $x < 21$
The solution set is
$\{x | x < 21\}$ or $(-\infty, 21)$.

27. $3(x-4) + 2(x+3) < 24$
Apply distributive property.
     $3x - 12 + 2x + 6 < 24$
Combine similar terms.
          $5x - 6 < 24$
Add 6 to both sides.
          $5x - 6 + 6 < 24 + 6$
             $5x < 30$
Divide both sides by 5.
          $\dfrac{5x}{5} < \dfrac{30}{5}$
             $x < 6$
The solution set is
$\{x | x < 6\}$ or $(-\infty, 6)$.

29. $5(n+1) - 3(n-1) > -9$

Apply distributive property.

$$5n + 5 - 3n + 3 > -9$$

Combine similar terms.

$$2n + 8 > -9$$

Subtract 8 from both sides.

$$2n + 8 - 8 > -9 - 8$$
$$2n > -17$$

Divide both sides by 2.

$$\frac{2n}{2} > \frac{-17}{2}$$
$$n > -\frac{17}{2}$$

The solution set is

$$\left\{ n \middle| n > -\frac{17}{2} \right\} \text{ or } \left(-\frac{17}{2}, \infty \right).$$

31. $\dfrac{1}{2}n - \dfrac{2}{3}n \geq -7$

Multiply both sides by 6.

$$6\left(\frac{1}{2}n - \frac{2}{3}n \right) \geq 6(-7)$$
$$3n - 4n \geq -42$$
$$-n \geq -42$$

Multiply both sides by -1,
which reverses the inequality.

$$n \leq 42$$

The solution set is

$\{n | n \leq 42\}$ or $(-\infty, 42]$.

33. $\dfrac{3}{4}n - \dfrac{5}{6}n < \dfrac{3}{8}$

Multiply both sides by 24.

$$24\left(\frac{3}{4}n - \frac{5}{6}n \right) < 24\left(\frac{3}{8} \right)$$
$$18n - 20n < 9$$

Combine similar terms.

$$-2n < 9$$

Divide both sides by -2,
which reverses the inequality.

$$\frac{-2n}{-2} > \frac{9}{-2}$$
$$n > -\frac{9}{2}$$

The solution set is

$$\left\{ n \middle| n > -\frac{9}{2} \right\} \text{ or } \left(-\frac{9}{2}, \infty \right).$$

35. $\dfrac{3x}{5} - \dfrac{2}{3} > \dfrac{x}{10}$

Multiply both sides by 30.

$$30\left(\frac{3x}{5} - \frac{2}{3} \right) > 30\left(\frac{x}{10} \right)$$
$$18x - 20 > 3x$$
$$18x - 20 + 20 > 3x + 20$$
$$18x > 3x + 20$$
$$15x > 20$$

Divide both sides by 15.

$$\frac{15x}{15} > \frac{20}{15}$$
$$x > \frac{4}{3}$$

The solution set is

$$\left\{ x \middle| x > \frac{4}{3} \right\} \text{ or } \left(\frac{4}{3}, \infty \right).$$

37. $n \geq 3.4 + 0.15n$

Subtract $0.15n$ from both sides.

$$n - 0.15n \geq 3.4$$
$$0.85n \geq 3.4$$
$$n \geq 4$$

The solution set is

$\{n | n \geq 4\}$ or $[4, \infty)$.

Problem Set 3.6

39. $0.09t + 0.1(t + 200) > 77$

$\qquad 0.09t + 0.1t + 20 > 77$

$\qquad\qquad 0.19t + 20 > 77$

$\qquad\qquad\qquad 0.19t > 57$

$\qquad\qquad\qquad\qquad t > 300$

The solution set is

$\{t | t > 300\}$ or $(300, \infty)$.

41. $0.06x + 0.08(250 - x) \geq 19$

$\qquad 0.06x + 20 - 0.08x \geq 19$

$\qquad\qquad -0.02x + 20 \geq 19$

$\qquad\qquad\qquad -0.02x \geq -1$

$\qquad\qquad\qquad\qquad x \leq 50$

The solution set is

$\{x | x \leq 50\}$ or $(-\infty, 50]$.

43. $\dfrac{x - 1}{2} + \dfrac{x + 3}{5} > \dfrac{1}{10}$

Multiply both sides by 10.

$10\left(\dfrac{x - 1}{2} + \dfrac{x + 3}{5}\right) > 10\left(\dfrac{1}{10}\right)$

$\qquad 5(x - 1) + 2(x + 3) > 1$

$\qquad 5x - 5 + 2x + 6 > 1$

$\qquad\qquad\qquad 7x + 1 > 1$

$\qquad\qquad\qquad\qquad 7x > 0$

$\qquad\qquad\qquad\qquad x > 0$

The solution set is

$\{x | x > 0\}$ or $(0, \infty)$.

45. $\dfrac{x + 2}{6} - \dfrac{x + 1}{5} < -2$

Multiply both sides by 30.

$30\left(\dfrac{x + 2}{6} - \dfrac{x + 1}{5}\right) < 30(-2)$

$\qquad 5(x + 2) - 6(x + 1) < -60$

$\qquad 5x + 10 - 6x - 6 < -60$

$\qquad\qquad\qquad -x + 4 < -60$

$\qquad\qquad\qquad\qquad -x < -64$

Multiply both sides by -1,
which reverses the inequality.

$\qquad\qquad\qquad\qquad x > 64$

The solution set is

$\{x | x > 64\}$ or $(64, \infty)$.

47. $\dfrac{n + 3}{3} + \dfrac{n - 7}{2} > 3$

Multiply both sides by 6.

$6\left(\dfrac{n + 3}{3} + \dfrac{n - 7}{2}\right) > 6(3)$

$\qquad 2(n + 3) + 3(n - 7) > 18$

$\qquad 2n + 6 + 3n - 21 > 18$

$\qquad\qquad\qquad 5n - 15 > 18$

$\qquad\qquad\qquad\qquad 5n > 33$

$\qquad\qquad\qquad\qquad n > \dfrac{33}{5}$

The solution set is

$\left\{n | n > \dfrac{33}{5}\right\}$ or $\left(\dfrac{33}{5}, \infty\right)$.

49. $\dfrac{x - 3}{7} - \dfrac{x - 2}{4} \leq \dfrac{9}{14}$

Multiply both sides by 28.

$28\left(\dfrac{x - 3}{7} - \dfrac{x - 2}{4}\right) \leq 28\left(\dfrac{9}{14}\right)$

$\qquad 4(x - 3) - 7(x - 2) \leq 2(9)$

$\qquad 4x - 12 - 7x + 14 \leq 18$

$\qquad\qquad\qquad -3x + 2 \leq 18$

$\qquad\qquad\qquad\qquad -3x \leq 16$

$\qquad\qquad\qquad\qquad x \geq -\dfrac{16}{3}$

The solution set is

$\left\{x | x \geq -\dfrac{16}{3}\right\}$ or $\left[-\dfrac{16}{3}, -\infty\right)$.

51. Since it is an "and" statement, we need to satisfy both inequalities at the same time. Thus, all numbers between -1 and 2, but not including -1 and 2, are solutions.

62

53. Since it is an "or" statement, the solution set consists of all numbers less than (but not including) -2 along with all numbers greater than (but not including) 1.

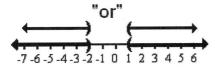

55. Since it is an "and" statement, we need to satisfy both inequalities at the same time. Thus, all numbers between -2 and 2, including 2 (but not including -2), are solutions.

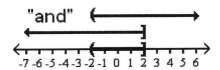

57. Since it is an "and" statement we are looking for all numbers that satisfy both inequalities at the same time. Thus, any number greater than 2 will work.

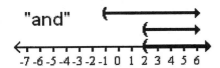

59. Since it is an "or" statement we are looking for all numbers that satisfy either inequality. Thus, any number greater than -4 will work.

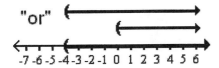

61. Since it is an "and" statement we are looking for all numbers that satisfy both inequalities at the same time. There are no numbers that are both greater than 3 and less than -1. So the solution set is $\emptyset$.

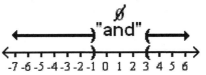

63. Since it is an "or" statement, the solution set consists of all numbers less than or equal to 0 along with all numbers greater than or equal to 2.

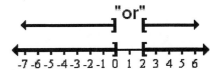

65. Since it is an "or" statement, we want all numbers greater than -4 along with all numbers less than 3. Thus, the solution set is the entire set of real numbers.

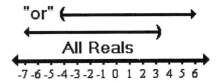

67. Let n represent the number.
$$3n + 5 > 26$$
$$3n > 21$$
$$n > 7$$
The numbers must be greater than 7.

69. Let w represent the width of the rectangle. Also remember that "length plus width equals one-half of the perimeter" of a rectangle.
$$w + 20 \leq 35$$
$$w \leq 15$$
Thus, 15 inches is the largest possible value for the width.

63

Problem Set 3.6

71. Let x be her score in the last game.
$$\frac{132 + 160 + x}{3} \geq 150$$
$$292 + x \geq 450$$
$$x \geq 158$$
She must bowl 158 or better in her last game.

73. Let x be his average on the last two exams.
$$\frac{96 + 90 + 94 + 2x}{5} > 92$$
$$280 + 2x > 460$$
$$2x > 180$$
$$x > 90$$
He must have an average better than 90 on the last two exams.

75. Let r represent the other rate.
$$500(0.08) + 500r > 100$$
$$40 + 500r > 100$$
$$500r > 60$$
$$r > 0.12$$
The other rate must be more than 12%.

77. Let x be his score on the final round.
$$\frac{82 + 84 + 78 + 79 + x}{5} \leq 80$$
$$323 + x \leq 400$$
$$x \leq 77$$
He must shoot 77 or less.

CHAPTER 3 **Review Problem Set**

1.
$$9x - 2 = -29$$
$$9x - 2 + 2 = -29 + 2$$
$$9x = -27$$
$$x = -3$$
The solution set is $\{-3\}$.

2.
$$-3 = -4y + 1$$
$$-4 = -4y$$
$$1 = y$$
The solution set is $\{1\}$.

3.
$$7 - 4x = 10$$
$$7 - 4x - 7 = 10 - 7$$
$$-4x = 3$$
$$x = -\frac{3}{4}$$
The solution set is $\left\{-\frac{3}{4}\right\}$.

4.
$$6y - 5 = 4y + 13$$
$$6y - 5 - 4y = 4y + 13 - 4y$$
$$2y - 5 = 13$$
$$2y = 18$$
$$y = 9$$
The solution set is $\{9\}$.

5.
$$4n - 3 = 7n + 9$$
$$4n - 3 - 7n = 7n + 9 - 7n$$
$$-3n - 3 = 9$$
$$-3n - 3 + 3 = 9 + 3$$
$$-3n = 12$$
$$n = -4$$
The solution set is $\{-4\}$.

6.
$$7(y - 4) = 4(y + 3)$$
$$7y - 28 = 4y + 12$$
$$3y - 28 = 12$$
$$3y = 40$$
$$y = \frac{40}{3}$$
The solution set is $\left\{\frac{40}{3}\right\}$.

7.
$$2(x + 1) + 5(x - 3) = 11(x - 2)$$
$$2x + 2 + 5x - 15 = 11x - 22$$
$$7x - 13 = 11x - 22$$
$$7x - 13 - 11x = 11x - 22 - 11x$$
$$-4x - 13 = -22$$
$$-4x - 13 + 13 = -22 + 13$$
$$-4x = -9$$
$$x = \frac{9}{4}$$
The solution set is $\left\{ \dfrac{9}{4} \right\}$.

8.
$$-3(x + 6) = 5x - 3$$
$$-3x - 18 = 5x - 3$$
$$-8x - 18 = -3$$
$$-8x = 15$$
$$x = -\frac{15}{8}$$
The solution set is $\left\{ -\dfrac{15}{8} \right\}$.

9.
$$\frac{2}{5}n - \frac{1}{2}n = \frac{7}{10}$$
$$10\left(\frac{2}{5}n - \frac{1}{2}n \right) = 10\left(\frac{7}{10} \right)$$
$$4n - 5n = 7$$
$$-n = 7$$
$$n = -7$$
The solution set is $\{ -7 \}$.

10.
$$\frac{3n}{4} + \frac{5n}{7} = \frac{1}{14}$$
$$28\left(\frac{3n}{4} + \frac{5n}{7} \right) = 28\left(\frac{1}{14} \right)$$
$$21n + 20n = 2$$
$$41n = 2$$
$$n = \frac{2}{41}$$
The solution set is $\left\{ \dfrac{2}{41} \right\}$.

11.
$$\frac{x - 3}{6} + \frac{x + 5}{8} = \frac{11}{12}$$
$$24\left(\frac{x - 3}{6} + \frac{x + 5}{8} \right) = 24\left(\frac{11}{12} \right)$$
$$4(x - 3) + 3(x + 5) = 2(11)$$
$$4x - 12 + 3x + 15 = 22$$
$$7x + 3 = 22$$
$$7x = 19$$
$$x = \frac{19}{7}$$
The solution set is $\left\{ \dfrac{19}{7} \right\}$.

12.
$$\frac{n}{2} - \frac{n - 1}{4} = \frac{3}{8}$$
$$8\left(\frac{n}{2} - \frac{n - 1}{4} \right) = 8\left(\frac{3}{8} \right)$$
$$4n - 2(n - 1) = 3$$
$$4n - 2n + 2 = 3$$
$$2n + 2 = 3$$
$$2n = 1$$
$$n = \frac{1}{2}$$
The solution set is $\left\{ \dfrac{1}{2} \right\}$.

13.
$$-2(x - 4) = -3(x + 8)$$
$$-2x + 8 = -3x - 24$$
$$-2x + 8 + 3x = -3x - 24 + 3x$$
$$x + 8 = -24$$
$$x = -32$$
The solution set is $\{ -32 \}$.

14.
$$3x - 4x - 2 = 7x - 14 - 9x$$
$$-x - 2 = -2x - 14$$
$$-x - 2 + 2x = -2x - 14 + 2x$$
$$x - 2 = -14$$
$$x = -12$$
The solution set is $\{ -12 \}$.

15.
$$5(n - 1) - 4(n + 2) = -3(n - 1) + 3n + 5$$
$$5n - 5 - 4n - 8 = -3n + 3 + 3n + 5$$
$$n - 13 = 8$$
$$n = 21$$
The solution set is $\{ 21 \}$.

16.
$$\frac{x-3}{9} = \frac{x+4}{8}$$
$$8(x-3) = 9(x+4)$$
$$8x - 24 = 9x + 36$$
$$-x = 60$$
$$x = -60$$
The solution set is $\{-60\}$.

17.
$$\frac{x-1}{-3} = \frac{x+2}{-4}$$
$$-4(x-1) = -3(x+2)$$
$$-4x + 4 = -3x - 6$$
$$-x + 4 = -6$$
$$-x = -10$$
$$x = 10$$
The solution set is $\{10\}$.

18.
$$-(t-3) - (2t+1) = 3(t+5) - 2(t+1)$$
$$-t + 3 - 2t - 1 = 3t + 15 - 2t - 2$$
$$-3t + 2 = t + 13$$
$$-4t = 11$$
$$t = -\frac{11}{4}$$
The solution set is $\left\{ -\frac{11}{4} \right\}$.

19.
$$\frac{2x-1}{3} = \frac{3x+2}{2}$$
$$2(2x-1) = 3(3x+2)$$
$$4x - 2 = 9x + 6$$
$$-5x - 2 = 6$$
$$-5x = 8$$
$$x = -\frac{8}{5}$$
The solution set is $\left\{ -\frac{8}{5} \right\}$.

20.
$$3(2t-4) + 2(3t+1) = -2(4t+3) - (t-1)$$
$$6t - 12 + 6t + 2 = -8t - 6 - t + 1$$
$$12t - 10 = -9t - 5$$
$$21t - 10 = -5$$
$$21t = 5$$
$$t = \frac{5}{21}$$
The solution set is $\left\{ \frac{5}{21} \right\}$.

21.
$$3x - 2 > 10$$
$$3x > 12$$
$$x > 4$$
The solution set is
$\{x | x > 4\}$ or $(4, \infty)$.

22.
$$-2x - 5 < 3$$
$$-2x < 8$$
$$x > -4$$
The solution set is
$\{x | x > -4\}$ or $(-4, \infty)$.

23.
$$2x - 9 \geq x + 4$$
$$x - 9 \geq 4$$
$$x \geq 13$$
The solution set is
$\{x | x \geq 13\}$ or $[13, \infty)$.

24.
$$3x + 1 \leq 5x - 10$$
$$-2x + 1 \leq -10$$
$$-2x \leq -11$$
$$x \geq \frac{11}{2}$$
The solution set is
$\left\{ x | x \geq \frac{11}{2} \right\}$ or $\left[\frac{11}{2}, \infty \right)$.

25.
$$6(x-3) > 4(x+13)$$
$$6x - 18 > 4x + 52$$
$$2x - 18 > 52$$
$$2x > 70$$
$$x > 35$$
The solution set is
$\{x | x > 35\}$ or $(35, \infty)$.

26.
$$2(x+3) + 3(x-6) < 14$$
$$2x + 6 + 3x - 18 < 14$$
$$5x - 12 < 14$$
$$5x < 26$$
$$x < \frac{26}{5}$$
The solution set is
$\left\{ x | x < \frac{26}{5} \right\}$ or $\left(-\infty, \frac{26}{5} \right)$.

27.
$$\frac{2n}{5} - \frac{n}{4} < \frac{3}{10}$$
$$20\left(\frac{2n}{5} - \frac{n}{4}\right) < 20\left(\frac{3}{10}\right)$$
$$8n - 5n < 6$$
$$3n < 6$$
$$n < 2$$
The solution set is
$\{n|n < 2\}$ or $(-\infty, 2)$.

28.
$$\frac{n+4}{5} + \frac{n-3}{6} > \frac{7}{15}$$
$$30\left(\frac{n+4}{5} + \frac{n-3}{6}\right) > 30\left(\frac{7}{15}\right)$$
$$6(n+4) + 5(n-3) > 14$$
$$6n + 24 + 5n - 15 > 14$$
$$11n + 9 > 14$$
$$11n > 5$$
$$n > \frac{5}{11}$$
The solution set is
$\left\{n|n > \frac{5}{11}\right\}$ or $\left(\frac{5}{11}, \infty\right)$.

29.
$$-16 < 8 + 2y - 3y$$
$$-16 < 8 - y$$
$$-24 < -y$$
$$24 > y$$
$$y < 24$$
The solution set is
$\{y|y < 24\}$ or $(-\infty, 24)$.

30.
$$-24 > 5x - 4 - 7x$$
$$-24 > -2x - 4$$
$$-20 > -2x$$
$$10 < x$$
$$x > 10$$
The solution set is
$\{x|x > 10\}$ or $(10, \infty)$.

31.
$$-3(n-4) > 5(n+2) + 3n$$
$$-3n + 12 > 5n + 10 + 3n$$
$$-3n + 12 > 8n + 10$$
$$-11n + 12 > 10$$
$$-11n > -2$$
$$n < \frac{2}{11}$$
The solution set is
$\left\{n|n < \frac{2}{11}\right\}$ or $\left(-\infty, \frac{2}{11}\right)$.

32.
$$-4(n-2) - (n-1) < -4(n+6)$$
$$-4n + 8 - n + 1 < -4n - 24$$
$$-5n + 9 < -4n - 24$$
$$-n + 9 < -24$$
$$-n < -33$$
$$n > 33$$
The solution set is
$\{n|n > 33\}$ or $(33, \infty)$.

33.
$$\frac{3}{4}n - 6 \le \frac{2}{3}n + 4$$
$$12\left(\frac{3}{4}n - 6\right) \le 12\left(\frac{2}{3}n + 4\right)$$
$$9n - 72 \le 8n + 48$$
$$n - 72 \le 48$$
$$n \le 120$$
The solution set is
$\{n|n \le 120\}$ or $(-\infty, 120]$.

34.
$$\frac{1}{2}n - \frac{1}{3}n - 4 \ge \frac{3}{5}n + 2$$
$$30\left(\frac{1}{2}n - \frac{1}{3}n - 4\right) \ge 30\left(\frac{3}{5}n + 2\right)$$
$$15n - 10n - 120 \ge 18n + 60$$
$$5n - 120 \ge 18n + 60$$
$$-13n \ge 180$$
$$n \le -\frac{180}{13}$$
The solution set is
$\left\{n|n \le -\frac{180}{13}\right\}$ or $\left(-\infty, -\frac{180}{13}\right]$.

35.
$$-12 > -4(x-1)+2$$
$$-12 > -4x+4+2$$
$$-12 > -4x+6$$
$$-18 > -4x$$
$$\frac{18}{4} < x$$
$$x > \frac{9}{2}$$
The solution set is
$$\left\{x \mid x > \frac{9}{2}\right\} \text{ or } \left(\frac{9}{2}, \infty\right).$$

36.
$$36 < -3(x+2)-1$$
$$36 < -3x-6-1$$
$$36 < -3x-7$$
$$43 < -3x$$
$$-\frac{43}{3} > x$$
$$x < -\frac{43}{3}$$
The solution set is
$$\left\{x \mid x < -\frac{43}{3}\right\} \text{ or } \left(-\infty, -\frac{43}{3}\right).$$

37. Since it is an "and" statement, we need to satisfy both inequalities at the same time. Thus, all numbers between -3 and 2, but not including -3 and 2, are solutions.

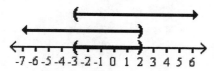

38. Since it is an "or" statement, the solution set consists of all numbers less than (but not including) -1 along with all numbers greater than (but not including) 4.

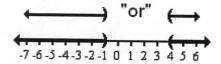

39. Since it is an "or" statement, we want all numbers less than 2 along with all numbers greater than 0. Thus, the solution set is the entire set of real numbers.

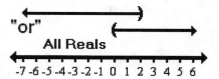

40. Since it is an "and" statement, we are looking for all numbers that satisfy both inequalities at the same time. Thus, any number greater than 1 will work.

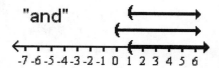

41. Let n represent the number.
$$\frac{3}{4}n = 18$$
$$4\left(\frac{3}{4}n\right) = 4(18)$$
$$3n = 72$$
$$n = 24$$
The number is 24.

42. Let n represent the number.
$$19 = 3n - 2$$
$$21 = 3n$$
$$7 = n$$
The number is 7.

43. Let n represent the larger number.
$$n - 21 = 12$$
$$n = 33$$
The larger number is 33.

44. Let n represent the number.
$$9n - 1 = 7n + 15$$
$$2n - 1 = 15$$
$$2n = 16$$
$$n = 8$$
The number is 8.

45. Let x represent the score on the fifth exam.
$$\frac{83 + 89 + 78 + 86 + x}{5} \geq 85$$
$$336 + x \geq 425$$
$$x \geq 89$$
She must have a score of 89 or better.

46. Let n represent the smaller number; then $40 - n$ represents the larger number.
$$6n = 4(40 - n)$$
$$6n = 160 - 4n$$
$$10n = 160$$
$$n = 16$$
The smaller number is 16 and the larger number is 24.

47. Let n represent the number.
$$\frac{2}{3}n - 2 = \frac{1}{2}n + 1$$
$$6\left(\frac{2}{3}n - 2\right) = 6\left(\frac{1}{2}n + 1\right)$$
$$4n - 12 = 3n + 6$$
$$n - 12 = 6$$
$$n = 18$$
The number is 18.

48. Let x represent the score on the fourth exam. The total score from the first three exams was $3(84) = 252$.
$$\frac{252 + x}{4} \geq 85$$
$$252 + x \geq 340$$
$$x \geq 88$$
She must have a score of 88 or better.

49. Let x represent the number of nickels, then $30 - x$ represents the number of dimes. The value in cents of the nickels is $5x$ and $10(30 - x)$ is the value in cents of the dimes.
$$5x + 10(30 - x) = 260$$
$$5x + 300 - 10x = 260$$
$$-5x + 300 = 260$$
$$-5x = -40$$
$$x = 8$$
There would be 8 nickels and $30 - (8) = 22$ dimes.

50. Let n represent the number of nickels, $3n + 1$ the number of dimes, and $2(3n + 1)$ the number of quarters. The value in cents of the nickels is $5n$, $10(3n + 1)$ is the value in cents of the dimes, and $25(6n + 2)$ is the value in cents of the quarters.
$$5n + 10(3n + 1) + 25(6n + 2) = 1540$$
$$5n + 30n + 10 + 150n + 50 = 1540$$
$$185n + 60 = 1540$$
$$185n = 1480$$
$$n = 8$$
There would be 8 nickels, $3(8) + 1 = 25$ dimes, and $2(25) = 50$ quarters.

51. Let a represent the measure of the angle; then $180 - a$ represents its supplement and $90 - a$ represents its complement.
$$180 - a = 3(90 - a) + 14$$
$$180 - a = 270 - 3a + 14$$
$$180 - a = 284 - 3a$$
$$180 + 2a = 284$$
$$2a = 104$$
$$a = 52$$
The measure of the angle is $52°$.

Chapter 3 Review Problem Set

52. Let m represent the number of miles that she drove.
$$3(25) + 0.20m = 215$$
$$75 + 0.20m = 215$$
$$0.20m = 140$$
$$20m = 14,000$$
$$m = 700$$
She drove 700 miles.

CHAPTER 3 | Test

1.
$$7x - 3 = 11$$
$$7x = 14$$
$$x = 2$$
The solution set is $\{2\}$.

2.
$$-7 = -3x + 2$$
$$-9 = -3x$$
$$3 = x$$
The solution set is $\{3\}$.

3.
$$4n + 3 = 2n - 15$$
$$2n + 3 = -15$$
$$2n = -18$$
$$n = -9$$
The solution set is $\{-9\}$.

4.
$$3n - 5 = 8n + 20$$
$$-5n - 5 = 20$$
$$-5n = 25$$
$$n = -5$$
The solution set is $\{-5\}$.

5.
$$4(x - 2) = 5(x + 9)$$
$$4x - 8 = 5x + 45$$
$$-x - 8 = 45$$
$$-x = 53$$
$$x = -53$$
The solution set is $\{-53\}$.

6.
$$9(x + 4) = 6(x - 3)$$
$$9x + 36 = 6x - 18$$
$$3x + 36 = -18$$
$$3x = -54$$
$$x = -18$$
The solution set is $\{-18\}$.

7.
$$5(y - 2) + 2(y + 1) = 3(y - 6)$$
$$5y - 10 + 2y + 2 = 3y - 18$$
$$7y - 8 = 3y - 18$$
$$4y - 8 = -18$$
$$4y = -10$$
$$y = -\frac{10}{4} = -\frac{5}{2}$$
The solution set is $\left\{ -\dfrac{5}{2} \right\}$.

8.
$$\frac{3}{5}x - \frac{2}{3} = \frac{1}{2}$$
$$30\left(\frac{3}{5}x - \frac{2}{3} \right) = 30\left(\frac{1}{2} \right)$$
$$18x - 20 = 15$$
$$18x = 35$$
$$x = \frac{35}{18}$$
The solution set is $\left\{ \dfrac{35}{18} \right\}$.

9.

$$\frac{x-2}{4} = \frac{x+3}{6}$$

$$12\left(\frac{x-2}{4}\right) = 12\left(\frac{x+3}{6}\right)$$

$$3(x-2) = 2(x+3)$$

$$3x - 6 = 2x + 6$$

$$x - 6 = 6$$

$$x = 12$$

The solution set is $\{12\}$.

10.

$$\frac{x+2}{3} + \frac{x-1}{2} = 2$$

$$6\left(\frac{x+2}{3} + \frac{x-1}{2}\right) = 6(2)$$

$$2(x+2) + 3(x-1) = 12$$

$$2x + 4 + 3x - 3 = 12$$

$$5x + 1 = 12$$

$$5x = 11$$

$$x = \frac{11}{5}$$

The solution set is $\left\{\dfrac{11}{5}\right\}$.

11.

$$\frac{x-3}{6} - \frac{x-1}{8} = \frac{13}{24}$$

$$24\left(\frac{x-3}{6} - \frac{x-1}{8}\right) = 24\left(\frac{13}{24}\right)$$

$$4(x-3) - 3(x-1) = 13$$

$$4x - 12 - 3x + 3 = 13$$

$$x - 9 = 13$$

$$x = 22$$

The solution set is $\{22\}$.

12.

$$-5(n-2) = -3(n+7)$$

$$-5n + 10 = -3n - 21$$

$$-2n + 10 = -21$$

$$-2n = -31$$

$$n = \frac{31}{2}$$

The solution set is $\left\{\dfrac{31}{2}\right\}$.

13.

$$3x - 2 < 13$$

$$3x < 15$$

$$x < 5$$

The solution set is
$\{x | x < 5\}$ or $(-\infty, 5)$.

14.

$$-2x + 5 \geq 3$$

$$-2x \geq -2$$

$$x \leq 1$$

The solution set is
$\{x | x \leq 1\}$ or $(-\infty, 1]$.

15.

$$3(x-1) \leq 5(x+3)$$

$$3x - 3 \leq 5x + 15$$

$$-2x - 3 \leq 15$$

$$-2x \leq 18$$

$$x \geq -9$$

The solution set is
$\{x | x \geq -9\}$ or $[-9, \infty)$.

16.

$$-4 > 7(x-1) + 3$$

$$-4 > 7x - 7 + 3$$

$$-4 > 7x - 4$$

$$0 > 7x$$

$$0 > x$$

$$x < 0$$

The solution set is
$\{x | x < 0\}$ or $(-\infty, 0)$.

17.

$$-2(x-1) + 5(x-2) < 5(x+3)$$

$$-2x + 2 + 5x - 10 < 5x + 15$$

$$3x - 8 < 5x + 15$$

$$-2x - 8 < 15$$

$$-2x < 23$$

$$x > -\frac{23}{2}$$

The solution set is

$\left\{x | x > -\dfrac{23}{2}\right\}$ or $\left(-\dfrac{23}{2}, \infty\right)$.

Chapter 3 Test

18.
$$\frac{1}{2}n + 2 \le \frac{3}{4}n - 1$$
$$8\left(\frac{1}{2}n + 2\right) \le 8\left(\frac{3}{4}n - 1\right)$$
$$4n + 16 \le 6n - 8$$
$$-2n + 16 \le -8$$
$$-2n \le -24$$
$$n \ge 12$$
The solution set is
$\{n | n \ge 12\}$ or $[12, \infty)$.

19. Since it is an "and" statement, we need to satisfy both inequalities at the same time. Thus, all numbers between and including -2 and 4 are solutions.

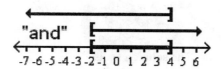

20. Since it is an "or" statement, the solution set consists of all numbers less than (but not including) 1 along with all numbers greater than (but not including) 3.

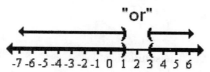

21. Let h represent the hourly rate for the labor.
$$53 + 4h = 127$$
$$4h = 74$$
$$h = 18.50$$
The hourly rate was $18.50 per hour.

22. Let s represent the length of the shortest side, $2s$ the longest side, and $s + 10$ the third side.
$$s + 2s + (s + 10) = 70$$
$$4s + 10 = 70$$
$$4s = 60$$
$$s = 15$$
The shortest side is 15 meters, the longest side is 30 meters and the third side is 25 meters.

23. Let x represent the score on the fifth exam.
$$\frac{86 + 88 + 89 + 91 + x}{5} \ge 90$$
$$354 + x \ge 450$$
$$x \ge 96$$
She would need a score of 96 or better.

24. Let n represent the number of nickels, $2n - 1$ the number of dimes, and $3n + 2$ the number of quarters.
$$n + (2n - 1) + (3n + 2) = 103$$
$$6n + 1 = 103$$
$$6n = 102$$
$$n = 17$$
There are 17 nickels, $2(17) - 1 = 33$ dimes, and $3(17) + 2 = 53$ quarters.

25. Let a represent the measure of the angle; then $180 - a$ represents its supplement and $90 - a$ represents its complement.
$$90 - a = \frac{1}{2}(180 - a) - 10$$
$$2(90 - a) = 2\left[\frac{1}{2}(180 - a) - 10\right]$$
$$180 - 2a = (180 - a) - 20$$
$$180 - 2a = 160 - a$$
$$180 - a = 160$$
$$-a = -20$$
$$a = 20$$
The measure of the angle is 20°.

72

Chapter 4 Formulas and Problem Solving

PROBLEM SET **4.1** **Ratio, Proportion, and Percent**

1. $\dfrac{x}{6} = \dfrac{3}{2}$

$2x = 18$ Cross products are equal.

$x = 9$

The solution set is $\{9\}$.

3. $\dfrac{5}{12} = \dfrac{n}{24}$

$120 = 12n$ Cross products are equal.

$10 = n$

The solution set is $\{10\}$.

5. $\dfrac{x}{3} = \dfrac{5}{2}$

$2x = 15$ Cross products are equal.

$x = \dfrac{15}{2}$

The solution set is $\left\{\dfrac{15}{2}\right\}$.

7. $\dfrac{x-2}{4} = \dfrac{x+4}{3}$

Cross products are equal.

$3(x-2) = 4(x+4)$

$3x - 6 = 4x + 16$

$-x = 22$

$x = -22$

The solution set is $\{-22\}$.

9. $\dfrac{x+1}{6} = \dfrac{x+2}{4}$

Cross products are equal.

$4(x+1) = 6(x+2)$

$4x + 4 = 6x + 12$

$-2x = 8$

$x = -4$

The solution set is $\{-4\}$.

11. $\dfrac{h}{2} - \dfrac{h}{3} = 1$

Be careful, this is not a proportion.

Multiply both sides by 6.

$6\left(\dfrac{h}{2} - \dfrac{h}{3}\right) = 6(1)$

$3h - 2h = 6$

$h = 6$

The solution set is $\{6\}$.

13. $\dfrac{x+1}{3} - \dfrac{x+2}{2} = 4$

Be careful, this is not a proportion.

Multiply both sides by 6.

$6\left(\dfrac{x+1}{3} - \dfrac{x+2}{2}\right) = 6(4)$

$2(x+1) - 3(x+2) = 24$

$2x + 2 - 3x - 6 = 24$

$-x - 4 = 24$

$-x = 28$

$x = -28$

The solution set is $\{-28\}$.

15. $\dfrac{-4}{x+2} = \dfrac{-3}{x-7}$

Cross products are equal.

$-4(x-7) = -3(x+2)$

$-4x + 28 = -3x - 6$

$-x = -34$

$x = 34$

The solution set is $\{34\}$.

73

Problem Set 4.1

17.
$$\frac{-1}{x-7} = \frac{5}{x-1}$$
Cross products are equal.
$$-1(x-1) = 5(x-7)$$
$$-x+1 = 5x-35$$
$$-6x = -36$$
$$x = 6$$
The solution set is $\{6\}$.

19.
$$\frac{3}{2x-1} = \frac{2}{3x+2}$$
Cross products are equal.
$$3(3x+2) = 2(2x-1)$$
$$9x+6 = 4x-2$$
$$5x = -8$$
$$x = -\frac{8}{5}$$
The solution set is $\left\{-\frac{8}{5}\right\}$.

21.
$$\frac{n+1}{n} = \frac{8}{7}$$
Cross products are equal.
$$7(n+1) = 8n$$
$$7n+7 = 8n$$
$$7 = n$$
The solution set is $\{7\}$.

23.
$$\frac{x-1}{2} - 1 = \frac{3}{4}$$
Be careful, this is not a proportion.
Multiply both sides by 8.
$$8\left(\frac{x-1}{2} - 1\right) = 8\left(\frac{3}{4}\right)$$
$$4(x-1) - 8 = 2(3)$$
$$4x - 4 - 8 = 6$$
$$4x - 12 = 6$$
$$4x = 18$$
$$x = \frac{18}{4} = \frac{9}{2}$$
The solution set is $\left\{\frac{9}{2}\right\}$.

25.
$$-3 - \frac{x+4}{5} = \frac{3}{2}$$
Be careful, this is not a proportion.
Multiply both sides by 10.
$$10\left(-3 - \frac{x+4}{5}\right) = 10\left(\frac{3}{2}\right)$$
$$-30 - 2(x+4) = 5(3)$$
$$-30 - 2x - 8 = 15$$
$$-2x - 38 = 15$$
$$-2x = 53$$
$$x = -\frac{53}{2}$$
The solution set is $\left\{\frac{-53}{2}\right\}$.

27.
$$\frac{n}{150-n} = \frac{1}{2}$$
Cross products are equal.
$$2n = 150 - n$$
$$3n = 150$$
$$n = 50$$
The solution set is $\{50\}$.

29.
$$\frac{300-n}{n} = \frac{3}{2}$$
Cross products are equal.
$$2(300 - n) = 3n$$
$$600 - 2n = 3n$$
$$600 = 5n$$
$$120 = n$$
The solution set is $\{120\}$.

31.
$$\frac{-1}{5x-1} = \frac{-2}{3x+7}$$
Cross products are equal.
$$-1(3x+7) = -2(5x-1)$$
$$-3x - 7 = -10x + 2$$
$$7x - 7 = 2$$
$$7x = 9$$
$$x = \frac{9}{7}$$
The solution set is $\left\{\frac{9}{7}\right\}$.

33. $\dfrac{11}{20} = \dfrac{n}{100}$

$20n = 1100$

$n = 55$

Therefore, $\dfrac{11}{20} = \dfrac{55}{100} = 55\%$

35. $\dfrac{3}{5} = \dfrac{n}{100}$

$5n = 300$

$n = 60$

Therefore, $\dfrac{3}{5} = \dfrac{60}{100} = 60\%$

37. $\dfrac{1}{6} = \dfrac{n}{100}$

$6n = 100$

$n = \dfrac{100}{6} = 16\dfrac{2}{3}$

Therefore, $\dfrac{1}{6} = \dfrac{16\frac{2}{3}}{100} = 16\dfrac{2}{3}\%$

39. $\dfrac{3}{8} = \dfrac{n}{100}$

$8n = 300$

$n = \dfrac{300}{8} = 37\dfrac{1}{2}$

Therefore, $\dfrac{3}{8} = \dfrac{37\frac{1}{2}}{100} = 37\dfrac{1}{2}\%$

41. $\dfrac{3}{2} = \dfrac{n}{100}$

$2n = 300$

$n = 150$

Therefore, $\dfrac{3}{2} = \dfrac{150}{100} = 150\%$

43. $\dfrac{12}{5} = \dfrac{n}{100}$

$5n = 1200$

$n = 240$

Therefore, $\dfrac{12}{5} = \dfrac{240}{100} = 240\%$

45. Let n represent the number.

$n = (7\%)(38)$

$n = 0.07(38)$

$n = 2.66$

Therefore, 2.66 is 7% of 38.

47. Let n represent the number.

$(15\%)(n) = 6.3$

$0.15n = 6.3$

Multiply both sides by 100.

$15n = 630$

$n = 42$

Therefore, 15% of 42 is 6.3.

49. Let r represent the percent to be found.

$76 = r(95)$

$\dfrac{76}{95} = r$

$0.80 = r$

Therefore, 76 is 80% of 95.

51. Let n represent the number.

$n = (120\%)(50)$

$n = 1.2(50)$

$n = 60$

Therefore, 60 is 120% of 50.

53. Let r represent the percent to be found.

$46 = r(40)$

$\dfrac{46}{40} = r$

$1.15 = r$

Therefore, 46 is 115% of 40.

55. Let n represent the number.

$(160\%)(n) = 144$

$1.6n = 144$

Multiply both sides by 10.

$16n = 1440$

$n = \dfrac{1440}{16} = 90$

Therefore, 160% of 90 is 144.

Problem Set 4.1

57. Let l and w represent the length and width of the room measured in feet.

$$\begin{array}{c} \\ \text{map} \\ \overline{\text{room}} \end{array} \quad \begin{array}{cc} \text{inches} & \text{feet} \\ \dfrac{1}{2\frac{1}{2}} = \dfrac{6}{w} \end{array}$$

$$w = 6\left(2\frac{1}{2}\right)$$
$$w = 6\left(\frac{5}{2}\right)$$
$$w = 15$$

$$\begin{array}{c} \\ \text{map} \\ \overline{\text{room}} \end{array} \quad \begin{array}{cc} \text{inches} & \text{feet} \\ \dfrac{1}{3\frac{1}{4}} = \dfrac{6}{l} \end{array}$$

$$l = 6\left(3\frac{1}{4}\right)$$
$$l = 6\left(\frac{13}{4}\right)$$
$$l = \frac{39}{2} = 19\frac{1}{2}$$

The room measures 15 feet by $19\frac{1}{2}$ feet.

59. Let m represent the number of miles traveled.

$$\begin{array}{c} \text{miles} \\ \overline{\text{gallons}} \end{array} \quad \dfrac{264}{12} = \dfrac{m}{15}$$

$$12m = 3960$$
$$m = 330$$

The car will travel 330 miles.

61. Let l represent the length of the rectangle.

$$\begin{array}{c} \text{length} \\ \overline{\text{width}} \end{array} \quad \dfrac{5}{2} = \dfrac{l}{24}$$

$$2l = 120$$
$$l = 60$$

The length of the rectangle is 60 centimeters.

63. Let s represent the number of pounds of salt needed.

$$\begin{array}{c} \text{salt} \\ \overline{\text{water}} \end{array} \quad \dfrac{3}{10} = \dfrac{s}{25}$$

$$10s = 75$$
$$s = 7.5$$

It will take 7.5 pounds of salt.

65. Let p represent the number of pounds of fertilizer need.

$$\begin{array}{c} \text{fertilizer} \\ \overline{\text{lawn}} \end{array} \quad \dfrac{20}{1500} = \dfrac{p}{2500}$$

$$1500p = 50,000$$
$$p = 33\frac{1}{3}$$

It will take $33\frac{1}{3}$ pounds of fertilizer to cover the lawn.

67. Let n represent the number of people who are expected to vote.

$$\dfrac{3}{7} = \dfrac{n}{210,000}$$
$$7n = 630,000$$
$$n = 90,000$$

It is expected that 90,000 people will vote in the election.

69. Let x represent the length of the rectangle. The sum of the length and the width is equal to one-half of the perimeter. Therefore, $25 - x$ represents the width.

$$\begin{array}{c} \text{length} \\ \overline{\text{width}} \end{array} \quad \dfrac{3}{2} = \dfrac{x}{25 - x}$$

$$2x = 3(25 - x)$$
$$2x = 75 - 3x$$
$$5x = 75$$
$$x = 15$$

The dimensions of the rectangle are 15 inches by 10 inches.

71. Let x represent the additional money to be invested.

$$\begin{array}{c} \text{investment} \\ \overline{\text{earnings}} \end{array} \quad \dfrac{500}{45} = \dfrac{500 + x}{72}$$

$$45(500 + x) = 500(72)$$
$$22,500 + 45x = 36,000$$
$$45x = 13,500$$
$$x = 300$$

The additional investment would be $300.

73. Let x represent the money the child will receive; then, the cancer fund will receive $180,000 - x$.

$$\frac{\text{child}}{\text{fund}} \quad \frac{5}{1} = \frac{x}{180,000 - x}$$
$$x = 5(180,000 - x)$$
$$x = 900,000 - 5x$$
$$6x = 900,000$$
$$x = 150,000$$
The child would receive $150,000$.

77.
$$\frac{3}{x-2} = \frac{6}{2x-4}$$
$$2(2x-4) = 6(x-2)$$
$$6x - 12 = 6x - 12$$
$$-12 = -12$$
(Always true except when $x = 2$, because the denominator would become zero.)
The solution set is
{all real numbers except 2}.

79.
$$\frac{5}{x-3} = \frac{10}{x-6}$$
$$5(x-6) = 10(x-3)$$
$$5x - 30 = 10x - 30$$
$$-5x - 30 = -30$$
$$-5x = 0$$
$$x = 0$$
The solution set is $\{0\}$.

81.
$$\frac{x-2}{2} = \frac{x}{2} - 1$$
$$2\left(\frac{x-2}{2}\right) = 2\left(\frac{x}{2} - 1\right)$$
$$x - 2 = x - 2$$
$$-2 = -2 \quad \text{(Always true.)}$$
The solution set is {all real numbers}.

PROBLEM SET **4.2** **More on Percents and Problem Solving**

1.
$$x - 0.36 = 0.75$$
Add 0.36 to both sides.
$$x - 0.36 + 0.36 = 0.75 + 0.36$$
$$x = 1.11$$
The solution set is $\{1.11\}$.

3.
$$x + 7.6 = 14.2$$
Subtract 7.6 from both sides.
$$x + 7.6 - 7.6 = 14.2 - 7.6$$
$$x = 6.6$$
The solution set is $\{6.6\}$.

5.
$$0.62 - y = 0.14$$
Subtract 0.62 from both sides.
$$0.62 - y - 0.62 = 0.14 - 0.62$$
$$-y = -0.48$$
$$y = 0.48$$
The solution set is $\{0.48\}$.

7.
$$0.7t = 56$$
Multiply both sides by 10.
$$10(0.7t) = 10(56)$$
$$7t = 560$$
$$t = 80$$
The solution set is $\{80\}$.

9.
$$x = 3.36 - 0.12x$$
Multiply both sides by 100.
$$100(x) = 100(3.36 - 0.12x)$$
$$100x = 336 - 12x$$
$$112x = 336$$
$$x = 3$$
The solution set is $\{3\}$.

11.
$$s = 35 + 0.3s$$
Multiply both sides by 10.
$$10(s) = 10(35 + 0.3s)$$
$$10s = 350 + 3s$$
$$7s = 350$$
$$s = 50$$
The solution set is $\{50\}$.

Problem Set 4.2

13.
$$s = 42 + 0.4s$$
Multiply both sides by 10.
$$10(s) = 10(42 + 0.4s)$$
$$10s = 420 + 4s$$
$$6s = 420$$
$$s = 70$$
The solution set is $\{70\}$.

15.
$$0.07x + 0.08(x + 600) = 78$$
Multiply both sides by 100.
$$100[0.07x + 0.08(x + 600)] = 100(78)$$
$$7x + 8(x + 600) = 7800$$
$$7x + 8x + 4800 = 7800$$
$$15x + 4800 = 7800$$
$$15x = 3000$$
$$x = 200$$
The solution set is $\{200\}$.

17.
$$0.09x + 0.1(2x) = 130.5$$
Multiply both sides by 100.
$$100[0.09x + 0.1(2x)] = 100(130.5)$$
$$9x + 10(2x) = 13,050$$
$$9x + 20x = 13,050$$
$$29x = 13,050$$
$$x = 450$$
The solution set is $\{450\}$.

19.
$$0.08x + 0.11(500 - x) = 50.5$$
Multiply both sides by 100.
$$100[0.08x + 0.11(500 - x)] = 100(50.5)$$
$$8x + 11(500 - x) = 5050$$
$$8x + 5500 - 11x = 5050$$
$$-3x + 5500 = 5050$$
$$-3x = -450$$
$$x = 150$$
The solution set is $\{150\}$.

21.
$$0.09x = 550 - 0.11(5400 - x)$$
Multiply both sides by 100.
$$9x = 55,000 - 11(5400 - x)$$
$$9x = 55,000 - 59,400 + 11x$$
$$-2x = -4400$$
$$x = 2200$$
The solution set is $\{2200\}$.

23. Let p represent the original price of the trousers.
$$(100\%)(p) - (30\%)(p) = 35$$
$$(70\%)(p) = 35$$
$$0.7p = 35$$
$$7p = 350$$
$$p = 50$$
The original price of the trousers was $50.

25. Let d represent the discount sale price of the sweater. Since the sweater is on sale for 25% off, the discount price is 75% of the original price.
$$d = (75\%)(48)$$
$$d = 0.75(48)$$
$$d = 36$$
The discount sale price is $36.

27. Let d represent the discount sale price of the putter. Since the putter is on sale for 35% off, the discount price is 65% of the original price.
$$d = (65\%)(32)$$
$$d = 0.65(32)$$
$$d = 20.8$$
The discount sale price is $20.80.

29. Let r represent the rate of discount.
$$180 - r(180) = 126$$
$$180 - 180r = 126$$
$$-180r = -54$$
$$r = 0.3$$
The rate of discount is 30%.

31. Let s represent the selling price. Profit is a percent of cost.
Selling price = Cost + Profit
$$s = 5 + (70\%)(5)$$
$$s = 5 + 0.7(5)$$
$$s = 5 + 3.5$$
$$s = 8.5$$
The selling price would be $8.50.

33. Let s represent the selling price.
Profit is a percent of cost.
Selling price = Cost + Profit
$$s = 3 + (55\%)(3)$$
$$s = 3 + 0.55(3)$$
$$s = 3 + 1.65$$
$$s = 4.65$$
The selling price would be $4.65.

35. Let s represent the selling price.
Profit is a percent of selling price.
Selling price = Cost + Profit
$$s = 400 + (60\%)(s)$$
$$s = 400 + 0.6s$$
$$10s = 4000 + 6s$$
$$4s = 4000$$
$$s = 1000$$
The selling price would be $1000.

37. Let r represent the rate of profit.
Profit is a percent of cost.
Selling price = Cost + Profit
$$44.8 = 32 + r(32)$$
$$448 = 320 + 320r$$
$$128 = 320r$$
$$0.4 = r$$
The rate of profit would be 40%
of the cost.

39.
$$i = Prt$$
$$560 = 3500(r)(2)$$
$$560 = 7000r$$
$$\frac{560}{7000} = r$$
$$0.8 = r$$
The annual interest rate will be 8%.

41.
$$i = Prt$$
$$1000 = P(0.08)(3)$$
$$1000 = 0.24P$$
$$\frac{1000}{0.24} = P$$
$$4166.67 = P$$
The principal will be $4166.67.

43.
$$i = Prt$$
$$i = (5000)(0.068)(10)$$
$$i = 3400$$
The interest earned will be $3400.

45.
$$i = Prt$$
Remember time must be in years.
So 1 month = $\frac{1}{12}$ year.
$$i = 95000(0.08)\left(\frac{1}{12}\right)$$
$$i = 633.33$$
The interest for a month will be $633.33.

49. The profit in dollars on the item is
$50 − $40 = $10. Since 10 is 20% of 50,
his claim would be correct that he makes a
20% profit if he calculates the profit as a
percent of the selling price.

51. Use problem 50 and reverse the
calculations to make a comparison.
$$d = (90\%)(60) = 0.9(60) = 54$$
$$a = (60\%)(54) = 0.6(54) = 32.40$$
Yes, both discount patterns would give
the same result. This is the result of the
commutative property of multiplication.

53.
$$2.4x + 5.7 = 9.6$$
$$2.4x = 3.9$$
$$x = 1.625$$
The solution set is $\{1.625\}$.

55.
$$0.08x + 0.09(800 − x) = 68.5$$
$$8x + 9(800 − x) = 6850$$
$$8x + 7200 − 9x = 6850$$
$$-x + 7200 = 6850$$
$$-x = -350$$
$$x = 350$$
The solution set is $\{350\}$.

Problem Set 4.2

57. $7x - 0.39 = 0.03$
$$7x = 0.42$$
$$x = 0.06$$
The solution set is $\{0.06\}$.

59. $0.2(t + 1.6) = 3.4$
$$2(t + 1.6) = 34$$
$$2t + 3.2 = 34$$
$$2t = 30.8$$
$$t = 15.4$$
The solution set is $\{15.4\}$.

PROBLEM SET | **4.3** **Formulas**

1. $d = 336, r = 48 : d = rt$
$$336 = 48t$$
$$7 = t$$

3. $i = 200, r = 0.08, t = 5 : i = Prt$
$$200 = P(0.08)(5)$$
$$200 = 0.4P$$
$$500 = P$$

5. $F = 68 : F = \dfrac{9}{5}C + 32$
$$68 = \dfrac{9}{5}C + 32$$
Multiply both sides by 5.
$$340 = 9C + 160$$
$$180 = 9C$$
$$20 = C$$

7. $V = 112, h = 7 : V = \dfrac{1}{3}Bh$
$$112 = \dfrac{1}{3}B(7)$$
Multipy both sides by 3.
$$336 = 7B$$
$$48 = B$$

9. $A{=}652, P{=}400, r{=}0.07; A = P + Prt$
$$652 = 400 + 400(0.07)t$$
$$652 = 400 + 28t$$
$$252 = 28t$$
$$9 = t$$

11. Substitute 14 for l and 9 for w in the formula for finding perimeter of a rectangle.

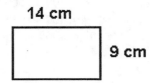

$$P = 2l + 2w$$
$$P = 2(14) + 2(9)$$
$$P = 28 + 18$$
$$P = 46$$
The perimeter of the rectangle is 46 centimeters.

13. Substitute $3\dfrac{1}{4}$ feet $= 39$ inches for l and 108 feet for P in the formula for finding perimeter of a rectangle.

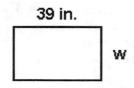

$$P = 2l + 2w$$
$$108 = 2(39) + 2w$$
$$108 = 78 + 2w$$
$$30 = 2w$$
$$15 = w$$
The width of the rectangle is 15 inches.

15. Subtract the area of the rectangular garden from the area of the garden plus the dirt path. The width of the larger rectangle is $17 + 4 + 4 = 25$ feet and the length of the larger rectangle is $38 + 4 + 4 = 46$ feet.

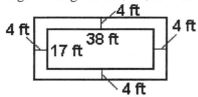

$A =$ large area $-$ garden area
$A = (46)(25) - (38)(17)$
$A = 1150 - 646$
$A = 504$
The area of the dirt path is 504 square feet.

17. Since the area is needed in square meters, it is easier to convert the measurements to meters first. Thus, the length of the wood piece is 60 centimeters $= 0.6$ meters and the width is 30 centimeters $= 0.3$ meters. The area of one piece is $(0.3)(0.6) = 0.18$ square meters. There are 50 pieces for a total area of $50(0.18) = 9$ square meters. One liter of paint would cover the total area for a cost of \$2.

19. Let h represent the length of the altitude of the trapezoid.

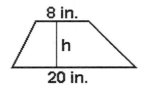

$A = \dfrac{1}{2}h(b_1 + b_2)$

$98 = \dfrac{1}{2}h(8 + 20)$

$98 = \dfrac{1}{2}h(28)$

$98 = 14h$

$7 = h$

The length of the altitude of the trapezoid is 7 inches.

21. The area of one washer can be computed by subtracting the area of the inner circle (diameter $= 2$ cm) from the area of the outer circle (diameter $= 4$ cm). The radius, which is one-half of the diameter, is used in the formula. See Figure 4.18 in the text for a drawing.
$A = \pi(2)^2 - \pi(1)^2 = 4\pi - 1\pi = 3\pi$
The area of one washer is 3π square centimeters, so 50 washers would have $50(3\pi) = 150\pi$ square centimeters of metal.

23. The radius of the circular region is $\dfrac{1}{2}$ yard.

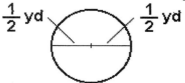

$A = \pi r^2$

$A = \pi \left(\dfrac{1}{2}\right)^2 = \dfrac{1}{4}\pi$

The area of the circular region is $\dfrac{1}{4}\pi$ square yards.

25. Substitute $r = 9$ into both formulas.
$$S = 4\pi r^2 = 4\pi(9)^2$$
$$= 324\pi \text{ square inches}$$
$$V = \dfrac{4}{3}\pi r^3 = \dfrac{4}{3}\pi(9)^3$$
$$= 972\pi \text{ cubic inches}$$

27. Substitute $r = 8$ and $h = 18$ into both formulas.
$$V = \pi r^2 h = \pi(8)^2(18)$$
$$= 1152\pi \text{ cubic inches}$$
$$S = 2\pi r^2 + 2\pi rh = 2\pi(8)^2 + 2\pi(8)(18)$$
$$= 128\pi + 288\pi$$
$$= 416\pi \text{ square inches}$$

Problem Set 4.3

29. Substitute $V = 324\pi$ and $r = 9$ in the formula.

$$V = \frac{1}{3}\pi r^2 h$$

$$324\pi = \frac{1}{3}\pi (9)^2 h$$

$$324\pi = \frac{81}{3}\pi h$$

$$324\pi = 27\pi h$$

$$12 = h$$

The height of the cone is 12 inches.

31. Substitute $S = 65\pi$ and $r = 5$ in the formula.

$$S = \pi r^2 + \pi r s$$

$$65\pi = \pi (5)^2 + \pi (5)s$$

$$65\pi = 25\pi + 5\pi s$$

$$40\pi = 5\pi s$$

$$8 = s$$

The slant height of the cone is 8 feet.

33. $V = Bh$

Divide both sides by B.

$$\frac{V}{B} = \frac{Bh}{B}$$

$$\frac{V}{B} = h$$

35. $V = \frac{1}{3}Bh$

Multiply both sides by 3.

$$3V = Bh$$

Divide both sides by h.

$$\frac{3V}{h} = B$$

37. $P = 2l + 2w$

Subtract $2l$ from both sides.

$$P - 2l = 2w$$

Divide both sides by 2.

$$\frac{P - 2l}{2} = w$$

39. $V = \frac{1}{3}\pi r^2 h$

Multiply both sides by 3.

$$3V = \pi r^2 h$$

Divide both sides by πr^2.

$$\frac{3V}{\pi r^2} = h$$

41. $$F = \frac{9}{5}C + 32$$

Subtract 32 from both sides.

$$F - 32 = \frac{9}{5}C$$

Multiply both sides by $\frac{5}{9}$.

$$\frac{5}{9}(F - 32) = C$$

43. $$A = 2\pi r^2 + 2\pi r h$$

Subtract $2\pi r^2$ from both sides.

$$A - 2\pi r^2 = 2\pi r h$$

Divide both sides by $2\pi r$.

$$\frac{A - 2\pi r^2}{2\pi r} = h$$

45. $3x + 7y = 9$

Subtract $7y$ from both sides.

$$3x = 9 - 7y$$

Divide both sides by 3.

$$x = \frac{9 - 7y}{3}$$

47. $9x - 6y = 13$

Subtracte $9x$ from both sides.

$$-6y = 13 - 9x$$

Multiply both sides by -1.

$$6y = -13 + 9x$$

Apply commutative property.

$$6y = 9x - 13$$

Divide both sides by 6.

$$y = \frac{9x - 13}{6}$$

49. $-2x + 11y = 14$
Subtract $11y$ from both sides.
$$-2x = 14 - 11y$$
Multiply both sides by -1.
$$2x = -14 + 11y$$
Apply commutative property.
$$2x = 11y - 14$$
Divide both sides by 2.
$$x = \frac{11y - 14}{2}$$

51. $y = -3x - 4$
Add 4 to both sides.
$$y + 4 = -3x$$
Multiply both sides by -1.
$$-y - 4 = 3x$$
Divide both sides by 3.
$$\frac{-y - 4}{3} = x$$

53. $\dfrac{x - 2}{4} = \dfrac{y - 3}{6}$
Cross products are equal.
$$4(y - 3) = 6(x - 2)$$
Distributive property.
$$4y - 12 = 6x - 12$$
Add 12 to both sides..
$$4y = 6x$$
Divide both sides by 4.
$$y = \frac{6x}{4} = \frac{3x}{2}$$

55. $ax - by - c = 0$
Add by to both sides.
$$ax - c = by$$
Divide both sides by b.
$$\frac{ax - c}{b} = y$$

57. $\dfrac{x + 6}{2} = \dfrac{y + 4}{5}$
Cross products are equal.
$$5(x + 6) = 2(y + 4)$$
$$5x + 30 = 2y + 8$$
Subtract 30 from both sides..
$$5x = 2y - 22$$
Divide both sides by 5.
$$x = \frac{2y - 22}{5}$$

59. $m = \dfrac{y - b}{x}$
Multiply both sides by x.
$$mx = y - b$$
Add b to both sides.
$$mx + b = y$$

65. Find the larger area by substituting $r = 7$ into the area formula. Then substitute $r = 3$ into the formula to find the smaller area. Finally, subtract the smaller area from the larger area to find the area of the shaded ring.
Larger Area $= \pi r^2 = (3.14)(7)^2 = 153.86$
Smaller Area $= \pi r^2 = (3.14)(3)^2 = 28.26$
Shaded Area $= 153.86 - 28.26 = 125.6$ square centimeters

67. Right Circular Cylinder with $r = 3$ and $h = 10$:
$$S = 2\pi r^2 + 2\pi rh$$
$$= 2(3.14)(3)^2 + 2(3.14)(3)(10)$$
$$= 56.52 + 188.4$$
$$= 244.92 \text{ rounds to } 245 \text{ square centimeters}$$

69. Sphere with diameter of 5 ($r = 2.5$):
$$V = \frac{4}{3}\pi r^3$$
$$= \frac{4}{3}(3.14)(2.5)^3$$
$$= 65.41666... \text{ rounds to } 65 \text{ cubic inches}$$

Problem Set 4.4

4.4 **Problem Solving**

1. $950(0.12)t = 950$
Divide both sides by 950.
$$0.12t = 1$$
Multiply both sides by 100.
$$12t = 100$$
Divide both sides by 12 and reduce.
$$t = \frac{100}{12} = 8\frac{1}{3}$$
The solution set is $\left\{8\frac{1}{3}\right\}$.

3. $l + \frac{1}{4}l - 1 = 19$
Add 1 to both sides.
$$l + \frac{1}{4}l = 20$$
Multiply both sides by 4.
$$4l + l = 80$$
$$5l = 80$$
Divide both sides by 5.
$$l = 16$$
The solution set is $\{16\}$.

5. $500(0.08)t = 1000$
Simplify the left side.
$$40t = 1000$$
Divide both sides by 40.
$$t = 25$$
The solution set is $\{25\}$.

7. $s + (2s - 1) + (3s - 4) = 37$
Combine like terms.
$$6s - 5 = 37$$
Add 5 to both sides.
$$6s = 42$$
Divide both sides by 6.
$$s = 7$$
The solution set is $\{7\}$.

9. $\frac{5}{2}r + \frac{5}{2}(r + 6) = 135$
Multiply both sides by 2.
$$2\left[\frac{5}{2}r + \frac{5}{2}(r + 6)\right] = 2(135)$$
$$5r + 5(r + 6) = 270$$
Apply distributive property.
$$5r + 5r + 30 = 270$$
Combine like terms.
$$10r + 30 = 270$$
Subtract 30 from both sides.
$$10r = 240$$
Divide both sides by 10.
$$r = 24$$
The solution set is $\{24\}$.

11. $24\left(t - \frac{2}{3}\right) = 18t + 8$
Apply distributive property.
$$24t - 16 = 18t + 8$$
Subtract $18t$ from both sides.
$$6t - 16 = 8$$
Add 16 to both sides.
$$6t = 24$$
Divide both sides by 6.
$$t = 4$$
The solution set is $\{4\}$.

13. To "double" the interest earned, i would equal 750 when $P = 750$ and $r = 8\%$.
$$i = Prt$$
$$750 = 750(0.08)t$$
Divide both sides by 750.
$$1 = 0.08t$$
Multiply by 100.
$$100 = 8t$$
Divide both sides by 8.
$$12\frac{1}{2} = t$$
It would take $12\frac{1}{2}$ years for $750 to double itself.

84

15. To "triple itself" the interest earned, i would equal 1600 when $P = 800$ and $r = 10\%$.

$1600 = 800(10\%)t$

Divide both sides by 800.

$2 = 0.1t$

Multiply by 100.

$200 = 10t$

Divide both sides by 10.

$20 = t$

It would take 20 years for $800 to triple itself.

17. Let w represent the width of the rectangle. Then $3w$ represents the length and the perimeter is 112 inches.

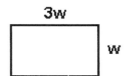

$P = 2l + 2w$
$112 = 2(3w) + 2w$
$112 = 6w + 2w$
$112 = 8w$
$14 = w$

The width is 14 inches and the length is $3(14) = 42$ inches.

19. Let w represent the width of the rectangle. Then $3w - 2$ represents the length and the perimeter is 92 cm.

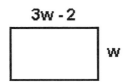

$P = 2l + 2w$
$92 = 2(3w - 2) + 2w$
$92 = 6w - 4 + 2w$
$96 = 8w$
$12 = w$

The width is 12 centimeters and the length is $3(12) - 2 = 34$ centimeters.

21. Let l represent the length of the rectangle; then $\frac{1}{2}l - 3$ represents the width when $P = 42$.

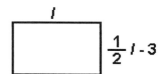

$P = 2l + 2w$
$42 = 2l + 2\left(\frac{1}{2}l - 3\right)$
$42 = 2l + l - 6$
$48 = 3l$
$16 = l$

The length is 16 inches and the width is $\frac{1}{2}(16) - 3 = 5$ inches. Therefore, the area $= lw = (16)(5) = 80$ square inches.

23. Let s represent the shortest side, $2s - 3$ the longest side, and $s + 7$ the third side of a triangle when the perimeter equals 100 feet.

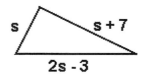

$s + (2s - 3) + (s + 7) = 100$
$4s + 4 = 100$
$4s = 96$
$s = 24$

The lengths of the sides are 24 feet, $2(24) - 3 = 45$ feet, and $24 + 7 = 31$ feet.

85

Problem Set 4.4

25. Let f represent the first side, $3f + 1$ the second side, and $3f + 3$ the third side of a triangle when the perimeter equals 46 centimeters.

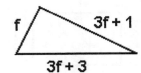

$$f + (3f + 1) + (3f + 3) = 46$$
$$7f + 4 = 46$$
$$7f = 42$$
$$f = 6$$

The lengths of the sides are 6 centimeters, 19 centimeters and 21 centimeters.

27. Let s represent each side of the equilateral triangle; then $s - 4$ represents each side of the square. The perimeter of the triangle is $3s$ and the perimeter of the rectangle is $4(s - 4)$. As stated in the problem, the perimeter of the triangle is 4 centimeters more than the perimeter of the rectangle. This gives the guideline for the equation to be solved.

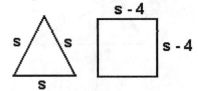

Triangle perimeter = Rectangle perimeter + 4
$$3s = 4(s - 4) + 4$$
$$3s = 4s - 16 + 4$$
$$-s = -12$$
$$s = 12$$

The length of a side of the triangle is 12 cm.

29. Let s represent the length of each side of the square; then s also represents the radius of the circle.

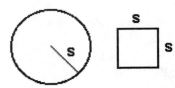

circle circumference = square perimeter + 15.96
$$2\pi r = 2s + 2s + 15.96$$
$$2(3.14)s = 4s + 15.96$$
$$6.28s = 4s + 15.96$$
$$2.28s = 15.96$$
$$s = 7$$

The length of the radius of the circle is 7cm.

31. Let t represent Monica's time; then $t + 1$ represents Sandy's time as she travels one hour longer. A chart of the information would be as follows:

	Rate	Time	Distance ($d = rt$)
Sandy	45	$t + 1$	$45(t + 1)$
Monica	50	t	$50t$

A diagram of the problem would be as follows:

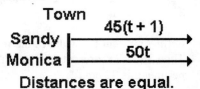

Distances are equal.

Set the distances equal to each other and solve.
$$45(t + 1) = 50t$$
$$45t + 45 = 50t$$
$$45 = 5t$$
$$9 = t$$

It would take 9 hours for Monica to overtake Sandy.

33. Let t represent the freight train's time; then t also represents the passenger train's time as both trains left at the same time. A chart of the information would be as follows:

	Rate	Time	Distance $(d = rt)$
Freight	40	t	$40t$
Passenger	90	t	$90t$

A diagram of the problem would be as follows:

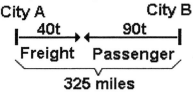

The sum of the two distances would equal the total distance of 325 miles.

$$40t + 90t = 325$$
$$130t = 325$$
$$t = 2\frac{1}{2}$$

It would take $2\frac{1}{2}$ hours for the two trains to meet.

35. Let r represent the second car's rate. Add two hours to the second car's time to find the first car's time. The time also needs to be converted into hours only. A chart of the information would be as follows:

	Rate	Time	Distance $(d = rt)$
First car	40	$7\frac{1}{3} = \frac{22}{3}$	$40\left(\frac{22}{3}\right) = \frac{880}{3}$
Second car	r	$5\frac{1}{3} = \frac{16}{3}$	$\frac{16}{3}r$

A diagram of the problem would be as follows:

Town

1st car $\qquad \dfrac{880/3}{\longrightarrow}$

2nd car $\qquad \dfrac{(16/3)r}{\longrightarrow}$

Distances are equal.

Set the two distances equal and solve.

$$\frac{16}{3}r = \frac{880}{3}$$
$$16r = 880$$
$$r = 55$$

The second car would be traveling 55 mph.

37. Let r represent the rate of the train traveling west; then $r + 8$ represents the rate of the train traveling east. The time is the same for both trains. A chart of the information would be as follows:

	Rate	Time	Distance $(d = rt)$
West train	r	$9\frac{1}{2} = \frac{19}{2}$	$\frac{19}{2}r$
East train	$r + 8$	$9\frac{1}{2} = \frac{19}{2}$	$\frac{19}{2}(r + 8)$

A diagram of the problem would be as follows:

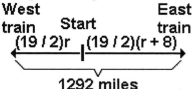

The sum of the two distances would be 1292 miles.

$$\frac{19}{2}r + \frac{19}{2}(r + 8) = 1292$$
$$19r + 19(r + 8) = 2(1292)$$
$$19r + 19r + 152 = 2584$$
$$38r = 2432$$
$$r = 64$$

The train traveling west has a rate of 64 mph and the train traveling east has a rate of 72 mph.

39. Let r represent Jeff's rate when he was leaving; then $r - 2$ represents his rate returning. A chart of the information would be as follows:

	Rate	Time	Distance $(d = rt)$
Leaving	r	3	$3r$
Returning	$r - 2$	$3\frac{3}{4} = \frac{15}{4}$	$\frac{15}{4}(r - 2)$

A diagram of the problem would be as follows:

Home

Leaving $\qquad \dfrac{3r}{\longrightarrow}$

Returning $\dfrac{(15/4)(r-2)}{\longleftarrow}$

Distances are equal.

Set the distances equal to each other and solve for r.

Problem Set 4.4

$$3r = \frac{15}{4}(r - 2)$$
$$12r = 15(r - 2)$$
$$12r = 15r - 30$$
$$-3r = -30$$
$$r = 10$$

Jeff's rate leaving was 10 mph. Therefore, he travelled 30 miles leaving and 30 miles returning for a total of 60 miles.

PROBLEM SET **4.5** **More about Problem Solving**

1. $0.3x + 0.7(20 - x) = 0.4(20)$
$$0.3x + 14 - 0.7x = 8$$
$$14 - 0.4x = 8$$
$$-0.4x = -6$$
$$x = 15$$
The solution set is $\{15\}$.

3. $0.2(20) + x = 0.3(20 + x)$
$$4 + x = 6 + 0.3x$$
$$4 + 0.7x = 6$$
$$0.7x = 2$$
$$7x = 20$$
$$x = \frac{20}{7}$$
The solution set is $\left\{\dfrac{20}{7}\right\}$.

5. $0.7(15) - x = 0.6(15 - x)$
$$10[0.7(15)] - 10x = 10[0.6(15 - x)]$$
$$7(15) - 10x = 6(15 - x)$$
$$105 - 10x = 90 - 6x$$
$$105 - 4x = 90$$
$$-4x = -15$$
$$x = \frac{15}{4}$$
The solution set is $\left\{\dfrac{15}{4}\right\}$.

7. $0.4(10) - 0.4x + x = 0.5(10)$
$$4 - 0.4x + x = 5$$
$$4 + 0.6x = 5$$
$$0.6x = 1$$
$$6x = 10$$
$$x = \frac{10}{6} = \frac{5}{3}$$
The solution set is $\left\{\dfrac{5}{3}\right\}$.

9. $20x + 12\left(4\dfrac{1}{2} - x\right) = 70$
$$20x + 12\left(\frac{9}{2} - x\right) = 70$$
$$20x + 54 - 12x = 70$$
$$8x + 54 = 70$$
$$8x = 16$$
$$x = 2$$
The solution set is $\{2\}$.

11. $3t = \dfrac{11}{2}\left(t - \dfrac{3}{2}\right)$
$$3t = \frac{11}{2}t - \frac{33}{4}$$
$$4(3t) = 4\left(\frac{11}{2}t - \frac{33}{4}\right)$$
$$12t = 22t - 33$$
$$-10t = -33$$
$$t = \frac{33}{10}$$
The solution set is $\left\{\dfrac{33}{10}\right\}$.

13. Let x represent the amount of pure acid to be added; then $100 + x$ represents the amount of final solution.

$$\left(\begin{matrix} \text{pure acid in} \\ \text{10\% solution} \end{matrix} \right) + \left(\begin{matrix} \text{pure acid} \\ \text{to be added} \end{matrix} \right) = \left(\begin{matrix} \text{pure acid in} \\ \text{final solution} \end{matrix} \right)$$

$$(10\%)(100) + x = (20\%)(100 + x)$$
$$0.10(100) + x = 0.20(100 + x)$$
$$10[0.10(100) + x] = 10[0.20(100 + x)]$$
$$1(100) + 10x = 2(100 + x)$$
$$100 + 10x = 200 + 2x$$
$$100 + 8x = 200$$
$$8x = 100$$
$$x = 12.5$$

We must add 12.5 milliliters of pure acid.

15. Let x represent the amount of distilled water to be added; then $10 + x$ represents the amount of final solution.

$$\left(\begin{matrix} \text{pure acid in} \\ \text{50\% solution} \end{matrix} \right) + \left(\begin{matrix} \text{no acid} \\ \text{to be added} \end{matrix} \right) = \left(\begin{matrix} \text{pure acid in} \\ \text{final solution} \end{matrix} \right)$$

$$(50\%)(10) + 0 = (20\%)(10 + x)$$
$$0.50(10) = 0.20(10 + x)$$
$$10[0.50(10)] = 10[0.20(10 + x)]$$
$$5(10) = 2(10 + x)$$
$$50 = 20 + 2x$$
$$30 = 2x$$
$$15 = x$$

We must add 15 centiliters of distilled water.

17. Let x represent the amount of 30% alcohol to be used; then $10 - x$ represents the amount of 50% solution.

$$\left(\begin{matrix} \text{pure alcohol in} \\ \text{30\% solution} \end{matrix} \right) + \left(\begin{matrix} \text{pure alcohol in} \\ \text{50\% solution} \end{matrix} \right) = \left(\begin{matrix} \text{pure alcohol in} \\ \text{final 35\% solution} \end{matrix} \right)$$

Problem Set 4.5

$$(30\%)(x) + (50\%)(10 - x) = (35\%)(10)$$
$$0.30(x) + 0.50(10 - x) = 0.35(10)$$
$$100[0.30(x) + 0.50(10 - x)] = 100[0.35(10)]$$
$$30(x) + 50(10 - x) = 35(10)$$
$$30x + 500 - 50x = 350$$
$$500 - 20x = 350$$
$$-20x = -150$$
$$x = \frac{150}{20} = 7\frac{1}{2}$$

We must add $7\frac{1}{2}$ quarts of 30% alcohol and $2\frac{1}{2}$ quarts of 50% alcohol.

19. Let x represent the amount of water to be removed; then
$20 - x$ represents the amount of final salt solution.

$$\left(\begin{array}{c} \text{pure salt in} \\ \text{30\% solution} \end{array} \right) - \left(\begin{array}{c} \text{no salt to} \\ \text{be removed} \end{array} \right) = \left(\begin{array}{c} \text{pure salt in} \\ \text{final solution} \end{array} \right)$$

$$(30\%)(20) - 0 = (40\%)(20 - x)$$
$$0.30(20) - 0 = 0.40(20 - x)$$
$$6 = 8 - 0.4x$$
$$-2 = -0.4x$$
$$-20 = -4x$$
$$5 = x$$

We must remove 5 gallons of water.

21. Let x represent the amount of pure antifreeze to be added; then
x also represents the amount of solution to be drained.

$$\left(\begin{array}{c} \text{antifreeze in} \\ \text{20\% solution} \end{array} \right) - \left(\begin{array}{c} \text{antifreeze to} \\ \text{be drained} \end{array} \right) + \left(\begin{array}{c} \text{antifreeze to} \\ \text{be added} \end{array} \right) = \left(\begin{array}{c} \text{antifreeze in} \\ \text{final solution} \end{array} \right)$$

$$(20\%)(12) - (20\%)(x) + x = (40\%)(12)$$
$$0.20(12) - 0.20x + x = 0.40(12)$$
$$10[0.20(12) - 0.20x + x] = 10[0.40(12)]$$
$$2(12) - 2(x) + 10x = 4(12)$$
$$24 - 2x + 10x = 48$$
$$24 + 8x = 48$$
$$8x = 24$$
$$x = 3$$

We must drain 3 quarts of the 20% solution and then replace with 3 quarts of pure antifreeze.

90

23. Let x represent the amount of 15% salt solution to be used; then $8 + x$ represents the amount of final solution.

$$\left(\begin{array}{c}\text{pure salt in}\\\text{15\% solution}\end{array}\right) + \left(\begin{array}{c}\text{pure salt in}\\\text{20\% solution}\end{array}\right) = \left(\begin{array}{c}\text{pure salt in}\\\text{final solution}\end{array}\right)$$

$$(15\%)(x) + (20\%)(8) = (17\%)(8 + x)$$
$$0.15(x) + 0.20(8) = 0.17(8 + x)$$
$$15x + 20(8) = 17(8 + x)$$
$$15x + 160 = 136 + 17x$$
$$-2x = -24$$
$$x = 12$$

We must add 12 gallons of the 15% salt solution.

25. Let p represent the percent of grapefruit juice in the resulting mixture.

$$\left(\begin{array}{c}\text{pure juice in}\\\text{10\% solution}\end{array}\right) + \left(\begin{array}{c}\text{pure juice in}\\\text{20\% solution}\end{array}\right) = \left(\begin{array}{c}\text{pure juice in}\\\text{final solution}\end{array}\right)$$

$$(10\%)(30) + (20\%)(50) = (p)(30 + 50)$$
$$0.10(30) + 0.20(50) = p(80)$$
$$3 + 10 = 80p$$
$$13 = 80p$$
$$\frac{13}{80} = p$$
$$p = 0.1625 = 16.25\%$$

The resulting mixture would be 16.25% grapefruit juice.

27. Let x represent the length of the side of the square; then $2x - 9$ represents the width of the rectangle and $2x - 3$ represents the length of the rectangle.

Rectangular perimeter = Square perimeter
$$2(2x - 9) + 2(2x - 3) = 4x$$
$$4x - 18 + 4x - 6 = 4x$$
$$8x - 24 = 4x$$
$$-24 = -4x$$
$$6 = x$$

Each side of the square is 6 inches, the width of the rectangle is $2(6) - 9 = 3$ inches, and the length of the rectangle is $2(6) - 3 = 9$ inches.

Problem Set 4.5

29. Let t represent Dick's time. Add one-half hour to Dick's time to represent Butch's time. A chart of the information would be as follows:

	Rate	Time	Distance $(d = rt)$
Butch	2	$t + \dfrac{1}{2}$	$2\left(t + \dfrac{1}{2}\right) = 2t + 1$
Dick	$3\dfrac{1}{2} = \dfrac{7}{2}$	t	$\dfrac{7}{2}t$

A diagram of the problem would be as follows:

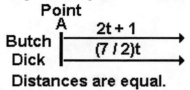

Set the two distances equal and solve.

$$2t + 1 = \frac{7}{2}t$$
$$4t + 2 = 7t$$
$$2 = 3t$$
$$t = \frac{2}{3}\text{hour} = 40 \text{ minutes}$$

Dick would catch up with Butch in 40 minutes.

31. Let the ages be represented as follows:

$$2x : \text{Bill's present age}$$
$$x : \text{Pam's present age}$$
$$2x - 6 : \text{Bill's age 6 years ago}$$
$$x - 6 : \text{Pam's age 6 years ago}$$

$$(\text{Bill's age 6 years ago}) = (4)(\text{Pam's age 6 years ago})$$
$$2x - 6 = 4(x - 6)$$
$$2x - 6 = 4x - 24$$
$$-2x - 6 = -24$$
$$-2x = -18$$
$$x = 9$$

Pam is 9 years old and Bill is 18 years old.

33. Let x represent the amount of money invested at 12% interest; then $12,000 - x$ represents the amount of money invested at 14%.

$$\begin{pmatrix} \text{Interest earned} \\ \text{at 12\%} \end{pmatrix} + \begin{pmatrix} \text{Interest earned} \\ \text{at 14\%} \end{pmatrix} = \begin{pmatrix} \text{Total interest} \\ \text{earned} \end{pmatrix}$$

$$(12\%)(x) + (14\%)(12,000 - x) = 1580$$
$$0.12x + 0.14(12,000 - x) = 1580$$
$$12x + 14(12,000 - x) = 158,000$$
$$12x + 168,000 - 14x = 158,000$$
$$-2x = -10,000$$
$$x = 5000$$

She invested $5000 at 12% and $7000 at 14%.

35. Let x represent the amount of money invested at 9% interest; then $2x$ represents the amount of money invested at 10% and $3x$ the amount at 11%.

$$\begin{pmatrix} \text{Interest earned} \\ \text{at 9\%} \end{pmatrix} + \begin{pmatrix} \text{Interest earned} \\ \text{at 10\%} \end{pmatrix} + \begin{pmatrix} \text{Interest earned} \\ \text{at 11\%} \end{pmatrix} = \begin{pmatrix} \text{Total interest} \\ \text{earned} \end{pmatrix}$$

$$(9\%)(x) + (10\%)(2x) + (11\%)(3x) = 310$$
$$0.09x + 0.10(2x) + 0.11(3x) = 310$$
$$9x + 10(2x) + 11(3x) = 31,000$$
$$9x + 20x + 33x = 31,000$$
$$62x = 31,000$$
$$x = 500$$

She invested $500 at 9%, $1000 at 10%, and $1500 at 11%.

37. Let x represent the amount of money invested at 10% interest; then $x + 250$ represents the amount invested at 11%.

$$\begin{pmatrix} \text{Interest earned} \\ \text{at 10\%} \end{pmatrix} + \begin{pmatrix} \text{Interest earned} \\ \text{at 11\%} \end{pmatrix} = \begin{pmatrix} \text{Total interest} \\ \text{earned} \end{pmatrix}$$

$$(10\%)(x) + (11\%)(x + 250) = 153.50$$
$$0.10(x) + 0.11(x + 250) = 153.50$$
$$10x + 11(x + 250) = 15,350$$
$$10x + 11x + 2750 = 15,350$$
$$21x + 2750 = 15,350$$
$$21x = 12,600$$
$$x = 600$$

She invested $600 at 10% and $850 at 11%.

Problem Set 4.5

39. Let x represent the amount of money invested at 12% interest; then $x + 3000$ represents the total amount of money invested at 11%.

$$\left(\begin{matrix}\text{Interest earned} \\ \text{at 9\%}\end{matrix}\right) + \left(\begin{matrix}\text{Interest earned} \\ \text{at 12\%}\end{matrix}\right) = \left(\begin{matrix}\text{Total interest} \\ \text{at 11\%}\end{matrix}\right)$$

$$(9\%)(3000) + (12\%)(x) = (11\%)(x + 3000)$$
$$0.09(3000) + 0.12x = 0.11(x + 3000)$$
$$9(3000) + 12x = 11(x + 3000)$$
$$27,000 + 12x = 11x + 33,000$$
$$x + 27,000 = 33,000$$
$$x = 6000$$

The amount invested must be $6000 at 12%.

41. Let x represent the amount of money invested at 9% interest; then $6000 - x$ represents the amount of money invested at 11%.

$$\left(\begin{matrix}\text{Interest earned} \\ \text{at 9\%}\end{matrix}\right) = \left(\begin{matrix}\text{Interest earned} \\ \text{at 11\%}\end{matrix}\right) - 160$$

$$(9\%)(x) = (11\%)(6000 - x) - 160$$
$$0.09x = 0.11(6000 - x) - 160$$
$$9x = 11(6000 - x) - 16,000$$
$$9x = 66,000 - 11x - 16,000$$
$$9x = 50,000 - 11x$$
$$20x = 50,000$$
$$x = 2500$$

The amounts invested are $2500 at 9% and $3500 at 11%.

CHAPTER 4 | **Review Problem Set**

1. $0.5x + 0.7x = 1.7$
Combine like terms.
$$1.2x = 1.7$$
Multiplied both sides by 10.
$$12x = 17$$
$$x = \frac{17}{12}$$
The solution set is $\left\{\dfrac{17}{12}\right\}$.

2. $0.07t + 0.12(t - 3) = 0.59$
Multiplied by 100.
$$7t + 12(t - 3) = 59$$
Distributive property
$$7t + 12t - 36 = 59$$
Combined like terms.
$$19t - 36 = 59$$
Added 36 to both sides.
$$19t = 95$$
$$t = 5$$
The solution set is $\{5\}$.

3. $0.1x + 0.12(1700 - x) = 188$
Multiplied both sides by 100.
$$10x + 12(1700 - x) = 18,800$$
$$10x + 20,400 - 12x = 18,800$$
$$-2x + 20,400 = 18,800$$
$$-2x = -1600$$
$$x = 800$$
The solution set is $\{800\}$.

4. $x - 0.25x = 12$
Multiplied both sides by 100.
$$100x - 25x = 1200$$
$$75x = 1200$$
$$x = 16$$
The solution set is $\{16\}$.

5. $0.2(x - 3) = 14$
Multiplied both sides by 10.
$$2(x - 3) = 140$$
$$2x - 6 = 140$$
$$2x = 146$$
$$x = 73$$
The solution set is $\{73\}$.

6. $P = 50, l = 19 : P = 2l + 2w$
$$50 = 2(19) + 2w$$
$$50 = 38 + 2w$$
$$12 = 2w$$
$$6 = w$$

7. $F = 77 : F = \frac{9}{5}C + 32$
$$77 = \frac{9}{5}C + 32$$
Subtracted 32 from both sides.
$$45 = \frac{9}{5}C$$
Multiply both sides by $\frac{5}{9}$.
$$\frac{5}{9}(45) = \frac{5}{9}\left(\frac{9}{5}\right)C$$
$$25 = C$$

8. $A = P + Prt$
$$A - P = Prt$$
Divide both sides by Pr.
$$\frac{A - P}{Pr} = t$$

9. $2x - 3y = 13$
Add $3y$ to both sides.
$$2x = 13 + 3y$$
Divide both sides by 2.
$$x = \frac{13 + 3y}{2}$$

10. Substitute $b_1 = 8$, $b_2 = 14$, and $h = 7$ into the formula for the area of a trapezoid.
$$A = \frac{1}{2}h(b_1 + b_2) = \frac{1}{2}(7)(8 + 14)$$
$$= \frac{1}{2}(7)(22) = 77 \text{ square inches}$$

11. Let h represent the altitude when $b = 9$ and $A = 27$.
$$A = \frac{1}{2}bh$$
$$27 = \frac{1}{2}(9)h$$
$$54 = 9h$$
$$6 = h$$
The altitude of the triangle is 6 centimeters.

12. Let h represent the height when $r = 4$ and $S = 152\pi$.
$$S = 2\pi r^2 + 2\pi rh$$
$$152\pi = 2\pi(4)^2 + 2\pi(4)h$$
$$152\pi = 32\pi + 8\pi h$$
$$120\pi = 8\pi h$$
$$15 = h$$
The height of the trapezoid is 15 feet.

13. Let r represent the percent to be found.
$$18 = r(30)$$
$$\frac{18}{30} = r$$
$$0.6 = r$$
Therefore, 18 is 60% of 30.

14. Let n represent one of the numbers, then $96 - n$ represents the other number.
$$\frac{5}{7} = \frac{n}{96 - n}$$
$$7n = 5(96 - n)$$
$$7n = 480 - 5n$$
$$12n = 480$$
$$n = 40$$
Therefore, 40 and 56 are the numbers.

15. Let n represent the number.
$$(15\%)(n) = 6$$
$$0.15n = 6$$
$$15n = 600$$
$$n = 40$$
Therefore, 15% of 40 is 6.

16. Let w represent the width of the rectangle, then $2w + 5$ represents the length when the perimeter is 46 meters.

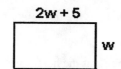

$$2(2w + 5) + 2w = 46$$
$$4w + 10 + 2w = 46$$
$$6w + 10 = 46$$
$$6w = 36$$
$$w = 6$$
The dimensions of the rectangle are 6 meters by 17 meters.

17. Let t represent the time for each airplane as both airplanes left at the same time. A chart of the information would be as follows:

	Rate	Time	Distance ($d = rt$)
Plane A	350	t	$350t$
Plane B	400	t	$400t$

A diagram of the problem would be as follows:

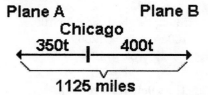

The sum of the two distances would equal the total distance of 1125 miles.
$$350t + 400t = 1125$$
$$750t = 1125$$
$$t = 1\frac{1}{2}$$
It would take $1\frac{1}{2}$ hours for the airplanes to be 1125 miles apart.

18. Let x represent the amount of pure alcohol to be added. Then $10 + x$ represents the amount of final solution.

$$\begin{pmatrix} \text{pure} \\ \text{alcohol} \\ \text{in 70\%} \\ \text{solution} \end{pmatrix} + \begin{pmatrix} \text{pure} \\ \text{alcohol} \\ \text{to be} \\ \text{added} \end{pmatrix} = \begin{pmatrix} \text{pure} \\ \text{alcohol} \\ \text{in final} \\ \text{solution} \end{pmatrix}$$
$$(70\%)(10) + x = (90\%)(10 + x)$$
$$0.70(10) + x = 0.90(10 + x)$$
$$10[0.70(10) + x] = 10[0.90(10 + x)]$$
$$7(10) + 10x = 9(10 + x)$$
$$70 + 10x = 90 + 9x$$
$$70 + x = 90$$
$$x = 20$$
We must add 20 liters of pure alcohol.

19. Let w represent the width of the rectangle, then $2w + 10$ represents the length when the perimeter is 110.

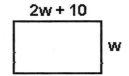

2w + 10

$$2(2w + 10) + 2w = 110$$
$$4w + 20 + 2w = 110$$
$$6w + 20 = 110$$
$$6w = 90$$
$$w = 15$$

The dimensions of the rectangle are 15 centimeters by 40 centimeters.

20. Let w represent the width of the rectangular garden, then $3w - 1$ represents the length when the perimeter is 78 yards.

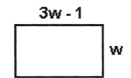

3w - 1

$$2(3w - 1) + 2w = 78$$
$$6w - 2 + 2w = 78$$
$$8w - 2 = 78$$
$$8w = 80$$
$$w = 10$$

The dimensions of the garden are 10 yards by 29 yards.

21. Let a represent the measure of the angle, then $90 - a$ represents the complement of the angle, and $180 - a$ represents the supplement.

$$\frac{\text{complement}}{\text{supplement}} \quad \frac{90 - a}{180 - a} = \frac{7}{16}$$
$$7(180 - a) = 16(90 - a)$$
$$1260 - 7a = 1440 - 16a$$
$$1260 + 9a = 1440$$
$$9a = 180$$
$$a = 20$$

The measure of the angle is $20°$.

22. Let g represent the number of gallons required for the trip.

$$\frac{\text{gallons}}{\text{miles}} \quad \frac{18}{369} = \frac{g}{615}$$
$$369g = 18(615)$$
$$369g = 11,070$$
$$g = 30$$

Therefore, 30 gallons would be needed for the 615-mile trip.

23. Let x represent the amount of money invested at 9% interest; then $2100 - x$ represents the amount of money invested at 11%.

$$\left(\begin{smallmatrix} \text{Interest earned} \\ \text{at 9\%} \end{smallmatrix} \right) + 51 = \left(\begin{smallmatrix} \text{Interest earned} \\ \text{at 11\%} \end{smallmatrix} \right)$$

$$(9\%)x + 51 = (11\%)(2100 - x)$$
$$0.09x + 51 = 0.11(2100 - x)$$
$$9x + 5100 = 11(2100 - x)$$
$$9x + 5100 = 23,100 - 11x$$
$$20x = 18,000$$
$$x = 900$$

He invested $900 at 9% and $1200 at 11%.

24. Let s represent the selling price. Profit is a percent of selling price.

Selling price = Cost + Profit
$$s = 28 + (30\%)(s)$$
$$s = 28 + 0.3s$$
$$10s = 280 + 3s$$
$$7s = 280$$
$$s = 40$$

The selling price would be $40.

25. Let r represent the rate of discount.
$$60 - r(60) = 39$$
$$60 - 60r = 39$$
$$-60r = -21$$
$$r = 0.35$$
The percent of discount that she received was 35%.

26. Let a represent the measure of the second angle, then $3a - 3$ represents the third angle. The sum of the measures of the angles of a triangle are $180°$
$$47 + a + (3a - 3) = 180$$
$$44 + 4a = 180$$
$$4a = 136$$
$$a = 34$$
The remaining angles are $34°$ and $3(34) - 3 = 99°$.

27. Let t represent Zak's time, then $t + 1$ represents Connie's time as she travels one hour longer. A chart of the information would be as follows:

	Rate	Time	Distance $(d = rt)$
Connie	10	$t + 1$	$10(t + 1)$
Zak	12	t	$12t$

A diagram of the problem would be as follows:

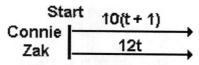

Start

Connie $\quad$ 10(t + 1)

Zak $\quad$ 12t

Distances are equal.

Set the distances equal to each other and solve.
$$10(t + 1) = 12t$$
$$10t + 10 = 12t$$
$$10 = 2t$$
$$5 = t$$
It would take 5 hours for Zak to catch up with Connie.

28. Let x represent the amount of 10% salt solution to be used, then $12 + x$ represents the amount of final solution.

$$\begin{pmatrix} \text{pure} \\ \text{salt} \\ \text{in 10\%} \\ \text{solution} \end{pmatrix} + \begin{pmatrix} \text{pure} \\ \text{salt} \\ \text{in 15\%} \\ \text{solution} \end{pmatrix} = \begin{pmatrix} \text{pure} \\ \text{salt} \\ \text{in final} \\ \text{solution} \end{pmatrix}$$

$$(10\%)(x) + (15\%)(12) = (12\%)(12 + x)$$
$$0.10(x) + 0.15(12) = 0.12(12 + x)$$
$$10x + 15(12) = 12(12 + x)$$
$$10x + 180 = 144 + 12x$$
$$-2x = -36$$
$$x = 18$$
We must add 18 gallons of the 10% salt solution.

29. Let p represent the percent of orange juice in the resulting mixture.

$$\begin{pmatrix} \text{pure} \\ \text{juice} \\ \text{in 20\%} \\ \text{solution} \end{pmatrix} + \begin{pmatrix} \text{pure} \\ \text{juice} \\ \text{in 30\%} \\ \text{solution} \end{pmatrix} = \begin{pmatrix} \text{pure} \\ \text{juice} \\ \text{in final} \\ \text{solution} \end{pmatrix}$$

$$(20\%)(20) + (30\%)(30) = (p)(20 + 30)$$
$$0.20(20) + 0.30(30) = p(50)$$
$$4 + 9 = 50p$$
$$13 = 50p$$
$$\frac{13}{50} = p$$
$$p = 0.26 = 26\%$$
The resulting mixture would be 26% orange juice.

30. $i = Prt$
$i = 3500(0.0525)(2)$
$i = 367.5$
The interest is \$367.50.

CHAPTER 4 Test

1. $\dfrac{x+2}{4} = \dfrac{x-3}{5}$

Cross products are equal.

$5(x+2) = 4(x-3)$

$5x + 10 = 4x - 12$

$x + 10 = -12$

$x = -22$

The solution set is $\{-22\}$.

2. $\dfrac{-4}{2x-1} = \dfrac{3}{3x+5}$

Cross products are equal.

$-4(3x+5) = 3(2x-1)$

$-12x - 20 = 6x - 3$

$-18x - 20 = -3$

$-18x = 17$

$x = -\dfrac{17}{18}$

The solution set is $\left\{-\dfrac{17}{18}\right\}$.

3. $\dfrac{x-1}{6} - \dfrac{x+2}{5} = 2$

This is NOT a proportion!

Multiply both sides by 30.

$30\left(\dfrac{x-1}{6} - \dfrac{x+2}{5}\right) = 30(2)$

$5(x-1) - 6(x+2) = 60$

$5x - 5 - 6x - 12 = 60$

$-x - 17 = 60$

$-x = 77$

$x = -77$

The solution set is $\{-77\}$.

4. $\dfrac{x+8}{7} - 2 = \dfrac{x-4}{4}$

This is NOT a proportion!

Multiply both sides by 28.

$28\left(\dfrac{x+8}{7} - 2\right) = 28\left(\dfrac{x-4}{4}\right)$

$4(x+8) - 28(2) = 7(x-4)$

$4x + 32 - 56 = 7x - 28$

$4x - 24 = 7x - 28$

$-3x - 24 = -28$

$-3x = -4$

$x = \dfrac{4}{3}$

The solution set is $\left\{\dfrac{4}{3}\right\}$.

5. $\dfrac{n}{20-n} = \dfrac{7}{3}$

Cross products are equal.

$7(20 - n) = 3n$

$140 - 7n = 3n$

$140 = 10n$

$14 = n$

The solution set is $\{14\}$.

6. $\dfrac{h}{4} + \dfrac{h}{6} = 1$

This is NOT a proportion!

Multiply both sides by 12.

$12\left(\dfrac{h}{4} + \dfrac{h}{6}\right) = 12(1)$

$3h + 2h = 12$

$5h = 12$

$h = \dfrac{12}{5}$

The solution set is $\left\{\dfrac{12}{5}\right\}$.

Chapter 4 Test

7. $0.05n + 0.06(400 - n) = 23$
Multiplied both sides by 100.
$$5n + 6(400 - n) = 2300$$
$$5n + 2400 - 6n = 2300$$
$$-n + 2400 = 2300$$
$$-n = -100$$
$$n = 100$$
The solution set is $\{100\}$.

8. $s = 35 + 0.5s$
Multiplied both sides by 10.
$$10s = 350 + 5s$$
$$5s = 350$$
$$s = 70$$
The solution set is $\{70\}$.

9. $0.07n = 45.5 - 0.08(600 - n)$
Multiplied both sides by 100.
$$7n = 4550 - 8(600 - n)$$
$$7n = 4550 - 4800 + 8n$$
$$7n = -250 + 8n$$
$$-n = -250$$
$$n = 250$$
The solution set is $\{250\}$.

10. $12t + 8\left(\dfrac{7}{2} - t\right) = 50$
Distributive property.
$$12t + 28 - 8t = 50$$
$$4t + 28 = 50$$
$$4t = 22$$
$$t = \frac{22}{4} = \frac{11}{2}$$
The solution set is $\left\{\dfrac{11}{2}\right\}$.

11. $F = \dfrac{9C + 160}{5}$
Multiplied both sides by 5.
$$5F = 9C + 160$$
Subtracted 160 from both sides.
$$5F - 160 = 9C$$
Divided both sides by 9.
$$\frac{5F - 160}{9} = C$$

12. $y = 2(x - 4)$
Distributive Property.
$$y = 2x - 8$$
Added 8 to both sides.
$$y + 8 = 2x$$
Divided both sides by 2.
$$\frac{y + 8}{2} = x$$

13. $\dfrac{x + 3}{4} = \dfrac{y - 5}{9}$
Cross products are equal.
$$4(y - 5) = 9(x + 3)$$
Distributive property.
$$4y - 20 = 9x + 27$$
Added 20 to both sides.
$$4y = 9x + 47$$
Divided both sides by 4.
$$y = \frac{9x + 47}{4}$$

14. Find the radius by solving the formula for circumference for r when $C = 16\pi$.
$$C = 2\pi r$$
$$16\pi = 2\pi r$$
$$8 = r$$
When the radius is 8 centimeters, the area will be $A = \pi(8)^2 = 64\pi$ square centimeters.

15. Let w represent the width of the rectangle when $l = 32$ and $P = 100$.

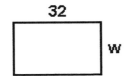

$$P = 2l + 2w$$
$$100 = 2(32) + 2w$$
$$100 = 64 + 2w$$
$$36 = 2w$$
$$18 = w$$

The width of the rectangle is 18 inches. Therefore, the area of the rectangle is $(32)(18) = 576$ square inches.

16. Let h represent the altitude of the triangular plot when $A = 133$ and $b = 19$.

$$A = \frac{1}{2}bh$$
$$133 = \frac{1}{2}(19)h$$
$$266 = 19h$$
$$14 = h$$

The altitude of the triangular plot is 14 yards.

17.
$$\frac{5}{4} = \frac{n}{100}$$
$$4n = 500$$
$$n = 125$$

Therefore, $\dfrac{5}{4} = \dfrac{125}{100} = 125\%$.

18. Let n represent the number.

$$(35\%)n = 24.5$$
$$0.35n = 24.5$$
$$35n = 2450$$
$$n = 70$$

Therefore, 35% of 70 is 24.5.

19. Let p represent the original price of the blouse.

$$(100\%)(p) - (30\%)(p) = 28$$
$$(70\%)(p) = 28$$
$$0.7p = 28$$
$$7p = 280$$
$$p = 40$$

The original price of the blouse was \$40.

20. Let s represent the selling price of the skirt. Profit is a percent of cost.
Selling price = Cost + Profit
$$s = 40 + (30\%)(40)$$
$$s = 40 + 0.3(40)$$
$$s = 40 + 12$$
$$s = 52$$

The selling price of the skirt should be \$52.

21. Let r represent the rate of discount he received.

$$80 - r(80) = 48$$
$$80 - 80r = 48$$
$$-80r = -32$$
$$r = 0.40$$

The rate of discount would be 40%.

22. Let f represent the number of females who voted, then $1500 - f$ represents the number of males who voted.

$$\frac{\text{female voters}}{\text{male voters}} \quad \frac{7}{5} = \frac{f}{1500 - f}$$
$$5f = 7(1500 - f)$$
$$5f = 10,500 - 7f$$
$$12f = 10,500$$
$$f = 875$$

There were 875 female voters.

23. Let t represent the time of the second car, then $t + 1$ represents the first car's time as it travels one hour longer. A chart of the information would be as follows:

	Rate	Time	Distance ($d = rt$)
First Car	50	$t + 1$	$50(t + 1)$
Second Car	55	t	$55t$

A diagram of the problem would be as follows:

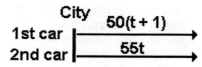

City

1st car $\xrightarrow{50(t+1)}$

2nd car $\xrightarrow{55t}$

Distances are equal.

Set the distances equal to each other and solve.

$$50(t + 1) = 55t$$
$$50t + 50 = 55t$$
$$50 = 5t$$
$$10 = t$$

It would take 10 hours for the second car to overtake the first car.

24. Let x represent the amount of pure acid to be added. Then $6 + x$ represents the amount of final solution.

$$\left(\begin{array}{c}\text{pure}\\\text{acid}\\\text{in 50\%}\\\text{solution}\end{array}\right) + \left(\begin{array}{c}\text{pure}\\\text{acid}\\\text{to be}\\\text{added}\end{array}\right) = \left(\begin{array}{c}\text{pure}\\\text{acid}\\\text{in final}\\\text{solution}\end{array}\right)$$

$$(50\%)(6) + x = (70\%)(6 + x)$$
$$0.50(6) + x = 0.70(6 + x)$$
$$10[0.50(6) + x] = 10[0.70(6 + x)]$$
$$5(6) + 10x = 7(6 + x)$$
$$30 + 10x = 42 + 7x$$
$$30 + 3x = 42$$
$$3x = 12$$
$$x = 4$$

We must add 4 centiliters of pure acid.

25.
$$i = Prt$$
$$4000 = 4000(0.09)t$$
$$4000 = 360t$$
$$\frac{4000}{360} = t$$
$$11.1 = t$$

It would take 11.1 years.

CHAPTERS 1-4 **Cumulative Review**

1. $7x - 9x - 14x =$
$(7 - 9 - 14)x =$
$-16x$

2. $-10a - 4 + 13a + a - 2 = 4a - 6$

3. $5(x - 3) + 7(x + 6) =$
$5x - 15 + 7x + 42 =$
$12x + 27$

4. $3(x - 1) - 4(2x - 1) =$
$3x - 3 - 8x + 4 =$
$-5x + 1$

5. $-3n - 2(n - 1) + 5(3n - 2) - n =$
$-3n - 2n + 2 + 15n - 10 - n =$
$9n - 8$

6. $6n + 3(4n - 2) - 2(2n - 3) - 5 =$
$6n + 12n - 6 - 4n + 6 - 5 =$
$14n - 5$

7. $\dfrac{1}{2}x - \dfrac{3}{4}x + \dfrac{2}{3}x - \dfrac{1}{6}x =$
$\dfrac{6}{12}x - \dfrac{9}{12}x + \dfrac{8}{12}x - \dfrac{2}{12}x =$
$\dfrac{3}{12}x = \dfrac{1}{4}x$

8. $\dfrac{1}{3}n - \dfrac{4}{15}n + \dfrac{5}{6}n - n =$
$\dfrac{10}{30}n - \dfrac{8}{30}n + \dfrac{25}{30}n - \dfrac{30}{30}n =$
$-\dfrac{3}{30}n = -\dfrac{1}{10}n$

9. $0.4x + 0.7x - 0.8x + 1.0x = 1.3x$

10. $0.5(x - 2) + 0.4(x + 3) - 0.2x =$
$0.5x - 1.0 + 0.4x + 1.2 - 0.2x =$
$0.7x + 0.2$

11. $x = -2, y = 5 : 5x - 7y + 2xy =$
$5(-2) - 7(5) + 2(-2)(5) =$
$-10 - 35 - 20 = -65$

12. $a = 3, b = -4 : 2ab - a + 6b =$
$2(3)(-4) - (3) + 6(-4) =$
$-24 - 3 - 24 = -51$

13. $-3(x - 1) + 2(x + 6) =$
$-3x + 3 + 2x + 12 = -x + 15 =$
$-(-5) + 15 = 5 + 15 = 20,$ when $x = -5$

14. $5(n + 3) - (n + 4) - n =$
$5n + 15 - n - 4 - n =$
$3n + 11 = 3(7) + 11,$ when $n = 7$
$= 21 + 11 = 32$

15. $x = 3, y = -6 : \dfrac{3x - 2y}{2x - 3y} =$

$\dfrac{3(3) - 2(-6)}{2(3) - 3(-6)} = \dfrac{9 + 12}{6 + 18} = \dfrac{21}{24} = \dfrac{7}{8}$

16. $n = -\dfrac{2}{3} : \dfrac{3}{4}n - \dfrac{1}{3}n + \dfrac{5}{6}n =$

$\dfrac{9}{12}n - \dfrac{4}{12}n + \dfrac{10}{12}n = \dfrac{15}{12}n = \dfrac{5}{4}n =$

$\dfrac{5}{4}\left(-\dfrac{2}{3}\right) = -\dfrac{5}{6}$

17. $a = 0.2, b = -0.3 : 2a^2 - 4b^2 =$
$2(0.2)^2 - 4(-0.3)^2 =$
$2(0.04) - 4(0.09) =$
$0.08 - 0.36 = -0.28$

18. $x = \dfrac{1}{2}, y = \dfrac{1}{4} : x^2 - 3xy - 2y^2 =$

$\left(\dfrac{1}{2}\right)^2 - 3\left(\dfrac{1}{2}\right)\left(\dfrac{1}{4}\right) - 2\left(\dfrac{1}{4}\right)^2 =$

$\dfrac{1}{4} - \dfrac{3}{8} - \dfrac{1}{8} = \dfrac{2}{8} - \dfrac{3}{8} - \dfrac{1}{8} = -\dfrac{2}{8} = -\dfrac{1}{4}$

19. $5x - 7y - 8x + 3y = -3x - 4y =$
$-3(9) - 4(-8),$ when $x = 9, y = -8$
$= -27 + 32 = 5$

20. $a = -1, b = 3 : \dfrac{3a - b - 4a + 3b}{a - 6b - 4b - 3a} =$

$\dfrac{-a + 2b}{-2a - 10b} = \dfrac{-(-1) + 2(3)}{-2(-1) - 10(3)} =$

$\dfrac{1 + 6}{2 - 30} = -\dfrac{7}{28} = -\dfrac{1}{4}$

21. $3^4 = 3 \cdot 3 \cdot 3 \cdot 3 = 81$

22. $-2^6 = -(2 \cdot 2 \cdot 2 \cdot 2 \cdot 2 \cdot 2) = -64$

23. $\left(\dfrac{2}{3}\right)^3 = \dfrac{2}{3} \cdot \dfrac{2}{3} \cdot \dfrac{2}{3} = \dfrac{8}{27}$

24. $\left(-\dfrac{1}{2}\right)^5 =$

$\left(-\dfrac{1}{2}\right)\left(-\dfrac{1}{2}\right)\left(-\dfrac{1}{2}\right)\left(-\dfrac{1}{2}\right)\left(-\dfrac{1}{2}\right) =$

$-\dfrac{1}{32}$

25. $\left(\dfrac{1}{2} + \dfrac{1}{3}\right)^2 = \left(\dfrac{3}{6} + \dfrac{2}{6}\right)^2 =$

$\left(\dfrac{5}{6}\right)^2 = \dfrac{5}{6} \cdot \dfrac{5}{6} = \dfrac{25}{36}$

26. $\left(\dfrac{3}{4} - \dfrac{7}{8}\right)^3 = \left(\dfrac{6}{8} - \dfrac{7}{8}\right)^3 = \left(-\dfrac{1}{8}\right)^3 =$

$\left(-\dfrac{1}{8}\right)\left(-\dfrac{1}{8}\right)\left(-\dfrac{1}{8}\right) = -\dfrac{1}{512}$

27. $-5x + 2 = 22$
$-5x = 20$
$x = -4$
The solution set is $\{-4\}$.

103

Chapters 1-4 Cumulative Review

28.
$$3x - 4 = 7x + 4$$
$$-4x - 4 = 4$$
$$-4x = 8$$
$$x = -2$$
The solution set is $\{-2\}$.

29.
$$7(n - 3) = 5(n + 7)$$
$$7n - 21 = 5n + 35$$
$$2n - 21 = 35$$
$$2n = 56$$
$$n = 28$$
The solution set is $\{28\}$.

30.
$$2(x - 1) - 3(x - 2) = 12$$
$$2x - 2 - 3x + 6 = 12$$
$$-x + 4 = 12$$
$$-x = 8$$
$$x = -8$$
The solution set is $\{-8\}$.

31.
$$\frac{2}{5}x - \frac{1}{3} = \frac{1}{3}x + \frac{1}{2}$$
Multiply both sides by 30.
$$30\left(\frac{2}{5}x - \frac{1}{3}\right) = 30\left(\frac{1}{3}x + \frac{1}{2}\right)$$
$$12x - 10 = 10x + 15$$
$$2x - 10 = 15$$
$$2x = 25$$
$$x = \frac{25}{2}$$
The solution set is $\left\{\frac{25}{2}\right\}$.

32.
$$\frac{t - 2}{4} + \frac{t + 3}{3} = \frac{1}{6}$$
$$12\left(\frac{t - 2}{4} + \frac{t + 3}{3}\right) = 12\left(\frac{1}{6}\right)$$
$$3(t - 2) + 4(t + 3) = 2$$
$$3t - 6 + 4t + 12 = 2$$
$$7t + 6 = 2$$
$$7t = -4$$
$$t = -\frac{4}{7}$$
The solution set is $\left\{-\frac{4}{7}\right\}$.

33.
$$\frac{2n - 1}{5} - \frac{n + 2}{4} = 1$$
Multiply both sides by 20.
$$20\left(\frac{2n - 1}{5} - \frac{n + 2}{4}\right) = 20(1)$$
$$4(2n - 1) - 5(n + 2) = 20$$
$$8n - 4 - 5n - 10 = 20$$
$$3n - 14 = 20$$
$$3n = 34$$
$$n = \frac{34}{3}$$
The solution set is $\left\{\frac{34}{3}\right\}$.

34.
$$0.09x + 0.12(500 - x) = 54$$
Multiply both sides by 100.
$$100[0.09x + 0.12(500 - x)] = 100(54)$$
$$9x + 12(500 - x) = 5400$$
$$9x + 6000 - 12x = 5400$$
$$6000 - 3x = 5400$$
$$-3x = -600$$
$$x = 200$$
The solution set is $\{200\}$.

35.
$$-5(n - 1) - (n - 2) = 3(n - 1) - 2n$$
Distributive property.
$$-5n + 5 - n + 2 = 3n - 3 - 2n$$
Combined like terms.
$$-6n + 7 = n - 3$$
$$-7n = -10$$
$$n = \frac{10}{7}$$
The solution set is $\left\{\frac{10}{7}\right\}$.

36.
$$\frac{-2}{x - 1} = \frac{-3}{x + 4}$$
Cross products are equal.
$$-2(x + 4) = -3(x - 1)$$
$$-2x - 8 = -3x + 3$$
$$x - 8 = 3$$
$$x = 11$$
The solution set is $\{11\}$.

37. $0.2x + 0.1(x - 4) = 0.7x - 1$

Multiplied both sides by 10.

$$2x + 1(x - 4) = 7x - 10$$
$$2x + x - 4 = 7x - 10$$
$$3x - 4 = 7x - 10$$
$$-4x - 4 = -10$$
$$-4x = -6$$
$$x = \frac{6}{4} = \frac{3}{2}$$

The solution set is $\left\{ \frac{3}{2} \right\}$.

38. $-(t - 2) + (t - 4) = 2\left(t - \frac{1}{2} \right) - 3\left(t + \frac{1}{3} \right)$

Distributive Property.

$$-t + 2 + t - 4 = 2t - 1 - 3t - 1$$
$$-2 = -t - 2$$
$$0 = -t$$
$$0 = t$$

The solution set is $\{0\}$.

39. $4x - 6 > 3x + 1$

$$x - 6 > 1$$
$$x > 7$$

The solution set is
$\{x | x > 7\}$ or $(7, \infty)$.

40. $-3x - 6 < 12$

$$-3x < 18$$

Reversed the inequality.

$$x > -6$$

The solution set is
$\{x | x > -6\}$ or $(-6, \infty)$.

41. $-2(n - 1) \leq 3(n - 2) + 1$

Distributive property.

$$-2n + 2 \leq 3n - 6 + 1$$
$$-2n + 2 \leq 3n - 5$$
$$-5n + 2 \leq -5$$
$$-5n \leq -7$$

Reversed the inequality.

$$n \geq \frac{7}{5}$$

The solution set is
$\left\{ n | n \geq \frac{7}{5} \right\}$ or $\left[\frac{7}{5}, \infty \right)$.

42. $\frac{2}{7}x - \frac{1}{4} \geq \frac{1}{4}x + \frac{1}{2}$

Multiply both sides by 28.

$$28\left(\frac{2}{7}x - \frac{1}{4} \right) \geq 28\left(\frac{1}{4}x + \frac{1}{2} \right)$$
$$8x - 7 \geq 7x + 14$$
$$x - 7 \geq 14$$
$$x \geq 21$$

The solution set is
$\{x | x \geq 21\}$ or $[21, \infty)$.

43. $0.08t + 0.1(300 - t) > 28$

Multiplied both sides by 100.

$$8t + 10(300 - t) > 2800$$
$$8t + 3000 - 10t > 2800$$
$$3000 - 2t > 2800$$
$$-2t > -200$$

Reversed the inequality.

$$t < 100$$

The solution set is
$\{t | t < 100\}$ or $(-\infty, 100)$.

44. $-4 > 5x - 2 - 3x$

$$-4 > 2x - 2$$
$$-2 > 2x$$
$$-1 > x \text{ means } x < -1$$

The solution set is
$\{x | x < -1\}$ or $(-\infty, -1)$.

45. $\dfrac{2}{3}n - 2 \geq \dfrac{1}{2}n + 1$

Multiply both sides by 6.

$6\left(\dfrac{2}{3}n - 2\right) \geq 6\left(\dfrac{1}{2}n + 1\right)$

$4n - 12 \geq 3n + 6$

$n - 12 \geq 6$

$n \geq 18$

The solution set is
$\{n \mid n \geq 18\}$ or $[18, \infty)$.

46. $-3 < -2(x - 1) - x$

$-3 < -2x + 2 - x$

$-3 < -3x + 2$

$-5 < -3x$

$\dfrac{5}{3} > x$ means $x < \dfrac{5}{3}$

The solution set is

$\left\{x \mid x < \dfrac{5}{3}\right\}$ or $\left(-\infty, \dfrac{5}{3}\right)$.

47. Let s represent her salary five years ago.

$2s + 2000 = 32,000$

$2s = 30,000$

$s = 15,000$

Five years ago, her salary was $15,000.

48. Let a represent the measure of one angle; then $180 - a$ represents the supplementary angle.

$a = 4(180 - a) - 45$

$a = 720 - 4a - 45$

$a = 675 - 4a$

$5a = 675$

$a = 135$

One angle is $135°$ and the other is $180 - (135) = 45°$.

49. Let n represent the number of nickels; then $25 - n$ represents the number of dimes. The value of the nickels in cents is represented by $5n$ and the value of the dimes in cents is $10(25 - n)$.

$5n + 10(25 - n) = 210$

$5n + 250 - 10n = 210$

$-5n + 250 = 210$

$-5n = -40$

$n = 8$

Jasmal has 8 nickels and $25 - 8 = 17$ dimes.

50. Let x represent what Hana's score in the third game.

$\dfrac{144 + 176 + x}{3} \geq 150$

$144 + 176 + x \geq 450$

$320 + x \geq 450$

$x \geq 130$

Hana must bowl 130 or higher in the third game.

51. Let x represent the shorter piece of the board; then $30 - x$ represents the longer piece.

$\dfrac{x}{30 - x} = \dfrac{2}{3}$

$3x = 2(30 - x)$

$3x = 60 - 2x$

$5x = 60$

$x = 12$

The pieces are 12 feet and $30 - 12 = 18$ feet.

52. Let x represent the selling price of the shoes. Profit is a percent of the selling price.

Selling price $=$ Cost $+$ Profit

$s = 32 + (20\%)(s)$

$s = 32 + 0.2s$

$10s = 320 + 2s$

$8s = 320$

$s = 40$

The selling price would be $40.

53. Let r represent the rate of one car; then $r + 5$ represents the other car's rate. Both cars travel for 6 hours. A chart of the information would be as follows:

	Rate	Time	Distance ($d = rt$)
Car A	r	6	$6r$
Car B	$r + 5$	6	$6(r + 5)$

A diagram of the problem would be as follows:

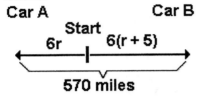

The sum of the two distances would equal the total distance of 570 miles.

$$6r + 6(r + 5) = 570$$
$$6r + 6r + 30 = 570$$
$$12r + 30 = 570$$
$$12r = 540$$
$$r = 45$$

The speeds for the cars were 45 mph and 50 mph.

54. Let x represent the amount of pure alcohol to be added. Then $15 + x$ represents the amount of final solution.

$$\begin{pmatrix} \text{pure} \\ \text{alcohol} \\ \text{in 20\%} \\ \text{solution} \end{pmatrix} + \begin{pmatrix} \text{pure} \\ \text{alcohol} \\ \text{to be} \\ \text{added} \end{pmatrix} = \begin{pmatrix} \text{pure} \\ \text{alcohol} \\ \text{in final} \\ \text{solution} \end{pmatrix}$$

$$(20\%)(15) + x = (40\%)(15 + x)$$
$$0.20(15) + x = 0.40(15 + x)$$
$$10[0.20(15) + x] = 10[0.40(15 + x)]$$
$$2(15) + 10x = 4(15 + x)$$
$$30 + 10x = 60 + 4x$$
$$30 + 6x = 60$$
$$6x = 30$$
$$x = 5$$

We must add 5 liters of pure alcohol.

Problem Set 5.1

Chapter 5 Exponents and Polynomials

PROBLEM SET **5.1** Addition and Subtraction of Polynomials

1. The degree of $7x^2y + 6xy$ is 3 because the degree of the term $7x^2y$ is 3.

3. The degree of $5x^2 - 9$ is 2 because the degree of the term $5x^2$ is 2.

5. The degree of $5x^3 - x^2 - x + 3$ is 3 because the degree of the term $5x^3$ is 3.

7. The degree of $5xy$ is 2 because the degree of the term $5xy$ is 2.

9. $(3x + 4) + (5x + 7) =$
$(3 + 5)x + (4 + 7) =$
$8x + 11$

11. $(-5y - 3) + (9y + 13) =$
$(-5 + 9)y + (-3 + 13) =$
$4y + 10$

13. $(-2x^2 + 7x - 9) + (4x^2 - 9x - 14) =$
$(-2 + 4)x^2 + (7 - 9)x + (-9 - 14) =$
$2x^2 - 2x - 23$

15. $(5x - 2) + (3x - 7) + (9x - 10) =$
$(5 + 3 + 9)x + (-2 - 7 - 10) =$
$17x - 19$

17. $(2x^2 - x + 4) + (-5x^2 - 7x - 2)$
$\quad + (9x^2 + 3x - 6) =$
$(2 - 5 + 9)x^2 + (-1 - 7 + 3)x$
$\quad + (4 - 2 - 6) =$
$6x^2 - 5x - 4$

19. $(-4n^2 - n - 1) + (4n^2 + 6n - 5) =$
$(-4 + 4)n^2 + (-1 + 6)n + (-1 - 5) =$
$5n - 6$

21. $(2x^2 - 7x - 10) + (-6x - 2) + (-9x^2 + 5) =$
$(2 - 9)x^2 + (-7 - 6)x + (-10 - 2 + 5) =$
$-7x^2 - 13x - 7$

23. $(12x + 6) - (7x + 1) =$
$12x + 6 - 7x - 1 =$
$(12 - 7)x + (6 - 1) =$
$5x + 5$

25. $(3x - 7) - (5x - 2) =$
$3x - 7 - 5x + 2 =$
$(3 - 5)x + (-7 + 2) =$
$-2x - 5$

27. $(-4x + 6) - (-x - 1) =$
$-4x + 6 + x + 1 =$
$(-4 + 1)x + (6 + 1) =$
$-3x + 7$

29. $(3x^2 + 8x - 4) - (x^2 - 7x + 2) =$
$3x^2 + 8x - 4 - x^2 + 7x - 2 =$
$(3 - 1)x^2 + (8 + 7)x + (-4 - 2) =$
$2x^2 + 15x - 6$

31. $(3n^2 - n + 7) - (-2n^2 - 3n + 4) =$
$3n^2 - n + 7 + 2n^2 + 3n - 4 =$
$(3 + 2)n^2 + (-1 + 3)n + (7 - 4) =$
$5n^2 + 2n + 3$

33. $(-7x^3 + x^2 + 6x - 12)$
$\quad - (-4x^3 - x^2 + 6x - 1) =$
$-7x^3 + x^2 + 6x - 12 + 4x^3$
$\quad + x^2 - 6x + 1 =$
$(-7 + 4)x^3 + (1 + 1)x^2 + (6 - 6)x$
$\quad + (-12 + 1) =$
$-3x^3 + 2x^2 - 11$

35. $12x - 4$ Add $12x - 4$
$\underline{3x - 2}$ the $\underline{-3x + 2}$
opposite. $9x - 2$

37. $-3a + 9$ Add $-3a + 9$
$\underline{-5a - 6}$ the $\underline{5a + 6}$
opposite. $2a + 15$

108

39.

$6x^2 - x + 11$	Add	$6x^2 - x + 11$
$8x^2 - x + \ 6$	the	$-8x^2 + x - \ 6$
	opposite.	$-2x^2 \qquad + \ 5$

41.
$$4x^3 + 6x^2 + 7x - 14$$
$$\underline{-2x^3 - 6x^2 + 7x - \ 9}$$
Add the opposite.

$$4x^3 + 6x^2 + 7x - 14$$
$$\underline{2x^3 + 6x^2 - 7x + \ 9}$$
$$6x^3 + 12x^2 \qquad - 5$$

43.
$$4x^3 - 6x^2 + 7x - \ 2$$
$$\underline{\qquad\ 2x^2 - 6x - 14}$$
Add the opposite.

$$4x^3 - 6x^2 + 7x - \ 2$$
$$\underline{\qquad -2x^2 + 6x + 14}$$
$$4x^3 - 8x^2 + 13x + 12$$

45. $(5x + 3) - (7x - 2) + (3x + 6) =$
$5x + 3 - 7x + 2 + 3x + 6 =$
$(5 - 7 + 3)x + (3 + 2 + 6) =$
$x + 11$

47. $(-x - 1) - (-2x + 6) + (-4x - 7) =$
$-x - 1 + 2x - 6 - 4x - 7 =$
$(-1 + 2 - 4)x + (-1 - 6 - 7) =$
$-3x - 14$

49. $(x^2 - 7x - 4) + (2x^2 - 8x - 9)$
$\qquad - (4x^2 - 2x - 1) =$
$x^2 - 7x - 4 + 2x^2 - 8x - 9$
$\qquad - 4x^2 + 2x + 1 =$
$(1 + 2 - 4)x^2 + (-7 - 8 + 2)x$
$\qquad + (-4 - 9 + 1) =$
$-x^2 - 13x - 12$

51. $(-x^2 - 3x + 4) + (-2x^2 - x - 2)$
$\qquad - (-4x^2 + 7x + 10) =$
$-x^2 - 3x + 4 - 2x^2 - x - 2$
$\qquad + 4x^2 - 7x - 10 =$
$(-1 - 2 + 4)x^2 + (-3 - 1 - 7)x$
$\qquad + (4 - 2 - 10) =$
$x^2 - 11x - 8$

53. $(3a - 2b) - (7a + 4b) - (6a - 3b) =$
$3a - 2b - 7a - 4b - 6a + 3b =$
$(3 - 7 - 6)a + (-2 - 4 + 3)b =$
$-10a - 3b$

55. $(n - 6) - (2n^2 - n + 4) + (n^2 - 7) =$
$n - 6 - 2n^2 + n - 4 + n^2 - 7 =$
$(-2 + 1)n^2 + (1 + 1)n + (-6 - 4 - 7) =$
$-n^2 + 2n - 17$

57. $7x + [3x - (2x - 1)] =$
$7x + [3x - 2x + 1] =$
$7x + [x + 1] =$
$7x + x + 1 =$
$(7 + 1)x + 1 =$
$8x + 1$

59. $-7n - [4n - (6n - 1)] =$
$-7n - [4n - 6n + 1] =$
$-7n - [-2n + 1] =$
$-7n + 2n - 1 =$
$(-7 + 2)n - 1 =$
$-5n - 1$

61. $(5a - 1) - [3a + (4a - 7)] =$
$(5a - 1) - [3a + 4a - 7] =$
$(5a - 1) - [7a - 7] =$
$5a - 1 - 7a + 7 =$
$(5 - 7)a + (-1 + 7) =$
$-2a + 6$

63. $13x - [5x - [4x - (x - 6)]] =$
$13x - [5x - [4x - x + 6]] =$
$13x - [5x - [3x + 6]] =$
$13x - [5x - 3x - 6] =$
$13x - [2x - 6] =$
$13x - 2x + 6 =$
$11x + 6$

65. $[(4x - 2) + (7x + 6)] - (5x - 3) =$
$[4x - 2 + 7x + 6] - (5x - 3) =$
$[4x + 7x - 2 + 6] - (5x - 3) =$
$[11x + 4] - (5x - 3) =$
$11x + 4 - 5x + 3 =$
$11x - 5x + 4 + 3 =$
$6x + 7$

Problem Set 5.1

67. $(-8n+9) - [(-2n-5) + (-n+7)] =$
$(-8n+9) - [-2n-5-n+7] =$
$(-8n+9) - [-2n-n-5+7] =$
$(-8n+9) - [-3n+2] =$
$-8n+9+3n-2 =$
$-8n+3n+9-2 =$
$-5n+7$

69. Use the formula $P = 2l + 2w$ with
$l = 3x + 5$ and $w = x - 3$.

$P = 2(3x+5) + 2(x-2)$
$= 6x + 10 + 2x - 4$
$= 6x + 2x + 10 - 4$
$= 8x + 6$

71. Use the formula $A = lw$ for all the figures.
Remember, $x(x) = x^2$
$A = 3x(x) + 4x(x) + 2x(2x) + 3x(3x)$
$= 3x^2 + 4x^2 + 4x^2 + 9x^2$
$= (3+4+4+9)x^2$
$= 20x^2$

PROBLEM SET **5.2** **Multiplying Monomials**

1. $(5x)(9x) = 5 \cdot 9 \cdot x \cdot x = 45x^{1+1} = 45x^2$

3. $(3x^2)(7x) = 3 \cdot 7 \cdot x^2 \cdot x = 21x^{2+1} = 21x^3$

5. $(-3xy)(2xy) = -3 \cdot 2 \cdot x \cdot x \cdot y \cdot y =$
$-6x^{1+1}y^{1+1} = -6x^2y^2$

7. $(-2x^2y)(-7x) =$
$(-2)(-7)(x^2)(x)(y) =$
$14x^{2+1}y^1 = 14x^3y$

9. $(4a^2b^2)(-12ab) =$
$(4)(-12)(a^2)(a)(b^2)(b) =$
$-48a^{2+1}b^{2+1} = -48a^3b^3$

11. $(-xy)(-5x^3) =$
$(-1)(-5)(x)(x^3)(y) =$
$5x^{1+3}y^1 = 5x^4y$

13. $(8ab^2c)(13a^2c) =$
$8 \cdot 13 \cdot a \cdot a^2 \cdot b^2 \cdot c \cdot c =$
$104a^{1+2}b^2c^{1+1} = 104a^3b^2c^2$

15. $(5x^2)(2x)(3x^3) = 5 \cdot 2 \cdot 3 \cdot x^2 \cdot x \cdot x^3 =$
$30x^{2+1+3} = 30x^6$

17. $(4xy)(-2x)(7y^2) =$
$(4)(-2)(7)(x)(x)(y)(y^2) =$
$-56x^{1+1}y^{1+2} = -56x^2y^3$

19. $(-2ab)(-ab)(-3b) =$
$(-2)(-1)(-3)(a^{1+1})(b^{1+1+1}) =$
$-6a^2b^3$

21. $(6cd)(-3c^2d)(-4d) =$
$(6)(-3)(-4)(c^{1+2}d^{1+1+1}) =$
$72c^3d^3$

23. $\left(\frac{2}{3}xy\right)\left(\frac{3}{5}x^2y^4\right) =$

$\left(\frac{2}{3}\right)\left(\frac{3}{5}\right)x^{1+2}y^{1+4} =$

$\frac{2}{5}x^3y^5$

25. $\left(-\frac{7}{12}a^2b\right)\left(\frac{8}{21}b^4\right) =$

$\left(-\frac{\overset{1}{\cancel{7}}}{\cancel{12}}_3\right)\left(\frac{\overset{2}{\cancel{8}}}{\cancel{21}}_3\right)a^2b^{1+4} = -\frac{2}{9}a^2b^5$

27. $(0.4x^5)(0.7x^3) = (0.4)(0.7)x^{5+3} = 0.28x^8$

29. $(-4ab)(1.6a^3b) =$
$(-4)(1.6)a^{1+3}b^{1+1} = -6.4a^4b^2$

31. $(2x^4)^2 = (2)^2(x^4)^2 = 4x^{4 \cdot 2} = 4x^8$

110

33. $(-3a^2b^3)^2 = (-3)^2(a^2)^2(b^3)^2 = 9a^4b^6$

35. $(3x^2)^3 = (3)^3(x^2)^3 = 27x^6$

37. $(-4x^4)^3 = (-4)^3(x^4)^3 = -64x^{12}$

39. $(9x^4y^5)^2 = (9)^2(x^4)^2(y^5)^2 = 81x^8y^{10}$

41. $(2x^2y)^4 = (2)^4(x^2)^4(y)^4 = 16x^8y^4$

43. $(-3a^3b^2)^4 = (-3)^4(a^3)^4(b^2)^4 =$
$81a^{12}b^8$

45. $(-x^2y)^6 = (-1)^6(x^2)^6(y)^6 =$
$1x^{12}y^6 = x^{12}y^6$

47. $5x(3x+2) = 5x(3x) + 5x(2) =$
$15x^2 + 10x$

49. $3x^2(6x-2) = 3x^2(6x) - 3x^2(2) =$
$18x^3 - 6x^2$

51. $-4x(7x^2-4) =$
$-4x(7x^2) - (-4x)(4) =$
$-28x^3 + 16x$

53. $2x(x^2-4x+6) =$
$2x(x^2) - 2x(4x) + 2x(6) =$
$2x^3 - 8x^2 + 12x$

55. $-6a(3a^2-5a-7) =$
$-6a(3a^2) - (-6a)(5a) - (-6a)(7) =$
$-18a^3 + 30a^2 + 42a$

57. $7xy(4x^2-x+5) =$
$7xy(4x^2) - 7xy(x) + 7xy(5) =$
$28x^3y - 7x^2y + 35xy$

59. $-xy(9x^2-2x-6) =$
$-xy(9x^2) - (-xy)(2x) - (-xy)(6) =$
$-9x^3y + 2x^2y + 6xy$

61. $5(x+2y) + 4(2x+3y) =$
$5x + 10y + 8x + 12y =$
$13x + 22y$

63. $4(x-3y) - 3(2x-y) =$
$4x - 12y - 6x + 3y =$
$-2x - 9y$

65. $2x(x^2-3x-4) + x(2x^2+3x-6) =$
$2x^3 - 6x^2 - 8x + 2x^3 + 3x^2 - 6x =$
$4x^3 - 3x^2 - 14x$

67. $3[2x-(x-2)] - 4(x-2) =$
$3[2x-x+2] - 4(x-2) =$
$3[x+2] - 4(x-2) =$
$3x + 6 - 4x + 8 =$
$-x + 14$

69. $-4(3x+2) - 5[2x-(3x+4)] =$
$-4(3x+2) - 5[2x-3x-4] =$
$-4(3x+2) - 5[-x-4] =$
$-12x - 8 + 5x + 20 =$
$-7x + 12$

71. $(3x)^2(2x^3) = (3)^2(x)^2(2)(x)^3 =$
$(9)(x^2)(2)(x^3) = 18x^5$

73. $(-3x)^3(-4x)^2 =$
$(-3)^3(x)^3(-4)^2(x)^2 =$
$(-27)(x^3)(16)(x^2) =$
$-432x^5$

75. $(5x^2y)^2(xy^2)^3 =$
$(5)^2(x^2)^2(y)^2(x)^3(y^2)^3 =$
$(25)(x^4)(y^2)(x^3)(y^6) =$
$25x^7y^8$

77. $(-a^2bc^3)^3(a^3b)^2 =$
$(-1)^3(a^2)^3(b)^3(c^3)^3(a^3)^2(b)^2 =$
$-1(a^6)(b^3)(c^9)(a^6)(b^2) =$
$-a^{12}b^5c^9$

79. $(-2x^2y^2)^4(-xy^3)^3 =$
$(-2)^4(x^2)^4(y^2)^4(-1)^3(x)^3(y^3)^3 =$
$(16)(x^8)(y^8)(-1)(x^3)(y^9) =$
$-16x^{11}y^{17}$

Problem Set 5.2

81. Use the formula $A = lw$ for both rectangles and then add.

Area = Left Area + Right Area
$$A = 3(x - 1) + 4(x + 2)$$
$$= 3x - 3 + 4x + 8$$
$$= 7x + 5$$

83. Use the formula $A = \pi r^2$ for the both circles, then subtract the area of the smaller circle from the area of the larger circle.

$$\text{Area} = \frac{\text{Area of}}{\text{Larger Circle}} - \frac{\text{Area of}}{\text{Smaller Circle}}$$
$$A = \pi(2x)^2 - \pi(x)^2$$
$$= \pi(4x^2) - \pi(x^2)$$
$$= 4\pi x^2 - \pi x^2$$
$$= 3\pi x^2$$

The area of the shaded region can be represented by $3\pi x^2$.

89. $(x^{2n})(x^{5n}) = x^{2n+5n} = x^{7n}$

91. $(x^{5n+2})(x^{n-1}) = x^{(5n+2)+(n-1)} = x^{6n+1}$

93. $(x^{6n-1})(x^4) = x^{(6n-1)+4} = x^{6n+3}$

95. $(4x^{3n})(-5x^{7n}) = -20x^{3n+7n} = -20x^{10n}$

97. $(-3x^{5n-2})(-4x^{2n+2}) = 12x^{(5n-2)+(2n+2)} = 12x^{7n}$

PROBLEM SET $\quad$ **5.3** $\quad$ **Multiplying Polynomials**

1. $(x + 2)(y + 3) =$
$x(y) + x(3) + 2(y) + 2(3) =$
$xy + 3x + 2y + 6$

3. $(x - 4)(y + 1) =$
$x(y) + x(1) - 4(y) - 4(1) =$
$xy + x - 4y - 4$

5. $(x - 5)(y - 6) =$
$x(y) + x(-6) - 5(y) - 5(-6) =$
$xy - 6x - 5y + 30$

7. $(x + 2)(y + z + 1) =$
$x(y) + x(z) + x(1) + 2(y) + 2(z) + 2(1) =$
$xy + xz + x + 2y + 2z + 2$

9. $(2x + 3)(3y + 1) =$
$2x(3y) + 2x(1) + 3(3y) + 3(1) =$
$6xy + 2x + 9y + 3$

11. $(x + 3)(x + 7) =$
$x(x) + x(7) + 3(x) + 3(7) =$
$x^2 + 7x + 3x + 21 =$
$x^2 + 10x + 21$

13. $(x + 8)(x - 3) =$
$x(x) + x(-3) + 8(x) + 8(-3) =$
$x^2 - 3x + 8x - 24 =$
$x^2 + 5x - 24$

15. $(x - 7)(x + 1) =$
$x(x) + x(1) - 7(x) - 7(1) =$
$x^2 + 1x - 7x - 7 =$
$x^2 - 6x - 7$

17. $(n - 4)(n - 6) =$
$n(n) + n(-6) - 4(n) - 4(-6) =$
$n^2 - 6n - 4n + 24 =$
$n^2 - 10n + 24$

19. $(3n + 1)(n + 6) =$
$3n(n) + 3n(6) + 1(n) + 1(6) =$
$3n^2 + 18n + 1n + 6 =$
$3n^2 + 19n + 6$

21. $(5x - 2)(3x + 7) =$
$5x(3x) + 5x(7) - 2(3x) - 2(7) =$
$15x^2 + 35x - 6x - 14 =$
$15x^2 + 29x - 14$

23. $(x+3)(x^2+4x+9) =$
$x(x^2) + x(4x) + x(9) + 3(x^2)$
$\quad + 3(4x) + 3(9) =$
$x^3 + 4x^2 + 9x + 3x^2 + 12x + 27 =$
$x^3 + 7x^2 + 21x + 27$

25. $(x+4)(x^2-x-6) =$
$x(x^2) + x(-x) + x(-6) + 4(x^2)$
$\quad + 4(-x) + 4(-6) =$
$x^3 - x^2 - 6x + 4x^2 - 4x - 24 =$
$x^3 + 3x^2 - 10x - 24$

27. $(x-5)(2x^2+3x-7) =$
$x(2x^2) + x(3x) + x(-7) - 5(2x^2)$
$\quad - 5(3x) - 5(-7) =$
$2x^3 + 3x^2 - 7x - 10x^2 - 15x + 35 =$
$2x^3 - 7x^2 - 22x + 35$

29. $(2a-1)(4a^2-5a+9) =$
$2a(4a^2) + 2a(-5a) + 2a(9) - 1(4a^2)$
$\quad - 1(-5a) - 1(9) =$
$8a^3 - 10a^2 + 18a - 4a^2 + 5a - 9 =$
$8a^3 - 14a^2 + 23a - 9$

31. $(3a+5)(a^2-a-1) =$
$3a(a^2) + 3a(-a) + 3a(-1) + 5(a^2)$
$\quad + 5(-a) + 5(-1) =$
$3a^3 - 3a^2 - 3a + 5a^2 - 5a - 5 =$
$3a^3 + 2a^2 - 8a - 5$

33. $(x^2+2x+3)(x^2+5x+4) =$
$x^2(x^2+5x+4) + 2x(x^2+5x+4)$
$\quad + 3(x^2+5x+4) =$
$x^4 + 5x^3 + 4x^2 + 2x^3 + 10x^2 + 8x$
$\quad + 3x^2 + 15x + 12 =$
$x^4 + 7x^3 + 17x^2 + 23x + 12$

35. $(x^2-6x-7)(x^2+3x-9) =$
$x^2(x^2+3x-9) - 6x(x^2+3x-9)$
$\quad - 7(x^2+3x-9) =$
$x^4 + 3x^3 - 9x^2 - 6x^3 - 18x^2 + 54x$
$\quad - 7x^2 - 21x + 63 =$
$x^4 - 3x^3 - 34x^2 + 33x + 63$

37. $(x+2)(x+9) =$
$x^2 + (2+9)x + 18 =$
$x^2 + 11x + 18$

39. $(x+6)(x-2) =$
$x^2 + (6-2)x - 12 =$
$x^2 + 4x - 12$

41. $(x+3)(x-11) =$
$x^2 + (3-11)x - 33 =$
$x^2 - 8x - 33$

43. $(n-4)(n-3) =$
$n^2 + (-4-3)n + 12 =$
$n^2 - 7n + 12$

45. $(n+6)(n+12) =$
$n^2 + (6+12)n + 72 =$
$n^2 + 18n + 72$

47. $(y+3)(y-7) =$
$y^2 + (3-7)y - 21 =$
$y^2 - 4y - 21$

49. $(y-7)(y-12) =$
$y^2 + (-7-12)y + 84 =$
$y^2 - 19y + 84$

51. $(x-5)(x+7) =$
$x^2 + (-5+7)x - 35 =$
$x^2 + 2x - 35$

53. $(x-14)(x+8) =$
$x^2 + (-14+8)x - 112 =$
$x^2 - 6x - 112$

55. $(a+10)(a-9) =$
$a^2 + (10-9)a - 90 =$
$a^2 + a - 90$

57. $(2a+1)(a+6) =$
$2a^2 + (12+1)a + 6 =$
$2a^2 + 13a + 6$

59. $(5x-2)(x+7) =$
$5x^2 + (35-2)x - 14 =$
$5x^2 + 33x - 14$

61. $(3x-7)(2x+1) =$
$6x^2 + (3-14)x - 7 =$
$6x^2 - 11x - 7$

Problem Set 5.3

63. $(4a+3)(3a-4) =$
$12a^2 + (-16+9)a - 12 =$
$12a^2 - 7a - 12$

65. $(6n-5)(2n-3) =$
$12n^2 + (-18-10)n + 15 =$
$12n^2 - 28n + 15$

67. $(7x-4)(2x+3) =$
$14x^2 + (21-8)x - 12 =$
$14x^2 + 13x - 12$

69. $(5-x)(9-2x) =$
$45 + (-10-9)x + 2x^2 =$
$45 - 19x + 2x^2$

71. $(-2x+3)(4x-5) =$
$-8x^2 + (10+12)x - 15 =$
$-8x^2 + 22x - 15$

73. $(-3x-1)(3x-4) =$
$-9x^2 + (12-3)x + 4 =$
$-9x^2 + 9x + 4$

75. $(8n+3)(9n-4) =$
$72n^2 + (-32+27)n - 12 =$
$72n^2 - 5n - 12$

77. $(3-2x)(9-x) =$
$27 + (-3-18)x + 2x^2 =$
$27 - 21x + 2x^2$

79. $(-4x+3)(-5x-2) =$
$20x^2 + (8-15)x - 6 =$
$20x^2 - 7x - 6$

81. Use the pattern
$(a+b)^2 = a^2 + 2ab + b^2$
$(x+7)^2 = x^2 + 2(x)(7) + (7)^2 =$
$x^2 + 14x + 49$

83. Use the pattern
$(a+b)(a-b) = a^2 - b^2$
$(5x-2)(5x+2) = (5x)^2 - (2)^2 =$
$25x^2 - 4$

85. Use the pattern
$(a-b)^2 = a^2 - 2ab + b^2$

$(x-1)^2 = x^2 - 2(x)(1) + (1)^2 =$
$x^2 - 2x + 1$

87. Use the pattern
$(a+b)^2 = a^2 + 2ab + b^2$
$(3x+7)^2 = (3x)^2 + 2(3x)(7) + (7)^2 =$
$9x^2 + 42x + 49$

89. Use the pattern
$(a-b)^2 = a^2 - 2ab + b^2$
$(2x-3)^2 = (2x)^2 - 2(2x)(3) + (3)^2 =$
$4x^2 - 12x + 9$

91. Use the pattern
$(a+b)(a-b) = a^2 - b^2$
$(2x+3y)(2x-3y) = (2x)^2 - (3y)^2 =$
$4x^2 - 9y^2$

93. Use the pattern
$(a-b)^2 = a^2 - 2ab + b^2$
$(1-5n)^2 = 1^2 - 2(1)(5n) + (5n)^2 =$
$1 - 10n + 25n^2$

95. Use the pattern
$(a+b)^2 = a^2 + 2ab + b^2$
$(3x+4y)^2 = (3x)^2 + 2(3x)(4y) + (4y)^2 =$
$9x^2 + 24xy + 16y^2$

97. Use the pattern
$(a+b)^2 = a^2 + 2ab + b^2$
$(3+4y)^2 = 3^2 + 2(3)(4y) + (4y)^2 =$
$9 + 24y + 16y^2$

99. Use the pattern
$(a+b)(a-b) = a^2 - b^2$
$(1+7n)(1-7n) = (1)^2 - (7n)^2 =$
$1 - 49n^2$

101. Use the pattern
$(a-b)^2 = a^2 - 2ab + b^2$
$(4a-7b)^2 = (4a)^2 - 2(4a)(7b) + (7b)^2 =$
$16a^2 - 56ab + 49b^2$

103. Use the pattern
$(a+b)^2 = a^2 + 2ab + b^2$
$(x+8y)^2 = x^2 + 2(x)(8y) + (8y)^2 =$
$x^2 + 16xy + 64y^2$

114

105. Use the pattern
$(a + b)(a - b) = a^2 - b^2$
$(5x - 11y)(5x + 11y) = (5x)^2 - (11y)^2 =$
$25x^2 - 121y^2$

107. Use the pattern
$(a + b)(a - b) = a^2 - b^2$
$x(8x + 1)(8x - 1) = x[(8x)^2 - (1)^2]$
$x[64x^2 - 1] = 64x^3 - x$

109. Use the pattern
$(a + b)(a - b) = a^2 - b^2$
$-2x(4x + y)(4x - y) = -2x[(4x)^2 - (y)^2]$
$-2x[16x^2 - y^2] = -32x^3 + 2xy^2$

111. $(x + 2)^3 =$
$(x + 2)(x + 2)(x + 2) =$
$(x + 2)(x^2 + 4x + 4) =$
$x(x^2 + 4x + 4) + 2(x^2 + 4x + 4) =$
$x^3 + 4x^2 + 4x + 2x^2 + 8x + 8 =$
$x^3 + 6x^2 + 12x + 8$

113. $(x - 3)^3 =$
$(x - 3)(x - 3)(x - 3) =$
$(x - 3)(x^2 - 6x + 9) =$
$x(x^2 - 6x + 9) - 3(x^2 - 6x + 9) =$
$x^3 - 6x^2 + 9x - 3x^2 + 18x - 27 =$
$x^3 - 9x^2 + 27x - 27$

115. $(2n + 1)^3 =$
$(2n + 1)(2n + 1)(2n + 1) =$
$(2n + 1)(4n^2 + 4n + 1) =$
$2n(4n^2 + 4n + 1) + 1(4n^2 + 4n + 1) =$
$8n^3 + 8n^2 + 2n + 4n^2 + 4n + 1 =$
$8n^3 + 12n^2 + 6n + 1$

117. $(3n - 2)^3 =$
$(3n - 2)(3n - 2)(3n - 2) =$
$(3n - 2)(9n^2 - 12n + 4) =$
$3n(9n^2 - 12n + 4) - 2(9n^2 - 12n + 4) =$
$27n^3 - 36n^2 + 12n - 18n^2 + 24n - 8 =$
$27n^3 - 54n^2 + 36n - 8$

119. Let $x + 3$ represent the width of the rectangle, then $x + 5$ represents the length. Therefore, the area of the figure is represented by

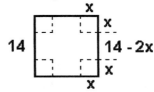

Area = (length)(width)
$\quad\quad = (x + 3)(x + 5)$
$\quad\quad = x^2 + 8x + 15$

Geometrically, the sum of the area of each section would be

Area $= \underset{\text{of A}}{\text{Area}} + \underset{\text{of B}}{\text{Area}} + \underset{\text{of C}}{\text{Area}} + \underset{\text{of D}}{\text{Area}}$
$A = \quad 3x \quad + \quad 15 \quad + \quad x^2 \quad + \quad 5x$
$\quad = x^2 + 8x + 15$

121. Each side of the box will be $14 - 2x$ and the height will be x.

The volume is $V = lwh$
$\quad\quad = (14 - 2x)(14 - 2x)x$
$\quad\quad = (196 - 56x + 4x^2)x$
$\quad\quad = 196x - 56x^2 + 4x^3$

The outside surface area is $S = \dfrac{\text{Original}}{\text{Area}} - 4\text{ Corners}$
$\quad\quad = (14)^2 - 4(x)^2$
$\quad\quad = 196 - 4x^2$

125. **111)** $(x + 2)^3 =$
$(x)^3 + 3(x)^2 2 + 3(x)(2)^2 + (2)^3 =$
$x^3 + 6x^2 + 12x + 8$

112) $(x + 4)^3 =$
$(x)^3 + 3(x)^2 4 + 3(x)(4)^2 + (4)^3 =$
$x^3 + 12x^2 + 48x + 64$

113) $(x - 3)^3 =$
$(x)^3 + 3(x)^2(-3) + 3(x)(-3)^2 + (-3)^3 =$
$x^3 - 9x^2 + 27x - 27$

115

Problem Set 5.3

114) $(x-1)^3 =$
$(x^3) - 3(x)^2(1) + 3(x)(1)^2 - (1)^3 =$
$x^3 - 3x^2 + 3x - 1$

115) $(2n+1)^3 =$
$(2n)^3 + 3(2n)^2(1) + 3(2n)(1)^2 + (1)^3 =$
$8n^3 + 12n^2 + 6n + 1$

116) $(3n+2)^3 =$
$(3n)^3 + 3(3n)^2(2) + 3(3n)(2)^2 + (2)^3 =$
$27n^3 + 54n^2 + 36n + 8$

117) $(3n-2)^3 =$
$(3n)^3 + 3(3n)^2(-2) + 3(3n)(-2)^2 + (-2)^3 =$

118) $(4n-3)^3 =$
$(4n)^3 - 3(4n)^2(3) + 3(4n)(3)^2 - (3)^3 =$
$64n^3 - 144n^2 + 108n - 27$

127a. $21^2 = (20+1)^2 = 20^2 + 2(20)(1) + 1 =$
$400 + 40 + 1 = 441$
b. $41^2 = (40+1)^2 = 40^2 + 2(40)(1) + 1 =$
$1600 + 80 + 1 = 1681$
c. $71^2 = (70+1)^2 = 70^2 + 2(70)(1) + 1 =$
$4900 + 140 + 1 = 5041$
d. $32^2 = (30+2)^2 = 30^2 + 2(30)(2) + 2^2 =$
$900 + 120 + 4 = 1024$
e. $52^2 = (50+2)^2 = 50^2 + 2(50)(2) + 2^2 =$
$2500 + 200 + 4 = 2704$
f. $82^2 = (80+2)^2 = 80^2 + 2(80)(2) + 2^2 =$
$6400 + 320 + 4 = 6724$

129 a. $15^2 = 1(2)(100) + 25 = 225$
b. $25^2 = 2(3)(100) + 25 = 625$
c. $45^2 = 4(5)(100) + 25 = 2025$
d. $55^2 = 5(6)(100) + 25 = 3025$
e. $65^2 = 6(7)(100) + 25 = 4225$
f. $75^2 = 7(8)(100) + 25 = 5625$
g. $85^2 = 8(9)(100) + 25 = 7225$
h. $95^2 = 9(10)(100) + 25 = 9025$
i. $105^2 = 10(11)(100) + 25 = 11025$

PROBLEM SET 5.4 **Dividing by Monomials**

1. $\dfrac{x^{10}}{x^2} = x^{10-2} = x^8$

11. $\dfrac{-91a^4b^6}{-13a^3b^4} = 7a^{4-3}b^{6-4} = 7ab^2$

3. $\dfrac{4x^3}{2x} = 2x^{3-1} = 2x^2$

13. $\dfrac{18x^2y^6}{xy^2} = 18x^{2-1}y^{6-2} = 18xy^4$

5. $\dfrac{-16n^6}{2n^2} = -8n^{6-2} = -8n^4$

15. $\dfrac{32x^6y^2}{-x} = -32x^{6-1}y^2 = -32x^5y^2$

7. $\dfrac{72x^3}{-9x^3} = -8(1) = -8$

17. $\dfrac{-96x^5y^7}{12y^3} = -8x^5y^{7-3} = -8x^5y^4$

9. $\dfrac{65x^2y^3}{5xy} = 13x^{2-1}y^{3-1} = 13xy^2$

19. $\dfrac{-ab}{ab} = -1(1)(1) = -1$

21. $\dfrac{56a^2b^3c^5}{4abc} = 14a^{2-1}b^{3-1}c^{5-1} = 14ab^2c^4$

23. $\dfrac{-80xy^2z^6}{-5xyz^2} = 16(1)y^{2-1}z^{6-2} = 16yz^4$

25. $\dfrac{8x^4 + 12x^5}{2x^2} = \dfrac{8x^4}{2x^2} + \dfrac{12x^5}{2x^2} = 4x^2 + 6x^3$

27. $\dfrac{9x^6 - 24x^4}{3x^3} = \dfrac{9x^6}{3x^3} - \dfrac{24x^4}{3x^3} = 3x^3 - 8x$

29. $\dfrac{-28n^5 + 36n^2}{4n^2} = \dfrac{-28n^5}{4n^2} + \dfrac{36n^2}{4n^2} =$
$-7n^3 + 9$

31. $\dfrac{35x^6 - 56x^5 - 84x^3}{7x^2} =$
$\dfrac{35x^6}{7x^2} - \dfrac{56x^5}{7x^2} - \dfrac{84x^3}{7x^2} =$
$5x^4 - 8x^3 - 12x$

33. $\dfrac{-24n^8 + 48n^5 - 78n^3}{-6n^3} =$
$\dfrac{-24n^8}{-6n^3} + \dfrac{48n^5}{-6n^3} - \dfrac{78n^3}{-6n^3} =$
$4n^5 - 8n^2 + 13$

35. $\dfrac{-60a^7 - 96a^3}{-12a} = \dfrac{-60a^7}{-12a} - \dfrac{96a^3}{-12a} =$
$5a^6 + 8a^2$

37. $\dfrac{27x^2y^4 - 45xy^4}{-9xy^3} =$

39. $\dfrac{48a^2b^2 + 60a^3b^4}{-6ab} =$
$\dfrac{48a^2b^2}{-6ab} + \dfrac{60a^3b^4}{-6ab} =$
$-8ab - 10a^2b^3$

41. $\dfrac{12a^2b^2c^2 - 52a^2b^3c^5}{-4a^2bc} =$
$\dfrac{12a^2b^2c^2}{-4a^2bc} - \dfrac{52a^2b^3c^5}{-4a^2bc} =$
$-3bc + 13b^2c^4$

43. $\dfrac{9x^2y^3 - 12x^3y^4}{-xy} = \dfrac{9x^2y^3}{-xy} - \dfrac{12x^3y^4}{-xy} =$
$-9xy^2 + 12x^2y^3$

37. (right column) $\dfrac{27x^2y^4}{-9xy^3} - \dfrac{45xy^4}{-9xy^3} =$
$-3xy + 5y$

45. $\dfrac{-42x^6 - 70x^4 + 98x^2}{14x^2} =$
$\dfrac{-42x^6}{14x^2} - \dfrac{70x^4}{14x^2} + \dfrac{98x^2}{14x^2} =$
$-3x^4 - 5x^2 + 7$

47. $\dfrac{15a^3b - 35a^2b - 65ab^2}{-5ab} =$
$\dfrac{15a^3b}{-5ab} - \dfrac{35a^2b}{-5ab} - \dfrac{65ab^2}{-5ab} =$
$-3a^2 + 7a + 13b$

49. $\dfrac{-xy + 5x^2y^3 - 7x^2y^6}{xy} =$
$\dfrac{-xy}{xy} + \dfrac{5x^2y^3}{xy} - \dfrac{7x^2y^6}{xy} =$
$-1 + 5xy^2 - 7xy^5$

Problem Set 5.5

1.
$$
\begin{array}{r}
x + 12 \\
x + 4 \overline{\smash{\big)}\ x^2 + 16x + 48} \\
\underline{x^2 + 4x} \\
12x + 48 \\
\underline{12x + 48}
\end{array}
$$

3.
$$
\begin{array}{r}
x + 2 \\
x - 7 \overline{\smash{\big)}\ x^2 - 5x - 14} \\
\underline{x^2 - 7x} \\
2x - 14 \\
\underline{2x - 14}
\end{array}
$$

5.
$$
\begin{array}{r}
x + 8 \\
x + 3 \overline{\smash{\big)}\ x^2 + 11x + 28} \\
\underline{x^2 + 3x} \\
8x + 28 \\
\underline{8x + 24} \\
4
\end{array}
$$

7.
$$
\begin{array}{r}
x + 4 \\
x - 8 \overline{\smash{\big)}\ x^2 - 4x - 39} \\
\underline{x^2 - 8x} \\
4x - 39 \\
\underline{4x - 32} \\
-7
\end{array}
$$

9.
$$
\begin{array}{r}
5n + 4 \\
n - 1 \overline{\smash{\big)}\ 5n^2 - n - 4} \\
\underline{5n^2 - 5n} \\
4n - 4 \\
\underline{4n - 4}
\end{array}
$$

11.
$$
\begin{array}{r}
8y - 3 \\
y + 7 \overline{\smash{\big)}\ 8y^2 + 53y - 19} \\
\underline{8y^2 + 56y} \\
-3y - 19 \\
\underline{-3y - 21} \\
2
\end{array}
$$

13.
$$
\begin{array}{r}
4x - 7 \\
5x + 1 \overline{\smash{\big)}\ 20x^2 - 31x - 7} \\
\underline{20x^2 + 4x} \\
-35x - 7 \\
\underline{-35x - 7}
\end{array}
$$

15.
$$
\begin{array}{r}
3x + 2 \\
2x + 7 \overline{\smash{\big)}\ 6x^2 + 25x + 8} \\
\underline{6x^2 + 21x} \\
4x + 8 \\
\underline{4x + 14} \\
-6
\end{array}
$$

17.
$$
\begin{array}{r}
2x^2 + 3x + 4 \\
x - 2 \overline{\smash{\big)}\ 2x^3 - x^2 - 2x - 8} \\
\underline{2x^3 - 4x^2} \\
3x^2 - 2x - 8 \\
\underline{3x^2 - 6x} \\
4x - 8 \\
\underline{4x - 8}
\end{array}
$$

19.
$$
\begin{array}{r}
5n^2 - 4n - 3 \\
n + 3 \overline{\smash{\big)}\ 5n^3 + 11n^2 - 15n - 9} \\
\underline{5n^3 + 15n^2} \\
-4n^2 - 15n - 9 \\
\underline{-4n^2 - 12n} \\
-3n - 9 \\
\underline{-3n - 9}
\end{array}
$$

21.
$$
\begin{array}{r}
n^2 + 6n - 4 \\
n - 6 \overline{\smash{\big)}\ n^3 + 0n^2 - 40n + 24} \\
\underline{n^3 - 6n^2} \\
6n^2 - 40n + 24 \\
\underline{6n^2 - 36n} \\
-4n + 24 \\
\underline{-4n + 24}
\end{array}
$$

23.

$$
\begin{array}{r}
x^2 + 3x + 9 \\
x - 3 \,\overline{\big)\, x^3 + 0x^2 + 0x - 27} \\
\underline{x^3 - 3x^2} \\
3x^2 - 0x - 27 \\
\underline{3x^2 - 9x} \\
9x - 27 \\
\underline{9x - 27}
\end{array}
$$

25.

$$
\begin{array}{r}
9x^2 + 12x + 16 \\
3x - 4 \,\overline{\big)\, 27x^3 + 0x^2 + 0x - 64} \\
\underline{27x^3 - 36x^2} \\
36x^2 + 0x - 64 \\
\underline{36x^2 - 48x} \\
48x - 64 \\
\underline{48x - 64}
\end{array}
$$

27.

$$
\begin{array}{r}
3n - 8 \\
n + 2 \,\overline{\big)\, 3n^2 - 2n + 1} \\
\underline{3n^2 + 6n} \\
-8n + 1 \\
\underline{-8n - 16} \\
17
\end{array}
$$

29.

$$
\begin{array}{r}
3t + 2 \\
3t - 1 \,\overline{\big)\, 9t^2 + 3t + 4} \\
\underline{9t^2 - 3t} \\
6t + 4 \\
\underline{6t - 2} \\
6
\end{array}
$$

31.

$$
\begin{array}{r}
3n^2 - n - 4 \\
2n - 1 \,\overline{\big)\, 6n^3 - 5n^2 - 7n + 4} \\
\underline{6n^3 - 3n^2} \\
-2n^2 - 7n + 4 \\
\underline{-2n^2 + 1n} \\
-8n + 4 \\
\underline{-8n + 4}
\end{array}
$$

33.

$$
\begin{array}{r}
4x^2 - 5x + 5 \\
x + 7 \,\overline{\big)\, 4x^3 + 23x^2 - 30x + 32} \\
\underline{4x^3 + 28x^2} \\
-5x^2 - 30x + 32 \\
\underline{-5x^2 - 35x} \\
5x + 32 \\
\underline{5x + 35} \\
-3
\end{array}
$$

35.

$$
\begin{array}{r}
x + 4 \\
x^2 - 2x \,\overline{\big)\, x^3 + 2x^2 - 3x - 1} \\
\underline{x^3 - 2x^2} \\
4x^2 - 3x - 1 \\
\underline{4x^2 - 8x} \\
5x - 1
\end{array}
$$

37.

$$
\begin{array}{r}
2x - 12 \\
x^2 + 4x \,\overline{\big)\, 2x^3 - 4x^2 + x - 5} \\
\underline{2x^3 + 8x^2} \\
-12x^2 + x - 5 \\
\underline{-12x^2 - 48x} \\
49x - 5
\end{array}
$$

39.

$$
\begin{array}{r}
x^3 - 2x^2 + 4x - 8 \\
x + 2 \,\overline{\big)\, x^4 + 0x^3 + 0x^2 + 0x - 16} \\
\underline{x^4 + 2x^3} \\
-2x^3 + 0x^2 + 0x - 16 \\
\underline{-2x^3 - 4x^2} \\
4x^2 + 0x - 16 \\
\underline{4x^2 + 8x} \\
-8x - 16 \\
\underline{-8x - 16}
\end{array}
$$

Problem Set 5.6

1. $\;$ $3^{-2} = \dfrac{1}{3^2} = \dfrac{1}{9}$

3. $\;$ $4^{-3} = \dfrac{1}{4^3} = \dfrac{1}{64}$

5. $\;$ $\left(\dfrac{3}{2}\right)^{-1} = \left(\dfrac{2}{3}\right)^1 = \dfrac{2}{3}$

7. $\;$ $\dfrac{1}{2^{-4}} = \dfrac{1}{\dfrac{1}{2^4}} = 2^4 = 16$

9. $\;$ $\left(-\dfrac{4}{3}\right)^0 = 1$

11. $\;$ $\left(-\dfrac{2}{3}\right)^{-3} = \left(-\dfrac{3}{2}\right)^3 = -\dfrac{27}{8}$

13. $\;$ $(-2)^{-2} = \dfrac{1}{(-2)^2} = \dfrac{1}{4}$

15. $\;$ $-\left(3^{-2}\right) = -\dfrac{1}{3^2} = -\dfrac{1}{9}$

17. $\;$ $\dfrac{1}{\left(\dfrac{3}{4}\right)^{-3}} = \left(\dfrac{3}{4}\right)^3 = \dfrac{27}{64}$

19. $\;$ $2^6 \cdot 2^{-9} = 2^{6-9} = 2^{-3} = \dfrac{1}{2^3} = \dfrac{1}{8}$

21. $\;$ $3^6 \cdot 3^{-3} = 3^{6-3} = 3^3 = 27$

23. $\;$ $\dfrac{10^2}{10^{-1}} = 10^{2-(-1)} = 10^{2+1} = 10^3 = 1000$

25. $\;$ $\dfrac{10^{-1}}{10^2} = 10^{-1-2} = 10^{-3} = \dfrac{1}{10^3} = \dfrac{1}{1000}$

27. $\;$ $(2^{-1} \cdot 3^{-2})^{-1} = (2^{-1})^{-1} \cdot (3^{-2})^{-1} =$ $2 \cdot 3^2 = 2 \cdot 9 = 18$

29. $\;$ $\left(\dfrac{4^{-1}}{3}\right)^{-2} = \dfrac{(4^{-1})^{-2}}{3^{-2}} = \dfrac{4^2}{3^{-2}} =$ $4^2 \cdot 3^2 = 16 \cdot 9 = 144$

31. $\;$ $x^6 x^{-1} = x^{6-1} = x^5$

33. $\;$ $n^{-4} n^2 = n^{-4+2} = n^{-2} = \dfrac{1}{n^2}$

35. $\;$ $a^{-2} a^{-3} = a^{-2-3} = a^{-5} = \dfrac{1}{a^5}$

37. $\;$ $(2x^3)(4x^{-2}) = 2 \cdot 4 \cdot x^{3-2} = 8x$

39. $\;$ $(3x^{-6})(9x^2) = 3 \cdot 9 \cdot x^{-6+2} = 27x^{-4} = \dfrac{27}{x^4}$

41. $\;$ $(5y^{-1})(-3y^{-2}) = (5)(-3)y^{-1-2} =$ $-15y^{-3} = -\dfrac{15}{y^3}$

43. $\;$ $(8x^{-4})(12x^4) = 8 \cdot 12 \cdot x^{-4+4} =$ $96x^0 = 96(1) = 96$

45. $\;$ $\dfrac{x^7}{x^{-3}} = x^{7-(-3)} = x^{7+3} = x^{10}$

47. $\;$ $\dfrac{n^{-1}}{n^3} = n^{-1-3} = n^{-4} = \dfrac{1}{n^4}$

49. $\;$ $\dfrac{4n^{-1}}{2n^{-3}} = 2n^{-1-(-3)} = 2n^{-1+3} = 2n^2$

51. $\;$ $\dfrac{-24x^{-6}}{8x^{-2}} = -3x^{-6-(-2)} =$ $-3x^{-6+2} = -3x^{-4} = -\dfrac{3}{x^4}$

53. $\;$ $\dfrac{-52y^{-2}}{-13y^{-2}} = 4y^{-2-(-2)} = 4y^{-2+2} =$ $4y^0 = 4(1) = 4$

55. $\;$ $(x^{-3})^{-2} = x^{-3(-2)} = x^6$

120

57. $\left(x^2\right)^{-2} = x^{2(-2)} = x^{-4} = \dfrac{1}{x^4}$

59. $\left(x^3 y^4\right)^{-1} = \left(x^3\right)^{-1}\left(y^4\right)^{-1} =$

$x^{-3}y^{-4} = \dfrac{1}{x^3 y^4}$

61. $\left(x^{-2} y^{-1}\right)^3 = \left(x^{-2}\right)^3 \left(y^{-1}\right)^3 =$

$x^{-6} y^{-3} = \dfrac{1}{x^6 y^3}$

63. $\left(2n^{-2}\right)^3 = (2)^3 \left(n^{-2}\right)^3 = 8n^{-6} = \dfrac{8}{n^6}$

65. $\left(4n^3\right)^{-2} = (4)^{-2}(n^3)^{-2} = 4^{-2}n^{-6} =$

$\dfrac{1}{4^2 n^6} = \dfrac{1}{16 n^6}$

67. $\left(3a^{-2}\right)^4 = 3^4 \left(a^{-2}\right)^4 = 81a^{-8} = \dfrac{81}{a^8}$

69. $\left(5x^{-1}\right)^{-2} = 5^{-2}\left(x^{-1}\right)^{-2} = 5^{-2}x^2 =$

$\dfrac{x^2}{5^2} = \dfrac{x^2}{25}$

71. $\left(2x^{-2}y^{-1}\right)^{-1} = 2^{-1}(x^{-2})^{-1}(y^{-1})^{-1} =$

$2^{-1}x^2 y^1 = \dfrac{x^2 y}{2}$

73. $\left(\dfrac{x^2}{y}\right)^{-1} = \dfrac{\left(x^2\right)^{-1}}{y^{-1}} = \dfrac{x^{-2}}{y^{-1}} = \dfrac{y}{x^2}$

75. $\left(\dfrac{a^{-1}}{b^2}\right)^{-4} = \dfrac{a^4}{b^{-8}} = a^4 b^8$

77. $\left(\dfrac{x^{-1}}{y^{-3}}\right)^{-2} = \dfrac{x^2}{y^6}$

79. $\left(\dfrac{x^2}{x^3}\right)^{-1} = \left(x^{2-3}\right)^{-1} = \left(x^{-1}\right)^{-1} = x$

81. $\left(\dfrac{2x^{-1}}{x^{-2}}\right)^{-3} = \left(2x^{-1-(-2)}\right)^{-3} =$

$\left(2x^{-1+2}\right)^{-3} = (2x)^{-3} = \dfrac{1}{(2x)^3} = \dfrac{1}{8x^3}$

83. $\left(\dfrac{18x^{-1}}{9x}\right)^{-2} = \left(2x^{-1-1}\right)^{-2} =$

$\left(2x^{-2}\right)^{-2} = 2^{-2}x^4 = \dfrac{x^4}{2^2} = \dfrac{x^4}{4}$

85. $321 = (3.21)(10^2)$
Number > 10, positive exponent.

87. $8000 = (8)(10^3)$
Number > 10, positive exponent.

89. $0.00246 = (2.46)(10^{-3})$
Number < 1, negative exponent.

91. $0.0000179 = (1.79)(10^{-5})$
Number < 1, negative exponent.

93. $87{,}000{,}000 = (8.7)(10^7)$
Number > 10, positive exponent.

95. $(8)(10^3) = 8000$
Positive exponent, move decimal right.

97. $(5.21)(10^4) = 52{,}100$
Positive exponent, move decimal right.

99. $(1.14)(10^7) = 11{,}400{,}000$
Positive exponent, move decimal right.

101. $(7)(10^{-2}) = 0.07$
Negative exponent, move decimal left.

103. $(9.87)(10^{-4}) = 0.000987$
Negative exponent, move decimal left.

105. $(8.64)(10^{-6}) = 0.00000864$
Negative exponent, move decimal left.

107. $(0.007)(120) =$
$(7)(10^{-3})(1.2)(10^2) =$
$(7)(1.2)(10^{-3})(10^2) =$
$(8.4)(10^{-1}) = 0.84$

Problem Set 5.6

109. $(5,000,000)(0.00009) =$
$(5)(10^6)(9)(10^{-5}) =$
$(5)(9)(10^6)(10^{-5}) =$
$(45)(10^1) = 450$

111. $\dfrac{6000}{0.0015} = \dfrac{(6)(10^3)}{(1.5)(10^{-3})} =$
$(4)\left(10^{3-(-3)}\right) = (4)(10^6) =$
$4,000,000$

113. $\dfrac{0.00086}{4300} = \dfrac{(8.6)(10^{-4})}{(4.3)(10^3)} =$
$(2)(10^{-4-3}) = (2)(10^{-7}) =$
0.0000002

115. $\dfrac{0.00039}{0.0013} = \dfrac{(3.9)(10^{-4})}{(1.3)(10^{-3})} =$
$(3)\left(10^{-4-(-3)}\right) = (3)(10^{-1}) = 0.3$

117. $\dfrac{(0.0008)(0.07)}{(20,000)(0.0004)} =$
$\dfrac{(8)(10^{-4})(7)(10^{-2})}{(2)(10^4)(4)(10^{-4})} =$
$\dfrac{(56)(10^{-6})}{(8)(10^0)} =$
$(7)(10^{-6}) = 0.000007$

CHAPTER 5 **Review Problem Set**

1. $(5x^2 - 6x + 4) + (3x^2 - 7x - 2) =$
$(5+3)x^2 + (-6-7)x + (4-2) =$
$8x^2 - 13x + 2$

2. $(7y^2 + 9y - 3) - (4y^2 - 2y + 6) =$
$7y^2 + 9y - 3 - 4y^2 + 2y - 6 =$
$3y^2 + 11y - 9$

3. $(2x^2 + 3x - 4) + (4x^2 - 3x - 6)$
$\quad - (3x^2 - 2x - 1) =$
$2x^2 + 3x - 4 + 4x^2 - 3x - 6 - 3x^2$
$\quad + 2x + 1 = 3x^2 + 2x - 9$

4. $(-3x^2 - 2x + 4) - (x^2 - 5x - 6)$
$\quad - (4x^2 + 3x - 8) =$
$-3x^2 - 2x + 4 - x^2 + 5x + 6$
$\quad - 4x^2 - 3x + 8 =$
$-8x^2 + 18$

5. $5(2x - 1) + 7(x + 3) - 2(3x + 4) =$
$10x - 5 + 7x + 21 - 6x - 8 =$
$11x + 8$

6. $3(2x^2 - 4x - 5) - 5(3x^2 - 4x + 1) =$
$6x^2 - 12x - 15 - 15x^2 + 20x - 5 =$
$-9x^2 + 8x - 20$

7. $6(y^2 - 7y - 3) - 4(y^2 + 3y - 9) =$
$6y^2 - 42y - 18 - 4y^2 - 12y + 36 =$
$2y^2 - 54y + 18$

8. $3(a - 1) - 2(3a - 4) - 5(2a + 7) =$
$3a - 3 - 6a + 8 - 10a - 35 =$
$-13a - 30$

9. $-(a + 4) + 5(-a - 2) - 7(3a - 1) =$
$-a - 4 - 5a - 10 - 21a + 7 =$
$-27a - 7$

10. $-2(3n - 1) - 4(2n + 6) + 5(3n + 4) =$
$-6n + 2 - 8n - 24 + 15n + 20 =$
$n - 2$

11. $3(n^2 - 2n - 4) - 4(2n^2 - n - 3) =$
$3n^2 - 6n - 12 - 8n^2 + 4n + 12 =$
$-5n^2 - 2n$

12. $-5(-n^2 + n - 1) + 3(4n^2 - 3n - 7) =$
$5n^2 - 5n + 5 + 12n^2 - 9n - 21 =$
$17n^2 - 14n - 16$

13. $(5x^2)(7x^4) = 5 \cdot 7 \cdot x^2 \cdot x^4$
$\quad = 35x^{2+4} = 35x^6$

14. $(-6x^3)(9x^5) = -6 \cdot 9 \cdot x^3 \cdot x^5 =$
$-54x^{3+5} = -54x^8$

122

15. $(-4xy^2)(-6x^2y^3) =$
$(-4)(-6)(x^{1+2})(y^{2+3}) =$
$24x^3y^5$

16. $(2a^3b^4)(-3ab^5) =$
$(2)(-3)(a^{3+1})(b^{4+5}) =$
$-6a^4b^9$

17. $(2a^2b^3)^3 = (2)^3(a^2)^3(b^3)^3 = 8a^6b^9$

18. $(-3xy^2)^2 = (-3)^2(x)^2(y^2)^2 = 9x^2y^4$

19. $5x(7x+3) = 35x^2 + 15x$

20. $(-3x^2)(8x-1) = -24x^3 + 3x^2$

21. $(x+9)(x+8) =$
$x^2 + (9+8)x + 72 =$
$x^2 + 17x + 72$

22. $(3x+7)(x+1) =$
$3x^2 + (3+7)x + 7 =$
$3x^2 + 10x + 7$

23. $(x-5)(x+2) =$
$x^2 + (-5+2)x - 10 =$
$x^2 - 3x - 10$

24. $(y-4)(y-9) =$
$y^2 + (-9-4)y + 36 =$
$y^2 - 13y + 36$

25. $(2x-1)(7x+3) =$
$14x^2 + (6-7)x - 3 =$
$14x^2 - x - 3$

26. $(4a-7)(5a+8) =$
$20a^2 + (32-35)a - 56 =$
$20a^2 - 3a - 56$

27. $(3a-5)^2 =$
$(3a)^2 - 2(3a)(5) + (5)^2 =$
$9a^2 - 30a + 25$

28. $(x+6)(2x^2+5x-4) =$
$x(2x^2+5x-4) + 6(2x^2+5x-4) =$
$2x^3 + 5x^2 - 4x + 12x^2 + 30x - 24 =$
$2x^3 + 17x^2 + 26x - 24$

29. $(5n-1)(6n+5) =$
$30n^2 + (25-6)n - 5 =$
$30n^2 + 19n - 5$

30. $(3n+4)(4n-1) =$
$12n^2 + (-3+16)n - 4 =$
$12n^2 + 13n - 4$

31. $(2n+1)(2n-1) =$
$(2n)^2 - (1)^2 =$
$4n^2 - 1$

32. $(4n-5)(4n+5) =$
$(4n)^2 - (5)^2 =$
$16n^2 - 25$

33. $(2a+7)^2 =$
$(2a)^2 + 2(2a)(7) + (7)^2 =$
$4a^2 + 28a + 49$

34. $(3a+5)^2 =$
$(3a)^2 + 2(3a)(5) + (5)^2 =$
$9a^2 + 30a + 25$

35. $(x-2)(x^2-x+6) =$
$x(x^2-x+6) - 2(x^2-x+6) =$
$x^3 - x^2 + 6x - 2x^2 + 2x - 12 =$
$x^3 - 3x^2 + 8x - 12$

36. $(2x-1)(x^2+4x+7) =$
$2x(x^2+4x+7) - 1(x^2+4x+7) =$
$2x^3 + 8x^2 + 14x - x^2 - 4x - 7 =$
$2x^3 + 7x^2 + 10x - 7$

37. $(a+5)^3 = (a+5)(a+5)(a+5) =$
$(a+5)(a^2+10a+25) =$
$a(a^2+10a+25) + 5(a^2+10a+25) =$
$a^3 + 10a^2 + 25a + 5a^2 + 50a + 125 =$
$a^3 + 15a^2 + 75a + 125$

38. $(a-6)^3 = (a-6)(a-6)(a-6) =$
$(a-6)(a^2-12a+36) =$
$a(a^2-12a+36) - 6(a^2-12a+36) =$
$a^3 - 12a^2 + 36a - 6a^2 + 72a - 216 =$
$a^3 - 18a^2 + 108a - 216$

39. $(x^2 - x - 1)(x^2 + 2x + 5) =$
$x^2(x^2 + 2x + 5) - x(x^2 + 2x + 5)$
$\quad - 1(x^2 + 2x + 5) =$
$x^4 + 2x^3 + 5x^2 - x^3 - 2x^2 - 5x$
$\quad - x^2 - 2x - 5 =$
$x^4 + x^3 + 2x^2 - 7x - 5$

40. $(n^2 + 2n + 4)(n^2 - 7n - 1) =$
$n^2(n^2 - 7n - 1) + 2n(n^2 - 7n - 1)$
$\quad + 4(n^2 - 7n - 1) =$
$n^4 - 7n^3 - n^2 + 2n^3 - 14n^2 - 2n$
$\quad + 4n^2 - 28n - 4 =$
$n^4 - 5n^3 - 11n^2 - 30n - 4$

41. $\dfrac{36x^4y^5}{-3xy^2} = -12x^{4-1}y^{5-2} = -12x^3y^3$

42. $\dfrac{-56a^5b^7}{-8a^2b^3} = 7a^{5-2}b^{7-3} = 7a^3b^4$

43. $\dfrac{-18x^4y^3 - 54x^6y^2}{6x^2y^2} =$

$\dfrac{-18x^4y^3}{6x^2y^2} - \dfrac{54x^6y^2}{6x^2y^2} =$

$-3x^2y - 9x^4$

44. $\dfrac{-30a^5b^{10} + 39a^4b^8}{-3ab} =$

$\dfrac{-30a^5b^{10}}{-3ab} + \dfrac{39a^4b^8}{-3ab} =$

$10a^4b^9 - 13a^3b^7$

45. $\dfrac{56x^4 - 40x^3 - 32x^2}{4x^2} =$

$\dfrac{56x^4}{4x^2} - \dfrac{40x^3}{4x^2} - \dfrac{32x^2}{4x^2} =$

$14x^2 - 10x - 8$

46.
$$
\begin{array}{r}
x + 4 \\
x + 5 \overline{\smash{\big)}\ x^2 + 9x - 1} \\
\underline{x^2 + 5x} \\
4x - 1 \\
\underline{4x + 20} \\
-21
\end{array}
$$

47.
$$
\begin{array}{r}
7x - 6 \\
3x + 2 \overline{\smash{\big)}\ 21x^2 - 4x - 12} \\
\underline{21x^2 + 14x} \\
-18x - 12 \\
\underline{-18x - 12}
\end{array}
$$

48.
$$
\begin{array}{r}
2x^2 + x + 4 \\
x - 2 \overline{\smash{\big)}\ 2x^3 - 3x^2 + 2x - 4} \\
\underline{2x^3 - 4x^2} \\
x^2 + 2x - 4 \\
\underline{x^2 - 2x} \\
4x - 4 \\
\underline{4x - 8} \\
4
\end{array}
$$

49. $3^2 + 2^2 = 9 + 4 = 13$

50. $(3 + 2)^2 = (5)^2 = 25$

51. $2^{-4} = \dfrac{1}{2^4} = \dfrac{1}{16}$

52. $(-5)^0 = 1$

53. $-5^0 = -1$

54. $\dfrac{1}{3^{-2}} = 3^2 = 9$

55. $\left(\dfrac{3}{4}\right)^{-2} = \left(\dfrac{4}{3}\right)^2 = \dfrac{16}{9}$

56. $\dfrac{1}{\left(\dfrac{1}{4}\right)^{-1}} = \left(\dfrac{1}{4}\right)^1 = \dfrac{1}{4}$

57. $\dfrac{1}{(-2)^{-3}} = (-2)^3 = -8$

58. $2^{-1} + 3^{-2} = \dfrac{1}{2} + \dfrac{1}{3^2} = \dfrac{1}{2} + \dfrac{1}{9} =$

$\dfrac{9}{18} + \dfrac{2}{18} = \dfrac{11}{18}$

59. $3^0 + 2^{-2} = 1 + \dfrac{1}{2^2} =$

$1 + \dfrac{1}{4} = \dfrac{4}{4} + \dfrac{1}{4} = \dfrac{5}{4}$

60. $(2 + 3)^{-2} = (5)^{-2} = \dfrac{1}{5^2} = \dfrac{1}{25}$

61. $x^5 x^{-8} = x^{5-8} = x^{-3} = \dfrac{1}{x^3}$

62. $(3x^5)(4x^{-2}) = 12x^{5-2} = 12x^3$

63. $\dfrac{x^{-4}}{x^{-6}} = x^{-4-(-6)} = x^{-4+6} = x^2$

64. $\dfrac{x^{-6}}{x^{-4}} = x^{-6-(-4)} = x^{-6+4} = x^{-2} = \dfrac{1}{x^2}$

65. $\dfrac{24a^5}{3a^{-1}} = 8a^{5-(-1)} = 8a^6$

66. $\dfrac{48n^{-2}}{12n^{-1}} = 4n^{-2-(-1)} =$

$4n^{-2+1} = 4n^{-1} = \dfrac{4}{n}$

67. $\left(x^{-2}y\right)^{-1} = \left(x^{-2}\right)^{-1}(y)^{-1} = x^2 y^{-1} = \dfrac{x^2}{y}$

68. $\left(a^2 b^{-3}\right)^{-2} = (a^2)^{-2}(b^{-3})^{-2}$

$= a^{-4} b^6 = \dfrac{b^6}{a^4}$

69. $(2x)^{-1} = \dfrac{1}{2x}$

70. $(3n^2)^{-2} = 3^{-2}(n^2)^{-2} = 3^{-2}n^{-4} =$

$\dfrac{1}{3^2 n^4} = \dfrac{1}{9n^4}$

71. $\left(2n^{-1}\right)^{-3} = \left(2^{-3}\right)\left(n^{-1}\right)^{-3} =$

$2^{-3}n^3 = \dfrac{n^3}{2^3} = \dfrac{n^3}{8}$

72. $(4ab^{-1})(-3a^{-1}b^2) = -12a^{1-1}b^{-1+2} =$

$-12a^0 b^1 = -12b$

73. $(6.1)(10^2) = 610$

74. $(5.6)(10^4) = 56,000$

75. $(8)(10^{-2}) = 0.08$

76. $(9.2)(10^{-4}) = 0.00092$

77. $9000 = (9)(10^3)$

78. $47 = (4.7)(10^1)$

79. $0.047 = (4.7)(10^{-2})$

80. $0.00021 = (2.1)(10^{-4})$

81. $(0.00004)(12,000) =$
$(4)(10^{-5})(1.2)(10^4) =$
$(4)(1.2)(10^{-5})(10^4) =$
$(4.8)(10^{-1}) = 0.48$

82. $(0.0021)(2000) =$
$(2.1)(10^{-3})(2)(10^3) =$
$(2.1)(2)(10^{-3})(10^3) =$
$(4.2)(10^0) = 4.2$

83. $\dfrac{0.0056}{0.0000028} = \dfrac{(5.6)(10^{-3})}{(2.8)(10^{-6})} =$
$(2)\left(10^{-3-(-6)}\right) = (2)(10^3) = 2000$

84. $\dfrac{0.00078}{39,000} = \dfrac{(7.8)(10^{-4})}{(3.9)(10^4)} =$
$(2)(10^{-4-4}) = (2)(10^{-8}) = 0.00000002$

Chapter 5 Test

1. $(-7x^2 + 6x - 2) + (5x^2 - 8x + 7) =$
$-2x^2 - 2x + 5$

2. $(-4x^2 + 3x + 6) - (-x^2 + 9x - 14) =$
$-4x^2 + 3x + 6 + x^2 - 9x + 14 =$
$-3x^2 - 6x + 20$

3. $3(2x - 1) - 6(3x - 2) - (x + 7) =$
$6x - 3 - 18x + 12 - x - 7 =$
$-13x + 2$

4. $(-4xy^2)(7x^2y^3) =$
$(-4)(7)(x^{1+2})(y^{2+3}) =$
$-28x^3y^5$

5. $(2x^2y)^2(3xy^3) =$
$(4x^4y^2)(3xy^3) =$
$12x^5y^5$

6. $(x - 9)(x + 2) =$
$x^2 + (2 - 9)x - 18 =$
$x^2 - 7x - 18$

7. $(n + 14)(n - 7) =$
$n^2 + (14 - 7)n - 98 =$
$n^2 + 7n - 98$

8. $(5a + 3)(8a + 7) =$
$40a^2 + (35 + 24)a + 21 =$
$40a^2 + 59a + 21$

9. $(3x - 7y)^2 =$
$9x^2 - 2(3x)(7y) + 49y^2 =$
$9x^2 - 42xy + 49y^2$

10. $(x + 3)(2x^2 - 4x - 7) =$
$x(2x^2 - 4x - 7) + 3(2x^2 - 4x - 7) =$
$2x^3 - 4x^2 - 7x + 6x^2 - 12x - 21 =$
$2x^3 + 2x^2 - 19x - 21$

11. $(9x - 5y)(9x + 5y) = 81x^2 - 25y^2$

12. $(3x - 7)(5x - 11) =$
$15x^2 + (-33 - 35)x + 77 =$
$15x^2 - 68x + 77$

13. $\dfrac{-96x^4y^5}{-12x^2y} = 8x^{4-2}y^{5-1} = 8x^2y^4$

14. $\dfrac{56x^2y - 72xy^2}{-8xy} = \dfrac{56x^2y}{-8xy} - \dfrac{72xy^2}{-8xy} =$
$-7x + 9y$

15.
$$
\begin{array}{r}
x^2 + 4x - 5 \\
2x - 3 \overline{\smash{\big)}\, 2x^3 + 5x^2 - 22x + 15} \\
\underline{2x^3 - 3x^2} \\
8x^2 - 22x + 15 \\
\underline{8x^2 - 12x} \\
-10x + 15 \\
\underline{-10x + 15}
\end{array}
$$

16.
$$
\begin{array}{r}
4x^2 - x + 6 \\
x + 6 \overline{\smash{\big)}\, 4x^3 + 23x^2 + 0x + 36} \\
\underline{4x^3 + 24x^2} \\
-1x^2 - 0x + 36 \\
\underline{-1x^2 - 6x} \\
6x + 36 \\
\underline{6x + 36}
\end{array}
$$

17. $\left(\dfrac{2}{3}\right)^{-3} = \left(\dfrac{3}{2}\right)^3 = \dfrac{27}{8}$

18. $4^{-2} + 4^{-1} + 4^0 = \dfrac{1}{4^2} + \dfrac{1}{4} + 1 =$
$\dfrac{1}{16} + \dfrac{1}{4} + 1 = \dfrac{1}{16} + \dfrac{4}{16} + \dfrac{16}{16} = \dfrac{21}{16}$

19. $\dfrac{1}{2^{-4}} = 2^4 = 16$

20. $(-6x^{-4})(4x^2) = -24x^{-2} = -\dfrac{24}{x^2}$

21. $\left(\dfrac{8x^{-1}}{2x^2}\right)^{-1} = (4x^{-1-2})^{-1} =$
$(4x^{-3})^{-1} = 4^{-1}x^3 = \dfrac{x^3}{4}$

126

22. $(x^{-3}y^5)^{-2} = (x^{-3})^{-2}(y^5)^{-2} =$

$x^6 y^{-10} = \dfrac{x^6}{y^{10}}$

23. $0.00027 = (2.7)(10^{-4})$

24. $(9.2)(10^6) = 9,200,000$

25. $(0.000002)(3000) =$
$(2)(10^{-6})(3)(10^3) =$
$(6)(10^{-3}) = 0.006$

CHAPTERS 1-5 | **Cumulative Review**

1. $5 + 3(2-7)^2 \div 3 \cdot 5$
$5 + 3(-5)^2 \div 3 \cdot 5$
$5 + 3(25) \div 3 \cdot 5$
$5 + 75 \div 3 \cdot 5$
$5 + 25 \cdot 5$
$5 + 125$
130

2. $8 \div 2 \cdot (-1) + 3$
$4 \cdot (-1) + 3$
$-4 + 3$
-1

3. $7 - 2^2 \cdot 5 \div (-1)$
$7 - 4 \cdot 5 \div (-1)$
$7 - 20 \div (-1)$
$7 + 20$
27

4. $4 + (-2) - 3(6)$
$4 + (-2) - 18$
$2 - 18$
-16

5. $(-3)^4 = (-3)(-3)(-3)(-3) = 81$

6. $-2^5 = -(2 \cdot 2 \cdot 2 \cdot 2 \cdot 2) = -32$

7. $\left(\dfrac{2}{3}\right)^{-1} = \dfrac{2^{-1}}{3^{-1}} = \dfrac{3^1}{2^1} = \dfrac{3}{2}$

8. $\dfrac{1}{4^{-2}} = 4^2 = 16$

9. $\left(\dfrac{1}{2} - \dfrac{1}{3}\right)^{-2} = \left(\dfrac{3}{6} - \dfrac{2}{6}\right)^{-2} = \left(\dfrac{1}{6}\right)^{-2} =$

$\dfrac{1^{-2}}{6^{-2}} = \dfrac{6^2}{1^2} = \dfrac{36}{1} = 36$

10. $2^0 + 2^{-1} + 2^{-2} = 1 + \dfrac{1}{2} + \dfrac{1}{2^2} =$

$1 + \dfrac{1}{2} + \dfrac{1}{4} = \dfrac{4}{4} + \dfrac{2}{4} + \dfrac{1}{4} = \dfrac{7}{4}$

11. $x = \dfrac{1}{2}, y = -\dfrac{1}{3} : \dfrac{2x + 3y}{x - y} =$

$\dfrac{2\left(\dfrac{1}{2}\right) + 3\left(-\dfrac{1}{3}\right)}{\dfrac{1}{2} - \left(-\dfrac{1}{3}\right)} = \dfrac{1 - 1}{\dfrac{3}{6} + \dfrac{2}{6}} =$

$\dfrac{0}{\dfrac{5}{6}} = 0$

12. $n = -\dfrac{3}{4} : \dfrac{2}{5}n - \dfrac{1}{3}n - n + \dfrac{1}{2}n =$

$\dfrac{12}{30}n - \dfrac{10}{30}n - \dfrac{30}{30}n + \dfrac{15}{30}n =$

$\dfrac{-13}{30}n = -\dfrac{13}{30}\left(-\dfrac{3}{4}\right) = \dfrac{13}{40}$

13. $a = -1, b = -\dfrac{1}{3} : \dfrac{3a - 2b - 4a + 7b}{-a - 3a + b - 2b} =$

$\dfrac{-a + 5b}{-4a - b} = \dfrac{-(-1) + 5\left(-\dfrac{1}{3}\right)}{-4(-1) - \left(-\dfrac{1}{3}\right)} =$

$\dfrac{1 - \dfrac{5}{3}}{4 + \dfrac{1}{3}} = \dfrac{\dfrac{3}{3} - \dfrac{5}{3}}{\dfrac{12}{3} + \dfrac{1}{3}} = \dfrac{-\dfrac{2}{3}}{\dfrac{13}{3}} = -\dfrac{2}{13}$

Chapters 1-5 Cumulative Review

14. $x = -2 : -2(x-4) + 3(2x-1) - (3x-2) =$
$-2x + 8 + 6x - 3 - 3x + 2 =$
$x + 7 = (-2) + 7 = 5$

15. $x = -1 : (x^2 + 2x - 4) - (x^2 - x - 2)$
$\qquad + (2x^2 - 3x - 1) =$
$x^2 + 2x - 4 - x^2 + x + 2 + 2x^2 - 3x - 1 =$
$2x^2 - 3 =$
$2(-1)^2 - 3 = 2(1) - 3 = 2 - 3 = -1$

16. $n = 3 : 2(n^2 - 3n - 1) - (n^2 + n + 4)$
$\qquad - 3(2n-1) =$
$2n^2 - 6n - 2 - n^2 - n - 4 - 6n + 3 =$
$n^2 - 13n - 3 =$
$(3)^2 - 13(3) - 3 = 9 - 39 - 3 = -33$

17. $(3x^2 y^3)(-5xy^4) =$
$(3)(-5)(x^{2+1})(y^{3+4}) =$
$-15x^3 y^7$

18. $(-6ab^4)(-2b^3) =$
$(-6)(-2)(a^1)(b^{4+3}) =$
$12ab^7$

19. $(-2x^2 y^5)^3 =$
$-(2)^3 (x^2)^3 (y^5)^3 =$
$-8x^6 y^{15}$

20. $-3xy(2x - 5y) = -6x^2 y + 15xy^2$

21. $(5x - 2)(3x - 1) =$
$15x^2 + (-5-6)x + 2 =$
$15x^2 - 11x + 2$

22. $(7x - 1)(3x + 4) =$
$21x^2 + (28 - 3)x - 4 =$
$21x^2 + 25x - 4$

23. $(-x - 2)(2x + 3) =$
$-2x^2 + (-3-4)x - 6 =$
$-2x^2 - 7x - 6$

24. $(7 - 2y)(7 + 2y) =$
$49 + (14 - 14)y - 4y^2 =$
$49 - 4y^2$

25. $(x - 2)(3x^2 - x - 4) =$
$x(3x^2 - x - 4) - 2(3x^2 - x - 4) =$
$3x^3 - x^2 - 4x - 6x^2 + 2x + 8 =$
$3x^3 - 7x^2 - 2x + 8$

26. $(2x - 5)(x^2 + x - 4) =$
$2x(x^2 + x - 4) - 5(x^2 + x - 4) =$
$2x^3 + 2x^2 - 8x - 5x^2 - 5x + 20 =$
$2x^3 - 3x^2 - 13x + 20$

27. $(2n + 3)^3 = (2n + 3)(2n + 3)(2n + 3) =$
$(2n + 3)(4n^2 + 12n + 9) =$
$2n(4n^2 + 12n + 9) + 3(4n^2 + 12n + 9) =$
$8n^3 + 24n^2 + 18n + 12n^2 + 36n + 27 =$
$8n^3 + 36n^2 + 54n + 27$

28. $(1 - 2n)^3 = (1 - 2n)(1 - 2n)(1 - 2n) =$
$(1 - 2n)(1 - 4n + 4n^2) =$
$1(1 - 4n + 4n^2) - 2n(1 - 4n + 4n^2) =$
$1 - 4n + 4n^2 - 2n + 8n^2 - 8n^3 =$
$1 - 6n + 12n^2 - 8n^3$

29. $(x^2 - 2x + 6)(2x^2 + 5x - 6) =$
$x^2(2x^2 + 5x - 6) - 2x(2x^2 + 5x - 6)$
$\qquad + 6(2x^2 + 5x - 6) =$
$2x^4 + 5x^3 - 6x^2 - 4x^3 - 10x^2 + 12x$
$\qquad + 12x^2 + 30x - 36 =$
$2x^4 + x^3 - 4x^2 + 42x - 36$

30. $\dfrac{-52x^3 y^4}{13xy^2} = -4x^{3-1} y^{4-2} = -4x^2 y^2$

31. $\dfrac{-126a^3 b^5}{-9a^2 b^3} = 14a^{3-2} b^{5-3}$

$\qquad = 14a^1 b^2 = 14ab^2$

32. $\dfrac{56xy^2 - 64x^3 y - 72x^4 y^4}{8xy} =$

$\dfrac{56xy^2}{8xy} - \dfrac{64x^3 y}{8xy} - \dfrac{72x^4 y^4}{8xy} =$

$7y - 8x^2 - 9x^3 y^3$

33.

$$
\begin{array}{r}
2x^2 - \ 4x - 7 \\
x+3\overline{\smash{\big)}\ 2x^3 + 2x^2 - 19x - 21} \\
\underline{2x^3 + 6x^2} \\
-4x^2 - 19x - 21 \\
\underline{-4x^2 - 12x} \\
-7x - 21 \\
\underline{-7x - 21}
\end{array}
$$

34.

$$
\begin{array}{r}
x^2 + \ 6x + 4 \\
3x-1\overline{\smash{\big)}\ 3x^3 + 17x^2 + 6x - 4} \\
\underline{3x^3 - \ x^2} \\
18x^2 + 6x - 4 \\
\underline{18x^2 - 6x} \\
12x - 4 \\
\underline{12x - 4}
\end{array}
$$

35. $(-2x^3)(3x^{-4}) = -6x^{3-4} =$

$-6x^{-1} = -\dfrac{6}{x}$

36. $\dfrac{4x^{-2}}{2x^{-1}} = 2x^{-2-(-1)} = 2x^{-2+1} =$

$2x^{-1} = \dfrac{2}{x}$

37. $(3x^{-1}y^{-2})^{-1} = 3^{-1}(x^{-1})^{-1}(y^{-2})^{-1} =$

$3^{-1}x^1y^2 = \dfrac{xy^2}{3}$

38. $(xy^2z^{-1})^{-2} = x^{-2}(y^2)^{-2}(z^{-1})^{-2} =$

$x^{-2}y^{-4}z^2 = \dfrac{z^2}{x^2y^4}$

39. $(0.00003)(4000) =$
$(3)(10^{-5})(4)(10^3) =$
$(12)(10^{-5+3}) =$
$(12)(10^{-2}) = 0.12$

40. $(0.0002)(0.003)^2 =$
$(2)(10^{-4})(3)^2(10^{-3})^2 =$
$(2)(9)(10^{-4})(10^{-6}) =$
$(18)(10^{-4-6}) =$
$(18)(10^{-10}) = 0.0000000018$

41. $\dfrac{0.00034}{0.0000017} = \dfrac{(34)(10^{-4})}{(17)(10^{-6})} =$

$(2)(10^{-4-(-6)}) = (2)(10^2) = 200$

42.
$$5x + 8 = 6x - 3$$
$$-x + 8 = -3$$
$$-x = -11$$
$$x = 11$$
The solution set is $\{11\}$.

43.
$$-2(4x - 1) = -5x + 3 - 2x$$
$$-8x + 2 = -5x + 3 - 2x$$
$$-8x + 2 = -7x + 3$$
$$-x + 2 = 3$$
$$-x = 1$$
$$x = -1$$
The solution set is $\{-1\}$.

44.
$$\dfrac{y}{2} - \dfrac{y}{3} = 8$$
$$6\left(\dfrac{y}{2} - \dfrac{y}{3}\right) = 6(8)$$
$$6\left(\dfrac{y}{2}\right) - 6\left(\dfrac{y}{3}\right) = 48$$
$$3y - 2y = 48$$
$$y = 48$$
The solution set is $\{48\}$.

45.
$$6x + 8 - 4x = 10(3x + 2)$$
$$6x + 8 - 4x = 30x + 20$$
$$2x + 8 = 30x + 20$$
$$-28x + 8 = 20$$
$$-28x = 12$$
$$x = \dfrac{12}{-28} = -\dfrac{3}{7}$$
The solution set is $\left\{-\dfrac{3}{7}\right\}$.

Chapters 1-5 Cumulative Review

46. $1.6 - 2.4x = 5x - 65$

$1.6 - 7.4x = -65$

$-7.4x = -66.6$

$x = \dfrac{-66.6}{-7.4} = 9$

The solution set is $\{9\}$.

47. $-3(x-1) + 2(x+3) = -4$

$-3x + 3 + 2x + 6 = -4$

$-x + 9 = -4$

$-x = -13$

$x = 13$

The solution set is $\{13\}$.

48.

$$\dfrac{3n+1}{5} + \dfrac{n-2}{3} = \dfrac{2}{15}$$

$$15\left(\dfrac{3n+1}{5} + \dfrac{n-2}{3}\right) = 15\left(\dfrac{2}{15}\right)$$

$$15\left(\dfrac{3n+1}{5}\right) + 15\left(\dfrac{n-2}{3}\right) = 2$$

$$3(3n+1) + 5(n-2) = 2$$

$$9n + 3 + 5n - 10 = 2$$

$$14n - 7 = 2$$

$$14n = 9$$

$$n = \dfrac{9}{14}$$

The solution set is $\left\{\dfrac{9}{14}\right\}$.

49. $0.06x + 0.08(1500 - x) = 110$

$0.06x + 120 - 0.08x = 110$

$-0.02x + 120 = 110$

$-0.02x = -10$

$x = \dfrac{-10}{-0.02} = 500$

The solution set is $\{500\}$.

50. $2x - 7 \leq -3(x+4)$

$2x - 7 \leq -3x - 12$

$5x - 7 \leq -12$

$5x \leq -5$

$x \leq -1$

The solution set is
$\{x | x \leq -1\}$ or $(-\infty, -1]$.

51. $6x + 5 - 3x > 5$

$3x + 5 > 5$

$3x > 0$

$x > \dfrac{0}{3}$

$x > 0$

The solution set is
$\{x | x > 0\}$ or $(0, \infty)$.

52. $4(x-5) + 2(3x+6) < 0$

$4x - 20 + 6x + 12 < 0$

$10x - 8 < 0$

$10x < 8$

$x < \dfrac{8}{10}$

$x < \dfrac{4}{5}$

The solution set is

$\left\{x | x < \dfrac{4}{5}\right\}$ or $\left(-\infty, \dfrac{4}{5}\right)$.

53. $-5x + 3 > -4x + 5$

$-x + 3 > 5$

$-x > 2$

$\dfrac{-x}{-1} < \dfrac{2}{-1}$

$x < -2$

The solution set is
$\{x | x < -2\}$ or $(-\infty, -2)$.

54.

$$\frac{3x}{4} - \frac{x}{2} \leq \frac{5x}{6} - 1$$

$$12\left(\frac{3x}{4} - \frac{x}{2}\right) \leq 12\left(\frac{5x}{6} - 1\right)$$

$$12\left(\frac{3x}{4}\right) - 12\left(\frac{x}{2}\right) \leq 12\left(\frac{5x}{6}\right) - 12(1)$$

$$3(3x) - 6(x) \leq 2(5x) - 12$$

$$9x - 6x \leq 10x - 12$$

$$3x \leq 10x - 12$$

$$-7x \leq -12$$

$$\frac{-7x}{-7} \geq \frac{-12}{-7}$$

$$x \geq \frac{12}{7}$$

The solution set is

$$\left\{x \mid x \geq \frac{12}{7}\right\} \text{ or } \left[\frac{12}{7}, \infty\right).$$

55.
$$0.08(700 - x) + 0.11x \geq 65$$
$$56 - 0.08x + 0.11x \geq 65$$
$$56 + 0.03x \geq 65$$
$$0.03 \geq 9$$
$$x \geq \frac{9}{0.03}$$
$$x \geq 300$$

The solution set is
$\{x \mid x \geq 300\}$ or $[300, \infty)$.

56. Let n represent the number.
$$4 + 3n = n + 10$$
$$4 + 2n = 10$$
$$2n = 6$$
$$n = 3$$
The number is 3.

57. Let n represent the number.
$$(15\%)(n) = 6$$
$$0.15n = 6$$
$$15n = 600$$
$$n = 40$$
Therefore, 15% of 40 = 6.

58. Let d represent the number of dimes, then $18 - d$ represents the number of quarters. The value of the dimes in cents is 10d and the value of the quarters in cents is $25(18 - d)$.

$$10d + 25(18 - d) = 330$$
$$10d + 450 - 25d = 330$$
$$-15d + 450 = 330$$
$$-15d = -120$$
$$d = 8$$

He would have 8 dimes and 10 quarters.

59. Let x represent the amount of money invested at 8% interest; then $1500 - x$ represents the amount of money invested at 9%.

$$\begin{pmatrix} \text{Interest} \\ \text{earned} \\ \text{at} \\ 8\% \end{pmatrix} + \begin{pmatrix} \text{Interest} \\ \text{earned} \\ \text{at} \\ 9\% \end{pmatrix} = \begin{pmatrix} \text{Total} \\ \text{interest} \\ \text{earned} \end{pmatrix}$$

$$(8\%)(x) + (9\%)(1500 - x) = 128$$
$$0.08x + 0.09(1500 - x) = 128$$
$$8x + 9(1500 - x) = 12{,}800$$
$$8x + 13{,}500 - 9x = 12{,}800$$
$$13{,}500 - x = 12{,}800$$
$$-x = -700$$
$$x = 700$$

The amounts invested are $700 at 8% and $800 at 9%.

60. Let x represent the amount of water to be added, then $15 + x$ represents the amount of final salt solution.

$$\begin{pmatrix} \text{Pure} \\ \text{salt in} \\ \text{water} \\ \text{added} \end{pmatrix} + \begin{pmatrix} \text{Pure} \\ \text{salt} \\ \text{in 12\%} \\ \text{solution} \end{pmatrix} = \begin{pmatrix} \text{Pure} \\ \text{salt} \\ \text{in final} \\ \text{solution} \end{pmatrix}$$

$$0 + (12\%)(15) = (10\%)(15 + x)$$
$$0 + 0.12(15) = 0.10(15 + x)$$
$$12(15) = 10(15 + x)$$
$$180 = 150 + 10x$$
$$30 = 10x$$
$$3 = x$$

The amount of water to be added would be 3 gallons.

61. Let t represent the time for both airplanes as they left Atlanta at the same time. A chart of the information would be as follows:

	Rate	Time	Distance ($d = rt$)
Plane A	400	t	$400t$
Plane B	450	t	$450t$

A diagram of the problem would be as follows:

Plane A   **Plane B**

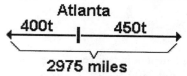

Atlanta

400t   **450t**

2975 miles

The sum of the two distances would be 2975 miles.

$$400t + 450t = 2975$$
$$850t = 2975$$
$$t = \frac{2975}{850} = 3\frac{1}{2} \text{ hours}$$

It would take $3\frac{1}{2}$ hours for the airplanes to be 2975 miles apart.

62. Let w represent the width of the rectangle, then $2w + 1$ represents the length of the rectangle.

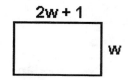

$$2(w) + 2(2w + 1) = 44$$
$$2w + 4w + 2 = 44$$
$$6w + 2 = 44$$
$$6w = 42$$
$$w = 7$$

The width of the rectangle is 7 meters and the length is 15 meters.

Chapter 6 Factoring and Solving Equations

PROBLEM SET **6.1** Factoring by Using the Distributive Property

1. $24y = 2 \cdot 2 \cdot 2 \cdot 3 \cdot y$
$30xy = 2 \cdot 3 \cdot 5 \cdot x \cdot y$
The greatest common factor is
$2 \cdot 3 \cdot y = 6y$.

3. $60x^2y = 2 \cdot 2 \cdot 3 \cdot 5 \cdot x \cdot x \cdot y$
$84xy^2 = 2 \cdot 2 \cdot 3 \cdot 7 \cdot x \cdot y \cdot y$
The greatest common factor is
$2 \cdot 2 \cdot 3 \cdot x \cdot y = 12xy$.

5. $42ab^3 = 2 \cdot 3 \cdot 7 \cdot a \cdot b \cdot b \cdot b$
$70a^2b^2 = 2 \cdot 5 \cdot 7 \cdot a \cdot a \cdot b \cdot b$
The greatest common factor is
$2 \cdot 7 \cdot a \cdot b \cdot b = 14ab^2$.

7. $6x^3 = 2 \cdot 3 \cdot x \cdot x \cdot x$
$8x = 2 \cdot 2 \cdot 2 \cdot x$
$24x^2 = 2 \cdot 2 \cdot 2 \cdot 3 \cdot x \cdot x$
The greatest common factor is
$2 \cdot x = 2x$.

9. $16a^2b^2 = 2 \cdot 2 \cdot 2 \cdot 2 \cdot a \cdot a \cdot b \cdot b$
$40a^2b^3 = 2 \cdot 2 \cdot 2 \cdot 5 \cdot a \cdot a \cdot b \cdot b \cdot b$
$56a^3b^4 = 2 \cdot 2 \cdot 2 \cdot 7 \cdot a \cdot a \cdot a \cdot b \cdot b \cdot b \cdot b$
The greatest common factor is
$2 \cdot 2 \cdot 2 \cdot a \cdot a \cdot b \cdot b = 8a^2b^2$.

11. $8x + 12y =$
$4(2x) + 4(3y) =$
$4(2x + 3y)$

13. $14xy - 21y =$
$7y(2x) - 7y(3) =$
$7y(2x - 3)$

15. $18x^2 + 45x =$
$9x(2x) + 9x(5) =$
$9x(2x + 5)$

17. $12xy^2 - 30x^2y =$
$6xy(2y) - 6xy(5x) =$

$6xy(2y - 5x)$

19. $36a^2b - 60a^3b^4 =$
$12a^2b(3) - 12a^2b(5ab^3) =$
$12a^2b(3 - 5ab^3)$

21. $16xy^3 + 25x^2y^2 =$
$xy^2(16y) + xy^2(25x) =$
$xy^2(16y + 25x)$

23. $64ab - 72cd =$
$8(8ab) - 8(9cd) =$
$8(8ab - 9cd)$

25. $9a^2b^4 - 27a^2b =$
$9a^2b(b^3) - 9a^2b(3) =$
$9a^2b(b^3 - 3)$

27. $52x^4y^2 + 60x^6y =$
$4x^4y(13y) + 4x^4y(15x^2) =$
$4x^4y(13y + 15x^2)$

29. $40x^2y^2 + 8x^2y =$
$8x^2y(5y) + 8x^2y(1) =$
$8x^2y(5y + 1)$

31. $12x + 15xy + 21x^2 =$
$3x(4) + 3x(5y) + 3x(7x) =$
$3x(4 + 5y + 7x)$

33. $2x^3 - 3x^2 + 4x =$
$x(2x^2) - x(3x) + x(4) =$
$x(2x^2 - 3x + 4)$

35. $44y^5 - 24y^3 - 20y^2 =$
$4y^2(11y^3) - 4y^2(6y) - 4y^2(5) =$
$4y^2(11y^3 - 6y - 5)$

37. $14a^2b^3 + 35ab^2 - 49a^3b =$
$7ab(2ab^2) + 7ab(5b) - 7ab(7a^2) =$
$7ab(2ab^2 + 5b - 7a^2)$

Problem Set 6.1

39. $x(y+1) + z(y+1) =$
$(y+1)(x+z)$

41. $a(b-4) - c(b-4) =$
$(b-4)(a-c)$

43. $x(x+3) + 6(x+3) =$
$(x+3)(x+6)$

45. $2x(x+1) - 3(x+1) =$
$(x+1)(2x-3)$

47. $5x + 5y + bx + by =$
$5(x+y) + b(x+y) =$
$(x+y)(5+b)$

49. $bx - by - cx + cy =$
$b(x-y) - c(x-y) =$
$(x-y)(b-c)$

51. $ac + bc + a + b =$
$c(a+b) + 1(a+b) =$
$(a+b)(c+1)$

53. $x^2 + 5x + 12x + 60 =$
$x(x+5) + 12(x+5) =$
$(x+5)(x+12)$

55. $x^2 - 2x - 8x + 16 =$
$x(x-2) - 8(x-2) =$
$(x-2)(x-8)$

57. $2x^2 + x - 10x - 5 =$
$x(2x+1) - 5(2x+1) =$
$(2x+1)(x-5)$

59. $6n^2 - 3n - 8n + 4 =$
$3n(2n-1) - 4(2n-1) =$
$(2n-1)(3n-4)$

61. $x^2 - 8x = 0$
$x(x-8) = 0$
$x = 0$ or $x - 8 = 0$
$x = 0$ or $x = 8$
The solution set is $\{0, 8\}$.

63. $x^2 + x = 0$
$x(x+1) = 0$

$x = 0$ or $x + 1 = 0$
$x = 0$ or $x = -1$
The solution set is $\{-1, 0\}$.

65. $n^2 = 5n$
$n^2 - 5n = 0$
$n(n-5) = 0$
$n = 0$ or $n - 5 = 0$
$n = 0$ or $n = 5$
The solution set is $\{0, 5\}$.

67. $2y^2 - 3y = 0$
$y(2y-3) = 0$
$y = 0$ or $2y - 3 = 0$
$y = 0$ or $2y = 3$
$y = 0$ or $y = \dfrac{3}{2}$
The solution set is $\left\{0, \dfrac{3}{2}\right\}$.

69. $7x^2 = -3x$
$7x^2 + 3x = 0$
$x(7x+3) = 0$
$x = 0$ or $7x + 3 = 0$
$x = 0$ or $7x = -3$
$x = 0$ or $x = -\dfrac{3}{7}$
The solution set is $\left\{-\dfrac{3}{7}, 0\right\}$.

71. $3n^2 + 15n = 0$
$3n(n+5) = 0$
$3n = 0$ or $n + 5 = 0$
$n = 0$ or $n = -5$
The solution set is $\{-5, 0\}$.

73. $4x^2 = 6x$
$2x^2 = 3x$
$2x^2 - 3x = 0$
$x(2x-3) = 0$
$x = 0$ or $2x - 3 = 0$
$x = 0$ or $2x = 3$
$x = 0$ or $x = \dfrac{3}{2}$
The solution set is $\left\{0, \dfrac{3}{2}\right\}$.

75. $7x - x^2 = 0$
$x(7 - x) = 0$
$x = 0$ or $7 - x = 0$
$x = 0$ or $7 = x$
The solution set is $\{0,\ 7\}$.

77. $13x = x^2$
$13x - x^2 = 0$
$x(13 - x) = 0$
$x = 0$ or $13 - x = 0$
$x = 0$ or $13 = x$
The solution set is $\{0,\ 13\}$.

79. $5x = -2x^2$
$2x^2 + 5x = 0$
$x(2x + 5) = 0$
$x = 0$ or $2x + 5 = 0$
$x = 0$ or $2x = -5$
$x = 0$ or $x = -\dfrac{5}{2}$
The solution set is $\left\{ -\dfrac{5}{2},\ 0 \right\}$.

81. $x(x + 5) - 4(x + 5) = 0$
$(x + 5)(x - 4) = 0$
$x + 5 = 0$ or $x - 4 = 0$
$x = -5$ or $x = 4$
The solution set is $\{-5,\ 4\}$.

83. $4(x - 6) - x(x - 6) = 0$
$(x - 6)(4 - x) = 0$
$x - 6 = 0$ or $4 - x = 0$
$x = 6$ or $4 = x$
The solution set is $\{4,\ 6\}$.

85. Let n represent the number, then n^2 represents the square of the number.
$n^2 = 9n$
$n^2 - 9n = 0$
$n(n - 9) = 0$
$n = 0$ or $n - 9 = 0$
$n = 0$ or $n = 9$
The number is 0 or 9.

87. Let s represent the length of a side of the square.

$\begin{aligned} \text{Area} &= (5)(\text{Perimeter}) \\ s^2 &= 5(4s) \\ s^2 &= 20s \\ s^2 - 20s &= 0 \\ s(s - 20) &= 0 \end{aligned}$
$s = 0$ or $s - 20 = 0$
$s = 0$ or $s = 20$
Since 0 is not a reasonable answer to the problem, the length of the side of the square must be 20 units.

89. Let s represent the length of a side of the square, then s also represents the length of a radius of the circle.

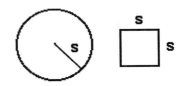

Circle Area = Square Perimeter
$\begin{aligned} \pi s^2 &= 4s \\ \pi s^2 - 4s &= 0 \end{aligned}$
$s(\pi s - 4) = 0$
$s = 0$ or $\pi s - 4 = 0$
$s = 0$ or $\pi s = 4$
$s = 0$ or $s = \dfrac{4}{\pi}$
The answer of 0 must be discarded, so the length of a side of the square is $\dfrac{4}{\pi}$ units.

Problem Set 6.1

91. Let w represent the width of the rectangle, then w also represents the length of a side of the square.

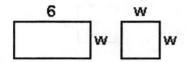

Rectangle Area $= (2)$(Square Area)
$$6w = 2\left(w^2\right)$$
$$6w - 2w^2 = 0$$
$$2w(3 - w) = 0$$
$$2w = 0 \quad \text{or} \quad 3 - w = 0$$
$$w = 0 \quad \text{or} \quad 3 = w$$
The answer of 0 must be discarded, so the dimensions of the rectangle are 3 inches by 6 inches and the dimensions of the square are 3 inches by 3 inches.

95. $A = P(1 + rt)$
a. $P = \$100$, $r = 8\%$, $t = 2$:
$$A = 100[1 + 0.08(2)] =$$
$$100(1 + 0.16) =$$
$$100(1.16) = \$116$$

b. $P = \$200$, $r = 9\%$, $t = 3$:
$$A = 200[1 + 0.09(3)] =$$
$$200(1 + 0.27) =$$
$$200(1.27) = \$254$$

c. $P = \$500$, $r = 10\%$, $t = 5$:
$$A = 500[1 + 0.10(5)] =$$
$$500(1 + 0.50) =$$
$$500(1.5) = \$750$$

d. $P = \$1000$, $r = 10\%$, $t = 10$:
$$A = 1000[1 + 0.10(10)] =$$
$$1000(1 + 1.00) =$$
$$1000(2) = \$2000$$

97. $b^2x^2 - cx = 0$
$$x\left(b^2x - c\right) = 0$$
$$x = 0 \quad \text{or} \quad b^2x - c = 0$$
$$x = 0 \quad \text{or} \quad b^2x = c$$
$$x = 0 \quad \text{or} \quad x = \frac{c}{b^2}$$

99. $y + ay - by - c = 0$
$$y + ay - by = c$$
$$y(1 + a - b) = c$$
$$y = \frac{c}{1 + a - b}$$

PROBLEM SET **6.2** **Factoring the Difference of Two Squares**

1. $x^2 - 1 =$
$(x)^2 - (1)^2 =$
$(x - 1)(x + 1)$

3. $x^2 - 100 =$
$(x)^2 - (10)^2 =$
$(x - 10)(x + 10)$

5. $x^2 - 4y^2 =$
$(x)^2 - (2y)^2 =$
$(x - 2y)(x + 2y)$

7. $9x^2 - y^2 =$
$(3x)^2 - (y)^2 =$
$(3x - y)(3x + y)$

9. $36a^2 - 25b^2 =$
$(6a)^2 - (5b)^2 =$
$(6a - 5b)(6a + 5b)$

11. $1 - 4n^2 =$
$(1)^2 - (2n)^2 =$
$(1 - 2n)(1 + 2n)$

13. $5x^2 - 20 =$
$5(x^2 - 4) =$
$5(x - 2)(x + 2)$

15. $8x^2 + 32 = 8(x^2 + 4)$

17. $2x^2 - 18y^2 =$
$2(x^2 - 9y^2) =$
$2(x - 3y)(x + 3y)$

19. $x^3 - 25x =$
$x(x^2 - 25) = x(x - 5)(x + 5)$

21. $x^2 + 9y^2$ is not factorable.

23. $45x^2 - 36xy = 9x(5x - 4y)$

25. $36 - 4x^2 =$
$4(9 - x^2) =$
$4(3 - x)(3 + x)$

27. $4a^4 + 16a^2 = 4a^2(a^2 + 4)$

29. $x^4 - 81 =$
$(x^2 + 9)(x^2 - 9) =$
$(x^2 + 9)(x + 3)(x - 3)$

31. $x^4 + x^2 = x^2(x^2 + 1)$

33. $3x^3 + 48x = 3x(x^2 + 16)$

35. $5x - 20x^3 =$
$5x(1 - 4x^2) =$
$5x(1 - 2x)(1 + 2x)$

37. $4x^2 - 64 =$
$4(x^2 - 16) =$
$4(x - 4)(x + 4)$

39. $75x^3y - 12xy^3 =$
$3xy(25x^2 - 4y^2) =$
$3xy(5x + 2y)(5x - 2y)$

41. $$x^2 = 9$$
$$x^2 - 9 = 0$$
$$(x - 3)(x + 3) = 0$$
$$x - 3 = 0 \quad \text{or} \quad x + 3 = 0$$
$$x = 3 \quad \text{or} \quad x = -3$$
The solution set is $\{-3, 3\}$.

43. $$4 = n^2$$
$$4 - n^2 = 0$$
$$(2 - n)(2 + n) = 0$$
$$2 - n = 0 \quad \text{or} \quad 2 + n = 0$$
$$2 = n \quad \text{or} \quad n = -2$$
The solution set is $\{-2, 2\}$.

45. $$9x^2 = 16$$
$$9x^2 - 16 = 0$$
$$(3x - 4)(3x + 4) = 0$$
$$3x - 4 = 0 \quad \text{or} \quad 3x + 4 = 0$$
$$3x = 4 \quad \text{or} \quad 3x = -4$$
$$x = \frac{4}{3} \quad \text{or} \quad x = -\frac{4}{3}$$
The solution set is $\left\{-\frac{4}{3}, \frac{4}{3}\right\}$.

47. $$n^2 - 121 = 0$$
$$(n - 11)(n + 11) = 0$$
$$n - 11 = 0 \quad \text{or} \quad n + 11 = 0$$
$$n = 11 \quad \text{or} \quad n = -11$$
The solution set is $\{-11, 11\}$.

49. $$25x^2 = 4$$
$$25x^2 - 4 = 0$$
$$(5x - 2)(5x + 2) = 0$$
$$5x - 2 = 0 \quad \text{or} \quad 5x + 2 = 0$$
$$5x = 2 \quad \text{or} \quad 5x = -2$$
$$x = \frac{2}{5} \quad \text{or} \quad x = -\frac{2}{5}$$
The solution set is $\left\{-\frac{2}{5}, \frac{2}{5}\right\}$.

51. $$3x^2 = 75$$
Divide both sides by 3.
$$x^2 = 25$$
$$x^2 - 25 = 0$$
$$(x - 5)(x + 5) = 0$$
$$x - 5 = 0 \quad \text{or} \quad x + 5 = 0$$
$$x = 5 \quad \text{or} \quad x = -5$$
The solution set is $\{-5, 5\}$.

Problem Set 6.2

53. $3x^3 - 48x = 0$
Divide both sides by 3.
$$x^3 - 16x = 0$$
$$x(x^2 - 16) = 0$$
$$x(x - 4)(x + 4) = 0$$
$x = 0$ or $x - 4 = 0$ or $x + 4 = 0$
$x = 0$ or $x = 4$ or $x = -4$
The solution set is $\{-4, 0, 4\}$.

55. $n^3 = 16n$
$$n^3 - 16n = 0$$
$$n(n^2 - 16) = 0$$
$$n(n - 4)(n + 4) = 0$$
$n = 0$ or $n - 4 = 0$ or $n + 4 = 0$
$n = 0$ or $n = 4$ or $n = -4$
The solution set is $\{-4, 0, 4\}$.

57. $5 - 45x^2 = 0$
Divide by 5.
$$1 - 9x^2 = 0$$
$$(1 - 3x)(1 + 3x) = 0$$
$1 - 3x = 0$ or $1 + 3x = 0$
$1 = 3x$ or $3x = -1$
$x = \dfrac{1}{3}$ or $x = -\dfrac{1}{3}$
The solution set is $\left\{-\dfrac{1}{3}, \dfrac{1}{3}\right\}$.

59. $4x^3 - 400x = 0$
Divide by 4.
$$x^3 - 100x = 0$$
$$x(x^2 - 100) = 0$$
$$x(x - 10)(x + 10) = 0$$
$x = 0$ or $x - 10 = 0$ or $x + 10 = 0$
$x = 0$ or $x = 10$ or $x = -10$
The solution set is $\{-10, 0, 10\}$.

61. $64x^2 = 81$
$$64x^2 - 81 = 0$$
$$(8x - 9)(8x + 9) = 0$$

$8x - 9 = 0$ or $8x + 9 = 0$
$8x = 9$ or $8x = -9$
$x = \dfrac{9}{8}$ or $x = -\dfrac{9}{8}$
The solution set is $\left\{-\dfrac{9}{8}, \dfrac{9}{8}\right\}$.

63. $36x^3 = 9x$
Divide both sides by 9.
$$4x^3 = x$$
$$4x^3 - x = 0$$
$$x(4x^2 - 1) = 0$$
$$x(2x - 1)(2x + 1) = 0$$
$x = 0$ or $2x - 1 = 0$ or $2x + 1 = 0$
$x = 0$ or $2x = 1$ or $2x = -1$
$x = 0$ or $x = \dfrac{1}{2}$ or $x = -\dfrac{1}{2}$
The solution set is $\left\{-\dfrac{1}{2}, 0, \dfrac{1}{2}\right\}$.

65. Let n represent the number.
$$n^2 - 49 = 0$$
$$(n - 7)(n + 7) = 0$$
$n - 7 = 0$ or $n + 7 = 0$
$n = 7$ or $n = -7$
The number could be either -7 or 7.

67. Let n represent the number.
$$5n^3 = 80n$$
Divide both sides by 5.
$$n^3 = 16n$$
$$n^3 - 16n = 0$$
$$n(n^2 - 16) = 0$$
$$n(n - 4)(n + 4) = 0$$
$n = 0$ or $n - 4 = 0$ or $n + 4 = 0$
$n = 0$ or $n = 4$ or $n = -4$
The number could be -4, 0 or 4.

138

69. Let x represent the length of a side of the smaller square, then $5x$ represents the length of a side of the larger square.

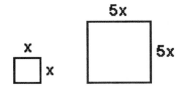

Larger Area + Smaller Area = 234
$$(5x)^2 + x^2 = 234$$
$$25x^2 + x^2 = 234$$
$$26x^2 = 234$$
$$x^2 = 9$$
$$x^2 - 9 = 0$$
$$(x - 3)(x + 3) = 0$$
$$x - 3 = 0 \quad \text{or} \quad x + 3 = 0$$
$$x = 3 \quad \text{or} \quad x = -3$$

The solution -3 must be discarded. Thus, the length of a side of the smaller square is 3 inches and the length of a side of the larger square is 15 inches.

71. Let w represent the width of the rectangle, then $2\frac{1}{2}w = \frac{5}{2}w$ represents the length of the rectangle.

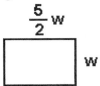

Rectangular Area = 160
$$w\left(\frac{5}{2}w\right) = 160$$
$$\frac{5}{2}w^2 = 160$$
$$w^2 = 64$$
$$w^2 - 64 = 0$$
$$(w - 8)(w + 8) = 0$$
$$w - 8 = 0 \quad \text{or} \quad w + 8 = 0$$
$$w = 8 \quad \text{or} \quad w = -8$$

The solution of -8 must be discarded. Thus, the width of the rectangle is 8 centimeters and the length is

$$\frac{5}{2}(8) = 20 \text{ centimeters.}$$

73. Let r represent the length of a radius of the smaller circle, then $2r$ represents the length of a radius of the larger circle.

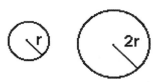

Smaller Area + Larger Area = 80π
$$\pi r^2 + \pi(2r)^2 = 80\pi$$
$$\pi r^2 + 4\pi r^2 = 80\pi$$
$$5\pi r^2 = 80\pi$$
$$r^2 = 16$$
$$r^2 - 16 = 0$$
$$(r - 4)(r + 4) = 0$$
$$r - 4 = 0 \quad \text{or} \quad r + 4 = 0$$
$$r = 4 \quad \text{or} \quad r = -4 \left(\begin{array}{c}\text{Discard this}\\\text{solution.}\end{array}\right)$$

The length of a radius of the smaller circle is 4 meters and the length of a radius of the larger circle is 8 meters.

75. Let x represent a radius of the base, then x also represents the altitude of the cylinder.

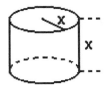

Surface Area = 100π
$$2\pi x^2 + 2\pi x(x) = 100\pi$$
$$2\pi x^2 + 2\pi x^2 = 100\pi$$
$$4\pi x^2 = 100\pi$$
$$x^2 = 25$$
$$x^2 - 25 = 0$$
$$(x - 5)(x + 5) = 0$$
$$x - 5 = 0 \quad \text{or} \quad x + 5 = 0$$
$$x = 5 \quad \text{or} \quad x = -5 \left(\begin{array}{c}\text{Discard this}\\\text{solution.}\end{array}\right)$$

The length of a radius is 5 centimeters.

Problem Set 6.2

81. $x^3 - 8 =$
$(x)^3 - (2)^3 =$
$(x-2)(x^2+2x+4)$

83. $n^3 + 64 =$
$(n)^3 + (4)^3 =$
$(n+4)(n^2-4n+16)$

85. $27a^3 - 64b^3 =$
$(3a)^3 - (4b)^3 =$
$(3a-4b)(9a^2+12ab+16b^2)$

87. $1 + 27a^3 =$
$(1)^3 + (3a)^3 =$
$(1+3a)(1-3a+9a^2)$

89. $8x^3 - y^3 =$
$(2x)^3 - (y)^3 =$
$(2x-y)(4x^2+2xy+y^2)$

91. $27x^3 - 8y^3 =$
$(3x)^3 - (2y)^3 =$
$(3x-2y)(9x^2+6xy+4y^2)$

93. $125x^3 + 8y^3 =$
$(5x)^3 + (2y)^3 =$
$(5x+2y)(25x^2-10xy+4y^2)$

95. $64 + x^3 =$
$(4)^3 + (x)^3 =$
$(4+x)(16-4x+x^2)$

PROBLEM SET **6.3** **Factoring Trinomials of the Form $x^2 + bx + c$**

1. We need two integers whose product is 24 and whose sum is 10. They are 4 and 6.
$x^2 + 10x + 24 = (x+4)(x+6)$

3. We need two integers whose product is 40 and whose sum is 13. They are 5 and 8.
$x^2 + 13x + 40 = (x+5)(x+8)$

5. We need two integers whose product is 18 and whose sum is -11. They are -2 and -9.
$x^2 - 11x + 18 = (x-2)(x-9)$

7. We need two integers whose product is 28 and whose sum is -11. They are -4 and -7.
$n^2 - 11n + 28 = (n-4)(n-7)$

9. We need two integers whose product is -27 and whose sum is 6. They are 9 and -3.
$n^2 + 6n - 27 = (n-3)(n+9)$

11. We need two integers whose product is -40 and whose sum is -6. They are 4 and -10.
$n^2 - 6n - 40 = (n+4)(n-10)$

13. We need two integers whose product is 24 and whose sum is 12. No such integers exist, therefore $t^2 + 12t + 24$ is not factorable.

15. We need two integers whose product is 72 and whose sum is -18. They are -6 and -12.
$x^2 - 18x + 72 = (x-6)(x-12)$

17. We need two integers whose product is -66 and whose sum is 5. They are -6 and 11.
$x^2 + 5x - 66 = (x-6)(x+11)$

19. We need two integers whose product is -72 and whose sum is -1. They are 8 and -9.
$y^2 - y - 72 = (y+8)(y-9)$

21. We need two integers whose product is 80 and whose sum is 21. They are 5 and 16.
$x^2 + 21x + 80 = (x+5)(x+16)$

23. We need two integers whose product is -72 and whose sum is 6. They are -6 and 12.
$x^2 + 6x - 72 = (x-6)(x+12)$

25. We need two integers whose product is -48 and whose sum is -10. No such integers exist, therefore $x^2 - 10x - 48$ is not factorable.

27. We need two integers whose product is -10 and whose sum is 3. They are -2 and 5.
$x^2 + 3xy - 10y^2 = (x - 2y)(x + 5y)$

29. We need two integers whose product is -32 and whose sum is -4. They are 4 and -8.
$a^2 - 4ab - 32b^2 = (a + 4b)(a - 8b)$

31. $x^2 + 10x + 21 = 0$
$(x + 3)(x + 7) = 0$
$x + 3 = 0$ or $x + 7 = 0$
$x = -3$ or $x = -7$
The solution set is $\{-7, -3\}$.

33. $x^2 - 9x + 18 = 0$
$(x - 3)(x - 6) = 0$
$x - 3 = 0$ or $x - 6 = 0$
$x = 3$ or $x = 6$
The solution set is $\{3, 6\}$.

35. $x^2 - 3x - 10 = 0$
$(x - 5)(x + 2) = 0$
$x - 5 = 0$ or $x + 2 = 0$
$x = 5$ or $x = -2$
The solution set is $\{-2, 5\}$.

37. $n^2 + 5n - 36 = 0$
$(n + 9)(n - 4) = 0$
$n + 9 = 0$ or $n - 4 = 0$
$n = -9$ or $n = 4$
The solution set is $\{-9, 4\}$.

39. $n^2 - 6n - 40 = 0$
$(n - 10)(n + 4) = 0$
$n - 10 = 0$ or $n + 4 = 0$
$n = 10$ or $n = -4$
The solution set is $\{-4, 10\}$.

41. $t^2 + t - 56 = 0$
$(t + 8)(t - 7) = 0$

$t + 8 = 0$ or $t - 7 = 0$
$t = -8$ or $t = 7$
The solution set is $\{-8, 7\}$.

43. $x^2 - 16x + 28 = 0$
$(x - 14)(x - 2) = 0$
$x - 14 = 0$ or $x - 2 = 0$
$x = 14$ or $x = 2$
The solution set is $\{2, 14\}$.

45. $x^2 + 11x = 12$
$x^2 + 11x - 12 = 0$
$(x + 12)(x - 1) = 0$
$x + 12 = 0$ or $x - 1 = 0$
$x = -12$ or $x = 1$
The solution set is $\{-12, 1\}$.

47. $x(x - 10) = -16$
$x^2 - 10x = -16$
$x^2 - 10x + 16 = 0$
$(x - 8)(x - 2) = 0$
$x - 8 = 0$ or $x - 2 = 0$
$x = 8$ or $x = 2$
The solution set is $\{2, 8\}$.

49. $-x^2 - 2x + 24 = 0$
Multiply both sides by -1.
$x^2 + 2x - 24 = 0$
$(x + 6)(x - 4) = 0$
$x + 6 = 0$ or $x - 4 = 0$
$x = -6$ or $x = 4$
The solution set is $\{-6, 4\}$.

51. Let n represent one integer, then $n + 1$ represents the next integer.
$n(n + 1) = 56$
$n^2 + n = 56$
$n^2 + n - 56 = 0$
$(n + 8)(n - 7) = 0$
$n + 8 = 0$ or $n - 7 = 0$
$n = -8$ or $n = 7$
If $n = -8$, then $n + 1 = -7$. If $n = 7$, then $n + 1 = 8$. Thus, the consecutive integers are either -8 and -7 or 7 and 8.

Problem Set 6.3

53. Let n represent one integer, then $n + 2$ represents the next consecutive even integer.

$$n(n + 2) = 168$$
$$n^2 + 2n = 168$$
$$n^2 + 2n - 168 = 0$$
$$(n + 14)(n - 12) = 0$$
$$n + 14 = 0 \quad \text{or} \quad n - 12 = 0$$
$$n = -14 \quad \text{or} \quad n = 12$$
$$n + 2 = -12 \quad \text{or} \quad n + 2 = 14$$

Since -14 and -12 are not whole numbers, the consecutive even whole numbers would be 12 and 14.

55. Let n represent the first integer, then $n + 1, n + 2,$ and $n + 3$ represent the other three consecutive integers.

$$(n + 2)(n + 3) = 2[n(n + 1)] - 22$$
$$n^2 + 5n + 6 = 2(n^2 + n) - 22$$
$$n^2 + 5n + 6 = 2n^2 + 2n - 22$$
$$-n^2 + 3n + 28 = 0$$
$$n^2 - 3n - 28 = 0$$
$$(n - 7)(n + 4) = 0$$
$$n - 7 = 0 \quad \text{or} \quad n + 4 = 0$$
$$n = 7 \quad \text{or} \quad n = -4$$

If the first integer is -4, then the four consecutive integers are $-4, -3, -2,$ and -1. If the first integer is 7, then the four consecutive integers are 7, 8, 9, and 10.

57. Let n represent the larger number, then $n - 3$ is the smaller number.

$$n^2 = 10(n - 3) + 9$$
$$n^2 = 10n - 30 + 9$$
$$n^2 = 10n - 21$$
$$n^2 - 10n + 21 = 0$$
$$(n - 7)(n - 3) = 0$$
$$n - 7 = 0 \quad \text{or} \quad n - 3 = 0$$
$$n = 7 \quad \text{or} \quad n = 3$$

If $n = 7$, then $n - 3 = 4$. If $n = 3$, then $n - 3 = 0$. Thus the numbers are 4 and 7 or 0 and 3.

59. Let l represent the length of the rectangle, then $l - 3$ represents the width of the rectangle.

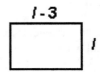

$$\text{Area} = (2)\text{Perimeter} - 6$$
$$l(l - 3) = 2[2l + 2(l - 3)] - 6$$
$$l^2 - 3l = 2[2l + 2l - 6] - 6$$
$$l^2 - 3l = 2[4l - 6] - 6$$
$$l^2 - 3l = 8l - 12 - 6$$
$$l^2 - 3l = 8l - 18$$
$$l^2 - 11l + 18 = 0$$
$$(l - 2)(l - 9) = 0$$
$$l - 2 = 0 \quad \text{or} \quad l - 9 = 0$$
$$l = 2 \quad \text{or} \quad l = 9$$

If $l = 2$, then $l - 3 = -1$ which is not possible. If $l = 9$, then $l - 3 = 6$. Therefore, the rectangle is 9 inches by 6 inches.

61. Let w represent the width of the rectangle, then one-half the perimeter less the width, $15 - w$, represents the length of the rectangle.

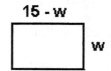

$$\text{Area} = 54$$
$$w(15 - w) = 54$$
$$15w - w^2 = 54$$
$$-w^2 + 15w - 54 = 0$$
$$w^2 - 15w + 54 = 0$$
$$(w - 6)(w - 9) = 0$$
$$w - 6 = 0 \quad \text{or} \quad w - 9 = 0$$
$$w = 6 \quad \text{or} \quad w = 9$$

If the width is 6, then $15 - 6 = 9$. If the width is 9, then $15 - 9 = 6$. Thus, the length and width of the rectangle must be 6 centimeters by 9 centimeters.

63. Let r represent the number of rows in the orchard, then $r + 5$ represents the number of trees per row.

(trees per row)(rows) = number of trees
$$(r + 5)(r) = 84$$
$$r^2 + 5r = 84$$
$$r^2 + 5r - 84 = 0$$
$$(r + 12)(r - 7) = 0$$
$$r + 12 = 0 \quad \text{or} \quad r - 7 = 0$$
$$r = -12 \quad \text{or} \quad r = 7$$

The number of rows cannot be negative, so $r = -12$ is discarded. Thus there would be 7 rows of trees.

65. Let x represent the length of one leg of the triangle, then $x - 7$ represents the other leg and $x + 2$ represents the length of the hypotenuse.

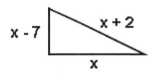

$$x^2 + (x - 7)^2 = (x + 2)^2$$
Pythagorean Theorem
$$x^2 + x^2 - 14x + 49 = x^2 + 4x + 4$$
$$2x^2 - 14x + 49 = x^2 + 4x + 4$$
$$x^2 - 18x + 45 = 0$$
$$(x - 3)(x - 15) = 0$$
$$x - 3 = 0 \quad \text{or} \quad x - 15 = 0$$
$$x = 3 \quad \text{or} \quad x = 15$$

If $x = 3$, then $x - 7 = -4$ which is not possible. If $x = 15$ then $x - 7 = 8$ and

$x + 2 = 17$. Thus, the sides of the right triangle are 8 feet, 15 feet and 17 feet.

67. Let x represent the length of one leg of the triangle, then $x - 2$ represents the other leg.

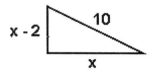

$$x^2 + (x - 2)^2 = 10^2$$
Pythagorean Theorem
$$x^2 + x^2 - 4x + 4 = 100$$
$$2x^2 - 4x + 4 = 100$$
$$2x^2 - 4x - 96 = 0$$
$$x^2 - 2x - 48 = 0$$
$$(x - 8)(x + 6) = 0$$
$$x - 8 = 0 \quad \text{or} \quad x + 6 = 0$$
$$x = 8 \quad \text{or} \quad x = -6$$

The solution of $x = -6$ is not possible. If $x = 8$, then $x - 2 = 6$. Thus, the legs of the triangle are 6 inches and 8 inches.

73. We need two integers whose product is 40 and whose sum is 13. They are 5 and 8.
$$x^{2a} + 13x^a + 40 = (x^a + 5)(x^a + 8)$$

75. We need two integers whose product is -27 and whose sum is 6. They are -3 and 9.
$$x^{2a} + 6x^a - 27 = (x^a - 3)(x^a + 9)$$

PROBLEM SET | **6.4** **Factoring Trinomials of the Form** $ax^2 + bx + c$

1. We need two integers whose product is 6 and whose sum is 7. They are 6 and 1.
$$3x^2 + 7x + 2 =$$
$$3x^2 + 1x + 6x + 2 =$$
$$x(3x + 1) + 2(3x + 1) =$$
$$(3x + 1)(x + 2)$$

3. We need two integers whose product is 60 and whose sum is 19. They are 4 and 15.
$$6x^2 + 19x + 10 =$$
$$6x^2 + 4x + 15x + 10 =$$
$$2x(3x + 2) + 5(3x + 2) =$$
$$(3x + 2)(2x + 5)$$

Problem Set 6.4

5. We need two integers whose product is 24 and whose sum is -25. They are -1 and -24.
$$4x^2 - 25x + 6 =$$
$$4x^2 - 1x - 24x + 6 =$$
$$x(4x - 1) - 6(4x - 1) =$$
$$(4x - 1)(x - 6)$$

7. We need two integers whose product is 240 and whose sum is -31. They are -15 and -16.
$$12x^2 - 31x + 20 =$$
$$12x^2 - 15x - 16x + 20 =$$
$$3x(4x - 5) - 4(4x - 5) =$$
$$(4x - 5)(3x - 4)$$

9. We need two integers whose product is -70 and whose sum is -33. They are 2 and -35.
$$5y^2 - 33y - 14 =$$
$$5y^2 + 2y - 35y - 14 =$$
$$y(5y + 2) - 7(5y + 2) =$$
$$(5y + 2)(y - 7)$$

11. We need two integers whose product is -48 and whose sum is 13. They are -3 and 16.
$$2n^2 + 13n - 24 =$$
$$2n^2 - 3n + 16n - 24 =$$
$$n(2n - 3) + 8(2n - 3) =$$
$$(2n - 3)(n + 8)$$

13. We need two integers whose product is 14 and whose sum is 1. No such integers exist, therefore $2x^2 + x + 7$ is not factorable.

15. We need two integers whose product is 126 and whose sum is 45. They are 3 and 42.
$$18x^2 + 45x + 7 =$$
$$18x^2 + 3x + 42x + 7 =$$
$$3x(6x + 1) + 7(6x + 1) =$$
$$(6x + 1)(3x + 7)$$

17. We need two integers whose product is 56 and whose sum is -30. They are -2 and -28.
$$7x^2 - 30x + 8 =$$

$$7x^2 - 2x - 28x + 8 =$$
$$x(7x - 2) - 4(7x - 2) =$$
$$(7x - 2)(x - 4)$$

19. We need two integers whose product is -168 and whose sum is 2. They are -12 and 14.
$$8x^2 + 2x - 21 =$$
$$8x^2 - 12x + 14x - 21 =$$
$$4x(2x - 3) + 7(2x - 3) =$$
$$(2x - 3)(4x + 7)$$

21. We need two integers whose product is -126 and whose sum is -15. They are 6 and -21.
$$9t^2 - 15t - 14 =$$
$$9t^2 + 6t - 21t - 14 =$$
$$3t(3t + 2) - 7(3t + 2) =$$
$$(3t + 2)(3t - 7)$$

23. We need two integers whose product is -420 and whose sum is 79. They are -5 and 84.
$$12y^2 + 79y - 35 =$$
$$12y^2 - 5y + 84y - 35 =$$
$$y(12y - 5) + 7(12y - 5) =$$
$$(12y - 5)(y + 7)$$

25. We need two integers whose product is -30 and whose sum is 2. No such integers exist, therefore $6n^2 + 2n - 5$ is not factorable.

27. We need two integers whose product is 294 and whose sum is 55. They are 6 and 49.
$$14x^2 + 55x + 21 =$$
$$14x^2 + 6x + 49x + 21 =$$
$$2x(7x + 3) + 7(7x + 3) =$$
$$(7x + 3)(2x + 7)$$

29. We need two integers whose product is 240 and whose sum is -31. They are -15 and -16.
$$20x^2 - 31x + 12 =$$
$$20x^2 - 15x - 16x + 12 =$$
$$5x(4x - 3) - 4(4x - 3) =$$
$$(4x - 3)(5x - 4)$$

144

31. We need two integers whose product is -240 and whose sum is -8. They are 12 and -20.
$16n^2 - 8n - 15 =$
$16n^2 + 12n - 20n - 15 =$
$4n(4n + 3) - 5(4n + 3) =$
$(4n + 3)(4n - 5)$

33. We need two integers whose product is 600 and whose sum is -50. They are -20 and -30.
$24x^2 - 50x + 25 =$
$24x^2 - 20x - 30x + 25 =$
$4x(6x - 5) - 5(6x - 5) = (6x - 5)(4x - 5)$

35. We need two integers whose product is 144 and whose sum is 25. They are 9 and 16.
$2x^2 + 25x + 72 =$
$2x^2 + 9x + 16x + 72 =$
$x(2x + 9) + 8(2x + 9) =$
$(2x + 9)(x + 8)$

37. We need two integers whose product is -42 and whose sum is 1. They are -6 and 7.
$21a^2 + a - 2 =$
$21a^2 - 6a + 7a - 2 =$
$3a(7a - 2) + 1(7a - 2) =$
$(7a - 2)(3a + 1)$

39. We need two integers whose product is -180 and whose sum is -31. They are 5 and -36.
$12a^2 - 31a - 15 =$
$12a^2 + 5a - 36a - 15 =$
$a(12a + 5) - 3(12a + 5) =$
$(12a + 5)(a - 3)$

41. We need two integers whose product is 36 and whose sum is 12. They are 6 and 6.
$4x^2 + 12x + 9 =$
$4x^2 + 6x + 6x + 9 =$
$2x(2x + 3) + 3(2x + 3) =$
$(2x + 3)(2x + 3)$

43. We need two integers whose product is 6 and whose sum is -5. They are -2 and -3.
$6x^2 - 5xy + y^2 =$
$6x^2 - 2xy - 3xy + y^2 =$
$2x(3x - y) - y(3x - y)$
$(3x - y)(2x - y)$

45. We need two integers whose product is -120 and whose sum is 7. They are -8 and 15.
$20x^2 + 7xy - 6y^2 =$
$20x^2 - 8xy + 15xy - 6y^2 =$
$4x(5x - 2y) + 3y(5x - 2y) =$
$(5x - 2y)(4x + 3y)$

47. We need two integers whose product is 60 and whose sum is -32. They are -2 and -30.
$5x^2 - 32x + 12 =$
$5x^2 - 30x - 2x + 12 =$
$5x(x - 6) - 2(x - 6) =$
$(5x - 2)(x - 6)$

49. We need two integers whose product is -56 and whose sum is -55. They are 1 and -56.
$8x^2 - 55x - 7 =$
$8x^2 + 1x - 56x - 7 =$
$x(8x + 1) - 7(8x + 1) =$
$(8x + 1)(x - 7)$

51. $2x^2 + 13x + 6 = 0$
$(2x + 1)(x + 6) = 0$
$2x + 1 = 0 \quad$ or $\quad x + 6 = 0$
$\quad 2x = -1 \quad$ or $\quad x = -6$
$\quad x = -\dfrac{1}{2} \quad$ or $\quad x = -6$
The solution set is $\left\{ -6, \ -\dfrac{1}{2} \right\}$.

53. $12x^2 + 11x + 2 = 0$
$(4x + 1)(3x + 2) = 0$
$4x + 1 = 0 \quad$ or $\quad 3x + 2 = 0$
$\quad 4x = -1 \quad$ or $\quad 3x = -2$
$\quad x = -\dfrac{1}{4} \quad$ or $\quad x = -\dfrac{2}{3}$
The solution set is $\left\{ -\dfrac{2}{3}, \ -\dfrac{1}{4} \right\}$.

Problem Set 6.4

55. $3x^2 - 25x + 8 = 0$
$(3x - 1)(x - 8) = 0$
$3x - 1 = 0 \quad$ or $\quad x - 8 = 0$
$3x = 1 \quad$ or $\qquad x = 8$
$x = \dfrac{1}{3} \quad$ or $\qquad x = 8$
The solution set is $\left\{ \dfrac{1}{3}, 8 \right\}$.

57. $15n^2 - 41n + 14 = 0$
$(3n - 7)(5n - 2) = 0$
$3n - 7 = 0 \quad$ or $\quad 5n - 2 = 0$
$3n = 7 \quad$ or $\qquad 5n = 2$
$n = \dfrac{7}{3} \quad$ or $\qquad n = \dfrac{2}{5}$
The solution set is $\left\{ \dfrac{2}{5}, \dfrac{7}{3} \right\}$.

59. $6t^2 + 37t - 35 = 0$
$(6t - 5)(t + 7) = 0$
$6t - 5 = 0 \quad$ or $\quad t + 7 = 0$
$6t = 5 \quad$ or $\qquad t = -7$
$t = \dfrac{5}{6} \quad$ or $\qquad t = -7$
The solution set is $\left\{ -7, \dfrac{5}{6} \right\}$.

61. $16y^2 - 18y - 9 = 0$
$(8y + 3)(2y - 3) = 0$
$8y + 3 = 0 \qquad$ or $\quad 2y - 3 = 0$
$8y = -3 \quad$ or $\qquad 2y = 3$
$y = -\dfrac{3}{8} \quad$ or $\qquad y = \dfrac{3}{2}$
The solution set is $\left\{ -\dfrac{3}{8}, \dfrac{3}{2} \right\}$.

63. $9x^2 - 6x - 8 = 0$
$(3x - 4)(3x + 2) = 0$
$3x - 4 = 0 \quad$ or $\quad 3x + 2 = 0$
$3x = 4 \quad$ or $\qquad 3x = -2$
$x = \dfrac{4}{3} \quad$ or $\qquad x = -\dfrac{2}{3}$
The solution set is $\left\{ -\dfrac{2}{3}, \dfrac{4}{3} \right\}$.

65. $10x^2 - 29x + 10 = 0$
$(2x - 5)(5x - 2) = 0$
$2x - 5 = 0 \quad$ or $\quad 5x - 2 = 0$
$2x = 5 \quad$ or $\qquad 5x = 2$
$x = \dfrac{5}{2} \quad$ or $\qquad x = \dfrac{2}{5}$
The solution set is $\left\{ \dfrac{2}{5}, \dfrac{5}{2} \right\}$.

67. $6x^2 + 19x = -10$
$6x^2 + 19x + 10 = 0$
$(2x + 5)(3x + 2) = 0$
$2x + 5 = 0 \qquad$ or $\quad 3x + 2 = 0$
$2x = -5 \quad$ or $\qquad 3x = -2$
$x = -\dfrac{5}{2} \quad$ or $\qquad x = -\dfrac{2}{3}$
The solution set is $\left\{ -\dfrac{5}{2}, -\dfrac{2}{3} \right\}$.

69. $16x(x + 1) = 5$
$16x^2 + 16x = 5$
$16x^2 + 16x - 5 = 0$
$(4x - 1)(4x + 5) = 0$
$4x - 1 = 0 \quad$ or $\quad 4x + 5 = 0$
$4x = 1 \quad$ or $\qquad 4x = -5$
$x = \dfrac{1}{4} \quad$ or $\qquad x = -\dfrac{5}{4}$
The solution set is $\left\{ -\dfrac{5}{4}, \dfrac{1}{4} \right\}$.

71. $35n^2 - 34n - 21 = 0$
$(7n + 3)(5n - 7) = 0$
$7n + 3 = 0 \qquad$ or $\quad 5n - 7 = 0$
$7n = -3 \quad$ or $\qquad 5n = 7$
$n = -\dfrac{3}{7} \quad$ or $\qquad n = \dfrac{7}{5}$
The solution set is $\left\{ -\dfrac{3}{7}, \dfrac{7}{5} \right\}$.

73. $4x^2 - 45x + 50 = 0$
$(4x - 5)(x - 10) = 0$
$4x - 5 = 0$ or $x - 10 = 0$
$4x = 5$ or $x = 10$
$x = \dfrac{5}{4}$ or $x = 10$
The solution set is $\left\{ \dfrac{5}{4}, 10 \right\}$.

75. $7x^2 + 46x - 21 = 0$
$(7x - 3)(x + 7) = 0$
$7x - 3 = 0$ or $x + 7 = 0$
$7x = 3$ or $x = -7$
$x = \dfrac{3}{7}$ or $x = -7$
The solution set is $\left\{ -7, \dfrac{3}{7} \right\}$.

77. $12x^2 - 43x - 20 = 0$
$(12x + 5)(x - 4) = 0$
$12x + 5 = 0$ or $x - 4 = 0$
$12x = -5$ or $x = 4$
$x = -\dfrac{5}{12}$ or $x = 4$
The solution set is $\left\{ -\dfrac{5}{12}, 4 \right\}$.

79. $18x^2 + 55x - 28 = 0$
$(9x - 4)(2x + 7) = 0$
$9x - 4 = 0$ or $2x + 7 = 0$
$9x = 4$ or $2x = -7$
$x = \dfrac{4}{9}$ or $x = -\dfrac{7}{2}$
The solution set is $\left\{ -\dfrac{7}{2}, \dfrac{4}{9} \right\}$.

PROBLEM SET **6.5**   **Factoring, Solving Equations, and Problem Solving**

1. $x^2 + 4x + 4 =$
$(x)^2 + 2(x)(2) + (2)^2 =$
$(x + 2)^2$

3. $x^2 - 10x + 25 =$
$(x)^2 - 2(x)(5) + (5)^2 =$
$(x - 5)^2$

5. $9n^2 + 12n + 4 =$
$(3n)^2 + 2(3n)(2) + (2)^2 =$
$(3n + 2)^2$

7. $16a^2 - 8a + 1 =$
$(4a)^2 - 2(4a)(1) + (1)^2 =$
$(4a - 1)^2$

9. $4 + 36x + 81x^2 =$
$(2)^2 + 2(2)(9x) + (9x)^2 =$
$(2 + 9x)^2$

11. $16x^2 - 24xy + 9y^2 =$
$(4x)^2 - 2(4x)(3y) + (3y)^2 =$
$(4x - 3y)^2$

13. We need two integers whose product is 16

and whose sum is 17. They are 1 and 16.
$2x^2 + 17x + 8 =$
$2x^2 + 1x + 16x + 8 =$
$x(2x + 1) + 8(2x + 1) =$
$(2x + 1)(x + 8)$

15. $2x^3 - 72x =$
$2x(x^2 - 36) =$
$2x(x - 6)(x + 6)$

17. We need two integers whose product is -60 and whose sum is -7. They are 5 and -12.
$n^2 - 7n - 60 = (n + 5)(n - 12)$

19. We need two integers whose product is -12 and whose sum is -7. No such integers exist, therefore $3a^2 - 7a - 4$ is not factorable.

21. $8x^2 + 72 = 8(x^2 + 9)$

23. $9x^2 + 30x + 25 =$
$(3x)^2 + 2(3x)(5) + (5)^2 =$
$(3x + 5)^2$

Problem Set 6.5

25. $15x^2 + 65x + 70 =$
$5(3x^2 + 13x + 14)$
We need two integers whose product is 42
and whose sum is 13. They are 6 and 7.
$5(3x^2 + 6x + 7x + 14) =$
$5[3x(x + 2) + 7(x + 2)] =$
$5(x + 2)(3x + 7)$

27. We need two integers whose product is
-360 and whose sum is 2. They are
-18 and 20.
$24x^2 + 2x - 15 =$
$24x^2 - 18x + 20x - 15 =$
$6x(4x - 3) + 5(4x - 3) =$
$(4x - 3)(6x + 5)$

29. $xy + 5y - 8x - 40 =$
$y(x + 5) - 8(x + 5) =$
$(x + 5)(y - 8)$

31. We need two integers whose product is
-140 and whose sum is 31. They are
-4 and 35.
$20x^2 + 31xy - 7y^2 =$
$20x^2 - 4xy + 35xy - 7y^2 =$
$4x(5x - y) + 7y(5x - y) =$
$(5x - y)(4x + 7y)$

33. $24x^2 + 18x - 81 =$
$3(8x^2 + 6x - 27)$
We need two integers whose product is
-216 and whose sum is 6. They are
-12 and 18.
$3(8x^2 + 6x - 27) =$
$3(8x^2 - 12x + 18x - 27) =$
$3[4x(2x - 3) + 9(2x - 3)] =$
$3(2x - 3)(4x + 9)$

35. $12x^2 + 6x + 30 = 6(2x^2 + x + 5)$
$[2x^2 + x + 5$ is not factorable.$]$

37. $5x^4 - 80 =$
$5(x^4 - 16) =$
$5[(x^2 - 4)(x^2 + 4)] =$
$5(x - 2)(x + 2)(x^2 + 4)$

39. $x^2 + 12xy + 36y^2 =$
$(x)^2 + 2(x)(6y) + (6y)^2 =$
$(x + 6y)^2$

41. $4x^2 - 20x = 0$
$4(x^2 - 5x) = 0$
$4x(x - 5) = 0$
$4x = 0$ or $x - 5 = 0$
$x = 0$ or $x = 5$
The solution set is $\{0, 5\}$.

43. $x^2 - 9x - 36 = 0$
$(x - 12)(x + 3) = 0$
$x - 12 = 0$ or $x + 3 = 0$
$x = 12$ or $x = -3$
The solution set is $\{-3, 12\}$.

45. $-2x^3 + 8x = 0$
Divide by -2.
$x^3 - 4x = 0$
$x(x^2 - 4) = 0$
$x(x - 2)(x + 2) = 0$
$x = 0$ or $x - 2 = 0$ or $x + 2 = 0$
$x = 0$ or $x = 2$ or $x = -2$
The solution set is $\{-2, 0, 2\}$.

47. $6n^2 - 29n - 22 = 0$
$(2n - 11)(3n + 2) = 0$
$2n - 11 = 0$ or $3n + 2 = 0$
$2n = 11$ or $3n = -2$
$n = \dfrac{11}{2}$ or $n = -\dfrac{2}{3}$
The solution set is $\left\{-\dfrac{2}{3}, \dfrac{11}{2}\right\}$.

49. $(3n - 1)(4n - 3) = 0$
$3n - 1 = 0$ or $4n - 3 = 0$
$3n = 1$ or $4n = 3$
$n = \dfrac{1}{3}$ or $n = \dfrac{3}{4}$
The solution set is $\left\{\dfrac{1}{3}, \dfrac{3}{4}\right\}$.

148

51. $(n-2)(n+6) = -15$
$n^2 + 4n - 12 = -15$
$n^2 + 4n + 3 = 0$
$(n+3)(n+1) = 0$
$n + 3 = 0 \quad$ or $\quad n + 1 = 0$
$\quad n = -3 \quad$ or $\quad n = -1$
The solution set is $\{-3, -1\}$.

53. $\qquad 2x^2 = 12x$
Divide by 2.
$\qquad x^2 = 6x$
$\quad x^2 - 6x = 0$
$\quad x(x - 6) = 0$
$\quad x = 0 \quad$ or $\quad x - 6 = 0$
$\quad x = 0 \quad$ or $\qquad x = 6$
The solution set is $\{0, 6\}$.

55. $t^3 - 2t^2 - 24t = 0$
$t(t^2 - 2t - 24) = 0$
$t(t - 6)(t + 4) = 0$
$t = 0 \quad$ or $\quad t - 6 = 0 \quad$ or $\quad t + 4 = 0$
$t = 0 \quad$ or $\qquad t = 6 \quad$ or $\qquad t = -4$
The solution set is $\{-4, 0, 6\}$.

57. $12 - 40x + 25x^2 = 0$
$(2 - 5x)(6 - 5x) = 0$
$2 - 5x = 0 \quad$ or $\quad 6 - 5x = 0$
$\quad -5x = -2 \quad$ or $\quad -5x = -6$
$\qquad x = \dfrac{2}{5} \quad$ or $\quad x = \dfrac{6}{5}$
The solution set is $\left\{\dfrac{2}{5}, \dfrac{6}{5}\right\}$.

59. $n^2 - 28n + 192 = 0$
$(n - 12)(n - 16) = 0$
$n - 12 = 0 \quad$ or $\quad n - 16 = 0$
$\quad n = 12 \quad$ or $\qquad n = 16$
The solution set is $\{12, 16\}$.

61. $(3n + 1)(n + 2) = 12$
$3n^2 + 7n + 2 = 12$
$3n^2 + 7n - 10 = 0$
$(3n + 10)(n - 1) = 0$
$3n + 10 = 0 \quad$ or $\quad n - 1 = 0$
$\quad 3n = -10 \quad$ or $\qquad n = 1$
$\qquad n = -\dfrac{10}{3} \quad$ or $\qquad n = 1$
The solution set is $\left\{-\dfrac{10}{3}, 1\right\}$.

63. $\qquad x^3 = 6x^2$
$\quad x^3 - 6x^2 = 0$
$\quad x^2(x - 6) = 0$
$\quad x(x)(x - 6) = 0$
$x = 0 \quad$ or $\quad x = 0 \quad$ or $\quad x - 6 = 0$
$x = 0 \quad$ or $\quad x = 0 \quad$ or $\qquad x = 6$
The solution set is $\{0, 6\}$.

65. $9x^2 - 24x + 16 = 0$
$(3x - 4)(3x - 4) = 0$
$3x - 4 = 0 \quad$ or $\quad 3x - 4 = 0$
$\quad 3x = 4 \quad$ or $\qquad 3x = 4$
$\qquad x = \dfrac{4}{3} \quad$ or $\qquad x = \dfrac{4}{3}$
The solution set is $\left\{\dfrac{4}{3}\right\}$.

67. $x^3 + 10x^2 + 25x = 0$
$x(x^2 + 10x + 25) = 0$
$x(x + 5)(x + 5) = 0$
$x = 0 \quad$ or $\quad x + 5 = 0 \qquad$ or $\quad x + 5 = 0$
$x = 0 \quad$ or $\qquad x = -5 \quad$ or $\qquad x = -5$
The solution set is $\{-5, 0\}$.

69. $24x^2 + 17x - 20 = 0$
$(3x + 4)(8x - 5) = 0$
$3x + 4 = 0 \quad$ or $\quad 8x - 5 = 0$
$\quad 3x = -4 \quad$ or $\qquad 8x = 5$
$\qquad x = -\dfrac{4}{3} \quad$ or $\quad x = \dfrac{5}{8}$
The solution set is $\left\{-\dfrac{4}{3}, \dfrac{5}{8}\right\}$.

Problem Set 6.5

71. Let n represent one of the numbers, then $4n + 7$ represents the other number.
$$n(4n + 7) = 15$$
$$4n^2 + 7n = 15$$
$$4n^2 + 7n - 15 = 0$$
$$(4n - 5)(n + 3) = 0$$
$$4n - 5 = 0 \quad \text{or} \quad n + 3 = 0$$
$$4n = 5 \quad \text{or} \quad n = -3$$
$$n = \frac{5}{4} \quad \text{or} \quad n = -3$$
If $n = \frac{5}{4}$, then $4n + 7 =$
$$4\left(\frac{5}{4}\right) + 7 = 12.$$
If $n = -3$, then $4n + 7 =$
$$4(-3) + 7 = -5.$$
Thus, then numbers are $\frac{5}{4}$ and 12
or -3 and -5.

73. Let n represent one number, then $2n + 3$ represents the other number.
$$n(2n + 3) = -1$$
$$2n^2 + 3n = -1$$
$$2n^2 + 3n + 1 = 0$$
$$(2n + 1)(n + 1) = 0$$
$$2n + 1 = 0 \quad \text{or} \quad n + 1 = 0$$
$$2n = -1 \quad \text{or} \quad n = -1$$
$$n = -\frac{1}{2} \quad \text{or} \quad n = -1$$
If $n = -\frac{1}{2}$, then $2n + 3 = 2\left(-\frac{1}{2}\right) + 3 = 2.$
If $n = -1$, then $2n + 3 = 2(-1) + 3 = 1.$
Thus, the numbers are $-\frac{1}{2}$ and 2 or
-1 and 1.

75. Let n represent one number, then $2n + 1$ represents the other number.
$$n^2 + (2n + 1)^2 = 97$$
$$n^2 + 4n^2 + 4n + 1 = 97$$
$$5n^2 + 4n + 1 = 97$$
$$5n^2 + 4n - 96 = 0$$
$$(5n + 24)(n - 4) = 0$$

$$5n + 24 = 0 \quad \text{or} \quad n - 4 = 0$$
$$5n = -24 \quad \text{or} \quad n = 4$$
$$n = -\frac{24}{5} \quad \text{or} \quad n = 4$$
If $n = -\frac{24}{5}$, then $2n + 1 =$
$$2\left(-\frac{24}{5}\right) + 1 = -\frac{48}{5} + \frac{5}{5} = -\frac{43}{5}.$$
If $n = 4$, then $2n + 1 = 2(4) + 1 = 9$
Thus, the numbers are $-\frac{24}{5}$ and $-\frac{43}{5}$
or 4 and 9.

77. Let r represent the number of rows, then $2r - 3$ represents the number of chairs per row.
$$(\text{rows})(\text{chairs per row}) = 54$$
$$r(2r - 3) = 54$$
$$2r^2 - 3r = 54$$
$$2r^2 - 3r - 54 = 0$$
$$(2r + 9)(r - 6) = 0$$
$$2r + 9 = 0 \quad \text{or} \quad r - 6 = 0$$
$$2r = -9 \quad \text{or} \quad r = 6$$
$$r = -\frac{9}{2} \quad \text{or} \quad r = 6$$
The solution of $r = -\frac{9}{2}$ is not reasonable.
If $r = 6$, then $2r - 3 = 2(6) - 3 = 9.$
Thus, there would be 6 rows with 9 chairs per row.

79. Let s represent the length of a side of the smaller square, then $3s$ represents the length of a side of the larger square.

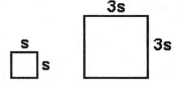

150

Smaller Area + Larger Area = 360

$$s^2 + (3s)^2 = 360$$
$$s^2 + 9s^2 = 360$$
$$10s^2 = 360$$
$$s^2 = 36$$
$$s^2 - 36 = 0$$
$$(s - 6)(s + 6) = 0$$
$$s - 6 = 0 \quad \text{or} \quad s + 6 = 0$$
$$s = 6 \quad \text{or} \quad s = -6$$

The negative solution must be discarded. Therefore, the smaller square is 6 feet by 6 feet and the larger square is 18 feet by 18 feet.

81. Let w represent the width of the rectangle, then $2w + 1$ represents the length.

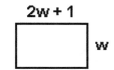

2w + 1

w

$$w(2w + 1) = 55$$
$$2w^2 + w = 55$$
$$2w^2 + w - 55 = 0$$
$$(2w + 11)(w - 5) = 0$$
$$2w + 11 = 0 \quad \text{or} \quad w - 5 = 0$$
$$2w = -11 \quad \text{or} \quad w = 5$$
$$w = -\frac{11}{2} \quad \text{or} \quad w = 5$$

The negative solution must be discarded. Therefore, the rectangle is 5 centimeters by 11 centimeters.

83. Let h represent the length of an altitude to a side of the triangle, then $3h - 1$ represents the side.

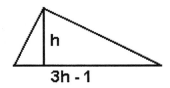

h

3h - 1

$$\frac{1}{2}h(3h - 1) = 51$$
$$h(3h - 1) = 102$$
$$3h^2 - h = 102$$
$$3h^2 - h - 102 = 0$$
$$(3h + 17)(h - 6) = 0$$
$$3h + 17 = 0 \quad \text{or} \quad h - 6 = 0$$
$$3h = -17 \quad \text{or} \quad h = 6$$
$$h = -\frac{17}{3} \quad \text{or} \quad h = 6$$

The negative solution must be discarded. Therefore, the length of the side is 17 inches and the altitude to that side is 6 inches long.

85. Let x represent the width of the strip, then $8 - 2x$ represents the reduced width of the paper, and $11 - 2x$ represents the reduced length of the paper.

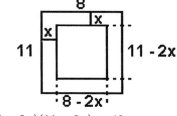

8

x

x

11

11 - 2x

8 - 2x

$$(8 - 2x)(11 - 2x) = 40$$
$$88 - 38x + 4x^2 = 40$$
$$48 - 38x + 4x^2 = 0$$
$$24 - 19x + 2x^2 = 0$$
$$(3 - 2x)(8 - x) = 0$$
$$3 - 2x = 0 \quad \text{or} \quad 8 - x = 0$$
$$-2x = -3 \quad \text{or} \quad 8 = x$$
$$x = \frac{3}{2} \quad \text{or} \quad x = 8$$

The solution $x = 8$ must be discarded since the width of the paper is 8 inches. Therefore, the width of the strip is $1\frac{1}{2}$ inches.

Problem Set 6.5

87. Let r represent a radius of the larger circle, then $r - 6$ represents a radius of the smaller circle.

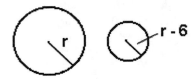

Larger Area + Smaller Area $= 180\pi$
$$\pi r^2 + \pi(r-6)^2 = 180\pi$$
$$\pi r^2 + \pi(r^2 - 12r + 36) = 180\pi$$
$$\pi r^2 + \pi r^2 - 12\pi r + 36\pi = 180\pi$$
$$2\pi r^2 - 12\pi r + 36\pi = 180\pi$$
$$2\pi r^2 - 12\pi r - 144\pi = 0$$
Divide both sides by 2π.
$$r^2 - 6r - 72 = 0$$
$$(r - 12)(r + 6) = 0$$
$$r - 12 = 0 \quad \text{or} \quad r + 6 = 0$$
$$r = 12 \quad \text{or} \quad r = -6$$
Discard the negative solution. The length of a radius of the larger circle is 12 inches and a radius of the smaller circle is 6 inches.

CHAPTER 6 Review Problem Set

1. We need two integers whose product is 14 and whose sum is -9. They are -2 and -7.
$$x^2 - 9x + 14 = (x - 2)(x - 7)$$

2. $3x^2 + 21x = 3x(x + 7)$

3. $9x^2 - 4 =$
$(3x)^2 - (2)^2 =$
$(3x - 2)(3x + 2)$

4. We need two integers whose product is -20 and whose sum is 8. They are -2 and 10.
$4x^2 + 8x - 5 =$
$4x^2 - 2x + 10x - 5 =$
$2x(2x - 1) + 5(2x - 1) =$
$(2x - 1)(2x + 5)$

5. $25x^2 - 60x + 36 =$
$(5x)^2 - 2(5x)(6) + (6)^2 =$
$(5x - 6)^2$

6. $n^3 + 13n^2 + 40n =$
$n(n^2 + 13n + 40)$
We need two integers whose product is 40 and whose sum is 13. They are 5 and 8.
$n(n^2 + 13n + 40) =$
$n(n + 5)(n + 8)$

7. We need two integers whose product is -12 and whose sum is 11. They are -1 and 12.
$$y^2 + 11y - 12 = (y - 1)(y + 12)$$

8. $3xy^2 + 6x^2y = 3xy(y + 2x)$

9. $x^4 - 1 =$
$(x^2)^2 - (1)^2 =$
$(x^2 - 1)(x^2 + 1) =$
$(x - 1)(x + 1)(x^2 + 1)$

10. We need two integers whose product is -90 and whose sum is 9. They are -6 and 15.
$18n^2 + 9n - 5 =$
$18n^2 - 6n + 15n - 5 =$
$6n(3n - 1) + 5(3n - 1) =$
$(3n - 1)(6n + 5)$

11. We need two integers whose product is 24 and whose sum is 7. No such integers exist, therefore $x^2 + 7x + 24$ is not factorable.

12. We need two integers whose product is
-28 and whose sum is -3. They are
4 and -7.
$$4x^2 - 3x - 7 =$$
$$4x^2 + 4x - 7x - 7 =$$
$$4x(x+1) - 7(x+1) =$$
$$(x+1)(4x-7)$$

13. $3n^2 + 3n - 90 =$
$$3(n^2 + n - 30)$$
We need two integers whose product is
-30 and whose sum is 1. They are
-5 and 6.
$$3(n^2 + n - 30) =$$
$$3(n-5)(n+6)$$

14. $x^3 - xy^2 = x(x^2 - y^2) = x(x-y)(x+y)$

15. We need two integers whose product is
-4 and whose sum is 3. They are -1
and 4.
$$2x^2 + 3xy - 2y^2 =$$
$$2x^2 - xy + 4xy - 2y^2 =$$
$$x(2x-y) + 2y(2x-y) =$$
$$(2x-y)(x+2y)$$

16. $4n^2 - 6n - 40 =$
$$2(2n^2 - 3n - 20)$$
We need two integers whose product is
-40 and whose sum is -3. They are
5 and -8.
$$2(2n^2 - 3n - 20) =$$
$$2(2n^2 + 5n - 8n - 20) =$$
$$2[n(2n+5) - 4(2n+5)] =$$
$$2(2n+5)(n-4)$$

17. $5x + 5y + ax + ay =$
$$5(x+y) + a(x+y) =$$
$$(x+y)(5+a)$$

18. We need two integers whose product is
-84 and whose sum is -5. They are
7 and -12.
$$21t^2 - 5t - 4 =$$
$$21t^2 + 7t - 12t - 4 =$$
$$7t(3t+1) - 4(3t+1) =$$
$$(3t+1)(7t-4)$$

19. $2x^3 - 2x =$
$$2x(x^2 - 1) =$$
$$2x(x-1)(x+1)$$

20. $3x^3 - 108x =$
$$3x(x^2 - 36) =$$
$$3x(x-6)(x+6)$$

21. $16x^2 + 40x + 25 =$
$$(4x)^2 + 2(4x)(5) + (5)^2 =$$
$$(4x+5)^2$$

22. $xy - 3x - 2y + 6 =$
$$x(y-3) - 2(y-3) =$$
$$(y-3)(x-2)$$

23. We need two integers whose product is
-30 and whose sum is -7. They are
3 and -10.
$$15x^2 - 7xy - 2y^2 =$$
$$15x^2 + 3xy - 10xy - 2y^2 =$$
$$3x(5x+y) - 2y(5x+y) =$$
$$(5x+y)(3x-2y)$$

24. $6n^4 - 5n^3 + n^2 =$
$$n^2(6n^2 - 5n + 1)$$
We need two integers whose product is 6
and whose sum is -5. They are -2 and -3.
$$n^2(6n^2 - 5n + 1) =$$
$$n^2(6n^2 - 2n - 3n + 1) =$$
$$n^2[2n(3n-1) - 1(3n-1)] =$$
$$n^2(3n-1)(2n-1)$$

25. $x^2 + 4x - 12 = 0$
$$(x+6)(x-2) = 0$$
$$x + 6 = 0 \quad \text{or} \quad x - 2 = 0$$
$$x = -6 \quad \text{or} \quad x = 2$$
The solution set is $\{-6, 2\}$.

26. $x^2 = 11x$
$$x^2 - 11x = 0$$
$$x(x - 11) = 0$$
$$x = 0 \quad \text{or} \quad x - 11 = 0$$
$$x = 0 \quad \text{or} \quad x = 11$$
The solution set is $\{0, 11\}$.

Chapter 6 Review Problem Set

27. $2x^2 + 3x - 20 = 0$
$(2x - 5)(x + 4) = 0$
$2x - 5 = 0 \quad$ or $\quad x + 4 = 0$
$2x = 5 \quad$ or $\qquad x = -4$
$x = \dfrac{5}{2} \quad$ or $\qquad x = -4$
The solution set is $\left\{ -4, \dfrac{5}{2} \right\}$.

28. $9n^2 + 21n - 8 = 0$
$(3n - 1)(3n + 8) = 0$
$3n - 1 = 0 \quad$ or $\quad 3n + 8 = 0$
$3n = 1 \quad$ or $\qquad 3n = -8$
$n = \dfrac{1}{3} \quad$ or $\qquad n = -\dfrac{8}{3}$
The solution set is $\left\{ -\dfrac{8}{3}, \dfrac{1}{3} \right\}$.

29. $6n^2 = 24$
$6n^2 - 24 = 0$
$n^2 - 4 = 0$
$(n - 2)(n + 2) = 0$
$n - 2 = 0 \quad$ or $\quad n + 2 = 0$
$n = 2 \quad$ or $\qquad n = -2$
The solution set is $\{ -2, 2 \}$.

30. $16y^2 + 40y + 25 = 0$
$(4y + 5)(4y + 5) = 0$
$4y + 5 = 0 \qquad$ or $\quad 4y + 5 = 0$
$4y = -5 \quad$ or $\qquad 4y = -5$
$y = -\dfrac{5}{4} \quad$ or $\qquad y = -\dfrac{5}{4}$
The solution set is $\left\{ -\dfrac{5}{4} \right\}$.

31. $t^3 - t = 0$
$t(t^2 - 1) = 0$
$t(t - 1)(t + 1) = 0$
$t = 0 \quad$ or $\quad t - 1 = 0 \quad$ or $\quad t + 1 = 0$
$t = 0 \quad$ or $\qquad t = 1 \quad$ or $\qquad t = -1$
The solution set is $\{ -1, 0, 1 \}$.

32. $28x^2 + 71x + 18 = 0$
$(4x + 9)(7x + 2) = 0$
$4x + 9 = 0 \qquad$ or $\quad 7x + 2 = 0$
$4x = -9 \quad$ or $\qquad 7x = -2$
$x = -\dfrac{9}{4} \quad$ or $\qquad x = -\dfrac{2}{7}$
The solution set is $\left\{ -\dfrac{9}{4}, -\dfrac{2}{7} \right\}$.

33. $x^2 + 3x - 28 = 0$
$(x + 7)(x - 4) = 0$
$x + 7 = 0 \qquad$ or $\quad x - 4 = 0$
$x = -7 \quad$ or $\qquad x = 4$
The solution set is $\{ -7, 4 \}$.

34. $(x - 2)(x + 2) = 21$
$x^2 - 4 = 21$
$x^2 - 25 = 0$
$(x + 5)(x - 5) = 0$
$x + 5 = 0 \qquad$ or $\quad x - 5 = 0$
$x = -5 \quad$ or $\qquad x = 5$
The solution set is $\{ -5, 5 \}$.

35. $5n^2 + 27n = 18$
$5n^2 + 27n - 18 = 0$
$(5n - 3)(n + 6) = 0$
$5n - 3 = 0 \quad$ or $\quad n + 6 = 0$
$5n = 3 \quad$ or $\qquad n = -6$
$n = \dfrac{3}{5} \quad$ or $\qquad n = -6$
The solution set is $\left\{ -6, \dfrac{3}{5} \right\}$.

36. $4n^2 + 10n = 14$
$4n^2 + 10n - 14 = 0$
$2n^2 + 5n - 7 = 0$
$(2n + 7)(n - 1) = 0$
$2n + 7 = 0 \qquad$ or $\quad n - 1 = 0$
$2n = -7 \quad$ or $\qquad n = 1$
$n = -\dfrac{7}{2} \quad$ or $\qquad n = 1$
The solution set is $\left\{ -\dfrac{7}{2}, 1 \right\}$.

37.
$$2x^3 - 8x = 0$$
Divide both sides by 2.
$$x^3 - 4x = 0$$
$$x(x^2 - 4) = 0$$
$$x(x - 2)(x + 2) = 0$$
$x = 0$ or $x - 2 = 0$ or $x + 2 = 0$
$x = 0$ or $x = 2$ or $x = -2$
The solution set is $\{-2, 0, 2\}$.

38.
$$x^2 - 20x + 96 = 0$$
$$(x - 8)(x - 12) = 0$$
$x - 8 = 0$ or $x - 12 = 0$
$x = 8$ or $x = 12$
The solution set is $\{8, 12\}$.

39.
$$4t^2 + 17t - 15 = 0$$
$$(4t - 3)(t + 5) = 0$$
$4t - 3 = 0$ or $t + 5 = 0$
$4t = 3$ or $t = -5$
$t = \dfrac{3}{4}$ or $t = -5$
The solution set is $\left\{-5, \dfrac{3}{4}\right\}$.

40.
$$3(x + 2) - x(x + 2) = 0$$
$$(x + 2)(3 - x) = 0$$
$x + 2 = 0$ or $3 - x = 0$
$x = -2$ or $3 = x$
The solution set is $\{-2, 3\}$.

41.
$$(2x - 5)(3x + 7) = 0$$
$2x - 5 = 0$ or $3x + 7 = 0$
$2x = 5$ or $3x = -7$
$x = \dfrac{5}{2}$ or $t = -\dfrac{7}{3}$
The solution set is $\left\{-\dfrac{7}{3}, \dfrac{5}{2}\right\}$.

42.
$$(x + 4)(x - 1) = 50$$
$$x^2 + 3x - 4 = 50$$
$$x^2 + 3x - 54 = 0$$
$$(x + 9)(x - 6) = 0$$
$x + 9 = 0$ or $x - 6 = 0$
$x = -9$ or $x = 6$
The solution set is $\{-9, 6\}$.

43.
$$-7n - 2n^2 = -15$$
$$-2n^2 - 7n = -15$$
$$-2n^2 - 7n + 15 = 0$$
$$2n^2 + 7n - 15 = 0$$
$$(2n - 3)(n + 5) = 0$$
$2n - 3 = 0$ or $n + 5 = 0$
$2n = 3$ or $n = -5$
$n = \dfrac{3}{2}$ or $n = -5$
The solution set is $\left\{-5, \dfrac{3}{2}\right\}$.

44.
$$-23x + 6x^2 = -20$$
$$6x^2 - 23x = -20$$
$$6x^2 - 23x + 20 = 0$$
$$(2x - 5)(3x - 4) = 0$$
$2x - 5 = 0$ or $3x - 4 = 0$
$2x = 5$ or $3x = 4$
$x = \dfrac{5}{2}$ or $x = \dfrac{4}{3}$
The solution set is $\left\{\dfrac{4}{3}, \dfrac{5}{2}\right\}$.

45. Let n represent the smaller number, then $2n - 1$ is the larger number.
$$(2n - 1)^2 - n^2 = 33$$
$$4n^2 - 4n + 1 - n^2 = 33$$
$$3n^2 - 4n + 1 = 33$$
$$3n^2 - 4n - 32 = 0$$
$$(3n + 8)(n - 4) = 0$$
$3n + 8 = 0$ or $n - 4 = 0$
$3n = -8$ or $n = 4$
$n = -\dfrac{8}{3}$ or $n = 4$

Chapter 6 Review Problem Set

If $n = -\dfrac{8}{3}$, then $2n - 1 = 2\left(-\dfrac{8}{3}\right) - 1 =$

$-\dfrac{16}{3} - \dfrac{3}{3} = -\dfrac{19}{3}.$

If $n = 4$, then $2n - 1 = 2(4) - 1 = 7$

The numbers are $-\dfrac{8}{3}$ and $-\dfrac{19}{3}$

or 4 and 7.

46. Let w represent the width of the rectangle, then $5w - 2$ represents the length of the rectangle.

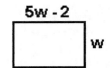

$$w(5w - 2) = 16$$
$$5w^2 - 2w = 16$$
$$5w^2 - 2w - 16 = 0$$
$$(5w + 8)(w - 2) = 0$$
$$5w + 8 = 0 \quad \text{or} \quad w - 2 = 0$$
$$5w = -8 \quad \text{or} \quad w = 2$$
$$w = -\dfrac{8}{5} \quad \text{or} \quad w = 2$$

Discard the negative solution. The width of the rectangle is 2 centimeters and the length is $5(2) - 2 = 8$ centimeters.

47. Let s represent the length of a side of the smaller square, then $5s$ represents the length of a side of the larger square.

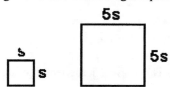

Smaller Area + Larger Area = 104
$$s^2 + (5s)^2 = 104$$
$$s^2 + 25s^2 = 104$$
$$26s^2 = 104$$
$$s^2 = 4$$
$$s^2 - 4 = 0$$
$$(s - 2)(s + 2) = 0$$
$$s - 2 = 0 \quad \text{or} \quad s + 2 = 0$$
$$s = 2 \quad \text{or} \quad s = -2$$

Discard the negative solution. The smaller square is 2 inches by 2 inches and the larger square is 10 inches by 10 inches.

48. Let x represent the length of the shorter leg, then $2x - 1$ represents the length of the longer leg, and $2x + 1$ represents the length of the hypotenuse.

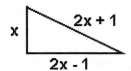

$$x^2 + (2x - 1)^2 = (2x + 1)^2$$
Pythagorean Theorem
$$x^2 + 4x^2 - 4x + 1 = 4x^2 + 4x + 1$$
$$5x^2 - 4x + 1 = 4x^2 + 4x + 1$$
$$x^2 - 8x = 0$$
$$x(x - 8) = 0$$
$$x = 0 \quad \text{or} \quad x - 8 = 0$$
$$x = 0 \quad \text{or} \quad x = 8$$

Discard the zero solution. The sides of the triangle are 8 units, 15 units, and 17 units.

49. Let n represent one number, then $6n + 1$ represents the other number.
$$n(6n + 1) = 26$$
$$6n^2 + n = 26$$
$$6n^2 + n - 26 = 0$$
$$(6n + 13)(n - 2) = 0$$

$$6n + 13 = 0 \quad \text{or} \quad n - 2 = 0$$
$$6n = -13 \quad \text{or} \quad n = 2$$
$$n = -\frac{13}{6} \quad \text{or} \quad n = 2$$

If $n = -\frac{13}{6}$, then $6n + 1 =$

$$6\left(-\frac{13}{6}\right) + 1 = -13 + 1 = -12.$$

If $n = 2$, then $6n + 1 = 6(2) + 1 = 13$.

The numbers are $-\frac{13}{6}$ and -12 or 2 and 13.

50. Let n represent the first whole number, then $n + 2$ and $n + 4$ represent the next two consecutive odd whole numbers.

$$n^2 + (n + 2)^2 = (n + 4)^2 + 9$$
$$n^2 + n^2 + 4n + 4 = n^2 + 8n + 16 + 9$$
$$2n^2 + 4n + 4 = n^2 + 8n + 25$$
$$n^2 - 4n - 21 = 0$$
$$(n + 3)(n - 7) = 0$$
$$n + 3 = 0 \quad \text{or} \quad n - 7 = 0$$
$$n = -3 \quad \text{or} \quad n = 7$$

Discard the negative solution because only positive whole numbers are wanted. Thus, the positive odd whole numbers are 7, 9, and 11.

51. Let s represent the number of shelves, then $9s - 1$ represents the number of books per shelf.

$$\left(\begin{array}{c}\text{Number of} \\ \text{shelves}\end{array}\right)\left(\begin{array}{c}\text{books per} \\ \text{shelf}\end{array}\right) = 140$$
$$s(9s - 1) = 140$$
$$9s^2 - s = 140$$
$$9s^2 - s - 140 = 0$$
$$(9s + 35)(s - 4) = 0$$
$$9s + 35 = 0 \quad \text{or} \quad s - 4 = 0$$
$$9s = -35 \quad \text{or} \quad s = 4$$
$$s = -\frac{35}{9} \quad \text{or} \quad s = 4$$

Discard the negative solution. There would be 4 shelves in the bookcase.

52. Let w represent the width of the rectangle, then $8w$ represents the length of the rectangle and w represents a side of the square.

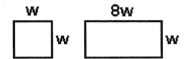

Square Area + Rectangle Area = 225
$$w^2 + w(8w) = 225$$
$$w^2 + 8w^2 = 225$$
$$9w^2 = 225$$
$$w^2 = 25$$
$$w^2 - 25 = 0$$
$$(w + 5)(w - 5) = 0$$
$$w + 5 = 0 \quad \text{or} \quad w - 5 = 0$$
$$w = -5 \quad \text{or} \quad w = 5$$

Discard the negative solution. The square would be 5 yards by 5 yards and the rectangle would be 5 yards by 40 yards.

53. Let n represent the first integer, then $n + 1$ represents the next integer.

$$n^2 + (n + 1)^2 = 613$$
$$n^2 + n^2 + 2n + 1 = 613$$
$$2n^2 + 2n - 612 = 0$$
$$n^2 + n - 306 = 0$$
$$(n + 18)(n - 17) = 0$$
$$n + 18 = 0 \quad \text{or} \quad n - 17 = 0$$
$$n = -18 \quad \text{or} \quad n = 17$$

The numbers would be -18 and -17 or 17 and 18.

54. Let x represent the length of a side of the cube.

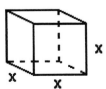

Volume of Cube = Surface Area
$$x^3 = 2x(x) + 2x(x) + 2x(x)$$
$$x^3 = 2x^2 + 2x^2 + 2x^2$$
$$x^3 = 6x^2$$
$$x^3 - 6x^2 = 0$$
$$x^2(x - 6) = 0$$
$$x = 0 \quad \text{or} \quad x = 0 \quad \text{or} \quad x - 6 = 0$$
$$x = 0 \quad \text{or} \quad x = 0 \quad \text{or} \quad x = 6$$
Discard the zero solution. The length of a side of the cube is 6 units.

55. Let r represent a radius of the smaller circle, then $3r + 1$ represents a radius of the larger circle.

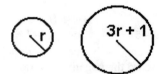

Smaller Area + Larger Area $= 53\pi$
$$\pi r^2 + \pi(3r + 1)^2 = 53\pi$$
$$\pi r^2 + \pi(9r^2 + 6r + 1) = 53\pi$$
$$\pi r^2 + 9\pi r^2 + 6\pi r + \pi = 53\pi$$
$$10\pi r^2 + 6\pi r + \pi = 53\pi$$
$$10\pi r^2 + 6\pi r - 52\pi = 0$$
Divide both sides by 2π.
$$5r^2 + 3r - 26 = 0$$
$$(5r + 13)(r - 2) = 0$$
$$5r + 13 = 0 \quad \text{or} \quad r - 2 = 0$$
$$5r = -13 \quad \text{or} \quad r = 2$$
$$r = -\frac{13}{5} \quad \text{or} \quad r = 2$$
Discard the negative solution. The length of a radius of the smaller circle is 2 meters and a radius of the larger circle is 7 meters.

56. Let n represent the first odd whole number, then $n + 2$ represents the next odd whole number. The sum of the two numbers is $n + (n + 2) = 2n + 2$.

$$n(n + 2) = 5(2n + 2) - 1$$
$$n^2 + 2n = 10n + 10 - 1$$
$$n^2 + 2n = 10n + 9$$
$$n^2 - 8n - 9 = 0$$
$$(n + 1)(n - 9) = 0$$
$$n + 1 = 0 \quad \text{or} \quad n - 9 = 0$$
$$n = -1 \quad \text{or} \quad n = 9$$
Discard the negative solution because whole numbers are positive. The numbers would be 9 and 11.

57. Let x represent the amount that the width and length are both reduced. The original area of the photograph is $8(14) = 112$ square centimeters. The new area is $112 - 40 = 72$ square centimeters.

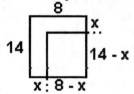

$$(8 - x)(14 - x) = 72$$
$$112 - 22x + x^2 = 72$$
$$40 - 22x + x^2 = 0$$
$$(20 - x)(2 - x) = 0$$
$$20 - x = 0 \quad \text{or} \quad 2 - x = 0$$
$$20 = x \quad \text{or} \quad 2 = x$$
The solution $x = 20$ is not reasonable. Thus, the length and the width must be reduced by 2 centimeters.

58. Let x represent the width of the plowed strip. Then $120 - 2x$ represents the length of the unplowed garden, and $90 - 2x$ represents the width. The original area of the garden is $90(120) = 10,800$ square feet. Thus, the unplowed garden is 5400 square feet.

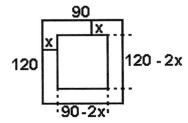

$$(90 - 2x)(120 - 2x) = 5400$$
$$10,800 - 420x + 4x^2 = 5400$$
$$5400 - 420x + 4x^2 = 0$$
$$1350 - 105x + x^2 = 0$$
$$(90 - x)(15 - x) = 0$$
$$90 - x = 0 \quad \text{or} \quad 15 - x = 0$$
$$90 = x \quad \text{or} \quad 15 = x$$

The solution $x = 90$ is not reasonable.
Thus, the width of the strip must be 15 feet.

CHAPTER 6 Test

1. We need two integers whose product is
-10 and whose sum is 3. They are
-2 and 5.
$$x^2 + 3x - 10 = (x - 2)(x + 5)$$

2. We need two integers whose product is
-24 and whose sum is -5. They are 3
and -8.
$$x^2 - 5x - 24 = (x + 3)(x - 8)$$

3. $2x^3 - 2x =$
$2x(x^2 - 1) =$
$2x(x - 1)(x + 1)$

4. We need two integers whose product is
108 and whose sum is 21. They are 9 and 12.
$$x^2 + 21x + 108 = (x + 9)(x + 12)$$

5. $18n^2 + 21n + 6 =$
$3(6n^2 + 7n + 2)$
We need two integers whose product is 12
and whose sum is 7. They are 3 and 4.
$3(6n^2 + 7n + 2) =$
$3(6n^2 + 3n + 4n + 2) =$
$3[3n(2n + 1) + 2(2n + 1)] =$
$3(2n + 1)(3n + 2)$

6. $ax + ay + 2bx + 2by =$
$a(x + y) + 2b(x + y) =$
$(x + y)(a + 2b)$

7. We need two integers whose product is -60
and whose sum is 17. They are -3 and 20.
$4x^2 + 17x - 15 =$
$4x^2 - 3x + 20x - 15 =$

$x(4x - 3) + 5(4x - 3) =$
$(4x - 3)(x + 5)$

8. $6x^2 + 24 = 6(x^2 + 4)$

9. $30x^3 - 76x^2 + 48x =$
$2x(15x^2 - 38x + 24)$
We need two integers whose product is
360 and whose sum is -38. They are
-18 and -20.
$2x(15x^2 - 38x + 24) =$
$2x(15x^2 - 18x - 20x + 24) =$
$2x[3x(5x - 6) - 4(5x - 6)] =$
$2x(5x - 6)(3x - 4)$

10. We need two integers whose product is
-168 and whose sum is 13. They are
-8 and 21.
$28 + 13x - 6x^2 =$
$28 - 8x + 21x - 6x^2 =$
$4(7 - 2x) + 3x(7 - 2x) =$
$(7 - 2x)(4 + 3x)$

11. $$7x^2 = 63$$
Divide both sides by 7.
$$x^2 = 9$$
$$x^2 - 9 = 0$$
$$(x - 3)(x + 3) = 0$$
$$x - 3 = 0 \quad \text{or} \quad x + 3 = 0$$
$$x = 3 \quad \text{or} \quad x = -3$$
The solution set is $\{-3, 3\}$.

159

Chapter 6 Test

12.
$$x^2 + 5x - 6 = 0$$
$$(x-1)(x+6) = 0$$
$$x - 1 = 0 \quad \text{or} \quad x + 6 = 0$$
$$x = 1 \quad \text{or} \quad x = -6$$
The solution set is $\{-6, 1\}$.

13.
$$4n^2 = 32n$$
$$4n^2 - 32n = 0$$
$$n^2 - 8n = 0$$
$$n(n-8) = 0$$
$$n = 0 \quad \text{or} \quad n - 8 = 0$$
$$n = 0 \quad \text{or} \quad n = 8$$
The solution set is $\{0, 8\}$.

14.
$$(3x-2)(2x+5) = 0$$
$$3x - 2 = 0 \quad \text{or} \quad 2x + 5 = 0$$
$$3x = 2 \quad \text{or} \quad 2x = -5$$
$$x = \frac{2}{3} \quad \text{or} \quad x = -\frac{5}{2}$$
The solution set is $\left\{-\frac{5}{2}, \frac{2}{3}\right\}$.

15.
$$(x-3)(x+7) = -9$$
$$x^2 + 4x - 21 = -9$$
$$x^2 + 4x - 12 = 0$$
$$(x+6)(x-2) = 0$$
$$x + 6 = 0 \quad \text{or} \quad x - 2 = 0$$
$$n = -6 \quad \text{or} \quad x = 2$$
The solution set is $\{-6, 2\}$.

16.
$$x^3 + 16x^2 + 48x = 0$$
$$x(x^2 + 16x + 48) = 0$$
$$x(x+4)(x+12) = 0$$
$$x = 0 \quad \text{or} \quad x + 4 = 0 \quad \text{or} \quad x + 12 = 0$$
$$x = 0 \quad \text{or} \quad x = -4 \quad \text{or} \quad x = -12$$
The solution set is $\{-12, -4, 0\}$.

17.
$$9(x-5) - x(x-5) = 0$$
$$(x-5)(9-x) = 0$$
$$x - 5 = 0 \quad \text{or} \quad 9 - x = 0$$
$$x = 5 \quad \text{or} \quad 9 = x$$
The solution set is $\{5, 9\}$.

18.
$$3t^2 + 35t = 12$$
$$3t^2 + 35t - 12 = 0$$
$$(3t-1)(t+12) = 0$$
$$3t - 1 = 0 \quad \text{or} \quad t + 12 = 0$$
$$3t = 1 \quad \text{or} \quad t = -12$$
$$t = \frac{1}{3} \quad \text{or} \quad t = -12$$
The solution set is $\left\{-12, \frac{1}{3}\right\}$.

19.
$$8 - 10x - 3x^2 = 0$$
$$(4+x)(2-3x) = 0$$
$$4 + x = 0 \quad \text{or} \quad 2 - 3x = 0$$
$$x = -4 \quad \text{or} \quad -3x = -2$$
$$x = -4 \quad \text{or} \quad x = \frac{2}{3}$$
The solution set is $\left\{-4, \frac{2}{3}\right\}$.

20.
$$3x^3 = 75x$$
$$x^3 = 25x$$
$$x^3 - 25x = 0$$
$$x(x^2 - 25) = 0$$
$$x(x+5)(x-5) = 0$$
$$x = 0 \quad \text{or} \quad x + 5 = 0 \quad \text{or} \quad x - 5 = 0$$
$$x = 0 \quad \text{or} \quad x = -5 \quad \text{or} \quad x = 5$$
The solution set is $\{-5, 0, 5\}$.

21.
$$25n^2 - 70n + 49 = 0$$
$$(5n-7)(5n-7) = 0$$
$$5n - 7 = 0 \quad \text{or} \quad 5n - 7 = 0$$
$$5n = 7 \quad \text{or} \quad 5n = 7$$
$$x = \frac{7}{5} \quad \text{or} \quad x = \frac{7}{5}$$
The solution set is $\left\{\frac{7}{5}\right\}$.

22. Let w represent the width of the rectangle, then $2w - 2$ represents the length of the rectangle.

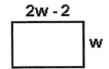

$$w(2w - 2) = 112$$
$$2w^2 - 2w = 112$$
$$2w^2 - 2w - 112 = 0$$
$$w^2 - w - 56 = 0$$
$$(w - 8)(w + 7) = 0$$
$$w - 8 = 0 \quad \text{or} \quad w + 7 = 0$$
$$w = 8 \quad \text{or} \qquad w = -7$$

Discard the negative solution. If $w = 8$, then the length of the rectangle is $2(8) - 2 = 14$ in.

23. Let x represent the length of the shorter leg, then $x + 4$ represents the length of the longer leg and $x + 8$ represents the length of the hypotenuse.

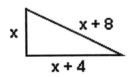

$$x^2 + (x + 4)^2 = (x + 8)^2$$
Pythagorean Theorem
$$x^2 + x^2 + 8x + 16 = x^2 + 16x + 64$$
$$2x^2 + 8x + 16 = x^2 + 16x + 64$$
$$x^2 - 8x - 48 = 0$$
$$(x - 12)(x + 4) = 0$$
$$x - 12 = 0 \quad \text{or} \quad x + 4 = 0$$
$$x = 12 \quad \text{or} \qquad x = -4$$

Discard the negative solution. The length of the shorter leg is 12 centimeters.

24. Let r represent the number of rows, then $3r - 5$ represents the number of chairs per row.

$$(\text{rows})(\text{chairs per rows}) = 112$$
$$r(3r - 5) = 112$$
$$3r^2 - 5r = 112$$
$$3r^2 - 5r - 112 = 0$$
$$(3r + 16)(r - 7) = 0$$
$$3r + 16 = 0 \qquad \text{or} \quad r - 7 = 0$$
$$3r = -16 \quad \text{or} \qquad r = 7$$
$$r = -\frac{16}{3} \quad \text{or} \qquad r = 7$$

Discard the negative solution. There are 7 rows with 16 chairs per row.

25. Let x represent the length of a side of the cube.

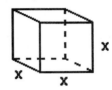

$$\text{Volume of Cube} = 2[\text{Surface Area}]$$
$$x^3 = 2[2x(x) + 2x(x) + 2x(x)]$$
$$x^3 = 2[2x^2 + 2x^2 + 2x^2]$$
$$x^3 = 2[6x^2]$$
$$x^3 = 12x^2$$
$$x^3 - 12x^2 = 0$$
$$x^2(x - 12) = 0$$
$$x = 0 \quad \text{or} \quad x = 0 \quad \text{or} \quad x - 12 = 0$$
$$x = 0 \quad \text{or} \quad x = 0 \quad \text{or} \qquad x = 12$$

Discard the zero solution. The length of a side of the cube is 12 units.

Chapter 7 Algebraic Fractions

PROBLEM SET **7.1** Simplifying Algebraic Fractions

1. $\dfrac{6x}{14y} = \dfrac{\cancel{2}\cdot 3\cdot x}{\cancel{2}\cdot 7\cdot y} = \dfrac{3x}{7y}$

3. $\dfrac{9xy}{24x} = \dfrac{3\cdot \cancel{3}\cdot \cancel{x}\cdot y}{\cancel{3}\cdot 8\cdot \cancel{x}} = \dfrac{3y}{8}$

5. $\dfrac{-15x^2y}{25x} = -\dfrac{3\cdot \cancel{5}\cdot \overset{x}{\cancel{x^2}}\cdot y}{5\cdot \cancel{5}\cdot \cancel{x}} = -\dfrac{3xy}{5}$

7. $\dfrac{-36x^4y^3}{-48x^6y^2} = +\dfrac{\overset{3}{\cancel{36}}\cdot \cancel{x^4}\cdot \overset{y}{\cancel{y^3}}}{\underset{4}{\cancel{48}}\cdot \underset{x^2}{\cancel{x^6}}\cdot \cancel{y^2}} = \dfrac{3y}{4x^2}$

9. $\dfrac{12a^2b^5}{-54a^2b^3} = -\dfrac{\cancel{6}\cdot 2\cdot \cancel{a^2}\cdot \overset{b^2}{\cancel{b^5}}}{\cancel{6}\cdot 9\cdot \cancel{a^2}\cdot \cancel{b^3}} = -\dfrac{2b^2}{9}$

11. $\dfrac{32xy^2z^3}{72yz^4} = \dfrac{\overset{4}{\cancel{32}}\cdot x\cdot \overset{y}{\cancel{y^2}}\cdot \cancel{z^3}}{\underset{9}{\cancel{72}}\cdot \cancel{y}\cdot \underset{z}{\cancel{z^4}}} = \dfrac{4xy}{9z}$

13. $\dfrac{xy}{x^2 - 2x} = \dfrac{\cancel{x}y}{\cancel{x}(x-2)} = \dfrac{y}{x-2}$

15. $\dfrac{8x + 12y}{12} = \dfrac{\overset{4}{\cancel{4}}(2x+3y)}{\underset{3}{\cancel{12}}} = \dfrac{2x+3y}{3}$

17. $\dfrac{x^2 + 2x}{x^2 - 7x} = \dfrac{\cancel{x}(x+2)}{\cancel{x}(x-7)} = \dfrac{x+2}{x-7}$

19. $\dfrac{7-x}{x-7} = -1$ because $7-x$ and $x-7$
     are opposites.

21. $\dfrac{15 - 3n}{n - 5} = \dfrac{3(5-n)}{n-5} = 3(-1) = -3$
     Remember $\dfrac{5-n}{n-5} = -1$

23. $\dfrac{4x^3 - 4x}{1 - x^2} = \dfrac{4x(x^2 - 1)}{1 - x^2} = 4x(-1) = -4x$
     Remember $\dfrac{x^2 - 1}{1 - x^2} = -1$

25. $\dfrac{x^2 - 1}{3x^2 - 3x} = \dfrac{\cancel{(x-1)}(x+1)}{3x\cancel{(x-1)}} = \dfrac{x+1}{3x}$

27. $\dfrac{x^2 + xy}{x^2} = \dfrac{\cancel{x}(x+y)}{\underset{x}{\cancel{x^2}}} = \dfrac{x+y}{x}$

29. $\dfrac{6x^3 - 15x^2y}{6x^2 + 24xy} = \dfrac{\overset{x}{\cancel{3}\cancel{x^2}}(2x - 5y)}{\underset{2}{\cancel{6}\cancel{x}(x+4y)}} =$
     $\dfrac{x(2x-5y)}{2(x+4y)}$

31. $\dfrac{n^2 + 2n}{n^2 + 3n + 2} = \dfrac{n\cancel{(n+2)}}{(n+1)\cancel{(n+2)}} =$
     $\dfrac{n}{n+1}$

33. $\dfrac{2n^2 + 5n - 3}{n^2 - 9} = \dfrac{(2n-1)\cancel{(n+3)}}{(n-3)\cancel{(n+3)}} =$
     $\dfrac{2n-1}{n-3}$

35. $\dfrac{2x^2 + 17x + 35}{3x^2 + 19x + 20} = \dfrac{(2x+7)\cancel{(x+5)}}{(3x+4)\cancel{(x+5)}} =$
     $\dfrac{2x+7}{3x+4}$

37. $\dfrac{9(x-1)^2}{12(x-1)^3} =$

$\dfrac{\overset{3}{\cancel{9}}\,(\cancel{x-1})\,(\cancel{x-1})}{\underset{4}{\cancel{12}}\,(\cancel{x-1})\,(\cancel{x-1})\,(x-1)} =$

$\dfrac{3}{4(x-1)}$

39. $\dfrac{7x^2 + 61x - 18}{7x^2 + 19x - 6} =$

$\dfrac{(\cancel{7x-2})(x+9)}{(\cancel{7x-2})(x+3)} =$

$\dfrac{x+9}{x+3}$

41. $\dfrac{10a^2 + a - 3}{15a^2 + 4a - 3} =$

$\dfrac{(\cancel{5a+3})(2a-1)}{(\cancel{5a+3})(3a-1)} =$

$\dfrac{2a-1}{3a-1}$

43. $\dfrac{x^2 + 2xy - 3y^2}{2x^2 - xy - y^2} =$

$\dfrac{(\cancel{x-y})\,(x+3y)}{(\cancel{x-y})\,(2x+y)} =$

$\dfrac{x+3y}{2x+y}$

45. $\dfrac{x^2 - 9}{-x^2 - 3x} =$

$\dfrac{(x-3)(\cancel{x+3})}{-x(\cancel{x+3})} =$

$-\dfrac{x-3}{x}$

47. $\dfrac{n^2 + 14n + 49}{8n + 56} =$

$\dfrac{(n+7)(\cancel{n+7})}{8(\cancel{n+7})} =$

$\dfrac{n+7}{8}$

49. $\dfrac{4n^2 - 12n + 9}{2n^2 - n - 3} =$

$\dfrac{(\cancel{2n-3})(2n-3)}{(\cancel{2n-3})(n+1)} =$

$\dfrac{2n-3}{n+1}$

51. $\dfrac{y^2 - 6y - 72}{y^2 - 8y - 84} =$

$\dfrac{(y-12)(\cancel{y+6})}{(y-14)(\cancel{y+6})} =$

$\dfrac{y-12}{y-14}$

53. $\dfrac{1 - x^2}{x - x^2} =$

$\dfrac{(\cancel{1-x})\,(1+x)}{x(\cancel{1-x})} =$

$\dfrac{1+x}{x}$

55. $\dfrac{6 - x - 2x^2}{12 + 7x - 10x^2} =$

$\dfrac{(2+x)(\cancel{3-2x})}{(4+5x)(\cancel{3-2x})} =$

$\dfrac{2+x}{4+5x}$

57. $\dfrac{x^2 + 7x - 18}{12 - 4x - x^2} = \dfrac{(x+9)(x-2)}{(6+x)(2-x)}$

$= -\dfrac{x+9}{x+6}$

Remember $\dfrac{x-2}{2-x} = -1$

59. $\dfrac{5x - 40}{80 - 10x} = \dfrac{\cancel{5}(x-8)}{\underset{2}{\cancel{10}}\,(8-x)} = -\dfrac{1}{2}$

Remember $\dfrac{x-8}{8-x} = -1$

Problem Set 7.1

63. $\dfrac{xy - 3x + 2y - 6}{xy + 5x + 2y + 10} =$

$\dfrac{x(y-3) + 2(y-3)}{x(y+5) + 2(y+5)} =$

$\dfrac{(y-3)\cancel{(x+2)}}{(y+5)\cancel{(x+2)}} = \dfrac{y-3}{y+5}$

65. $\dfrac{xy - 6x + y - 6}{xy - 6x + 5y - 30} =$

$\dfrac{x(y-6) + 1(y-6)}{x(y-6) + 5(y-6)} =$

$\dfrac{\cancel{(y-6)}\,(x+1)}{\cancel{(y-6)}\,(x+5)} = \dfrac{x+1}{x+5}$

67. $\dfrac{x^3}{x^9} = x^{3-9} = x^{-6} = \dfrac{1}{x^6}$

69. $\dfrac{x^4 y^3}{x^7 y^5} = x^{4-7} y^{3-5} = x^{-3} y^{-2} = \dfrac{1}{x^3 y^2}$

71. $\dfrac{28 a^2 b^3}{-7 a^5 b^3} = -4 a^{2-5} b^{3-3} =$

$-4 a^{-3} b^0 = -\dfrac{4}{a^3}$

PROBLEM SET | **7.2**   **Multiplying and Dividing Algebraic Fractions**

1. $\dfrac{5}{9} \cdot \dfrac{3}{10} = \dfrac{\cancel{5} \cdot \cancel{3}}{\underset{3}{\cancel{9}} \cdot \underset{2}{\cancel{10}}} = \dfrac{1}{6}$

3. $\left(-\dfrac{3}{4}\right)\left(\dfrac{6}{7}\right) = -\dfrac{3 \cdot \overset{3}{\cancel{6}}}{\underset{2}{\cancel{4}} \cdot 7} = -\dfrac{9}{14}$

5. $\left(\dfrac{17}{9}\right) \div \left(-\dfrac{19}{9}\right) =$

$\left(\dfrac{17}{9}\right) \cdot \left(-\dfrac{9}{19}\right) =$

$-\dfrac{17 \cdot \cancel{9}}{\cancel{9} \cdot 19} = -\dfrac{17}{19}$

7. $\dfrac{8xy}{12y} \cdot \dfrac{6x}{14y} = \dfrac{\overset{2}{\underset{}{\cancel{8}}} \cdot \overset{}{\cancel{6}} \cdot x \cdot x \cdot \cancel{y}}{\underset{2}{\cancel{12}} \cdot \underset{7}{\cancel{14}} \cdot y \cdot \cancel{y}} = \dfrac{2x^2}{7y}$

9. $\left(-\dfrac{5n^2}{18n}\right)\left(\dfrac{27n}{25}\right) =$

$-\dfrac{\cancel{5} \cdot \overset{3}{\cancel{27}} \cdot n^2 \cdot \cancel{n}}{\underset{2}{\cancel{18}} \cdot \underset{5}{\cancel{25}} \cdot \cancel{n}} =$

$-\dfrac{3n^2}{10}$

11. $\dfrac{3a^2}{7} \div \dfrac{6a}{28} = \dfrac{3a^2}{7} \cdot \dfrac{28}{6a} =$

$\dfrac{\cancel{3} \cdot \overset{\overset{2}{\cancel{4}}}{\cancel{28}} \cdot \overset{a}{\cancel{a^2}}}{7 \cdot \underset{\underset{2}{\cancel{2}}}{\cancel{6}} \cdot \cancel{a}} = 2a$

13. $\dfrac{18a^2 b^2}{-27a} \div \dfrac{-9a}{5b} = \dfrac{18a^2 b^2}{-27a} \cdot \dfrac{5b}{-9a} =$

$\dfrac{\overset{2}{\cancel{18}} \cdot 5 \cdot \cancel{a^2} \cdot b^2 \cdot b}{27 \cdot \cancel{9} \cdot \cancel{a} \cdot \cancel{a}} = \dfrac{10b^3}{27}$

15. $24x^3 \div \dfrac{16x}{y} = \dfrac{24x^3}{1} \cdot \dfrac{y}{16x} =$

$$\dfrac{\overset{3}{\cancel{24}} \cdot \overset{x^2}{\cancel{x^3}} \cdot y}{\underset{2}{\cancel{16}} \cdot \cancel{x}} = \dfrac{3x^2 y}{2}$$

17. $\dfrac{1}{15ab^3} \div \dfrac{-1}{12a} = \dfrac{1}{15ab^3} \cdot \dfrac{12a}{-1} =$

$$-\dfrac{\overset{4}{\cancel{12}} \cdot \cancel{a}}{\underset{5}{\cancel{15}} \cdot \cancel{a} \cdot b^3} = -\dfrac{4}{5b^3}$$

19. $\dfrac{18rs}{34} \div \dfrac{9r}{1} = \dfrac{18rs}{34} \cdot \dfrac{1}{9r} =$

$$\dfrac{\overset{\cancel{2}}{\cancel{18}} \cdot \cancel{r} \cdot s}{\underset{17}{\cancel{34}} \cdot \cancel{9} \cdot \cancel{r}} = \dfrac{s}{17}$$

21. $\dfrac{y}{x+y} \cdot \dfrac{x^2 - y^2}{xy} =$

$$\dfrac{\cancel{y}(x-y)\cancel{(x+y)}}{\cancel{(x+y)}(x)\cancel{(y)}} = \dfrac{x-y}{x}$$

23. $\dfrac{2x^2 + xy}{xy} \cdot \dfrac{y}{10x + 5y} =$

$$\dfrac{\cancel{x}\cancel{(2x+y)}\cancel{(y)}}{\cancel{(x)}\cancel{(y)}(5)\cancel{(2x+y)}} = \dfrac{1}{5}$$

25. $\dfrac{6ab}{4ab + 4b^2} \div \dfrac{7a - 7b}{a^2 - b^2} =$

$$\dfrac{6ab}{4ab + 4b^2} \cdot \dfrac{a^2 - b^2}{7a - 7b} =$$

$$\dfrac{\overset{3}{\cancel{6}}a\cancel{b}\cancel{(a-b)}(a+b)}{\underset{2}{\cancel{4}}\cancel{b}\cancel{(a+b)}(7)\cancel{(a-b)}} = \dfrac{3a}{14}$$

27. $\dfrac{x^2 + 11x + 30}{x^2 + 4} \cdot \dfrac{5x^2 + 20}{x^2 + 14x + 45} =$

$$\dfrac{\cancel{(x+5)}(x+6)(5)\cancel{(x^2+4)}}{\cancel{(x^2+4)}\cancel{(x+5)}(x+9)} =$$

$$\dfrac{5(x+6)}{x+9}$$

29. $\dfrac{2x^2 - 3xy + y^2}{4x^2 y} \div \dfrac{x^2 - y^2}{6x^2 y^2} =$

$$\dfrac{2x^2 - 3xy + y^2}{4x^2 y} \cdot \dfrac{6x^2 y^2}{x^2 - y^2} =$$

$$\dfrac{(2x-y)\cancel{(x-y)}(\overset{3}{\cancel{6}}\cancel{x^2}\overset{y}{\cancel{y^2}})}{\underset{2}{\cancel{4}}\cancel{x^2}\cancel{y}\cancel{(x-y)}(x+y)} =$$

$$\dfrac{3y(2x-y)}{2(x+y)}$$

31. $\dfrac{a + a^2}{15a^2 + 11a + 2} \cdot \dfrac{1 - a}{1 - a^2} =$

$$\dfrac{a\cancel{(1+a)}\cancel{(1-a)}}{(5a+2)(3a+1)\cancel{(1-a)}\cancel{(1+a)}} =$$

$$\dfrac{a}{(5a+2)(3a+1)}$$

33. $\dfrac{2x^2 - 2xy}{x^2 + 4x - 32} \cdot \dfrac{x^2 - 16}{5xy - 5y^2} =$

$$\dfrac{2x\cancel{(x-y)}\cancel{(x-4)}(x+4)}{(x+8)\cancel{(x-4)}(5y)\cancel{(x-y)}} =$$

$$\dfrac{2x(x+4)}{5y(x+8)}$$

Problem Set 7.2

35. $\dfrac{2x^2 - xy - 3y^2}{(x+y)^2} \div \dfrac{4x^2 - 12xy + 9y^2}{10x - 15y} =$

$\dfrac{2x^2 - xy - 3y^2}{(x+y)^2} \cdot \dfrac{10x - 15y}{4x^2 - 12xy + 9y^2} =$

$\dfrac{(2x - 3y)(x + y)\,(5)(2x - 3y)}{(x+y)\,(x+y)(2x - 3y)(2x - 3y)} =$

$\dfrac{5}{x+y}$

37. $\dfrac{(3t-1)^2}{45t - 15} \div \dfrac{12t^2 + 5t - 3}{20t + 5} =$

$\dfrac{(3t-1)^2}{45t - 15} \cdot \dfrac{20t + 5}{12t^2 + 5t - 3} =$

$\dfrac{(3t - 1)(3t - 1)(5)(4t + 1)}{\underset{3}{15}\,(3t - 1)(3t - 1)(4t + 3)} =$

$\dfrac{4t + 1}{3(4t + 3)}$

39. $\dfrac{n^3 - n}{n^2 + 7n + 6} \cdot \dfrac{4n + 24}{n^2 - n} =$

$\dfrac{n(n - 1)(n + 1)(4)(n + 6)}{(n + 1)(n + 6)(n)(n - 1)} = 4$

41. $\dfrac{6}{9y} \div \dfrac{30x}{12y^2} \cdot \dfrac{5xy}{4} =$

$\left(\dfrac{6}{9y} \cdot \dfrac{12y^2}{30x} \right) \cdot \dfrac{5xy}{4} =$

$\left(\dfrac{\overset{}{6} \cdot \overset{4}{12} \cdot \overset{y}{y^2}}{\underset{3}{9} \cdot \underset{5}{30} \cdot x \cdot y} \right) \cdot \dfrac{5xy}{4} =$

$\dfrac{\overset{4y}{\underset{3}{15}}}{} \not{} \cdot \dfrac{5\not{x}y}{4} = \dfrac{y^2}{3}$

43. $\dfrac{8x^2}{xy - xy^2} \cdot \dfrac{x - 1}{8x^2 - 8y^2} \div \dfrac{xy}{x + y} =$

$\left[\dfrac{\overset{x}{8x^2}\,(x - 1)}{xy(1 - y)(8)(x^2 - y^2)} \right] \cdot \dfrac{x + y}{xy} =$

$\dfrac{x(x - 1)(x + y)}{y(1 - y)(x - y)(x + y)(xy)} =$

$\dfrac{x - 1}{y^2(1 - y)(x - y)}$

45. $\dfrac{x^2 + 9x + 18}{x^2 + 3x} \cdot \dfrac{x^2 + 5x}{x^2 - 25} \div \dfrac{x^2 + 8x}{x^2 + 3x - 40} =$

$\dfrac{(x + 3)(x + 6)(x)(x + 5)}{(x)(x + 3)(x - 5)(x + 5)} \div \dfrac{x^2 + 8x}{x^2 + 3x - 40} =$

$\dfrac{(x + 6)}{(x - 5)} \cdot \dfrac{(x + 8)(x - 5)}{x(x + 8)} = \dfrac{x + 6}{x}$

PROBLEM SET 7.3   **Adding and Subtracting Algebraic Fractions**

1. $\dfrac{5}{x} + \dfrac{12}{x} = \dfrac{5 + 12}{x} = \dfrac{17}{x}$

3. $\dfrac{7}{3x} - \dfrac{5}{3x} = \dfrac{7 - 5}{3x} = \dfrac{2}{3x}$

5. $\dfrac{7}{2n} + \dfrac{1}{2n} = \dfrac{7 + 1}{2n} = \dfrac{8}{2n} = \dfrac{4}{n}$

7. $\dfrac{9}{4x^2} - \dfrac{13}{4x^2} = \dfrac{9 - 13}{4x^2} = \dfrac{-4}{4x^2} = -\dfrac{1}{x^2}$

9. $\dfrac{x+1}{x} + \dfrac{3}{x} = \dfrac{(x+1)+(3)}{x} = \dfrac{x+4}{x}$

11. $\dfrac{3}{x-1} - \dfrac{6}{x-1} = \dfrac{3-6}{x-1} =$

$\dfrac{-3}{x-1} = -\dfrac{3}{x-1}$

13. $\dfrac{x+1}{x} - \dfrac{1}{x} = \dfrac{(x+1)-(1)}{x} = \dfrac{x}{x} = 1$

15. $\dfrac{3t-1}{4} + \dfrac{2t+3}{4} =$

$\dfrac{(3t-1)+(2t+3)}{4} =$

$\dfrac{5t+2}{4}$

17. $\dfrac{7a+2}{3} - \dfrac{4a-6}{3} =$

$\dfrac{(7a+2)-(4a-6)}{3} =$

$\dfrac{7a+2-4a+6}{3} =$

$\dfrac{3a+8}{3}$

19. $\dfrac{4n+3}{8} + \dfrac{6n+5}{8} =$

$\dfrac{(4n+3)+(6n+5)}{8} =$

$\dfrac{10n+8}{8} = \dfrac{2(5n+4)}{8} =$

$\dfrac{5n+4}{4}$

21. $\dfrac{3n-7}{6} - \dfrac{9n-1}{6} =$

$\dfrac{(3n-7)-(9n-1)}{6} =$

$\dfrac{3n-7-9n+1}{6} =$

$\dfrac{-6n-6}{6} = \dfrac{6(-n-1)}{6} = -n-1$

23. $\dfrac{5x-2}{7x} - \dfrac{8x+3}{7x} =$

$\dfrac{(5x-2)-(8x+3)}{7x} =$

$\dfrac{5x-2-8x-3}{7x} =$

$\dfrac{-3x-5}{7x}$

25. $\dfrac{3(x+2)}{4x} + \dfrac{6(x-1)}{4x} =$

$\dfrac{3(x+2)+6(x-1)}{4x} =$

$\dfrac{3x+6+6x-6}{4x} =$

$\dfrac{9x}{4x} = \dfrac{9}{4}$

27. $\dfrac{6(n-1)}{3n} + \dfrac{3(n+2)}{3n} =$

$\dfrac{6(n-1)+3(n+2)}{3n} =$

$\dfrac{6n-6+3n+6}{3n} =$

$\dfrac{9n}{3n} = 3$

29. $\dfrac{2(3x-4)}{7x^2} - \dfrac{7x-8}{7x^2} =$

$\dfrac{2(3x-4)-(7x-8)}{7x^2} =$

$\dfrac{6x-8-7x+8}{7x^2} =$

$\dfrac{-x}{7x^2} = -\dfrac{1}{7x}$

31. $\dfrac{a^2}{a+2} - \dfrac{4}{a+2} = \dfrac{a^2-4}{a+2} =$

$\dfrac{(a-2)(a+2)}{a+2} = a-2$

167

Problem Set 7.3

33. $\dfrac{3x}{(x-6)^2} - \dfrac{18}{(x-6)^2} =$

$\dfrac{3x-18}{(x-6)^2} =$

$\dfrac{3(x-6)}{(x-6)(x-6)} = \dfrac{3}{x-6}$

35. $\dfrac{3x}{8} + \dfrac{5x}{4} =$

$\dfrac{3x}{8} + \left(\dfrac{5x}{4}\right)\left(\dfrac{2}{2}\right) =$

$\dfrac{3x}{8} + \dfrac{2(5x)}{8} =$

$\dfrac{3x+10x}{8} = \dfrac{13x}{8}$

37. $\dfrac{7n}{12} - \dfrac{4n}{3} =$

$\dfrac{7n}{12} - \left(\dfrac{4n}{3}\right)\left(\dfrac{4}{4}\right) =$

$\dfrac{7n}{12} - \dfrac{16n}{12} = \dfrac{7n-16n}{12} =$

$\dfrac{-9n}{12} = -\dfrac{3n}{4}$

39. $\dfrac{y}{6} + \dfrac{3y}{4} =$

$\left(\dfrac{y}{6}\right)\left(\dfrac{2}{2}\right) + \left(\dfrac{3y}{4}\right)\left(\dfrac{3}{3}\right) =$

$\dfrac{2y}{12} + \dfrac{9y}{12} =$

$\dfrac{2y+9y}{12} = \dfrac{11y}{12}$

41. $\dfrac{8x}{3} - \dfrac{3x}{7} =$

$\left(\dfrac{8x}{3}\right)\left(\dfrac{7}{7}\right) - \left(\dfrac{3x}{7}\right)\left(\dfrac{3}{3}\right) =$

$\dfrac{56x}{21} - \dfrac{9x}{21} = \dfrac{56x-9x}{21} = \dfrac{47x}{21}$

43. $\dfrac{2x}{6} + \dfrac{3x}{5} =$

$\left(\dfrac{2x}{6}\right)\left(\dfrac{5}{5}\right) + \left(\dfrac{3x}{5}\right)\left(\dfrac{6}{6}\right) =$

$\dfrac{10x}{30} + \dfrac{18x}{30} = \dfrac{10x+18x}{30} =$

$\dfrac{28x}{30} = \dfrac{14x}{15}$

45. $\dfrac{7n}{8} - \dfrac{3n}{9} =$

$\left(\dfrac{7n}{8}\right)\left(\dfrac{9}{9}\right) - \left(\dfrac{3n}{9}\right)\left(\dfrac{8}{8}\right) =$

$\dfrac{63n}{72} - \dfrac{24n}{72} = \dfrac{63n-24n}{72} =$

$\dfrac{39n}{72} = \dfrac{13n}{24}$

47. $\dfrac{x+3}{5} + \dfrac{x-4}{2} =$

$\left(\dfrac{x+3}{5}\right)\left(\dfrac{2}{2}\right) + \left(\dfrac{x-4}{2}\right)\left(\dfrac{5}{5}\right) =$

$\dfrac{2(x+3)}{10} + \dfrac{5(x-4)}{10} =$

$\dfrac{2(x+3)+5(x-4)}{10} =$

$\dfrac{2x+6+5x-20}{10} = \dfrac{7x-14}{10}$

49. $\dfrac{x-6}{9} + \dfrac{x+2}{3} =$

$\dfrac{x-6}{9} + \left(\dfrac{x+2}{3}\right)\left(\dfrac{3}{3}\right) =$

$\dfrac{x-6}{9} + \dfrac{3(x+2)}{9} =$

$\dfrac{x-6+3(x+2)}{9} =$

$\dfrac{x-6+3x+6}{9} = \dfrac{4x}{9}$

51. $\dfrac{3n-1}{3} + \dfrac{2n+5}{4} =$

$\left(\dfrac{3n-1}{3}\right)\left(\dfrac{4}{4}\right) + \left(\dfrac{2n+5}{4}\right)\left(\dfrac{3}{3}\right) =$

$\dfrac{4(3n-1)}{12} + \dfrac{3(2n+5)}{12} =$

$\dfrac{4(3n-1) + 3(2n+5)}{12} =$

$\dfrac{12n - 4 + 6n + 15}{12} = \dfrac{18n + 11}{12}$

53. $\dfrac{4n-3}{6} - \dfrac{3n+5}{18} =$

$\left(\dfrac{4n-3}{6}\right)\left(\dfrac{3}{3}\right) - \dfrac{3n+5}{18} =$

$\dfrac{3(4n-3)}{18} - \dfrac{3n+5}{18} =$

$\dfrac{3(4n-3) - (3n+5)}{18} =$

$\dfrac{12n - 9 - 3n - 5}{18} = \dfrac{9n - 14}{18}$

55. $\dfrac{3x}{4} + \dfrac{x}{6} - \dfrac{5x}{8} =$

$\left(\dfrac{3x}{4}\right)\left(\dfrac{6}{6}\right) + \left(\dfrac{x}{6}\right)\left(\dfrac{4}{4}\right) - \left(\dfrac{5x}{8}\right)\left(\dfrac{3}{3}\right) =$

$\dfrac{18x}{24} + \dfrac{4x}{24} - \dfrac{15x}{24} =$

$\dfrac{18x + 4x - 15x}{24} = \dfrac{7x}{24}$

57. $\dfrac{x}{5} - \dfrac{3}{10} - \dfrac{7x}{12} =$

$\left(\dfrac{x}{5}\right)\left(\dfrac{12}{12}\right) - \left(\dfrac{3}{10}\right)\left(\dfrac{6}{6}\right) - \left(\dfrac{7x}{12}\right)\left(\dfrac{5}{5}\right) =$

$\dfrac{12x}{60} - \dfrac{18}{60} - \dfrac{35x}{60} =$

$\dfrac{12x - 18 - 35x}{60} = \dfrac{-23x - 18}{60}$

59. $\dfrac{5}{8x} + \dfrac{1}{6x} =$

$\left(\dfrac{5}{8x}\right)\left(\dfrac{3}{3}\right) + \left(\dfrac{1}{6x}\right)\left(\dfrac{4}{4}\right) =$

$\dfrac{15}{24x} + \dfrac{4}{24x} = \dfrac{15 + 4}{24x} = \dfrac{19}{24x}$

61. $\dfrac{5}{6y} - \dfrac{7}{9y} =$

$\left(\dfrac{5}{6y}\right)\left(\dfrac{3}{3}\right) - \left(\dfrac{7}{9y}\right)\left(\dfrac{2}{2}\right) =$

$\dfrac{15}{18y} - \dfrac{14}{18y} = \dfrac{15 - 14}{18y} = \dfrac{1}{18y}$

63. $\dfrac{5}{12x} - \dfrac{11}{16x^2} =$

$\left(\dfrac{5}{12x}\right)\left(\dfrac{4x}{4x}\right) - \left(\dfrac{11}{16x^2}\right)\left(\dfrac{3}{3}\right) =$

$\dfrac{20x}{48x^2} - \dfrac{33}{48x^2} = \dfrac{20x - 33}{48x^2}$

65. $\dfrac{3}{2x} - \dfrac{2}{3x} + \dfrac{5}{4x} =$

$\left(\dfrac{3}{2x}\right)\left(\dfrac{6}{6}\right) - \left(\dfrac{2}{3x}\right)\left(\dfrac{4}{4}\right) + \left(\dfrac{5}{4x}\right)\left(\dfrac{3}{3}\right) =$

$\dfrac{18 - 8 + 15}{12x} = \dfrac{25}{12x}$

67. $\dfrac{3}{x-5} + \dfrac{7}{x} =$

$\left(\dfrac{3}{x-5}\right)\left(\dfrac{x}{x}\right) + \left(\dfrac{7}{x}\right)\left(\dfrac{x-5}{x-5}\right) =$

$\dfrac{3x}{x(x-5)} + \dfrac{7(x-5)}{x(x-5)} =$

$\dfrac{3x + 7(x-5)}{x(x-5)} =$

$\dfrac{3x + 7x - 35}{x(x-5)} = \dfrac{10x - 35}{x(x-5)}$

Problem Set 7.3

69. $\dfrac{2}{n-1} - \dfrac{3}{n} =$

$$\left(\dfrac{2}{n-1}\right)\left(\dfrac{n}{n}\right) - \left(\dfrac{3}{n}\right)\left(\dfrac{n-1}{n-1}\right) =$$

$$\dfrac{2n}{n(n-1)} - \dfrac{3(n-1)}{n(n-1)} =$$

$$\dfrac{2n - 3(n-1)}{n(n-1)} =$$

$$\dfrac{2n - 3n + 3}{n(n-1)} = \dfrac{-n+3}{n(n-1)}$$

71. $\dfrac{4}{n} - \dfrac{6}{n+4} =$

$$\left(\dfrac{4}{n}\right)\left(\dfrac{n+4}{n+4}\right) - \left(\dfrac{6}{n+4}\right)\left(\dfrac{n}{n}\right) =$$

$$\dfrac{4(n+4)}{n(n+4)} - \dfrac{6n}{n(n+4)} = \dfrac{4(n+4) - 6n}{n(n+4)} =$$

$$\dfrac{4n + 16 - 6n}{n(n+4)} = \dfrac{-2n+16}{n(n+4)}$$

73. $\dfrac{6}{x} - \dfrac{12}{2x+1} =$

$$\left(\dfrac{6}{x}\right)\left(\dfrac{2x+1}{2x+1}\right) - \left(\dfrac{12}{2x+1}\right)\left(\dfrac{x}{x}\right) =$$

$$\dfrac{6(2x+1)}{x(2x+1)} - \dfrac{12x}{x(2x+1)} =$$

$$\dfrac{6(2x+1) - 12x}{x(2x+1)} = \dfrac{12x + 6 - 12x}{x(2x+1)} =$$

$$\dfrac{6}{x(2x+1)}$$

75. $\dfrac{4}{x+4} + \dfrac{6}{x-3} =$

$$\left(\dfrac{4}{x+4}\right)\left(\dfrac{x-3}{x-3}\right) + \left(\dfrac{6}{x-3}\right)\left(\dfrac{x+4}{x+4}\right) =$$

$$\dfrac{4(x-3)}{(x+4)(x-3)} + \dfrac{6(x+4)}{(x-3)(x+4)} =$$

$$\dfrac{4(x-3) + 6(x+4)}{(x+4)(x-3)} =$$

$$\dfrac{4x - 12 + 6x + 24}{(x+4)(x-3)} = \dfrac{10x + 12}{(x+4)(x-3)}$$

77. $\dfrac{3}{x-2} - \dfrac{9}{x+1} =$

$$\left(\dfrac{3}{x-2}\right)\left(\dfrac{x+1}{x+1}\right) - \left(\dfrac{9}{x+1}\right)\left(\dfrac{x-2}{x-2}\right) =$$

$$\dfrac{3(x+1)}{(x-2)(x+1)} - \dfrac{9(x-2)}{(x+1)(x-2)} =$$

$$\dfrac{3(x+1) - 9(x-2)}{(x-2)(x+1)} =$$

$$\dfrac{3x + 3 - 9x + 18}{(x-2)(x+1)} = \dfrac{-6x + 21}{(x-2)(x+1)}$$

79. $\dfrac{3}{2x-1} - \dfrac{4}{3x+1} =$

$$\left(\dfrac{3}{2x-1}\right)\left(\dfrac{3x+1}{3x+1}\right) - \left(\dfrac{4}{3x+1}\right)\left(\dfrac{2x-1}{2x-1}\right) =$$

$$\dfrac{3(3x+1)}{(2x-1)(3x+1)} - \dfrac{4(2x-1)}{(3x+1)(2x-1)} =$$

$$\dfrac{3(3x+1) - 4(2x-1)}{(2x-1)(3x+1)} =$$

$$\dfrac{9x + 3 - 8x + 4}{(2x-1)(3x+1)} = \dfrac{x+7}{(2x-1)(3x+1)}$$

85. $\dfrac{5}{x-3} + \dfrac{1}{3-x} =$

$$\dfrac{5}{x-3} + \left(\dfrac{1}{3-x}\right)\left(\dfrac{-1}{-1}\right) =$$

$$\dfrac{5}{x-3} + \dfrac{-1}{x-3} = \dfrac{5 + (-1)}{x-3} = \dfrac{4}{x-3}$$

87. $\dfrac{-4}{a-1} + \dfrac{2}{1-a} =$

$$\dfrac{-4}{a-1} + \left(\dfrac{2}{1-a}\right)\left(\dfrac{-1}{-1}\right) =$$

$$\dfrac{-4}{a-1} + \dfrac{-2}{a-1} = \dfrac{-4 + (-2)}{a-1} =$$

$$\dfrac{-6}{a-1} = -\dfrac{6}{a-1}$$

170

89. $\dfrac{n}{2n-1} - \dfrac{3}{1-2n} =$

$\dfrac{n}{2n-1} - \left(\dfrac{3}{1-2n}\right)\left(\dfrac{-1}{-1}\right) =$

$\dfrac{n}{2n-1} - \dfrac{-3}{2n-1} =$

$\dfrac{n-(-3)}{2n-1} = \dfrac{n+3}{2n-1}$

PROBLEM SET | **7.4**   More on Addition and Subtraction of Algebraic Fractions

1. $x^2 - 4x = x(x-4)$

$x = x$

LCD is $x(x-4)$.

$\dfrac{4}{x^2-4x} + \dfrac{3}{x} = \dfrac{4}{x(x-4)} + \dfrac{3}{x} =$

$\dfrac{4}{x(x-4)} + \left(\dfrac{3}{x}\right)\left(\dfrac{x-4}{x-4}\right) =$

The following step will be eliminated in future problems.

$\dfrac{4}{x(x-4)} + \dfrac{3(x-4)}{x(x-4)} =$

$\dfrac{4+3(x-4)}{x(x-4)} = \dfrac{4+3x-12}{x(x-4)} = \dfrac{3x-8}{x(x-4)}$

3. $x^2 + 2x = x(x+2)$

$x = x$

LCD is $x(x+2)$.

$\dfrac{7}{x^2+2x} - \dfrac{5}{x} = \dfrac{7}{x(x+2)} - \dfrac{5}{x} =$

$\dfrac{7}{x(x+2)} - \left(\dfrac{5}{x}\right)\left(\dfrac{x+2}{x+2}\right) =$

$\dfrac{7-5(x+2)}{x(x+2)} = \dfrac{7-5x-10}{x(x+2)} = \dfrac{-5x-3}{x(x+2)}$

5. $n = n$

$n^2 - 6n = n(n-6)$

LCD is $n(n-6)$.

$\dfrac{8}{n} - \dfrac{2}{n^2-6n} = \dfrac{8}{n} - \dfrac{2}{n(n-6)} =$

$\left(\dfrac{8}{n}\right)\left(\dfrac{n-6}{n-6}\right) - \dfrac{2}{n(n-6)} =$

$\dfrac{8(n-6)-2}{n(n-6)} = \dfrac{8n-48-2}{n(n-6)} = \dfrac{8n-50}{n(n-6)}$

7. $n^2 + n = n(n+1)$

$n = n$

LCD is $n(n+1)$.

$\dfrac{4}{n^2+n} - \dfrac{4}{n} = \dfrac{4}{n(n+1)} - \dfrac{4}{n} =$

$\dfrac{4}{n(n+1)} - \left(\dfrac{4}{n}\right)\left(\dfrac{n+1}{n+1}\right) =$

$\dfrac{4-4(n+1)}{n(n+1)} = \dfrac{4-4n-4}{n(n+1)} =$

$\dfrac{-4n}{n(n+1)} = -\dfrac{4}{n+1}$

9. $2x = 2x$

$x^2 - x = x(x-1)$

LCD is $2x(x-1)$.

$\dfrac{7}{2x} - \dfrac{x}{x^2-x} = \dfrac{7}{2x} - \dfrac{x}{x(x-1)} =$

$\left(\dfrac{7}{2x}\right)\left(\dfrac{x-1}{x-1}\right) - \left[\dfrac{x}{x(x-1)}\right]\left(\dfrac{2}{2}\right) =$

$\dfrac{7(x-1)-2x}{2x(x-1)} = \dfrac{7x-7-2x}{2x(x-1)} =$

$\dfrac{5x-7}{2x(x-1)}$

Problem Set 7.4

11. $x^2 - 16 = (x - 4)(x + 4)$
$x + 4 = x + 4$
LCD is $(x - 4)(x + 4)$.

$$\frac{3}{x^2 - 16} + \frac{5}{x + 4} =$$

$$\frac{3}{(x - 4)(x + 4)} + \frac{5}{x + 4} =$$

$$\frac{3}{(x - 4)(x + 4)} + \left(\frac{5}{x + 4}\right)\left(\frac{x - 4}{x - 4}\right) =$$

$$\frac{3 + 5(x - 4)}{(x - 4)(x + 4)} = \frac{3 + 5x - 20}{(x - 4)(x + 4)} =$$

$$\frac{5x - 17}{(x - 4)(x + 4)}$$

13. $x^2 - 1 = (x - 1)(x + 1)$
$x - 1 = x - 1$
LCD is $(x - 1)(x + 1)$.

$$\frac{8x}{x^2 - 1} - \frac{4}{x - 1} =$$

$$\frac{8x}{(x - 1)(x + 1)} - \frac{4}{x - 1} =$$

$$\frac{8x}{(x - 1)(x + 1)} - \left(\frac{4}{x - 1}\right)\left(\frac{x + 1}{x + 1}\right) =$$

$$\frac{8x - 4(x + 1)}{(x - 1)(x + 1)} = \frac{8x - 4x - 4}{(x - 1)(x + 1)} =$$

$$\frac{4x - 4}{(x - 1)(x + 1)} = \frac{4(x - 1)}{(x - 1)(x + 1)} = \frac{4}{x + 1}$$

15. $a^2 - 2a = a(a - 2)$
$a^2 + 2a = a(a + 2)$
LCD is $a(a - 2)(a + 2)$.

$$\frac{4}{a^2 - 2a} + \frac{7}{a^2 + 2a} =$$

$$\frac{4}{a(a - 2)} + \frac{7}{a(a + 2)} =$$

$$\left[\frac{4}{a(a - 2)}\right]\left(\frac{a + 2}{a + 2}\right) + \left[\frac{7}{a(a + 2)}\right]\left(\frac{a - 2}{a - 2}\right) =$$

$$\frac{4(a + 2) + 7(a - 2)}{a(a - 2)(a + 2)} =$$

$$\frac{4a + 8 + 7a - 14}{a(a - 2)(a + 2)} =$$

$$\frac{11a - 6}{a(a - 2)(a + 2)}$$

17. $x^2 - 6x = x(x - 6)$
$x^2 + 6x = x(x + 6)$
LCD is $x(x - 6)(x + 6)$.

$$\frac{1}{x^2 - 6x} - \frac{1}{x^2 + 6x} =$$

$$\frac{1}{x(x - 6)} - \frac{1}{x(x + 6)} =$$

$$\left[\frac{1}{x(x - 6)}\right]\left(\frac{x + 6}{x + 6}\right) - \left[\frac{1}{x(x + 6)}\right]\left(\frac{x - 6}{x - 6}\right) =$$

$$\frac{1(x + 6) - 1(x - 6)}{x(x - 6)(x + 6)} = \frac{x + 6 - x + 6}{x(x - 6)(x + 6)} =$$

$$\frac{12}{x(x - 6)(x + 6)} =$$

19. $n^2 - 16 = (n - 4)(n + 4)$
$3n + 12 = 3(n + 4)$
LCD is $3(n - 4)(n + 4)$.

$$\frac{n}{n^2 - 16} - \frac{2}{3n + 12} =$$

$$\frac{n}{(n - 4)(n + 4)} - \frac{2}{3(n + 4)} =$$

$$\left[\frac{n}{(n - 4)(n + 4)}\right]\left(\frac{3}{3}\right) - \left[\frac{2}{3(n + 4)}\right]\left(\frac{n - 4}{n - 4}\right) =$$

$$\frac{3n - 2(n - 4)}{3(n - 4)(n + 4)} = \frac{3n - 2n + 8}{3(n - 4)(n + 4)} =$$

$$\frac{n + 8}{3(n - 4)(n + 4)}$$

21. $6x + 4 = 2(3x + 2)$
$9x + 6 = 3(3x + 2)$
LCD is $6(3x + 2)$.

$$\frac{5x}{6x + 4} + \frac{2x}{9x + 6} =$$

$$\frac{5x}{2(3x + 2)} + \frac{2x}{3(3x + 2)} =$$

$$\left[\frac{5x}{2(3x + 2)}\right]\left(\frac{3}{3}\right) + \left[\frac{2x}{3(3x + 2)}\right]\left(\frac{2}{2}\right) =$$

$$\frac{15x + 4x}{6(3x + 2)} = \frac{19x}{6(3x + 2)}$$

23. $5x + 5 = 5(x + 1)$
$3x + 3 = 3(x + 1)$
LCD is $15(x + 1)$.

$$\frac{x - 1}{5x + 5} - \frac{x - 4}{3x + 3} =$$

$$\frac{x - 1}{5(x + 1)} - \frac{x - 4}{3(x + 1)} =$$

$$\left[\frac{x - 1}{5(x + 1)}\right]\left(\frac{3}{3}\right) - \left[\frac{x - 4}{3(x + 1)}\right]\left(\frac{5}{5}\right) =$$

$$\frac{3(x - 1) - 5(x - 4)}{15(x + 1)} =$$

$$\frac{3x - 3 - 5x + 20}{15(x + 1)} = \frac{-2x + 17}{15(x + 1)}$$

25. $x^2 + 7x + 12 = (x + 3)(x + 4)$
$x^2 - 9 = (x - 3)(x + 3)$
LCD is $(x + 3)(x + 4)(x - 3)$.

$$\frac{2}{x^2 + 7x + 12} + \frac{3}{x^2 - 9} =$$

$$\frac{2}{(x + 3)(x + 4)} + \frac{3}{(x - 3)(x + 3)} =$$

$$\left[\frac{2}{(x + 3)(x + 4)}\right]\left(\frac{x - 3}{x - 3}\right)$$

$$+ \left[\frac{3}{(x - 3)(x + 3)}\right]\left(\frac{x + 4}{x + 4}\right) =$$

$$\frac{2(x - 3) + 3(x + 4)}{(x + 3)(x + 4)(x - 3)} =$$

$$\frac{2x - 6 + 3x + 12}{(x + 3)(x + 4)(x - 3)} =$$

$$\frac{5x + 6}{(x + 3)(x + 4)(x - 3)}$$

27. $x^2 + 6x + 8 = (x + 2)(x + 4)$
$x^2 - 3x - 10 = (x - 5)(x + 2)$
LCD is $(x + 2)(x + 4)(x - 5)$.

$$\frac{x}{x^2 + 6x + 8} - \frac{5}{x^2 - 3x - 10} =$$

$$\frac{x}{(x + 2)(x + 4)} - \frac{5}{(x - 5)(x + 2)} =$$

$$\left[\frac{x}{(x + 2)(x + 4)}\right]\left(\frac{x - 5}{x - 5}\right)$$

$$- \left[\frac{5}{(x - 5)(x + 2)}\right]\left(\frac{x + 4}{x + 4}\right) =$$

$$\frac{x(x - 5) - 5(x + 4)}{(x + 2)(x + 4)(x - 5)} =$$

$$\frac{x^2 - 5x - 5x - 20}{(x + 2)(x + 4)(x - 5)} =$$

$$\frac{x^2 - 10x - 20}{(x + 2)(x + 4)(x - 5)}$$

29. $ab + b^2 = b(a + b)$
$a^2 + ab = a(a + b)$
LCD is $ab(a + b)$.

$$\frac{a}{ab + b^2} - \frac{b}{a^2 + ab} =$$

$$\frac{a}{b(a + b)} - \frac{b}{a(a + b)} =$$

$$\left[\frac{a}{b(a + b)}\right]\left(\frac{a}{a}\right) - \left[\frac{b}{a(a + b)}\right]\left(\frac{b}{b}\right) =$$

$$\frac{a^2 - b^2}{ab(a + b)} = \frac{(a - b)(a + b)}{ab(a + b)} =$$

$$\frac{a - b}{ab}$$

Problem Set 7.4

31. $x - 5 = x - 5$
$x^2 - 25 = (x - 5)(x + 5)$
$x + 5 = x + 5$
LCD is $(x - 5)(x + 5)$.

$$\frac{3}{x - 5} - \frac{4}{x^2 - 25} + \frac{5}{x + 5} =$$

$$\frac{3}{x - 5} - \frac{4}{(x - 5)(x + 5)} + \frac{5}{x + 5} =$$

$$\left(\frac{3}{x - 5}\right)\left(\frac{x + 5}{x + 5}\right) - \frac{4}{(x - 5)(x + 5)}$$

$$+ \left(\frac{5}{x + 5}\right)\left(\frac{x - 5}{x - 5}\right) =$$

$$\frac{3(x + 5) - 4 + 5(x - 5)}{(x - 5)(x + 5)} =$$

$$\frac{3x + 15 - 4 + 5x - 25}{(x - 5)(x + 5)} = \frac{8x - 14}{(x - 5)(x + 5)}$$

33. $x^2 - 2x = x(x - 2)$
$x^2 + 2x = x(x + 2)$
$x^2 - 4 = (x - 2)(x + 2)$
LCD is $x(x - 2)(x + 2)$.

$$\frac{10}{x^2 - 2x} + \frac{8}{x^2 + 2x} - \frac{3}{x^2 - 4} =$$

$$\frac{10}{x(x - 2)} + \frac{8}{x(x + 2)} - \frac{3}{(x - 2)(x + 2)} =$$

$$\left[\frac{10}{x(x - 2)}\right]\left(\frac{x + 2}{x + 2}\right)$$

$$+ \left[\frac{8}{x(x + 2)}\right]\left(\frac{x - 2}{x - 2}\right)$$

$$- \left[\frac{3}{(x - 2)(x + 2)}\right]\left(\frac{x}{x}\right) =$$

$$\frac{10(x + 2) + 8(x - 2) - 3(x)}{x(x - 2)(x + 2)} =$$

$$\frac{10x + 20 + 8x - 16 - 3x}{x(x - 2)(x + 2)} =$$

$$\frac{15x + 4}{x(x - 2)(x + 2)}$$

35. $x^2 + 7x + 10 = (x + 2)(x + 5)$
$x + 2 = x + 2$
$x + 5 = x + 5$
LCD is $(x + 2)(x + 5)$.

$$\frac{3x}{x^2 + 7x + 10} - \frac{2}{x + 2} + \frac{3}{x + 5} =$$

$$\frac{3x}{(x + 2)(x + 5)} - \frac{2}{x + 2} + \frac{3}{x + 5} =$$

$$\frac{3x}{(x + 2)(x + 5)}$$

$$- \left(\frac{2}{x + 2}\right)\left(\frac{x + 5}{x + 5}\right)$$

$$+ \left(\frac{3}{x + 5}\right)\left(\frac{x + 2}{x + 2}\right) =$$

$$\frac{3x - 2(x + 5) + 3(x + 2)}{(x + 2)(x + 5)} =$$

$$\frac{3x - 2x - 10 + 3x + 6}{(x + 2)(x + 5)} =$$

$$\frac{4x - 4}{(x + 2)(x + 5)}$$

37. $3x^2 + 7x - 20 = (3x - 5)(x + 4)$
$3x - 5 = 3x - 5$
$x + 4 = x + 4$
LCD is $(3x - 5)(x + 4)$.

$$\frac{5x}{3x^2 + 7x - 20} - \frac{1}{3x - 5} - \frac{2}{x + 4} =$$

$$\frac{5x}{(3x - 5)(x + 4)} - \frac{1}{3x - 5} - \frac{2}{x + 4} =$$

$$\frac{5x}{(3x - 5)(x + 4)}$$

$$- \left(\frac{1}{3x - 5}\right)\left(\frac{x + 4}{x + 4}\right)$$

$$- \left(\frac{2}{x + 4}\right)\left(\frac{3x - 5}{3x - 5}\right) =$$

$$\frac{5x - 1(x + 4) - 2(3x - 5)}{(3x - 5)(x + 4)} =$$

$$\frac{5x - x - 4 - 6x + 10}{(3x - 5)(x + 4)} = \frac{-2x + 6}{(3x - 5)(x + 4)}$$

174

39. $x + 4 = x + 4$

$x - 3 = x - 3$

$x^2 + x - 12 = (x + 4)(x - 3)$

LCD is $(x + 4)(x - 3)$.

$$\frac{2}{x + 4} - \frac{1}{x - 3} + \frac{2x + 1}{x^2 + x - 12} =$$

$$\frac{2}{x + 4} - \frac{1}{x - 3} + \frac{2x + 1}{(x + 4)(x - 3)} =$$

$$\left(\frac{2}{x + 4}\right)\left(\frac{x - 3}{x - 3}\right)$$

$$- \left(\frac{1}{x - 3}\right)\left(\frac{x + 4}{x + 4}\right)$$

$$+ \frac{2x + 1}{(x + 4)(x - 3)} =$$

$$\frac{2(x - 3) - 1(x + 4) + 2x + 1}{(x + 4)(x - 3)} =$$

$$\frac{2x - 6 - x - 4 + 2x + 1}{(x + 4)(x - 3)} =$$

$$\frac{3x - 9}{(x + 4)(x - 3)} = \frac{3(x - 3)}{(x + 4)(x - 3)} = \frac{3}{x + 4}$$

41. LCM of 2, 3, 4 and 6 is 12.

$$\frac{\dfrac{1}{2} - \dfrac{3}{4}}{\dfrac{1}{6} + \dfrac{1}{3}} = \left(\frac{12}{12}\right)\left[\frac{\dfrac{1}{2} - \dfrac{3}{4}}{\dfrac{1}{6} + \dfrac{1}{3}}\right] =$$

$$\frac{12\left(\dfrac{1}{2} - \dfrac{3}{4}\right)}{12\left(\dfrac{1}{6} + \dfrac{1}{3}\right)} = \frac{6 - 9}{2 + 4} = \frac{-3}{6} = -\frac{1}{2}$$

43. LCM of 3, 6 and 9 is 18.

$$\frac{\dfrac{2}{9} + \dfrac{1}{3}}{\dfrac{5}{6} - \dfrac{2}{3}} = \left(\frac{18}{18}\right)\left[\frac{\dfrac{2}{9} + \dfrac{1}{3}}{\dfrac{5}{6} - \dfrac{2}{3}}\right] =$$

$$\frac{18\left(\dfrac{2}{9} + \dfrac{1}{3}\right)}{18\left(\dfrac{5}{6} - \dfrac{2}{3}\right)} = \frac{4 + 6}{15 - 12} = \frac{10}{3}$$

45. LCM of 1, 3 and 4 is 12.

$$\frac{3 - \dfrac{2}{3}}{2 + \dfrac{1}{4}} = \left(\frac{12}{12}\right)\left[\frac{3 - \dfrac{2}{3}}{2 + \dfrac{1}{4}}\right] =$$

$$\frac{12\left(3 - \dfrac{2}{3}\right)}{12\left(2 + \dfrac{1}{4}\right)} = \frac{36 - 8}{24 + 3} = \frac{28}{27}$$

47. LCM of x and y is xy.

$$\frac{\dfrac{3}{x}}{\dfrac{9}{y}} = \left(\frac{xy}{xy}\right)\left[\frac{\dfrac{3}{x}}{\dfrac{9}{y}}\right] =$$

$$\frac{xy\left(\dfrac{3}{x}\right)}{xy\left(\dfrac{9}{y}\right)} = \frac{3y}{9x} = \frac{y}{3x}$$

49. LCM of x and y is xy.

$$\frac{\dfrac{2}{x} + \dfrac{3}{y}}{\dfrac{5}{x} - \dfrac{1}{y}} = \left(\frac{xy}{xy}\right)\left[\frac{\dfrac{2}{x} + \dfrac{3}{y}}{\dfrac{5}{x} - \dfrac{1}{y}}\right] =$$

$$\frac{xy\left(\dfrac{2}{x} + \dfrac{3}{y}\right)}{xy\left(\dfrac{5}{x} - \dfrac{1}{y}\right)} = \frac{2y + 3x}{5y - x}$$

51. LCM of x^2 and y is $x^2 y$.

$$\frac{\dfrac{1}{y} - \dfrac{4}{x^2}}{\dfrac{7}{x} - \dfrac{3}{y}} = \left(\frac{x^2 y}{x^2 y}\right)\left[\frac{\dfrac{1}{y} - \dfrac{4}{x^2}}{\dfrac{7}{x} - \dfrac{3}{y}}\right] =$$

$$\frac{x^2 y\left(\dfrac{1}{y} - \dfrac{4}{x^2}\right)}{x^2 y\left(\dfrac{7}{x} - \dfrac{3}{y}\right)} = \frac{x^2 - 4y}{7xy - 3x^2}$$

53. LCM of x.

$$\dfrac{\dfrac{6}{x}+2}{\dfrac{3}{x}+4} = \left(\dfrac{x}{x}\right)\left[\dfrac{\dfrac{6}{x}+2}{\dfrac{3}{x}+4}\right] =$$

$$\dfrac{x\left(\dfrac{6}{x}+2\right)}{x\left(\dfrac{3}{x}+4\right)} = \dfrac{6+2x}{3+4x}$$

55. LCM of 2, 3, and x^2 is $6x^2$.

$$\dfrac{\dfrac{3}{2x^2}-\dfrac{4}{x}}{\dfrac{5}{3x}+\dfrac{7}{x^2}} = \left(\dfrac{6x^2}{6x^2}\right)\left[\dfrac{\dfrac{3}{2x^2}-\dfrac{4}{x}}{\dfrac{5}{3x}+\dfrac{7}{x^2}}\right] =$$

$$\dfrac{6x^2\left(\dfrac{3}{2x^2}-\dfrac{4}{x}\right)}{6x^2\left(\dfrac{5}{3x}+\dfrac{7}{x^2}\right)} = \dfrac{9-24x}{10x+42}$$

57. LCM of 2, 4 and x is $4x$.

$$\dfrac{\dfrac{x+2}{4}}{\dfrac{1}{x}+\dfrac{3}{2}} = \left(\dfrac{4x}{4x}\right)\left[\dfrac{\dfrac{x+2}{4}}{\dfrac{1}{x}+\dfrac{3}{2}}\right] =$$

$$\dfrac{4x\left(\dfrac{x+2}{4}\right)}{4x\left(\dfrac{1}{x}+\dfrac{3}{2}\right)} = \dfrac{x(x+2)}{4+6x} =$$

$$\dfrac{x^2+2x}{6x+4}$$

59. LCM is $x-1$.

$$\dfrac{\dfrac{1}{x-1}-2}{\dfrac{3}{x-1}+4} = \left(\dfrac{x-1}{x-1}\right)\left[\dfrac{\dfrac{1}{x-1}-2}{\dfrac{3}{x-1}+4}\right] =$$

$$\dfrac{(x-1)\left(\dfrac{1}{x-1}-2\right)}{(x-1)\left(\dfrac{3}{x-1}+4\right)} = \dfrac{1-2(x-1)}{3+4(x-1)} =$$

$$\dfrac{1-2x+2}{3+4x-4} = \dfrac{-2x+3}{4x-1}$$

61. She would have completed
$$\dfrac{m\text{ minutes}}{40\text{ minutes}} = \dfrac{m}{40}\text{ of the course.}$$

63. Let t represent the time in hours. Use $rt = d$ as a guideline and solve for t.

$$\left(r\,^{\text{kilometers}}_{\text{per hour}}\right)(t) = k\text{ kilometers}$$

$$t = \dfrac{k\text{ kilometers}}{r\,^{\text{kilometers}}_{\text{per hour}}}$$

Her time would be $\dfrac{k}{r}$ hours.

65. The price per liter would be $\dfrac{d}{l}$ dollars per liter.

67. Let n represent the number, then $\dfrac{34}{n}$ represents the other number.

69. Let w represent the width of the rectangle. Use $A = lw$ as a guideline and solve for w.

$$lw = 47$$
$$w = \dfrac{47}{l}$$

The width of the rectangle would be $\dfrac{47}{l}$ inches.

71. Let h represent the altitude to the given side.

$$48 = \dfrac{1}{2}bh$$
$$96 = bh$$
$$\dfrac{96}{b} = h$$

The altitude to the given side would be $\dfrac{96}{b}$ feet.

73. $1 - \dfrac{n}{1 - \dfrac{1}{n}} = 1 - \dfrac{n(n)}{n\left(1 - \dfrac{1}{n}\right)} =$

$1 - \dfrac{n^2}{n-1} = \dfrac{n-1}{n-1} - \dfrac{n^2}{n-1} =$

$\dfrac{n-1-n^2}{n-1} = \dfrac{-n^2+n-1}{n-1}$

75. $\dfrac{3x}{4 - \dfrac{2}{x}} - 1 = \dfrac{x(3x)}{x\left(4 - \dfrac{2}{x}\right)} - 1 =$

$\dfrac{3x^2}{4x-2} - 1 = \dfrac{3x^2}{4x-2} - \dfrac{4x-2}{4x-2} =$

$\dfrac{3x^2-(4x-2)}{4x-2} = \dfrac{3x^2-4x+2}{4x-2}$

PROBLEM SET **7.5** **Fractional Equations and Problem Solving**

1. $\dfrac{x}{2} + \dfrac{x}{3} = 10$

Multiply both sides by 6.

$6\left(\dfrac{x}{2} + \dfrac{x}{3}\right) = 6(10)$

$3x + 2x = 60$

$5x = 60$

$x = 12$

The solution set is $\{12\}$.

3. $\dfrac{x}{6} - \dfrac{4x}{3} = \dfrac{1}{9}$

Multiply both sides by 18.

$18\left(\dfrac{x}{6} - \dfrac{4x}{3}\right) = 18\left(\dfrac{1}{9}\right)$

$3x - 24x = 2$

$-21x = 2$

$x = -\dfrac{2}{21}$

The solution set is $\left\{-\dfrac{2}{21}\right\}$.

5. $\dfrac{n}{2} + \dfrac{n-1}{6} = \dfrac{5}{2}$

Multiply both sides by 6.

$6\left(\dfrac{n}{2} + \dfrac{n-1}{6}\right) = 6\left(\dfrac{5}{2}\right)$

$3n + (n-1) = 15$

$4n - 1 = 15$

$4n = 16$

$n = 4$

The solution set is $\{4\}$.

7. $\dfrac{t-3}{4} + \dfrac{t+1}{9} = -1$

Multiply both sides by 36.

$36\left(\dfrac{t-3}{4} + \dfrac{t+1}{9}\right) = 36(-1)$

$9(t-3) + 4(t+1) = -36$

$9t - 27 + 4t + 4 = -36$

$13t - 23 = -36$

$13t = -13$

$t = -1$

The solution set is $\{-1\}$.

9. $\dfrac{2x+3}{3} + \dfrac{3x-4}{4} = \dfrac{17}{4}$

Multiply both sides by 12.

$12\left(\dfrac{2x+3}{3} + \dfrac{3x-4}{4}\right) = 12\left(\dfrac{17}{4}\right)$

$4(2x+3) + 3(3x-4) = 3(17)$

$8x + 12 + 9x - 12 = 51$

$17x = 51$

$x = 3$

The solution set is $\{3\}$.

Problem Set 7.5

11. $\dfrac{x-4}{8} - \dfrac{x+5}{4} = 3$

Multiply both sides by 8.

$$8\left(\dfrac{x-4}{8} - \dfrac{x+5}{4}\right) = 8(3)$$

$$x - 4 - 2(x+5) = 24$$

$$x - 4 - 2x - 10 = 24$$

$$-x - 14 = 24$$

$$-x = 38$$

$$x = -38$$

The solution set is $\{-38\}$.

13. $\dfrac{3x+2}{5} - \dfrac{2x-1}{6} = \dfrac{2}{15}$

Multiply both sides by 30.

$$30\left(\dfrac{3x+2}{5} - \dfrac{2x-1}{6}\right) = 30\left(\dfrac{2}{15}\right)$$

$$6(3x+2) - 5(2x-1) = 2(2)$$

$$18x + 12 - 10x + 5 = 4$$

$$8x + 17 = 4$$

$$8x = -13$$

$$x = -\dfrac{13}{8}$$

The solution set is $\left\{-\dfrac{13}{8}\right\}$.

15. $\dfrac{1}{x} + \dfrac{2}{3} = \dfrac{7}{6}$, $x \neq 0$

Multiply both sides by $6x$.

$$6x\left(\dfrac{1}{x} + \dfrac{2}{3}\right) = 6x\left(\dfrac{7}{6}\right)$$

$$6(1) + 2x(2) = x(7)$$

$$6 + 4x = 7x$$

$$6 = 3x$$

$$2 = x$$

The solution set is $\{2\}$.

17. $\dfrac{5}{3n} - \dfrac{1}{9} = \dfrac{1}{n}$, $n \neq 0$

Multiply both sides by $9n$.

$$9n\left(\dfrac{5}{3n} - \dfrac{1}{9}\right) = 9n\left(\dfrac{1}{n}\right)$$

$$3(5) - n(1) = 9(1)$$

$$15 - n = 9$$

$$-n = -6$$

$$n = 6$$

The solution set is $\{6\}$.

19. $\dfrac{1}{2x} + 3 = \dfrac{4}{3x}$, $x \neq 0$

Multiply both sides by $6x$.

$$6x\left(\dfrac{1}{2x} + 3\right) = 6x\left(\dfrac{4}{3x}\right)$$

$$3(1) + 6x(3) = 2(4)$$

$$3 + 18x = 8$$

$$18x = 5$$

$$x = \dfrac{5}{18}$$

The solution set is $\left\{\dfrac{5}{18}\right\}$.

21. $\dfrac{4}{5t} - 1 = \dfrac{3}{2t}$, $t \neq 0$

Multiply both sides by $10t$.

$$10t\left(\dfrac{4}{5t} - 1\right) = 10t\left(\dfrac{3}{2t}\right)$$

$$2(4) - 10t(1) = 5(3)$$

$$8 - 10t = 15$$

$$-10t = 7$$

$$t = -\dfrac{7}{10}$$

The solution set is $\left\{-\dfrac{7}{10}\right\}$.

23. $\dfrac{-5}{4h} + \dfrac{7}{6h} = \dfrac{1}{4}$, $h \neq 0$

Multiply both sides by $12h$.

$$12h\left(\dfrac{-5}{4h} + \dfrac{7}{6h}\right) = 12h\left(\dfrac{1}{4}\right)$$
$$3(-5) + 2(7) = 3h(1)$$
$$-15 + 14 = 3h$$
$$-1 = 3h$$
$$-\dfrac{1}{3} = h$$

The solution set is $\left\{-\dfrac{1}{3}\right\}$.

25. $\dfrac{90-n}{n} = 10 + \dfrac{2}{n}$, $n \neq 0$

Multiply both sides by n.

$$n\left(\dfrac{90-n}{n}\right) = n\left(10 + \dfrac{2}{n}\right)$$
$$90 - n = 10n + 2$$
$$90 - 11n = 2$$
$$-11n = -88$$
$$n = 8$$

The solution set is $\{8\}$.

27. $\dfrac{n}{49-n} = 3 + \dfrac{1}{49-n}$, $n \neq 49$

Multiply both sides by $(49-n)$.

$$(49-n)\left(\dfrac{n}{49-n}\right) = (49-n)\left(3 + \dfrac{1}{49-n}\right)$$
$$n = 3(49-n) + 1$$
$$n = 147 - 3n + 1$$
$$n = 148 - 3n$$
$$4n = 148$$
$$n = 37$$

The solution set is $\{37\}$.

29. $\dfrac{x}{x+3} - 2 = \dfrac{-3}{x+3}$, $x \neq -3$

Multiply both sides by $(x+3)$.

$$(x+3)\left(\dfrac{x}{x+3} - 2\right) = (x+3)\left(\dfrac{-3}{x+3}\right)$$
$$x - 2(x+3) = -3$$
$$x - 2x - 6 = -3$$
$$-x - 6 = -3$$
$$-x = 3$$
$$x = -3$$

Since the initial restriction was $x \neq -3$, the solution is $\emptyset$.

31. $\dfrac{7}{x+3} = \dfrac{5}{x-9}$, $x \neq -3$ and $x \neq 9$

Multiply both sides by $(x+3)(x-9)$.

$$(x+3)(x-9)\left(\dfrac{7}{x+3}\right) = (x+3)(x-9)\left(\dfrac{5}{x-9}\right)$$
$$7(x-9) = 5(x+3)$$
$$7x - 63 = 5x + 15$$
$$2x - 63 = 15$$
$$2x = 78$$
$$x = 39$$

The solution set is $\{39\}$.

33. $\dfrac{x}{x+2} + 3 = \dfrac{1}{x+2}$, $x \neq -2$

Multiply both sides by $(x+2)$.

$$(x+2)\left(\dfrac{x}{x+2} + 3\right) = (x+2)\left(\dfrac{1}{x+2}\right)$$
$$x + 3(x+2) = 1$$
$$x + 3x + 6 = 1$$
$$4x + 6 = 1$$
$$4x = -5$$
$$x = -\dfrac{5}{4}$$

The solution set is $\left\{-\dfrac{5}{4}\right\}$.

Problem Set 7.5

35.
$$-1 - \frac{5}{x-2} = \frac{3}{x-2}, \; x \neq 2$$
Multiply both sides by $(x-2)$.
$$(x-2)\left(-1-\frac{5}{x-2}\right) = (x-2)\left(\frac{3}{x-2}\right)$$
$$-1(x-2)-5 = 3$$
$$-x+2-5 = 3$$
$$-x-3 = 3$$
$$-x = 6$$
$$x = -6$$
The solution set is $\{-6\}$.

37.
$$1 + \frac{n+1}{2n} = \frac{3}{4}, \; n \neq 0$$
Multiply both sides by $4n$.
$$4n\left(1+\frac{n+1}{2n}\right) = 4n\left(\frac{3}{4}\right)$$
$$4n+2(n+1) = n(3)$$
$$4n+2n+2 = 3n$$
$$6n+2 = 3n$$
$$2 = -3n$$
$$-\frac{2}{3} = n$$
The solution set is $\left\{-\frac{2}{3}\right\}$.

39.
$$\frac{h}{2} - \frac{h}{4} + \frac{h}{3} = 1$$
Multiply both sides by 12.
$$12\left(\frac{h}{2}-\frac{h}{4}+\frac{h}{3}\right) = 12(1)$$
$$6h - 3h + 4h = 12$$
$$7h = 12$$
$$h = \frac{12}{7}$$
The solution set is $\left\{\frac{12}{7}\right\}$.

41. Let x represent the denominator of the fraction, then $x-8$ represents the numerator of the fraction.

$$\frac{x-8}{x} = \frac{5}{6}$$
Cross products are equal.
$$6(x-8) = 5x$$
$$6x - 48 = 5x$$
$$x - 48 = 0$$
$$x = 48$$
The fraction is $\dfrac{48-8}{48} = \dfrac{40}{48}$.

43. Let n represent the number to be added.
$$\frac{2+n}{5+n} = \frac{4}{5}$$
$$5(2+n) = 4(5+n)$$
$$10 + 5n = 20 + 4n$$
$$10 + n = 20$$
$$n = 10$$
The number to be added would be 10.

45. Let n represent the smaller number, then $65 - n$ represents the larger number.
$$\frac{65-n}{n} = 8 + \frac{2}{n}$$
$$n\left(\frac{65-n}{n}\right) = n\left(8+\frac{2}{n}\right)$$
$$65 - n = 8n + 2$$
$$65 - 9n = 2$$
$$-9n = -63$$
$$n = 7$$
The smaller number is 7 and the larger number is $65 - 7 = 58$.

47. Let x represent the numerator of the fraction, then $x-4$ represents the denominator.
$$\frac{x+6}{2(x-4)} = 1$$
$$2(x-4)\left[\frac{x+6}{2(x-4)}\right] = 2(x-4)(1)$$
$$x + 6 = 2x - 8$$
$$x + 14 = 2x$$
$$14 = x$$
The original fraction was $\dfrac{14}{14-4} = \dfrac{14}{10}$.

49. Let x represent the rate at which they both rode. Heidi's time was $3\frac{1}{3} = \frac{10}{3}$ hours more than Abby's time.

	Rate	Time	Distance
Heidi	x	$\dfrac{125}{x}$	125
Abby	x	$\dfrac{75}{x}$	75

Heidi's time = Abby's time plus $3\frac{1}{3}$

$$\frac{125}{x} = \frac{75}{x} + \frac{10}{3}$$
$$3x\left(\frac{125}{x}\right) = 3x\left(\frac{75}{x} + \frac{10}{3}\right)$$
$$3(125) = 3(75) + x(10)$$
$$375 = 225 + 10x$$
$$150 = 10x$$
$$15 = x$$

They both rode at a rate of 15 miles per hour.

51. Let r represent Dave's rate, then $r + 4$ represents Kent's rate. Their times are equal.

$$\text{time} = \frac{\text{distance Dave rides}}{\text{Dave's time}} = \frac{\text{distance Kent rides}}{\text{Kent's time}}$$
$$\frac{250}{r} = \frac{270}{r+4}$$

Cross products are equal.

$$250(r + 4) = 270r$$
$$250r + 1000 = 270r$$
$$1000 = 20r$$
$$50 = r$$

Dave drove at a rate of 50 miles per hour and Kent's rate is 54 miles per hour.

57.
$$\frac{1}{2n} + \frac{4}{n} = \frac{9}{2n}, \quad n \neq 0$$
$$2n\left(\frac{1}{2n} + \frac{4}{n}\right) = 2n\left(\frac{9}{2n}\right)$$
$$1 + 2(4) = 9$$
$$1 + 8 = 9$$
$$9 = 9 \text{ (always true)}$$

The solution set is {all real numbers except 0}.

Problem Set 7.5

59.
$$\frac{1}{n+2} + \frac{2}{n+3} = \frac{3n+7}{(n+2)(n+3)}, \quad n \neq -3 \text{ and } n \neq -2$$

$$(n+2)(n+3)\left(\frac{1}{n+2} + \frac{2}{n+3}\right) = (n+2)(n+3)\left[\frac{3n+7}{(n+2)(n+3)}\right]$$

$$1(n+3) + 2(n+2) = 3n+7$$

$$n + 3 + 2n + 4 = 3n + 7$$

$$3n + 7 = 3n + 7 \text{ (always true)}$$

The solution set is {all real numbers except -3 and -2}.

PROBLEM SET **7.6** **More Fractional Equations and Problem Solving**

1.
$$\frac{4}{x} + \frac{7}{6} = \frac{1}{x} + \frac{2}{3x}, \quad x \neq 0$$
Multiply both sides by $6x$.

$$6x\left(\frac{4}{x} + \frac{7}{6}\right) = 6x\left(\frac{1}{x} + \frac{2}{3x}\right)$$

$$6(4) + x(7) = 6(1) + 2(2)$$

$$24 + 7x = 6 + 4$$

$$24 + 7x = 10$$

$$7x = -14$$

$$x = -2$$

The solution set is $\{-2\}$.

3.
$$\frac{3}{2(x+1)} + \frac{4}{x+1} = \frac{11}{12}, \quad x \neq -1$$
Factor the first denominator.
Multiply both sides by $12(x+1)$.

$$12(x+1)\left[\frac{3}{2(x+1)} + \frac{4}{x+1}\right] = 12(x+1)\left(\frac{11}{12}\right)$$

$$6(3) + 12(4) = 11(x+1)$$

$$18 + 48 = 11x + 11$$

$$66 = 11x + 11$$

$$55 = 11x$$

$$5 = x$$

The solution set is $\{5\}$.

5.
$$\frac{5}{2(n-5)} - \frac{3}{n-5} = 1, \quad n \neq 5$$
Factor the first denominator.
Multiply both sides by $2(n-5)$.

$$2(n-5)\left[\frac{5}{2(n-5)} - \frac{3}{n-5}\right] = 2(n-5)(1)$$

$$5 - 2(3) = 2n - 10$$

$$5 - 6 = 2n - 10$$

$$-1 = 2n - 10$$

$$9 = 2n$$

$$\frac{9}{2} = n$$

The solution set is $\left\{\frac{9}{2}\right\}$.

7.
$$\frac{3}{2t} - \frac{5}{t} = \frac{7}{5t} + 1, \quad t \neq 0$$
Multiply by $10t$.

$$10t\left(\frac{3}{2t} - \frac{5}{t}\right) = 10t\left(\frac{7}{5t} + 1\right)$$

$$5(3) - 10(5) = 2(7) + 10t$$

$$15 - 50 = 14 + 10t$$

$$-35 = 14 + 10t$$

$$-49 = 10t$$

$$-\frac{49}{10} = t$$

The solution set is $\left\{-\frac{49}{10}\right\}$.

9.
$$\frac{x}{x-2} + \frac{4}{x+2} = 1, x \neq -2 \text{ and } x \neq 2$$

Multiply by the LCD.

$$(x-2)(x+2)\left(\frac{x}{x-2} + \frac{4}{x+2}\right) = (x-2)(x+2)(1)$$
$$(x+2)(x) + (x-2)(4) = x^2 - 4$$
$$x^2 + 2x + 4x - 8 = x^2 - 4$$
$$6x - 8 = -4$$
$$6x = 4$$
$$x = \frac{4}{6} = \frac{2}{3} \qquad \text{The solution set is } \left\{\frac{2}{3}\right\}.$$

11.
$$\frac{x}{x-4} - \frac{2x}{x+4} = -1, x \neq -4 \text{ and } x \neq 4$$

Multiply by the LCD.

$$(x-4)(x+4)\left(\frac{x}{x-4} - \frac{2x}{x+4}\right) = (x-4)(x+4)(-1)$$
$$x(x+4) - 2x(x-4) = -1(x^2 - 16)$$
$$x^2 + 4x - 2x^2 + 8x = -x^2 + 16$$
$$-x^2 + 12x = -x^2 + 16$$
$$12x = 16$$
$$x = \frac{16}{12} = \frac{4}{3} \qquad \text{The solution set is } \left\{\frac{4}{3}\right\}.$$

13.
$$\frac{3n}{n+3} - \frac{n}{n-3} = 2, n \neq -3 \text{ and } n \neq 3$$

Multiply by the LCD.

$$(n+3)(n-3)\left(\frac{3n}{n+3} - \frac{n}{n-3}\right) = (n+3)(n-3)(2)$$
$$3n(n-3) - n(n+3) = 2(n^2 - 9)$$
$$3n^2 - 9n - n^2 - 3n = 2n^2 - 18$$
$$2n^2 - 12n = 2n^2 - 18$$
$$-12n = -18$$
$$n = \frac{18}{12} = \frac{3}{2} \qquad \text{The solution set is } \left\{\frac{3}{2}\right\}.$$

Problem Set 7.6

15.
$$\frac{3}{t^2-4}+\frac{5}{t+2}=\frac{2}{t-2},\ t\neq -2 \text{ and } t\neq 2$$

Factor the denominator.
$$\frac{3}{(t-2)(t+2)}+\frac{5}{t+2}=\frac{2}{t-2}$$
$$(t-2)(t+2)\left[\frac{3}{(t-2)(t+2)}+\frac{5}{t+2}\right]=(t-2)(t+2)\left(\frac{2}{t-2}\right)$$
$$3+5(t-2)=2(t+2)$$
$$3+5t-10=2t+4$$
$$5t-7=2t+4$$
$$3t-7=4$$
$$3t=11$$
$$t=\frac{11}{3}\qquad \text{The solution set is } \left\{\frac{11}{3}\right\}.$$

17.
$$\frac{4}{x-1}-\frac{2x-3}{x^2-1}=\frac{6}{x+1},\ x\neq -1 \text{ and } x\neq 1$$

Factor the denominator.
$$\frac{4}{x-1}-\frac{2x-3}{(x-1)(x+1)}=\frac{6}{x+1}$$
$$(x-1)(x+1)\left[\frac{4}{x-1}-\frac{2x-3}{(x-1)(x+1)}\right]=(x-1)(x+1)\left(\frac{6}{x+1}\right)$$
$$4(x+1)-(2x-3)=6(x-1)$$
$$4x+4-2x+3=6x-6$$
$$2x+7=6x-6$$
$$-4x+7=-6$$
$$-4x=-13$$
$$x=\frac{13}{4}\qquad \text{The solution set is } \left\{\frac{13}{4}\right\}.$$

19.
$$8+\frac{5}{y^2+2y}=\frac{3}{y+2},\ y\neq -2 \text{ and } y\neq 0$$

Factor the denominator.
$$8+\frac{5}{y(y+2)}=\frac{3}{(y+2)}$$
$$y(y+2)\left[8+\frac{5}{y(y+2)}\right]=y(y+2)\left(\frac{3}{y+2}\right)$$
$$8y(y+2)+5=3y$$
$$8y^2+16y+5=3y$$
$$8y^2+13y+5=0$$
$$(8y+5)(y+1)=0$$

$$8y + 5 = 0 \quad \text{or} \quad y + 1 = 0$$
$$8y = -5 \quad \text{or} \quad y = -1$$
$$y = -\frac{5}{8} \quad \text{or} \quad y = -1$$

The solution set is $\left\{ -1, \ -\frac{5}{8} \right\}$.

21.
$$n + \frac{1}{n} = \frac{17}{4}, \ n \neq 0$$
$$4n\left(n + \frac{1}{n} \right) = 4n\left(\frac{17}{4} \right)$$
$$4n^2 + 4 = 17n$$

Subtracted $17n$ from both sides.

$$4n^2 - 17n + 4 = 0$$
$$(4n - 1)(n - 4) = 0$$
$$4n - 1 = 0 \quad \text{or} \quad n - 4 = 0$$
$$4n = 1 \quad \text{or} \quad n = 4$$
$$n = \frac{1}{4} \quad \text{or} \quad n = 4 \qquad \text{The solution set is } \left\{ \frac{1}{4}, 4 \right\}.$$

23.
$$\frac{15}{4n} + \frac{15}{4(n + 4)} = 1, \ n \neq -4 \text{ and } n \neq 0$$
$$4n(n + 4)\left[\frac{15}{4n} + \frac{15}{4(n + 4)} \right] = 4n(n + 4)(1)$$
$$15(n + 4) + 15n = 4n^2 + 16n$$
$$15n + 60 + 15n = 4n^2 + 16n$$
$$30n + 60 = 4n^2 + 16n$$
$$-4n^2 + 14n + 60 = 0$$

Divide by -2.

$$2n^2 - 7n - 30 = 0$$
$$(2n + 5)(n - 6) = 0$$
$$2n + 5 = 0 \quad \text{or} \quad n - 6 = 0$$
$$2n = -5 \quad \text{or} \quad n = 6$$
$$n = -\frac{5}{2} \quad \text{or} \quad n = 6$$

The solution set is $\left\{ -\frac{5}{2}, 6 \right\}$.

Problem Set 7.6

25.
$$x - \frac{5x}{x-2} = \frac{-10}{x-2}, x \neq 2$$
$$(x-2)\left(x - \frac{5x}{x-2}\right) = (x-2)\left(\frac{-10}{x-2}\right)$$
$$x(x-2) - 5x = -10$$
$$x^2 - 2x - 5x = -10$$
$$x^2 - 7x = -10$$
$$x^2 - 7x + 10 = 0$$
$$(x-5)(x-2) = 0$$
$$x - 5 = 0 \quad \text{or} \quad x - 2 = 0$$
$$x = 5 \quad \text{or} \quad x = 2$$
Since the original restriction was $x \neq 2$, the solution set is $\{5\}$.

27.
$$\frac{t}{4t-4} + \frac{5}{t^2-1} = \frac{1}{4}, t \neq -1 \text{ and } t \neq 1$$
$$\frac{t}{4(t-1)} + \frac{5}{(t-1)(t+1)} = \frac{1}{4}$$
$$4(t+1)(t-1)\left[\frac{t}{4(t-1)} + \frac{5}{(t-1)(t+1)}\right] = 4(t-1)(t+1)\left(\frac{1}{4}\right)$$
$$t(t+1) + 4(5) = (t-1)(t+1)$$
$$t^2 + t + 20 = t^2 - 1$$
$$t + 20 = -1$$
$$t = -21 \qquad \text{The solution set is } \{-21\}.$$

29.
$$\frac{3}{n-5} + \frac{4}{n+7} = \frac{2n+11}{n^2+2n-35}, n \neq -7 \text{ and } n \neq 5$$
$$\frac{3}{n-5} + \frac{4}{n+7} = \frac{2n+11}{(n-5)(n+7)}$$
$$(n-5)(n+7)\left(\frac{3}{n-5} + \frac{4}{n+7}\right) = (n-5)(n+7)\left[\frac{2n+11}{(n-5)(n+7)}\right]$$
$$3(n+7) + 4(n-5) = 2n+11$$
$$3n + 21 + 4n - 20 = 2n+11$$
$$7n + 1 = 2n+11$$
$$5n + 1 = 11$$
$$5n = 10$$
$$n = 2 \qquad\qquad \text{The solution set is } \{2\}.$$

31.
$$\frac{a}{a+2} + \frac{3}{a+4} = \frac{14}{a^2+6a+8}, \; a \neq -4 \text{ and } a \neq -2$$
$$\frac{a}{a+2} + \frac{3}{a+4} = \frac{14}{(a+2)(a+4)}$$
$$(a+2)(a+4)\left(\frac{a}{a+2} + \frac{3}{a+4}\right) = (a+2)(a+4)\left[\frac{14}{(a+2)(a+4)}\right]$$
$$a(a+4) + 3(a+2) = 14$$
$$a^2 + 4a + 3a + 6 = 14$$
$$a^2 + 7a + 6 = 14$$
$$a^2 + 7a - 8 = 0$$
$$(a+8)(a-1) = 0$$
$$a+8 = 0 \quad \text{or} \quad a-1 = 0$$
$$a = -8 \quad \text{or} \quad a = 1$$
The solution set is $\{-8, 1\}$.

33. Let n represent the number, then $\dfrac{1}{n}$ represents the reciprocal.
$$n + 2\left(\frac{1}{n}\right) = \frac{9}{2}$$
$$n + \frac{2}{n} = \frac{9}{2}, \; n \neq 0$$
$$2n\left(n + \frac{2}{n}\right) = 2n\left(\frac{9}{2}\right)$$
$$2n^2 + 4 = 9n$$
$$2n^2 - 9n + 4 = 0$$
$$(2n-1)(n-4) = 0$$
$$2n - 1 = 0 \quad \text{or} \quad n - 4 = 0$$
$$2n = 1 \quad \text{or} \quad n = 4$$
$$n = \frac{1}{2} \quad \text{or} \quad n = 4$$
The number is $\dfrac{1}{2}$ or 4.

35. Let n represent the number, then $\dfrac{1}{n}$ represents the reciprocal.
$$n = \frac{1}{n} + \frac{21}{10}, \; n \neq 0$$
$$10n(n) = 10n\left(\frac{1}{n} + \frac{21}{10}\right)$$
$$10n^2 = 10 + 21n$$
$$10n^2 - 21n - 10 = 0$$
$$(5n+2)(2n-5) = 0$$

Problem Set 7.6

$$5n + 2 = 0 \quad \text{or} \quad 2n - 5 = 0$$
$$5n = -2 \quad \text{or} \quad 2n = 5$$
$$n = -\frac{2}{5} \quad \text{or} \quad n = \frac{5}{2}$$

The number is $-\frac{2}{5}$ or $\frac{5}{2}$.

37. Let r represent Tom's rate, then $r + 3$ represents Celia's rate. The information is recorded in the following table:.

	Distance	Rate	Time $\left(t = \dfrac{d}{r}\right)$
Tom	85	r	$\dfrac{85}{r}$
Celia	60	$r + 3$	$\dfrac{60}{r + 3}$

Time of Celia = Time of Tom − 2

$$\frac{60}{r+3} = \frac{85}{r} - 2, \ r \neq -3 \text{ and } r \neq 0$$
$$r(r+3)\left(\frac{60}{r+3}\right) = r(r+3)\left(\frac{85}{r} - 2\right)$$
$$60r = 85(r+3) - 2r(r+3)$$
$$60r = 85r + 255 - 2r^2 - 6r$$
$$60r = -2r^2 + 79r + 255$$
$$0 = -2r^2 + 19r + 255$$
$$0 = 2r^2 - 19r - 255$$
$$0 = (2r + 15)(r - 17)$$
$$2r + 15 = 0 \quad \text{or} \quad r - 17 = 0$$
$$2r = -15 \quad \text{or} \quad r = 17$$
$$r = -\frac{15}{2} \quad \text{or} \quad r = 17$$

The negative solution needs to be discarded. Tom's rate is 17 miles per hour and Celia's rate is 20 miles per hour.

39. Let r represent Jeff's rate back from the country, then $r + 4$ represents his rate out to the country. The information is recorded in the following table:

	Distance	Rate	Time $\left(t = \dfrac{d}{r}\right)$
Trip Out	40	$r + 4$	$\dfrac{40}{r + 4}$
Trip Back	42	r	$\dfrac{42}{r}$

Time of Trip Back = Time of Trip Out + 1

$$\frac{42}{r} = \frac{40}{r+4} + 1, \ r \neq -4 \text{ and } r \neq 0$$

$$r(r+4)\left(\frac{42}{r}\right) = r(r+4)\left(\frac{40}{r+4} + 1\right)$$

$$42(r+4) = 40r + r(r+4)$$

$$42r + 168 = 40r + r^2 + 4r$$

$$42r + 168 = r^2 + 44r$$

$$0 = r^2 + 2r - 168$$

$$0 = (r+14)(r-12)$$

$$r + 14 = 0 \quad \text{or} \quad r - 12 = 0$$

$$r = -14 \quad \text{or} \quad r = 12$$

The negative solution needs to be discarded. The rate back is 12 miles per hour and the rate out is 16 miles per hour.

41. Let t represent the time for the tank to overflow. The information is recorded in the following table:

	Rate	Time	Quantity
Fill	$\frac{1}{5}$	t	$\frac{t}{5}$
Drain	$\frac{1}{6}$	t	$\frac{t}{6}$

Fill − Drain = 1 tank Filled

$$\frac{t}{5} - \frac{t}{6} = 1$$

$$30\left(\frac{t}{5} - \frac{t}{6}\right) = 30(1)$$

$$6t - 5t = 30$$

$$t = 30$$

It would take 30 minutes for the tank to overflow.

43. Let t represent Mike's time, then $2t$ represents Barry's time. The sum of the individual rates equal the rate working together.

Mike's rate + Barry's rate = Rate together

$$\frac{1}{t} + \frac{1}{2t} = \frac{1}{40}$$

$$40t\left(\frac{1}{t} + \frac{1}{2t}\right) = 40t\left(\frac{1}{40}\right)$$

$$40 + 20 = t$$

$$60 = t$$

It would take Mike 60 minutes and Barry 120 minutes to deliver the papers alone.

Problem Set 7.6

45. Let t represent Mike's time to do the job by himself. The information is recorded in the following table:

	Rate	Time	Quantity
Pat	$\dfrac{1}{12}$	$3 + 5 = 8$	$\dfrac{8}{12} = \dfrac{2}{3}$
Mike	$\dfrac{1}{t}$	5	$\dfrac{5}{t}$

Pat's Portion $+$ Mike's Portion $= 1$ Task

$$\frac{2}{3} + \frac{5}{t} = 1, \ t \neq 0$$
$$3t\left(\frac{2}{3} + \frac{5}{t}\right) = 3t(1)$$
$$2t + 15 = 3t$$
$$15 = t$$

It would take Mike 15 hours to complete the task.

47. Let t represent the time that Card reader B is used. The information is recorded in the following table:

	Rate	Time	Quantity
Reader A	600	$6 + t$	$600(6 + t)$
Reader B	850	t	$850t$

Reader A $+$ Reader B $= 9400$
$$600(6 + t) + 850t = 9400$$
$$3600 + 600t + 850t = 9400$$
$$3600 + 1450t = 9400$$
$$1450t = 5800$$
$$t = 4$$

Card reader B was used for 4 minutes.

49. Let t represent Paul's rate, then $r + 20$ represents Amelia's rate. The information is recorded in the following table:

	Quantity	Rate	Time $\left(t = \frac{Q}{r}\right)$
Paul	600	r	$\dfrac{600}{r}$
Amelia	600	$r + 20$	$\dfrac{600}{r + 20}$

Amelia's time $=$ Paul's time $- 5$

$$\frac{600}{r+20} = \frac{600}{r} - 5, \ r \neq -20 \ \text{and} \ r \neq 0$$

$$r(r+20)\left(\frac{600}{r+20}\right) = r(r+20)\left(\frac{600}{r} - 5\right)$$

$$600r = 600(r+20) - 5r(r+20)$$

$$600r = 600r + 12,000 - 5r^2 - 100r$$

$$600r = -5r^2 + 500r + 12,000$$

$$0 = -5r^2 - 100r + 12,000$$

$$0 = r^2 + 20r - 2400$$

$$0 = (r+60)(r-40)$$

$$r + 60 = 0 \qquad \text{or} \quad r - 40 = 0$$

$$r = -60 \quad \text{or} \qquad r = 40$$

Discard the negative solution. Paul types at a rate of 40 words per minute and Amelia types at a rate of 60 words per minute.

53.

$$\frac{x-2}{x^2-1} + \frac{3}{x+1} = \frac{-5}{x-1}, \ x \neq -1 \ \text{and} \ x \neq 1$$

$$\frac{x-2}{(x-1)(x+1)} + \frac{3}{x+1} = \frac{-5}{x-1}$$

$$(x-1)(x+1)\left[\frac{x-2}{(x-1)(x+1)} + \frac{3}{x+1}\right] = (x-1)(x+1)\frac{-5}{x-1}$$

$$(x-2) + 3(x-1) = -5(x+1)$$

$$x - 2 + 3x - 3 = -5x - 5$$

$$4x - 5 = -5x - 5$$

$$9x - 5 = -5$$

$$9x = 0$$

$$x = 0 \qquad\qquad \text{The solution set is } \{0\}$$

CHAPTER 7 **Review Problem Set**

1. $\dfrac{56x^3y}{72xy^3} = \dfrac{\overset{7}{\cancel{56}} \cdot \cancel{x^3} \cdot \cancel{y}}{\underset{9}{\cancel{72}} \cdot \cancel{x} \cdot \cancel{y^3}}_{y^2} = \dfrac{7x^2}{9y^2}$

2. $\dfrac{x^2 - 9x}{x^2 - 6x - 27} = $

$\dfrac{x(\cancel{x-9})}{(\cancel{x-9})\,(x+3)} = \dfrac{x}{x+3}$

3. $\dfrac{3n^2 - n - 10}{n^2 - 4} = $

$\dfrac{(3n+5)(\cancel{n-2})}{(\cancel{n-2})\,(n+2)} = \dfrac{3n+5}{n+2}$

4. $\dfrac{16a^2 + 24a + 9}{20a^2 + 7a - 6} = $

$\dfrac{(4a+3)(\cancel{4a+3})}{(\cancel{4a+3})(5a-2)} = \dfrac{4a+3}{5a-2}$

191

Chapter 7 Review Problem Set

5. $\dfrac{7x^2y^2}{12y^3} \cdot \dfrac{18y}{28x} = \dfrac{7 \cdot \overset{3}{\cancel{18}} \cdot \overset{x}{\cancel{x^2}} \cdot \cancel{y^2} \cdot \cancel{y}}{\underset{2}{\cancel{12}} \cdot \underset{4}{\cancel{28}} \cdot \underset{y}{\cancel{y^3}} \cdot \cancel{x}} = \dfrac{3x}{8}$

6. $\dfrac{x^2y}{x^2+2x} \cdot \dfrac{x^2-x-6}{y} =$

$\dfrac{\overset{x}{\cancel{x^2}} \cdot \cancel{y}(x-3)\cancel{(x+2)}}{\cancel{x}\cancel{(x+2)}\,\cancel{(y)}} = x(x-3)$

7. $\dfrac{n^2-2n-24}{n^2+11n+28} \div \dfrac{n^3-6n^2}{n^2-49} =$

$\dfrac{n^2-2n-24}{n^2+11n+28} \cdot \dfrac{n^2-49}{n^3-6n^2} =$

$\dfrac{\cancel{(n-6)}\,\cancel{(n+4)}}{\cancel{(n+4)}\,\cancel{(n+7)}} \cdot \dfrac{(n-7)\cancel{(n+7)}}{n^2\cancel{(n-6)}} =$

$\dfrac{n-7}{n^2}$

8. $\dfrac{4a^2+4a+1}{(a+6)^2} \div \dfrac{6a^2-5a-4}{3a^2+14a-24} =$

$\dfrac{4a^2+4a+1}{(a+6)^2} \cdot \dfrac{3a^2+14a-24}{6a^2-5a-4} =$

$\dfrac{\cancel{(2a+1)}(2a+1)}{\cancel{(a+6)}\,(a+6)} \cdot \dfrac{\cancel{(3a-4)}\cancel{(a+6)}}{\cancel{(2a+1)}\cancel{(3a-4)}} =$

$\dfrac{2a+1}{a+6}$

9. $\dfrac{3x+4}{5} + \dfrac{2x-7}{4} =$

$\left(\dfrac{3x+4}{4}\right)\left(\dfrac{4}{4}\right) + \left(\dfrac{2x-7}{4}\right)\left(\dfrac{5}{5}\right) =$

$\dfrac{4(3x+4)}{20} + \dfrac{5(2x-7)}{20} =$

$\dfrac{4(3x+4)+5(2x-7)}{20} =$

$\dfrac{12x+16+10x-35}{20} = \dfrac{22x-19}{20}$

10. $\dfrac{7}{3x} + \dfrac{5}{4x} - \dfrac{2}{8x^2} =$

$\dfrac{7}{3x} \cdot \dfrac{8x}{8x} + \dfrac{5}{4x} \cdot \dfrac{6x}{6x} - \dfrac{2}{8x^2} \cdot \dfrac{3}{3} =$

$\dfrac{7(8x)+5(6x)-2(3)}{24x^2} =$

$\dfrac{56x+30x-6}{24x^2} = \dfrac{86x-6}{24x^2} =$

$\dfrac{2(43x-3)}{24x^2} = \dfrac{43x-3}{12x^2}$

11. $\dfrac{7}{n} + \dfrac{3}{n-1} =$

$\left(\dfrac{7}{n}\right)\left(\dfrac{n-1}{n-1}\right) + \left(\dfrac{3}{n-1}\right)\left(\dfrac{n}{n}\right) =$

$\dfrac{7(n-1)}{n(n-1)} + \dfrac{3n}{n(n-1)} =$

$\dfrac{7(n-1)+3n}{n(n-1)} = \dfrac{7n-7+3n}{n(n-1)} =$

$\dfrac{10n-7}{n(n-1)}$

12. $\dfrac{2}{a-4} - \dfrac{3}{a-2} =$

$\dfrac{2}{a-4} \cdot \dfrac{a-2}{a-2} - \dfrac{3}{a-2} \cdot \dfrac{a-4}{a-4} =$

$\dfrac{2(a-2)-3(a-4)}{(a-2)(a-4)} =$

$\dfrac{2a-4-3a+12}{(a-2)(a-4)} = \dfrac{-a+8}{(a-2)(a-4)}$

13. $\dfrac{2x}{x^2-3x} - \dfrac{3}{4x} = \dfrac{2x}{x(x-3)} - \dfrac{3}{4x} =$

$\left[\dfrac{2x}{x(x-3)}\right]\left(\dfrac{4}{4}\right) - \left(\dfrac{3}{4x}\right)\left(\dfrac{x-3}{x-3}\right) =$

$\dfrac{2x(4)-3(x-3)}{4x(x-3)} = \dfrac{8x-3x+9}{4x(x-3)} =$

$\dfrac{5x+9}{4x(x-3)}$

14. $\dfrac{2}{x^2+7x+10}+\dfrac{3}{x^2-25}=$

$\dfrac{2}{(x+5)(x+2)}+\dfrac{3}{(x-5)(x+5)}=$

$\dfrac{2}{(x+5)(x+2)}\cdot\dfrac{x-5}{x-5}$

$+\dfrac{3}{(x-5)(x+5)}\cdot\dfrac{x+2}{x+2}=$

$\dfrac{2(x-5)+3(x+2)}{(x+5)(x+2)(x-5)}=$

$\dfrac{2x-10+3x+6}{(x+5)(x+2)(x-5)}=$

$\dfrac{5x-4}{(x+5)(x+2)(x-5)}$

15. $\dfrac{5x}{x^2-4x-21}-\dfrac{3}{x-7}+\dfrac{4}{x+3}=$

$\dfrac{5x}{(x-7)(x+3)}-\dfrac{3}{x-7}+\dfrac{4}{x+3}=$

$\dfrac{5x}{(x-7)(x+3)}-\left(\dfrac{3}{x-7}\right)\left(\dfrac{x+3}{x+3}\right)+$

$\left(\dfrac{4}{x+3}\right)\left(\dfrac{x-7}{x-7}\right)=$

$\dfrac{5x-3(x+3)+4(x-7)}{(x-7)(x+3)}=$

$\dfrac{5x-3x-9+4x-28}{(x-7)(x+3)}=\dfrac{6x-37}{(x-7)(x+3)}$

16. LCM of x, y and y^2 is xy^2.

$\dfrac{\dfrac{3}{x}-\dfrac{4}{y^2}}{\dfrac{4}{y}+\dfrac{5}{x}}=\left(\dfrac{xy^2}{xy^2}\right)\left(\dfrac{\dfrac{3}{x}-\dfrac{4}{y^2}}{\dfrac{4}{y}+\dfrac{5}{x}}\right)=$

$\dfrac{xy^2\left(\dfrac{3}{x}-\dfrac{4}{y^2}\right)}{xy^2\left(\dfrac{4}{y}+\dfrac{5}{x}\right)}=\dfrac{3y^2-4x}{4xy+5y^2}$

17. LCM of x and y is xy.

$\dfrac{\dfrac{2}{x}-1}{3+\dfrac{5}{y}}=\left(\dfrac{xy}{xy}\right)\left(\dfrac{\dfrac{2}{x}-1}{3+\dfrac{5}{y}}\right)=$

$\dfrac{xy\left(\dfrac{2}{x}-1\right)}{xy\left(3+\dfrac{5}{y}\right)}=\dfrac{2y-xy}{3xy+5x}$

18. $\dfrac{2x-1}{3}+\dfrac{3x-2}{4}=\dfrac{5}{6}$

$12\left(\dfrac{2x-1}{3}+\dfrac{3x-2}{4}\right)=12\left(\dfrac{5}{6}\right)$

$4(2x-1)+3(3x-2)=2(5)$

$8x-4+9x-6=10$

$17x-10=10$

$17x=20$

$x=\dfrac{20}{17}$

The solution set is $\left\{\dfrac{20}{17}\right\}$.

19. $\dfrac{5}{3x}-2=\dfrac{7}{2x}+\dfrac{1}{5x},\ x\neq0$

$30x\left(\dfrac{5}{3x}-2\right)=30x\left(\dfrac{7}{2x}+\dfrac{1}{5x}\right)$

$10(5)-60x=15(7)+6$

$50-60x=105+6$

$50-60x=111$

$-60x=61$

$x=-\dfrac{61}{60}$

The solution set is $\left\{-\dfrac{61}{60}\right\}$.

Chapter 7 Review Problem Set

20.
$$\frac{67-x}{x} = 6 + \frac{4}{x}, \; x \neq 0$$

$$x\left(\frac{67-x}{x}\right) = x\left(6 + \frac{4}{x}\right)$$

$$67 - x = 6x + 4$$

$$67 - 7x = 4$$

$$-7x = -63$$

$$x = 9$$

The solution set is $\{9\}$.

21.
$$\frac{5}{2n+3} = \frac{6}{3n-2}, \; n \neq -\frac{3}{2} \text{ and } n \neq \frac{2}{3}$$

Cross products are equal.

$$5(3n-2) = 6(2n+3)$$

$$15n - 10 = 12n + 18$$

$$3n - 10 = 18$$

$$3n = 28$$

$$n = \frac{28}{3}$$

The solution set is $\left\{\dfrac{28}{3}\right\}$.

22.
$$\frac{x}{x-3} + \frac{5}{x+3} = 1, \; n \neq -3 \text{ and } n \neq 3$$

$$(x-3)(x+3)\left(\frac{x}{x-3} + \frac{5}{x+3}\right) = (x-3)(x+3)(1)$$

$$x(x+3) + 5(x-3) = x^2 - 9$$

$$x^2 + 3x + 5x - 15 = x^2 - 9$$

$$x^2 + 8x - 15 = x^2 - 9$$

$$8x - 15 = -9$$

$$8x = 6$$

$$x = \frac{6}{8} = \frac{3}{4} \qquad \text{The solution set is } \left\{\dfrac{3}{4}\right\}.$$

23.
$$n + \frac{1}{n} = 2, \; n \neq 0$$

$$n\left(n + \frac{1}{n}\right) = n(2)$$

$$n^2 + 1 = 2n$$

$$n^2 - 2n + 1 = 0$$

$$(n-1)(n-1) = 0$$

$$n - 1 = 0 \quad \text{or} \quad n - 1 = 0$$

$$n = 1 \quad \text{or} \qquad n = 1 \qquad \text{The solution set is } \{1\}.$$

24.

$$\frac{n-1}{n^2+8n-9} - \frac{n}{n+9} = 4, \; n \neq -9 \text{ and } n \neq 1$$

$$\frac{(n-1)}{(n+9)(n-1)} - \frac{n}{n+9} = 4$$

$$\frac{1}{n+9} - \frac{n}{n+9} = 4$$

$$(n+9)\left(\frac{1-n}{n+9}\right) = (n+9)(4)$$

$$1 - n = 4n + 36$$
$$1 = 5n + 36$$
$$-35 = 5n$$
$$-7 = n \qquad \text{The solution set is } \{-7\}.$$

25.

$$\frac{6}{7x} - \frac{1}{6} = \frac{5}{6x}, \; x \neq 0$$

$$42x\left(\frac{6}{7x} - \frac{1}{6}\right) = 42x\left(\frac{5}{6x}\right)$$

$$6(6) - 7x = 7(5)$$
$$36 - 7x = 35$$
$$-7x = -1$$
$$x = \frac{1}{7} \qquad \text{The solution set is } \left\{\frac{1}{7}\right\}.$$

26.

$$n + \frac{1}{n} = \frac{5}{2}, \; n \neq 0$$

$$2n\left(n + \frac{1}{n}\right) = 2n\left(\frac{5}{2}\right)$$

$$2n^2 + 2 = 5n$$
$$2n^2 - 5n + 2 = 0$$
$$(2n-1)(n-2) = 0$$
$$2n - 1 = 0 \quad \text{or} \quad n - 2 = 0$$
$$2n = 1 \quad \text{or} \quad n = 2$$
$$n = \frac{1}{2} \quad \text{or} \quad n = 2 \quad \text{The solution set is } \left\{\frac{1}{2}, 2\right\}.$$

27.

$$\frac{n}{5} = \frac{10}{n-5}, \; n \neq 5$$

Cross products are equal.
$$n(n-5) = 5(10)$$
$$n^2 - 5n = 50$$
$$n^2 - 5n - 50 = 0$$
$$(n-10)(n+5) = 0$$
$$n - 10 = 0 \quad \text{or} \quad n + 5 = 0$$
$$n = 10 \quad \text{or} \quad n = -5 \quad \text{The solution set is } \{-5, 10\}.$$

195

28.

$$\frac{-1}{2x-5} + \frac{2x-4}{4x^2-25} = \frac{5}{6x+15}, \; x \neq -\frac{5}{2} \text{ and } x \neq \frac{5}{2}$$

$$\frac{-1}{2x-5} + \frac{2x-4}{(2x-5)(2x+5)} = \frac{5}{3(2x+5)}$$

$$3(2x-5)(2x+5)\left[\frac{-1}{2x-5} + \frac{2x-4}{(2x-5)(2x+5)}\right] = 3(2x-5)(2x+5)\left[\frac{5}{3(2x+5)}\right]$$

$$-3(2x+5) + 3(2x-4) = 5(2x-5)$$

$$-6x - 15 + 6x - 12 = 10x - 25$$

$$-27 = 10x - 25$$

$$-10x = 2$$

$$x = -\frac{2}{10} = -\frac{1}{5}$$

The solution set is $\left\{-\dfrac{1}{5}\right\}$.

29.

$$1 + \frac{1}{n-1} = \frac{1}{n^2-n}, \; n \neq 0 \text{ and } n \neq 1$$

$$1 + \frac{1}{n-1} = \frac{1}{n(n-1)}$$

$$n(n-1)\left(1 + \frac{1}{n-1}\right) = n(n-1)\left[\frac{1}{n(n-1)}\right]$$

$$n(n-1) + n = 1$$

$$n^2 - n + n = 1$$

$$n^2 = 1$$

$$n^2 - 1 = 0$$

$$(n-1)(n+1) = 0$$

$$n - 1 = 0 \quad \text{or} \quad n + 1 = 0$$

$$n = 1 \quad \text{or} \quad n = -1$$

Since $n \neq 1$ because of the original restriction, the solution set is $\{-1\}$.

30. Let n represent the smaller number, then $75 - n$ represents the larger number.

$$\frac{75-n}{n} = 9 + \frac{5}{n}, \; n \neq 0$$

$$n\left(\frac{75-n}{n}\right) = n\left(9 + \frac{5}{n}\right)$$

$$75 - n = 9n + 5$$

$$75 = 10n + 5$$

$$70 = 10n$$

$$7 = n$$

The numbers are 7 and 68.

31. Let t represent Becky's time, then $3t$ represents Nancy's time. The sum of the individual rates equal the rate working together.

$$\frac{1}{t} + \frac{1}{3t} = \frac{1}{2}$$

$$6t\left(\frac{1}{t} + \frac{1}{3t}\right) = 6t\left(\frac{1}{2}\right)$$

$$6 + 2 = 3t$$

$$8 = 3t$$

$$\frac{8}{3} = t$$

It would take Becky $\frac{8}{3} = 2\frac{2}{3}$ hours to complete the task and it would take Nancy $3\left(\frac{8}{3}\right) = 8$ hours to complete the task.

32. Let n represent the number, then $\frac{1}{n}$ represents the reciprocal.

$$n + (2)\frac{1}{n} = 3, \, n \neq 0$$

$$n\left(n + \frac{2}{n}\right) = n(3)$$

$$n^2 + 2 = 3n$$

$$n^2 - 3n + 2 = 0$$

$$(n - 2)(n - 1) = 0$$

$$n - 2 = 0 \quad \text{or} \quad n - 1 = 0$$

$$n = 2 \quad \text{or} \quad n = 1$$

The number could be either 1 or 2.

33. Let n represent the numerator of the fraction, then $2n$ represents the denominator.

$$\frac{n + 4}{2n + 18} = \frac{4}{9}, \, n \neq -9$$

Cross products are equal.

$$9(n + 4) = 4(2n + 18)$$

$$9n + 36 = 8n + 72$$

$$n + 36 = 72$$

$$n = 36$$

The original fraction is $\frac{36}{72}$.

Chapter 7 Review Problem Set

34. Let r represent Todd's rate, then $r + 7$ represents Lanette's rate.

	Distance	Rate	Time $\left(t = \dfrac{d}{r} \right)$
Todd	30	r	$\dfrac{30}{r}$
Lanette	44	$r + 7$	$\dfrac{44}{r + 7}$

Todd's time = Lanette's time

$$\frac{30}{r} = \frac{44}{r + 7}, \ r \neq -7 \text{ and } r \neq 0$$

$$r(r + 7)\left(\frac{30}{r} \right) = r(r + 7)\left(\frac{44}{r + 7} \right)$$

$$30(r + 7) = 44r$$

$$30r + 210 = 44r$$

$$210 = 14r$$

$$15 = r$$

Todd's rate is 15 miles per hour and Lanette's rate is 22 miles per hour.

35. Let r represent Jim's rate for the first 20 miles, then $r - 2$ represents his rate on the last 16 miles. The information is recorded in the following table:

	Distance	Rate	Time $\left(t = \dfrac{d}{r} \right)$
First part	20	r	$\dfrac{20}{r}$
Last part	16	$r - 2$	$\dfrac{16}{r - 2}$

Time of the first part + Time of the last part = 4

$$\frac{20}{r} + \frac{16}{r - 2} = 4, \ r \neq 2 \text{ and } r \neq 0$$

$$r(r - 2)\left(\frac{20}{r} + \frac{16}{r - 2} \right) = r(r - 2)(4)$$

$$20(r - 2) + 16r = 4r(r - 2)$$

$$20r - 40 + 16r = 4r^2 - 8r$$

$$36r - 40 = 4r^2 - 8r$$

$$0 = 4r^2 - 44r + 40$$

$$0 = r^2 - 11r + 10$$

$$0 = (r - 1)(r - 10)$$

$$r - 1 = 0 \quad \text{or} \quad r - 10 = 0$$

$$r = 1 \quad \text{or} \quad r = 10$$

If the rate for the first part is 1 mile per hour, the rate for the second part would be -1 mile per hour which is not reasonable. Thus, the rate for the first part would be 10 miles per hour and the rate for the second part would be 8 miles per hour.

36. Let r represent the time for the tank to overflow.

	Rate	Time	Quantity
Fill	$\dfrac{1}{10}$	t	$\dfrac{t}{10}$
Drain	$\dfrac{1}{12}$	t	$\dfrac{t}{12}$

Fill $-$ Drain $= 1$ Tank Filled

$$\frac{t}{10} - \frac{t}{12} = 1$$

$$120\left(\frac{t}{10} - \frac{t}{12}\right) = 120(1)$$

$$12t - 10t = 120$$

$$2t = 120$$

$$t = 60$$

It would take 60 minutes for the tank to overflow.

37. Let r represent Corinne's rate, then $r - 6$ represents Sue's rate. The information is recorded in the following table:

	Quantity	Rate	Time $\left(t = \dfrac{Q}{r}\right)$
Corinne	840	r	$\dfrac{840}{r}$
Sue	1000	$r - 6$	$\dfrac{1000}{r - 6}$

Sue's time $=$ Corinne's time $+ 5$

$$\frac{1000}{r-6} = \frac{840}{r} + 5, \ r \neq 0 \text{ and } r \neq 6$$

$$r(r-6)\left(\frac{1000}{r-6}\right) = r(r-6)\left(\frac{840}{r} + 5\right)$$

$$1000r = 840(r-6) + 5r(r-6)$$

$$1000r = 840r + 5040 + 5r^2 - 30r$$

$$1000r = 5r^2 + 810r + 5040$$

$$0 = 5r^2 - 190r + 5040$$

$$0 = r^2 - 38r + 1008$$

$$0 = (r+18)(r-56)$$

$$r + 18 = 0 \quad \text{or} \quad r - 56 = 0$$

$$r = -18 \quad \text{or} \quad r = 56$$

Discard the negative solution. Corinne types at a rate of 56 words per minute and Sue types at a rate of 50 words per minute.

Chapter 7 Test

1. $\dfrac{72x^4y^5}{81x^2y^4} = \dfrac{\overset{8}{\cancel{72}} \cdot \overset{x^2}{\cancel{x^4}} \cdot \overset{y}{\cancel{y^5}}}{\underset{9}{\cancel{81}} \cdot \cancel{x^2} \cdot \cancel{y^4}} = \dfrac{8x^2y}{9}$

2. $\dfrac{x^2 + 6x}{x^2 - 36} =$

$\dfrac{x\cancel{(x+6)}}{(x-6)\cancel{(x+6)}} =$

$\dfrac{x}{x-6}$

3. $\dfrac{2n^2 - 7n - 4}{3n^2 - 8n - 16} =$

$\dfrac{(2n+1)\cancel{(n-4)}}{(3n+4)\cancel{(n-4)}} =$

$\dfrac{2n+1}{3n+4}$

4. $\dfrac{2x^3 + 7x^2 - 15x}{x^3 - 25x} =$

$\dfrac{\cancel{x}(2x-3)\cancel{(x+5)}}{\cancel{x}(x-5)\cancel{(x+5)}} =$

$\dfrac{2x-3}{x-5}$

5. $\left(\dfrac{8x^2y}{7x}\right)\left(\dfrac{21xy^3}{12y^2}\right) =$

$\dfrac{\overset{2}{\cancel{8}} \cdot \overset{3}{\cancel{21}} \cdot x^2 \cdot \cancel{x} \cdot y \cdot \overset{y}{\cancel{y^3}}}{\underset{}{7} \cdot \underset{\underset{3}{12}}{} \cdot \cancel{x} \cdot \cancel{y^2}} = 2x^2y^2$

6. $\dfrac{x^2 - 49}{x^2 + 7x} \div \dfrac{x^2 - 4x - 21}{x^2 - 2x} =$

$\dfrac{x^2 - 49}{x^2 + 7x} \cdot \dfrac{x^2 - 2x}{x^2 - 4x - 21} =$

$\dfrac{\cancel{(x-7)}\;\cancel{(x+7)}}{\cancel{x}\cancel{(x+7)}} \cdot \dfrac{\cancel{x}(x-2)}{\cancel{(x-7)}(x+3)} =$

$\dfrac{x-2}{x+3}$

7. $\dfrac{x^2 - 5x - 36}{x^2 - 15x + 54} \cdot \dfrac{x^2 - 2x - 24}{x^2 + 7x} =$

$\dfrac{\cancel{(x-9)}\,(x+4)}{\cancel{(x-6)}\;\cancel{(x-9)}} \cdot \dfrac{\cancel{(x-6)}\,(x+4)}{x(x+7)} =$

$\dfrac{(x+4)(x+4)}{x(x+7)} = \dfrac{(x+4)^2}{x(x+7)}$

8. $\dfrac{3x-1}{6} - \dfrac{2x-3}{8} =$

$\left(\dfrac{3x-1}{6}\right)\left(\dfrac{4}{4}\right) - \left(\dfrac{2x-3}{8}\right)\left(\dfrac{3}{3}\right) =$

$\dfrac{4(3x-1) - 3(2x-3)}{24} =$

$\dfrac{12x - 4 - 6x + 9}{24} = \dfrac{6x+5}{24}$

9. $\dfrac{n+2}{3} - \dfrac{n-1}{5} + \dfrac{n-6}{6} =$

$\dfrac{n+2}{3}\left(\dfrac{10}{10}\right) - \dfrac{n-1}{5}\left(\dfrac{6}{6}\right) + \dfrac{n-6}{6}\left(\dfrac{5}{5}\right) =$

$\dfrac{10(n+2) - 6(n-1) + 5(n-6)}{30} =$

$\dfrac{10n + 20 - 6n + 6 + 5n - 30}{30} = \dfrac{9n - 4}{30}$

10. $\dfrac{3}{2x} - \dfrac{5}{6} + \dfrac{7}{9x} =$

$\left(\dfrac{3}{2x}\right)\left(\dfrac{9}{9}\right) - \left(\dfrac{5}{6}\right)\left(\dfrac{3x}{3x}\right) + \left(\dfrac{7}{9x}\right)\left(\dfrac{2}{2}\right) =$

$\dfrac{27 - 15x + 14}{18x} = \dfrac{-15x + 41}{18x}$

11. $\dfrac{6}{n} - \dfrac{4}{n-1} =$

$\dfrac{6}{n}\left(\dfrac{n-1}{n-1}\right) - \dfrac{4}{n-1}\left(\dfrac{n}{n}\right) =$

$\dfrac{6(n-1) - 4n}{n(n-1)} = \dfrac{6n - 6 - 4n}{n(n-1)} =$

$\dfrac{2n - 6}{n(n-1)}$

12. $\dfrac{2x}{x^2 + 6x} - \dfrac{3}{4x} = \dfrac{2x}{x(x+6)} - \dfrac{3}{4x} =$

$\left[\dfrac{2x}{x(x+6)}\right]\left(\dfrac{4}{4}\right) - \left(\dfrac{3}{4x}\right)\left(\dfrac{x+6}{x+6}\right) =$

$\dfrac{4(2x) - 3(x+6)}{4x(x+6)} = \dfrac{8x - 3x - 18}{4x(x+6)} =$

$\dfrac{5x - 18}{4x(x+6)}$

13. $\dfrac{9}{x^2 + 4x - 32} + \dfrac{5}{x+8} =$

$\dfrac{9}{(x+8)(x-4)} + \dfrac{5}{x+8} =$

$\dfrac{9}{(x+8)(x-4)} + \dfrac{5}{x+8}\left(\dfrac{x-4}{x-4}\right) =$

$\dfrac{9 + 5(x-4)}{(x+8)(x-4)} = \dfrac{9 + 5x - 20}{(x+8)(x-4)} =$

$\dfrac{5x - 11}{(x+8)(x-4)}$

14. $\dfrac{-3}{6x^2 - 7x - 20} - \dfrac{5}{3x^2 - 14x - 24} =$

$\dfrac{-3}{(3x+4)(2x-5)} - \dfrac{5}{(3x+4)(x-6)} =$

$\left[\dfrac{-3}{(3x+4)(2x-5)}\right]\left(\dfrac{x-6}{x-6}\right)$
$\quad - \left[\dfrac{5}{(3x+4)(x-6)}\right]\left(\dfrac{2x-5}{2x-5}\right) =$

$\dfrac{-3(x-6) - 5(2x-5)}{(3x+4)(2x-5)(x-6)} =$

$\dfrac{-3x + 18 - 10x + 25}{(3x+4)(2x-5)(x-6)} =$

$\dfrac{-13x + 43}{(3x+4)(2x-5)(x-6)}$

15. $\dfrac{x+3}{5} - \dfrac{x-2}{6} = \dfrac{23}{30}$

$30\left(\dfrac{x+3}{5} - \dfrac{x-2}{6}\right) = 30\left(\dfrac{23}{30}\right)$

$6(x+3) - 5(x-2) = 23$

$6x + 18 - 5x + 10 = 23$

$x + 28 = 23$

$x = -5$

The solution set is $\{-5\}$.

16. $\dfrac{5}{8x} - 2 = \dfrac{3}{x}, \; x \neq 0$

$8x\left(\dfrac{5}{8x} - 2\right) = 8x\left(\dfrac{3}{x}\right)$

$5 - 2(8x) = 8(3)$

$5 - 16x = 24$

$-16x = 19$

$x = -\dfrac{19}{16}$

The solution set is $\left\{-\dfrac{19}{16}\right\}$.

17.
$$n + \frac{4}{n} = \frac{13}{3}$$
$$3n\left(n + \frac{4}{n}\right) = 3n\left(\frac{13}{3}\right)$$
$$3n^2 + 12 = 13n$$
$$3n^2 - 13n + 12 = 0$$
$$(3n - 4)(n - 3) = 0$$
$$3n - 4 = 0 \quad \text{or} \quad n - 3 = 0$$
$$3n = 4 \quad \text{or} \quad n = 3$$
$$n = \frac{4}{3} \quad \text{or} \quad n = 3$$
The solution set is $\left\{\frac{4}{3}, 3\right\}$.

18.
$$\frac{x}{8} = \frac{6}{x - 2}, \; x \neq 2$$
Cross products are equal.
$$x(x - 2) = 6(8)$$
$$x^2 - 2x = 48$$
$$x^2 - 2x - 48 = 0$$
$$(x - 8)(x + 6) = 0$$
$$x - 8 = 0 \quad \text{or} \quad x + 6 = 0$$
$$x = 8 \quad \text{or} \quad x = -6$$
The solution set is $\{-6, 8\}$.

19.
$$\frac{x}{x - 1} + \frac{2}{x + 1} = \frac{8}{3}, \; x \neq -1 \text{ and } x \neq 1$$
$$3(x - 1)(x + 1)\left(\frac{x}{x - 1} + \frac{2}{x + 1}\right) = 3(x - 1)(x + 1)\left(\frac{8}{3}\right)$$
$$3x(x + 1) + 6(x - 1) = 8\left(x^2 - 1\right)$$
$$3x^2 + 3x + 6x - 6 = 8x^2 - 8$$
$$3x^2 + 9x - 6 = 8x^2 - 8$$
$$-5x^2 + 9x + 2 = 0$$
$$5x^2 - 9x - 2 = 0$$
$$(5x + 1)(x - 2) = 0$$
$$5x + 1 = 0 \quad \text{or} \quad x - 2 = 0$$
$$5x = -1 \quad \text{or} \quad x = 2$$
$$x = -\frac{1}{5} \quad \text{or} \quad x = 2$$
The solution set is $\left\{-\frac{1}{5}, 2\right\}$.

20. $\dfrac{3}{2x+1} = \dfrac{5}{3x-6}$, $x \neq -\dfrac{1}{2}$ and $x \neq 2$

Cross products are equal.

$3(3x-6) = 5(2x+1)$

$9x - 18 = 10x + 5$

$-x = 23$

$x = -23$

The solution set is $\{-23\}$.

21. $\dfrac{4}{n^2-n} - \dfrac{3}{n-1} = -1$, $n \neq 0$ and $n \neq 1$

$\dfrac{4}{n(n-1)} - \dfrac{3}{n-1} = -1$

$n(n-1)\left[\dfrac{4}{n(n-1)} - \dfrac{3}{n-1}\right] = n(n-1)(-1)$

$4 - 3n = -n(n-1)$

$4 - 3n = -n^2 + n$

$0 = -n^2 + 4n - 4$

$0 = n^2 - 4n + 4$

$0 = (n-2)(n-2)$

$n - 2 = 0$ or $n - 2 = 0$

$n = 2$ or $n = 2$

The solution set is $\{2\}$.

22. $\dfrac{3n-1}{3} + \dfrac{2n+5}{4} = \dfrac{4n-6}{9}$

$36\left(\dfrac{3n-1}{3} + \dfrac{2n+5}{4}\right) = 36\left(\dfrac{4n-6}{9}\right)$

$12(3n-1) + 9(2n+5) = 4(4n-6)$

$36n - 12 + 18n + 45 = 16n - 24$

$54n + 33 = 16n - 24$

$38n + 33 = -24$

$38x = -57$

$x = -\dfrac{57}{38} = -\dfrac{3}{2}$

The solution set is $\left\{-\dfrac{3}{2}\right\}$.

Chapter 7 Test

23. Let n represent the number, then $\dfrac{1}{n}$ represents the reciprocal.

$$n + (2)\frac{1}{n} = \frac{11}{3}$$

$$3n\left(n + \frac{2}{n}\right) = 3n\left(\frac{11}{3}\right)$$

$$3n^2 + 6 = 11n$$

$$3n^2 - 11n + 6 = 0$$

$$(3n - 2)(n - 3) = 0$$

$$3n - 2 = 0 \quad \text{or} \quad n - 3 = 0$$

$$3n = 2 \quad \text{or} \qquad n = 3$$

$$n = \frac{2}{3} \quad \text{or} \qquad n = 3 \quad \text{The number is either } \frac{2}{3} \text{ or } 3.$$

24. Let r represent Betty's rate, then $r + 2$ represents Wendy's rate.

	Distance	Rate	Time $\left(t = \dfrac{d}{r}\right)$
Betty	36	r	$\dfrac{36}{r}$
Wendy	42	$r+2$	$\dfrac{42}{r+2}$

Betty's time = Wendy's time

$$\frac{36}{r} = \frac{42}{r+2}, \ r \neq -2 \text{ and } r \neq 0$$

$$r(r+2)\left(\frac{36}{r}\right) = r(r+2)\left(\frac{42}{r+2}\right)$$

$$36(r+2) = 42r$$

$$36r + 72 = 42r$$

$$72 = 6r$$

$$12 = r$$

Wendy's rate is 14 miles per hour.

25. Let t represent the time working together. The sum of the individual rates equals the rate working together.

$$\frac{1}{20} + \frac{1}{30} = \frac{1}{t}$$

$$60t\left(\frac{1}{20} + \frac{1}{30}\right) = 60t\left(\frac{1}{t}\right)$$

$$3t + 2t = 60$$

$$5t = 60$$

$$t = 12$$

It takes them 12 minutes working together.

CHAPTERS 1-7 Cumulative Review

1. $3x - 2xy - 7x + 5xy = -4x + 3xy$

when $x = \dfrac{1}{2}$, $y = 3$: $-4\left(\dfrac{1}{2}\right) + 3\left(\dfrac{1}{2}\right)(3)$

$$= \dfrac{-4}{2} + \dfrac{9}{2}$$

$$= \dfrac{5}{2}$$

2. $a = -3$, $b = -5$:
$7(a - b) - 3(a - b) - (a - b) =$
$3(a - b) = 3[(-3) - (-5)] =$
$3[2] = 6$

3. $x = \dfrac{2}{3}$, $y = \dfrac{5}{6}$, $z = \dfrac{3}{4}$:

$\dfrac{xy + yz}{y} = \dfrac{xy}{y} + \dfrac{yz}{y} = x + z =$

$\dfrac{2}{3} + \dfrac{3}{4} = \dfrac{8}{12} + \dfrac{9}{12} = \dfrac{17}{12}$

4. $a = 0.4$, $b = 0.6$: $ab + b^2 =$
$(0.4)(0.6) + (0.6)^2 = 0.24 + 0.36 = 0.6$

5. $x = -6$, $y = 4$: $x^2 - y^2 =$
$(-6)^2 - (4)^2 = 36 - 16 = 20$

6. $x = -9$: $x^2 + 5x - 36 =$
$(x + 9)(x - 4) = (-9 + 9)(-9 - 4) =$
$0(-13) = 0$

7. $x = -6$: $\dfrac{x^2 + 2x}{x^2 + 5x + 6} =$

$\dfrac{x\cancel{(x + 2)}}{(x + 3)\cancel{(x + 2)}} = \dfrac{x}{x + 3} =$

$\dfrac{-6}{-6 + 3} = \dfrac{-6}{-3} = 2$

8. $x = 4$: $\dfrac{x^2 + 3x - 10}{x^2 - 9x + 14} = \dfrac{(x + 5)\cancel{(x - 2)}}{(x - 7)\cancel{(x - 2)}} =$

$\dfrac{x + 5}{x - 7} = \dfrac{4 + 5}{4 - 7} = \dfrac{9}{-3} = -3$

9. $3^{-3} = \dfrac{1}{3^3} = \dfrac{1}{27}$

10. $\left(\dfrac{2}{3}\right)^{-1} = \dfrac{3}{2}$

11. $\left(\dfrac{1}{2} + \dfrac{1}{3}\right)^0 = 1$

12. $\left(\dfrac{1}{3} + \dfrac{1}{4}\right)^{-1} = \left(\dfrac{1 \cdot 4}{3 \cdot 4} + \dfrac{1 \cdot 3}{4 \cdot 3}\right)^{-1} =$

$\left(\dfrac{4}{12} + \dfrac{3}{12}\right)^{-1} = \left(\dfrac{7}{12}\right)^{-1} = \dfrac{12}{7}$

13. $-4^{-2} = -\dfrac{1}{4^2} = -\dfrac{1}{16}$

14. $\left(\dfrac{2}{3}\right)^{-2} = \left(\dfrac{3}{2}\right)^2 = \dfrac{3^2}{2^2} = \dfrac{9}{4}$

15. $\dfrac{1}{\frac{2^{-2}}{5}} = \left(\dfrac{2}{5}\right)^2 = \dfrac{4}{25}$

16. $(-3)^{-3} = \dfrac{1}{(-3)^3} = \dfrac{1}{-27} = -\dfrac{1}{27}$

17. $\dfrac{7}{5x} + \dfrac{2}{x} - \dfrac{3}{2x} =$

$\dfrac{7}{5x} \cdot \dfrac{2}{2} + \dfrac{2}{x} \cdot \dfrac{10}{10} - \dfrac{3}{2x} \cdot \dfrac{5}{5} =$

$\dfrac{14 + 20 - 15}{10x} = \dfrac{19}{10x}$

18. $\dfrac{4x}{5y} \div \dfrac{12x^2}{10y^2} = \dfrac{4x}{5y} \cdot \dfrac{10y^2}{12x^2} =$

$$\dfrac{\overset{2}{\cancel{4}} \cdot \overset{2}{\cancel{10}} \cdot \cancel{x} \cdot \overset{y}{\cancel{y^2}}}{\underset{3}{\cancel{5}} \cdot \underset{3}{\cancel{12}} \cdot \underset{x}{\cancel{x^2}} \cdot \cancel{y}} = \dfrac{2y}{3x}$$

19. $\dfrac{4}{x-6} + \dfrac{3}{x+4} =$

$$\dfrac{4}{x-6}\left(\dfrac{x+4}{x+4}\right) + \dfrac{3}{x+4}\left(\dfrac{x-6}{x-6}\right) =$$

$$\dfrac{4(x+4) + 3(x-6)}{(x+4)(x-6)} =$$

$$\dfrac{4x + 16 + 3x - 18}{(x+4)(x-6)} = \dfrac{7x-2}{(x+4)(x-6)}$$

20. $\dfrac{2}{x^2-4x} - \dfrac{3}{x^2} = \dfrac{2}{x(x-4)} - \dfrac{3}{x^2} =$

$$\left[\dfrac{2}{x(x-4)}\right]\left(\dfrac{x}{x}\right) - \left(\dfrac{3}{x^2}\right)\left(\dfrac{x-4}{x-4}\right) =$$

$$\dfrac{2x - 3(x-4)}{x^2(x-4)} = \dfrac{2x - 3x + 12}{x^2(x-4)} =$$

$$\dfrac{-x+12}{x^2(x-4)}$$

21. $\dfrac{x^2-8x}{x^2-x-56} \cdot \dfrac{x^2-49}{3xy} =$

$$\dfrac{\cancel{x}(x-8)}{(x-8)(x+7)} \cdot \dfrac{(x-7)(x+7)}{3 \cdot \cancel{x} \cdot y} =$$

$$\dfrac{x-7}{3y}$$

22. $\dfrac{5}{x^2-x-12} - \dfrac{3}{x-4} =$

$$\dfrac{5}{(x-4)(x+3)} - \dfrac{3}{x-4} =$$

$$\dfrac{5}{(x-4)(x+3)} - \left(\dfrac{3}{x-4}\right)\left(\dfrac{x+3}{x+3}\right) =$$

$$\dfrac{5 - 3(x+3)}{(x-4)(x+3)} = \dfrac{5 - 3x - 9}{(x-4)(x+3)} =$$

$$\dfrac{-3x - 4}{(x-4)(x+3)}$$

23. $(-5x^2y)(7x^3y^4) =$
$(-5)(7)(x^{2+3})(y^{1+4}) = -35x^5y^5$

24. $(9ab^3)^2 = 9^2a^2(b^3)^2 = 81a^2b^6$

25. $(-3n^2)(5n^2 + 6n - 2) =$
$(-3n^2)(5n^2) + (-3n^2)(6n) - (-3n^2)(2) =$
$-15n^4 - 18n^3 + 6n^2$

26. $(5x-1)(3x+4) =$
$(5x)(3x) + (20 - 3)x - 4 =$
$15x^2 + 17x - 4$

27. $(2x+5)^2 = (2x)^2 + 2(2x)(5) + (5)^2 =$
$4x^2 + 20x + 25$

28. $(x+2)(2x^2 - 3x - 1) =$
$x(2x^2 - 3x - 1) + 2(2x^2 - 3x - 1) =$
$2x^3 - 3x^2 - x + 4x^2 - 6x - 2 =$
$2x^3 + x^2 - 7x - 2$

29. $(x^2 - x - 1)(x^2 + 2x - 3) =$

$$x^2(x^2 + 2x - 3) - x(x^2 + 2x - 3)$$
$$- 1(x^2 + 2x - 3) =$$

$$x^4 + 2x^3 - 3x^2 - x^3 - 2x^2 + 3x$$
$$- x^2 - 2x + 3 =$$

$$x^4 + x^3 - 6x^2 + x + 3$$

30. $(-2x - 1)(3x - 7) =$
$(-2x)(3x) + (14 - 3)x + 7 =$
$-6x^2 + 11x + 7$

31. $\dfrac{24x^2y^3 - 48x^4y^5}{8xy^2} =$

$\dfrac{24x^2y^3}{8xy^2} - \dfrac{48x^4y^5}{8xy^2} =$

$3xy - 6x^3y^3$

32.

$$\begin{array}{r}
7x + 4 \\
4x - 5 \overline{\smash{)}\,28x^2 - 19x - 20} \\
\underline{28x^2 - 35x} \\
16x - 20 \\
\underline{16x - 20}
\end{array}$$

33. $3x^3 + 15x + 27x = 3x(x^2 + 5x + 9)$
$(x^2 + 5x + 9$ is not factorable.)

34. $x^2 - 100 = (x - 10)(x + 10)$

35. $5x^2 - 22x + 8 =$
$5x^2 - 20x - 2x + 8 =$
$5x(x - 4) - 2(x - 4) =$
$(x - 4)(5x - 2)$

36. $8x^2 - 22x - 63 =$
$8x^2 + 14x - 36x - 63 =$
$2x(4x + 7) - 9(4x + 7) =$
$(4x + 7)(2x - 9)$

37. $n^2 + 25n + 144 =$
$n^2 + 16n + 9n + 144 =$
$n(n + 16) + 9(n + 16) =$
$(n + 16)(n + 9)$

38. $nx + ny - 2x - 2y =$
$n(x + y) - 2(x + y) =$
$(x + y)(n - 2)$

39. $3x^3 - 3x = 3x(x^2 - 1) =$
$3x(x - 1)(x + 1)$

40. $2x^3 - 6x^2 - 108x =$
$2x(x^2 - 3x - 54) = 2x(x - 9)(x + 6)$

41. $36x^2 - 60x + 25 =$
$(6x)^2 - 2(6x)(5) + (5)^2 =$
$(6x - 5)^2$

42. $3x^2 - 5xy - 2y^2 =$
$3x^2 - 6xy + xy - 2y^2 =$
$3x(x - 2y) + y(x - 2y) =$
$(x - 2y)(3x + y)$

43. $3(x - 2) - 2(x + 6) = -2(x + 1)$
$3x - 6 - 2x - 12 = -2x - 2$
$x - 18 = -2x - 2$
$3x - 18 = -2$
$3x = 16$
$x = \dfrac{16}{3}$
The solution set is $\left\{ \dfrac{16}{3} \right\}$.

44. $x^2 = -11x$
$x^2 + 11x = 0$
$x(x + 11) = 0$
$x = 0$ or $x + 11 = 0$
$x = 0$ or $x = -11$
The solution set is $\{-11, 0\}$.

45. $0.2x - 3(x - 0.4) = 1$
Multiply by 10.
$2x - 30(x - 0.4) = 10$
$2x - 30x + 12 = 10$
$-28x + 12 = 10$
$-28x = -2$
$x = \dfrac{2}{28} = \dfrac{1}{14}$.
The solution set is $\left\{ \dfrac{1}{14} \right\}$.

207

46. $\dfrac{3n-1}{4} = \dfrac{5n+2}{7}$

Cross products are equal.

$7(3n-1) = 4(5n+2)$

$21n - 7 = 20n + 8$

$n - 7 = 8$

$n = 15$

The solution set is $\{15\}$.

47. $5n^2 - 5 = 0$

Divide by 5.

$n^2 - 1 = 0$

$(n-1)(n+1) = 0$

$n - 1 = 0 \quad \text{or} \quad n + 1 = 0$

$n = 1 \quad \text{or} \qquad n = -1$

The solution set is $\{-1, 1\}$.

48. $x^2 + 5x - 6 = 0$

$(x-1)(x+6) = 0$

$x - 1 = 0 \quad \text{or} \quad x + 6 = 0$

$x = 1 \quad \text{or} \qquad x = -6$

The solution set is $\{-6, 1\}$.

49. $n + \dfrac{4}{n} = 4, \; n \neq 0$

$n\left(n + \dfrac{4}{n}\right) = n(4)$

$n^2 + 4 = 4n$

$n^2 - 4n + 4 = 0$

$(n-2)(n-2) = 0$

$n - 2 = 0 \quad \text{or} \quad n - 2 = 0$

$n = 2 \quad \text{or} \qquad n = 2$

The solution set is $\{2\}$.

50. $\dfrac{2x+1}{2} + \dfrac{3x-4}{3} = 1$

$6\left(\dfrac{2x+1}{2} + \dfrac{3x-4}{3}\right) = 6(1)$

$3(2x+1) + 2(3x-4) = 6$

$6x + 3 + 6x - 8 = 6$

$12x - 5 = 6$

$12x = 11$

$x = \dfrac{11}{12}$

The solution set is $\left\{\dfrac{11}{12}\right\}$.

51. $2(x-1) - x(x-1) = 0$

$(x-1)(2-x) = 0$

$x - 1 = 0 \quad \text{or} \quad 2 - x = 0$

$x = 1 \quad \text{or} \qquad 2 = x$

The solution set is $\{1, 2\}$.

52. $\dfrac{3}{2x} - 1 = \dfrac{5}{3x} + 2, \; x \neq 0$

$6x\left(\dfrac{3}{2x} - 1\right) = 6x\left(\dfrac{5}{3x} + 2\right)$

$3(3) - 6x = 2(5) + 12x$

$9 - 6x = 10 + 12x$

$9 - 18x = 10$

$-18x = 1$

$x = -\dfrac{1}{18}$

The solution set is $\left\{-\dfrac{1}{18}\right\}$.

53. $6t^2 + 19t - 7 = 0$

$(3t-1)(2t+7) = 0$

$3t - 1 = 0 \quad \text{or} \quad 2t + 7 = 0$

$3t = 1 \quad \text{or} \qquad 2t = -7$

$t = \dfrac{1}{3} \quad \text{or} \qquad t = -\dfrac{7}{2}$

The solution set is $\left\{-\dfrac{7}{2}, \dfrac{1}{3}\right\}$.

54. $(2x - 1)(x - 8) = 0$

$2x - 1 = 0$ or $x - 8 = 0$

$2x = 1$ or $x = 8$

$x = \dfrac{1}{2}$ or $x = 8$

The solution set is $\left\{\dfrac{1}{2}, 8\right\}$.

55. $(x + 1)(x + 6) = 24$

$x^2 + 7x + 6 = 24$

$x^2 + 7x - 18 = 0$

$(x + 9)(x - 2) = 0$

$x + 9 = 0$ or $x - 2 = 0$

$x = -9$ or $x = 2$

The solution set is $\{-9, 2\}$.

56.
$$\frac{x}{x - 2} - \frac{7}{x + 1} = 1,\ x \neq -1 \text{ and } x \neq 2$$

$$(x - 2)(x + 1)\left(\frac{x}{x - 2} - \frac{7}{x + 1}\right) = (x - 2)(x + 1)(1)$$

$$x(x + 1) - 7(x - 2) = x^2 - x - 2$$

$$x^2 + x - 7x + 14 = x^2 - x - 2$$

$$x^2 - 6x + 14 = x^2 - x - 2$$

$$-5x + 14 = -2$$

$$-5x = -16$$

$$x = \frac{16}{5}$$

The solution set is $\left\{\dfrac{16}{5}\right\}$.

57.
$$\frac{1}{n} - \frac{2}{n - 1} = \frac{3}{n},\ n \neq 0 \text{ and } n \neq 1$$

$$n(n - 1)\left(\frac{1}{n} - \frac{2}{n - 1}\right) = n(n - 1)\left(\frac{3}{n}\right)$$

$$1(n - 1) - 2n = 3(n - 1)$$

$$n - 1 - 2n = 3n - 3$$

$$-1 - n = 3n - 3$$

$$-1 - 4n = -3$$

$$-4n = -2$$

$$n = \frac{2}{4} = \frac{1}{2}$$

The solution set is $\left\{\dfrac{1}{2}\right\}$.

58.

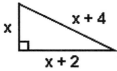

$x^2 + (x + 2)^2 = (x + 4)^2$

$x^2 + x^2 + 4x + 4 = x^2 + 8x + 16$

$2x^2 + 4x + 4 = x^2 + 8x + 16$

$x^2 - 4x - 12 = 0$

$(x - 6)(x + 2) = 0$

$x - 6 = 0$ or $x + 2 = 0$

$x = 6$ or $x = -2$

Since length can not be negative, discard the root $x = -2$. The length of the legs are 6 inches and 8 inches. The hypotenuse is 10 inches.

59. Let $x =$ the number

$$0.20(x) = 15$$

$$x = \frac{15}{0.20} = 75$$

The number is 75.

60. Let $x =$ milliliters of 65% HCl acid

$$0.65x + 0.30(40) = 0.55(x + 40)$$

$$0.65x + 12 = 0.55x + 22$$

$$0.10x = 10$$

$$x = \frac{10}{0.10} = 100$$

100 ml of 65% hydrochloric acid must be added.

61. Let $x =$ width of rectangle

then $x + 2 =$ length of rectangle

$$P = 2L + 2W$$

$$28 = 2(x + 2) + 2x$$

$$28 = 2x + 4 + 2x$$

$$28 = 4x + 4$$

$$24 = 4x$$

$$6 = x$$

The dimensions will be 6 feet by 8 feet.

62. Let $t =$ time traveling

$$55t + 65t = 300$$

$$120t = 300$$

$$t = \frac{300}{120} = 2.5$$

It will take 2.5 hours to be 300 miles apart.

63.
$$A = \frac{1}{2}h(b_1 + b_2)$$

$$120 = \frac{1}{2}h(10 + 22)$$

$$120 = \frac{1}{2}h(32)$$

$$120 = 16h$$

$$\frac{120}{16} = h$$

$$7.5 = h$$

The altitude is 7.5 cm.

64.
$$\frac{16}{352} = \frac{x}{594}$$

$$352x = 16(594)$$

$$352x = 9504$$

$$x = \frac{9504}{352} = 27$$

The consumption would be 27 gallons.

65. Let x represent the score of the fifth exam.

$$\frac{89 + 92 + 87 + 90 + x}{5} \geq 90$$

$$\frac{358 + x}{5} \geq 90$$

$$358 + x \geq 450$$

$$x \geq 92$$

Swati must have a score of 92 or better.

66. Let x represent a certain number.

$$3x - 2 < 10$$

$$3x < 12$$

$$x < 4$$

The numbers must be positive integers that are less than 4, which are 1, 2 and 3.

67. Let x represent a certain number.

$$1 + 4x > 15$$

$$4x > 14$$

$$x > \frac{14}{4}$$

$$x > \frac{7}{2}$$

$$x > 3\frac{1}{2}$$

The numbers must be all real numbers greater than $3\frac{1}{2}$.

Chapter 8 Coordinate Geometry and Linear Systems

PROBLEM SET | **8.1** Cartesian Coordinate System

1. $3x + 7y = 13$ for y

$7y = 13 - 3x$

$y = \dfrac{13 - 3x}{7}$

3. $x - 3y = 9$ for x

$x = 9 + 3y$

5. $-x + 5y = 14$ for y

$5y = x + 14$

$y = \dfrac{x + 14}{5}$

7. $-3x + y = 7$ for x

$-3x = -y + 7$

$x = \dfrac{-y + 7}{-3} \cdot \dfrac{-1}{-1} = \dfrac{y - 7}{3}$

9. $-2x + 3y = -5$ for y

$3y = 2x - 5$

$y = \dfrac{2x - 5}{3}$

11. $y = x + 1$

$x = -2:$ $y = -2 + 1 = -1$

$x = 0:$ $y = 0 + 1 = 1$

$x = 2:$ $y = 2 + 1 = 3$

Point
$(-2, -1)$
$(0, 1)$
$(2, 3)$

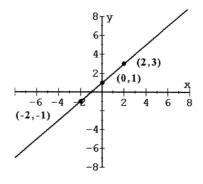

13. $y = x - 2$

$x = -2:$ $y = -2 - 2 = -4$

$x = 0:$ $y = 0 - 2 = -2$

$x = 2:$ $y = 2 - 2 = 0$

Point
$(-2, -4)$
$(0, -2)$
$(2, 0)$

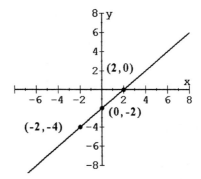

211

Problem Set 8.1

15. $y = (x-2)^2$

$x = -2 : \quad y = (-2-2)^2 = (-4)^2 = 16$

$x = 0 : \quad y = (0-2)^2 = (-2)^2 = 4$

$x = 2 : \quad y = (2-2)^2 = (0)^2 = 0$

$x = 4 : \quad y = (4-2)^2 = (2)^2 = 4$

$x = 6 : \quad y = (6-2)^2 = (4)^2 = 16$

Point
$(-2, 16)$
$(0, 4)$
$(2, 0)$
$(4, 4)$
$(6, 16)$

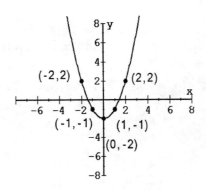

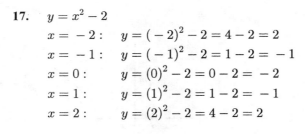

17. $y = x^2 - 2$

$x = -2 : \quad y = (-2)^2 - 2 = 4 - 2 = 2$

$x = -1 : \quad y = (-1)^2 - 2 = 1 - 2 = -1$

$x = 0 : \quad y = (0)^2 - 2 = 0 - 2 = -2$

$x = 1 : \quad y = (1)^2 - 2 = 1 - 2 = -1$

$x = 2 : \quad y = (2)^2 - 2 = 4 - 2 = 2$

Point
$(-2, 2)$
$(-1, -1)$
$(0, -2)$
$(1, -1)$
$(2, 2)$

19. $y = \dfrac{1}{2}x + 3$

$x = -2 : \quad y = \dfrac{1}{2}(-2) + 3 = -1 + 3 = 2$

$x = 0 : \quad y = \dfrac{1}{2}(0) + 3 = 0 + 3 = 3$

$x = 2 : \quad y = \dfrac{1}{2}(2) + 3 = 1 + 3 = 4$

Point
$(-2, -2)$
$(0, 3)$
$(2, 4)$

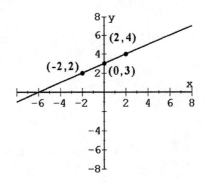

21. $x + 2y = 4$, so $y = \dfrac{4 - x}{2}$

(Use values of x that are divisible by 2.)

$x = -2: \quad y = \dfrac{4 - (-2)}{2} = \dfrac{6}{2} = 3$

$x = 0: \quad y = \dfrac{4 - (0)}{2} = \dfrac{4}{2} = 2$

$x = 2: \quad y = \dfrac{4 - (2)}{2} = \dfrac{2}{2} = 1$

Point
$\overline{}$
$(-2, 3)$
$(0, 2)$
$(2, 1)$

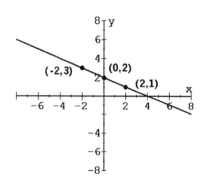

23. $2x - 5y = 10$, so $y = \dfrac{2x - 10}{5}$

(Use values of x that are divisible by 5.)

$x = -5: \quad y = \dfrac{2(-5) - 10}{5} = \dfrac{-10 - 10}{5}$

$\qquad \qquad = \dfrac{-20}{5} = -4$

$x = 0: \quad y = \dfrac{2(0) - 10}{5} = \dfrac{-10}{5} = -2$

$x = 5: \quad y = \dfrac{2(5) - 10}{5} = \dfrac{10 - 10}{5}$

$\qquad \qquad = \dfrac{0}{5} = 0$

Point
$\overline{}$
$(-5, -4)$
$(0, -2)$
$(5, 0)$

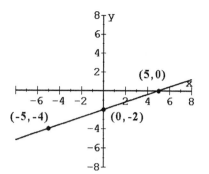

25. $y = x^3$

$x = -2: \quad y = (-2)^3 = -8$

$x = -1: \quad y = (-1)^3 = -1$

$x = 0: \quad y = (0)^3 = 0$

$x = 1: \quad y = (1)^3 = 1$

$x = 2: \quad y = (2)^3 = 8$

Point
$\overline{}$
$(-2, -8)$
$(-1, -1)$
$(0, 0)$
$(1, 1)$
$(2, 8)$

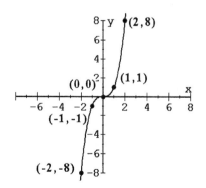

27. $y = -x^2$

$x = -2: \quad y = -(-2)^2 = -4$

$x = -1: \quad y = -(-1)^2 = -1$

$x = 0: \quad y = -(0)^2 = 0$

$x = 1: \quad y = -(1)^2 = -1$

$x = 2: \quad y = -(2)^2 = -4$

Problem Set 8.1

Point
$(-2, -4)$
$(-1, -1)$
$(0, 0)$
$(1, -1)$
$(2, -4)$

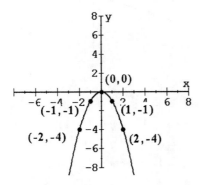

31. $y = -3x + 2$
$x = -2:$ $\quad y = -3(-2)+2=6+2=8$
$x = 0:$ $\quad y = -3(0)+2=0+2=2$
$x = 2:$ $\quad y = -3(2)+2 = -6+2 = -4$

Point
$(-2, 8)$
$(0, 2)$
$(2, -4)$

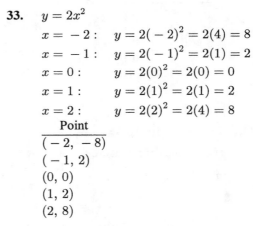

29. $y = x$
$x = -2:$ $\quad y = -2$
$x = 0:$ $\quad y = 0$
$x = 2:$ $\quad y = 2$

Point
$(-2, -2)$
$(0, 0)$
$(2, 2)$

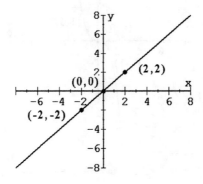

33. $y = 2x^2$
$x = -2:$ $\quad y = 2(-2)^2 = 2(4) = 8$
$x = -1:$ $\quad y = 2(-1)^2 = 2(1) = 2$
$x = 0:$ $\quad y = 2(0)^2 = 2(0) = 0$
$x = 1:$ $\quad y = 2(1)^2 = 2(1) = 2$
$x = 2:$ $\quad y = 2(2)^2 = 2(4) = 8$

Point
$(-2, -8)$
$(-1, 2)$
$(0, 0)$
$(1, 2)$
$(2, 8)$

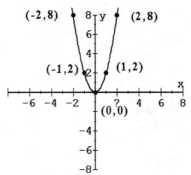

214

37 a. $y = x^2 + 2$

$x = -1: \quad y = (-1)^2 + 2 = 1 + 2 = 3$

$x = 0: \quad y = (0)^2 + 2 = 0 + 2 = 2$

$x = 1: \quad y = (1)^2 + 2 = 1 + 2 = 3$

$y = x^2 + 4$

$x = -1: \quad y = (-1)^2 + 4 = 1 + 4 = 5$

$x = 0: \quad y = (0)^2 + 4 = 0 + 4 = 4$

$x = 1: \quad y = (1)^2 + 4 = 1 + 4 = 5$

$y = x^2 - 3$

$x = -1: \quad y = (-1)^2 - 3 = 1 - 3 = -2$

$x = 0: \quad y = (0)^2 - 3 = 0 - 3 = -3$

$x = 1: \quad y = (1)^2 - 3 = 1 - 3 = -2$

$y = x^2 + 2$	$y = x^2 + 4$	$y = x^2 - 3$
Point	Point	Point
$(-1, 3)$	$(-1, 5)$	$(-1, -2)$
$(0, 2)$	$(0, 4)$	$(0, -3)$
$(1, 3)$	$(1, 5)$	$(1, -2)$

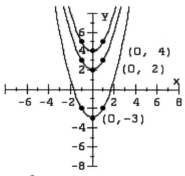

b. $y = x^2 - 1$

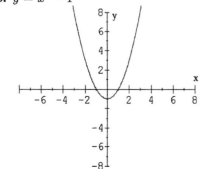

39a.

$y = (x - 1)^2 + 2$

$x = -1: \quad y = (-1-1)^2 + 2 = 4 + 2 = 6$

$x = 1: \quad y = (1-1)^2 + 2 = 0 + 2 = 2$

$x = 3: \quad y = (3-1)^2 + 2 = 4 + 2 = 6$

$y = (x - 3)^2 - 2$

$x = 1: \quad y = (1-3)^2 - 2 = 4 - 2 = 2$

$x = 3: \quad y = (3-3)^2 - 2 = 0 - 2 = -2$

$x = 5: \quad y = (5-3)^2 - 2 = 4 - 2 = 2$

$y = (x + 2)^2 + 3$

$x = -4: \quad y = (-4+2)^2 + 3 = 4 + 3 = 7$

$x = -2: \quad y = (-2+2)^2 + 3 = 0 + 3 = 3$

$x = 0: \quad y = (0+2)^2 + 3 = 4 + 3 = 7$

$y=(x-1)^2+2$	$y=(x-3)^2-2$	$y=(x+2)^2+3$
Point	Point	Point
$(-1, 6)$	$(1, 2)$	$(-4, 7)$
$(1, 2)$	$(3, -2)$	$(-2, 3)$
$(3, 6)$	$(5, 2)$	$(0, 7)$

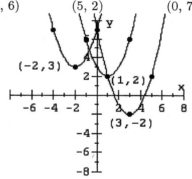

b. $y = (x + 1)^2 - 4$

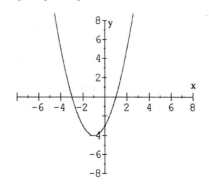

Problem Set 8.2

For Problems 1 — 36, the y-intercept, the x-intercept and one "check" point is given. If the line contains the origin or is parallel to an axis, two points in addition to the one intercept is given.

1.

$x + y = 2$		Point
$x = 0:$	$0 + y = 2$	$(0, 2)$
$y = 0:$	$x + 0 = 2$	$(2, 0)$
$x = 1:$	$1 + y = 2$	$(1, 1)$
	$y = 1$	

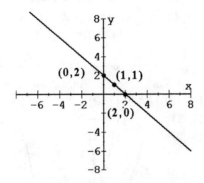

3.

$x - y = 3$		Point
$x = 0:$	$0 - y = 3$	$(0, -3)$
	$-y = 3$	$(3, 0)$
	$y = -3$	$(4, 1)$
$y = 0:$	$x - 0 = 3$	
$y = 1:$	$x - 1 = 3$	
	$x = 4$	

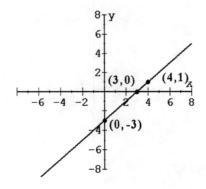

5.

$x - y = -4$		Point
$x = 0:$	$0 - y = -4$	$(0, 4)$
	$-y = -4$	$(-4, 0)$
	$y = 4$	$(-2, 2)$
$y = 0:$	$x - 0 = -4$	
$y = 2:$	$x - 2 = -4$	
	$x = -2$	

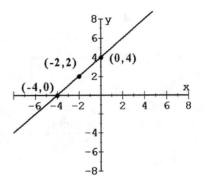

7.

$x + 2y = 2$		Point
$x = 0:$	$0 + 2y = 2$	$(0, 1)$
	$y = 1$	$(2, 0)$
$y = 0:$	$x + 0 = 2$	$(-2, 2)$
$y = 2:$	$x + 4 = 2$	
	$x = -2$	

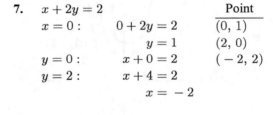

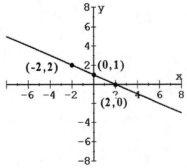

9. $3x - y = 6$ Point

$x = 0:$ $0 - y = 6$ $\overline{(0, -6)}$

               $-y = 6$ $(2, 0)$

                 $y = -6$ $(1, -3)$

$y = 0:$ $3x - 0 = 6$

                $x = 2$

$x = 1:$ $3 - y = 6$

                $-y = 3$

                $y = -3$

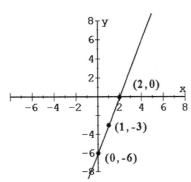

13. $x - y = 0$ Point

$x = 0:$ $0 - y = 0$ $\overline{(0, 0)}$

                 $y = 0$ $(2, 2)$

$x = 2:$ $2 - y = 0$ $(3, 3)$

                 $y = 2$

$y = 3:$ $x - 3 = 0$

                $x = 3$

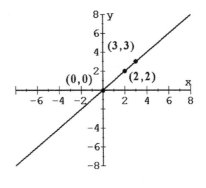

11. $3x - 2y = 6$ Point

$x = 0:$ $0 - 2y = 6$ $\overline{(0, -3)}$

                 $-2y = 6$ $(2, 0)$

                   $y = -3$ $(4, 3)$

$y = 0:$ $3x - 0 = 6$

                $x = 2$

$x = 4:$ $12 - 2y = 6$

                 $-2y = -6$

                 $y = 3$

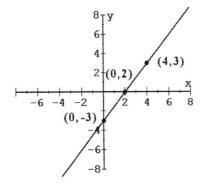

15. $y = 3x$

$x = 0:$ $y = 3(0) = 0$

$x = 1:$ $y = 3(1) = 3$

$x = -2:$ $y = 3(-2) = -6$

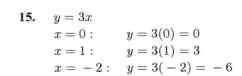

Point

$\overline{(0, 0)}$

$(1, 3)$

$(-2, -6)$

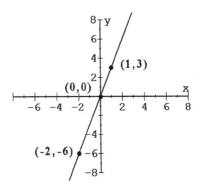

Problem Set 8.2

17. $x = -2$

		Point
$y = 0:$	$x = -2$	$(-2, 0)$
$y = 2:$	$x = -2$	$(-2, 2)$
$y = -3:$	$x = -2$	$(-2, -3)$

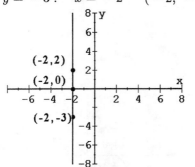

19. $y = 0$

		Point
$x = 0:$	$y = 0$	$(0, 0)$
$x = 2:$	$y = 0$	$(2, 0)$
$x = -3:$	$y = 0$	$(-3, 0)$

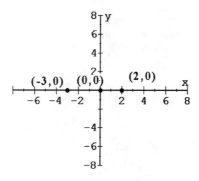

21. $y = -2x - 1$

		Point
$x = 0:$	$y = 0 - 1$	$(0, -1)$
	$y = -1$	
$y = 0:$	$0 = -2x - 1$	$\left(-\dfrac{1}{2}, 0\right)$
	$2x = -1$	$(2, -5)$
	$x = -\dfrac{1}{2}$	
$x = 2:$	$y = -2(2) - 1$	
	$y = -4 - 1$	
	$y = -5$	

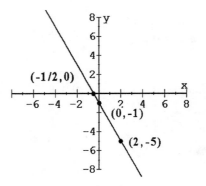

23. $y = \dfrac{1}{2}x + 1$

		Point
$x = 0:$	$y = 0 + 1$	$(0, 1)$
	$y = 1$	$(-2, 0)$
$y = 0:$	$0 = \dfrac{1}{2}x + 1$	$(2, 2)$
	$0 = x + 2$	
	$x = -2$	
$x = 2:$	$y = \dfrac{1}{2}(2) + 1$	
	$y = 1 + 1$	
	$y = 2$	

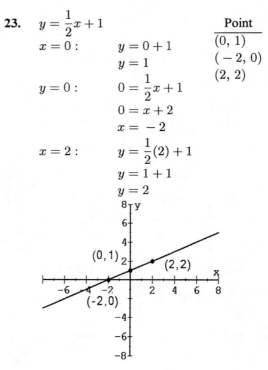

218

25. $y = -\dfrac{1}{3}x - 2$

$x = 0$: $y = 0 - 2$

               $y = -2$

$y = 0$: $0 = -\dfrac{1}{3}x - 2$

               $0 = -x - 6$

               $x = -6$

$x = 3$: $y = -\dfrac{1}{3}(3) - 2$

               $y = -1 - 2$

               $y = -3$

Point
$(0, -2)$
$(-6, 0)$
$(3, -3)$

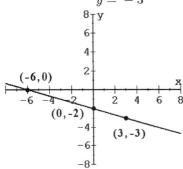

27. $4x + 5y = -10$

$x = 0$: $0 + 5y = -10$

               $y = -2$

$y = 0$: $4x + 0 = -10$

               $x = -\dfrac{10}{4}$

               $x = -\dfrac{5}{2}$

$x = 5$: $20 + 5y = -10$

               $5y = -30$

               $y = -6$

Point
$(0, -2)$
$\left(-\dfrac{5}{2}, 0\right)$
$(5, -6)$

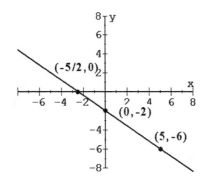

29. $-2x + y = -4$

$x = 0$: $0 + y = -4$

               $y = -4$

$y = 0$: $-2x + 0 = -4$

               $x = 2$

$x = 4$: $-8 + y = -4$

               $y = 4$

Point
$(0, -4)$
$(2, 0)$
$(4, 4)$

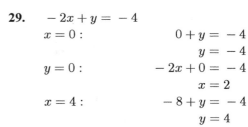

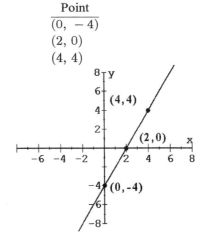

31. $3x - 4y = 7$

$x = 0$: $0 - 4y = 7$

               $y = -\dfrac{7}{4}$

$y = 0$: $3x - 0 = 7$

               $x = \dfrac{7}{3}$

$y = -7$: $3x + 28 = 7$

               $3x = -21$

               $x = -7$

Point
$\left(0, -\dfrac{7}{4}\right)$
$\left(\dfrac{7}{3}, 0\right)$
$(-7, -7)$

Problem Set 8.2

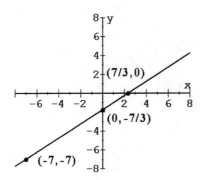

33. $y + 4x = 0$

		Point
$x = 0:$	$y + 0 = 0$	$(0, 0)$
	$y = 0$	$(1, -4)$
$x = 1:$	$y + 4 = 0$	$(-1, 4)$
	$y = -4$	
$x = -1:$	$y - 4 = 0$	
	$y = 4$	

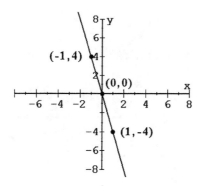

35. $x = 2y$

		Point
$x = 0:$	$0 = 2y$	$(0, 0)$
	$y = 0$	$(2, 1)$
$y = 1:$	$x = 2$	$(4, 2)$
$y = 2:$	$x = 4$	

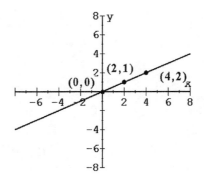

37. Use the points $(0, 1)$ and $(1, 0)$ to graph a dashed line for $x + y = 1$. Use $(0, 0)$ as a test point. The given inequality $x + y > 1$ becomes $0 + 0 > 1$ which is a false statement. Therefore, the solution set is the half-plane that does not contain the origin.

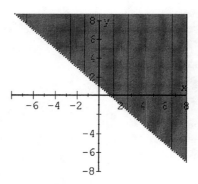

39. Use the points $(0, 3)$ and $(2, 0)$ to graph a dashed line for $3x + 2y = 6$. Use $(0, 0)$ as a test point. The given inequality $3x + 2y < 1$ becomes $0 + 0 < 1$ which is a true statement. Therefore, the solution set is the half-plane that does contain the origin.

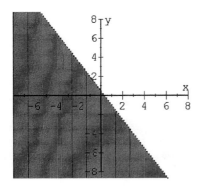

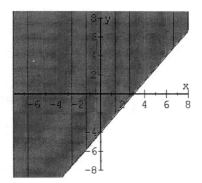

41. Use the points $(0, -4)$ and $(2, 0)$ to graph a solid line for $2x - y = 4$. Use $(0, 0)$ as a test point. The given inequality $2x - y \geq 4$ becomes $0 - 0 \geq 4$ which is a false statement. Therefore, the solution set is the half-plane that does not contain the origin.

45. Use the points $(0, 0)$ and $(1, -1)$ to graph a dashed line for $y = -x$. Since the origin is on the line use $(2, 1)$ as a test point. The given inequality $y > -x$ becomes $1 > -2$ which is a true statement. Therefore, the solution set is the half-plane that contains the point $(2, 1)$.

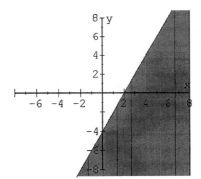

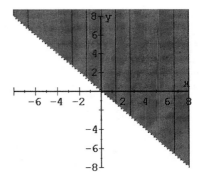

43. Use the points $(0, -4)$ and $(3, 0)$ to graph a solid line for $4x - 3y = 12$. Use $(0, 0)$ as a test point. The given inequality $4x - 3y \leq 12$ becomes $0 - 0 \leq 12$ which is a true statement. Therefore, the solution set is the half-plane that does contain the origin.

47. Use the points $(0, 0)$ and $(1, 2)$ to graph a solid line for $2x - y = 0$. Since the origin is on the line use $(2, -1)$ as a test point. The given inequality $2x - y \geq 0$ becomes $4 + 1 \geq 0$ which is a true statement. Therefore, the solution set is the half-plane that contains the point $(2, -1)$.

Problem Set 8.2

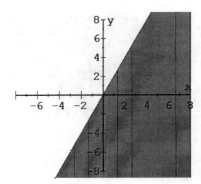

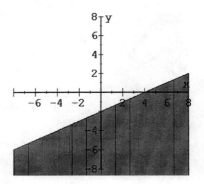

49. Use the points $(0, -1)$ and $(2, 0)$ to graph a dashed line for $-x + 2y = -2$. Use $(0, 0)$ as a test point. The given inequality $-x + 2y < -2$ becomes $0 + 0 < -2$ which is a false statement. Therefore, the solution set is the half-plane that does not contain the origin.

53. Use the points $(0, 4)$ and $(4, 0)$ to graph a solid line for $y = -x + 4$. Use $(0, 0)$ as a test point. The given inequality $y \geq -x + 4$ becomes $0 \geq 0 + 4$ which is a false statement. Therefore, the solution set is the half-plane that does not contain the origin.

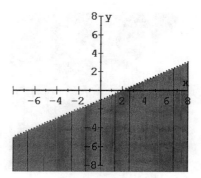

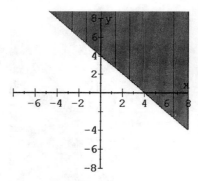

51. Use the points $(0, -2)$ and $(4, 0)$ to graph a solid line for $y = \frac{1}{2}x - 2$. Use $(0, 0)$ as a test point. The given inequality $y \leq \frac{1}{2}x - 2$ becomes $0 \leq 0 - 2$ which is a false statement. Therefore, the solution set is the half-plane that does not contain the origin.

55. Use the points $(0, -3)$ and $(-4, 0)$ to graph a dashed line for $3x + 4y = -12$. Use $(0, 0)$ as a test point. The given inequality $3x + 4y > -12$ becomes $0 + 0 > -12$ which is a true statement. Therefore, the solution set is the half-plane that does contain the origin.

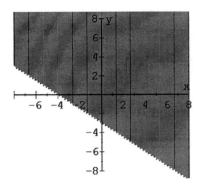

PROBLEM SET 8.3   Slope of a Line

1. Let $(7, 5)$ be P_1 and $(3, 2)$ be P_2.
$$m = \frac{y_2 - y_1}{x_2 - x_1} = \frac{2 - 5}{3 - 7} = \frac{-3}{-4} = \frac{3}{4}$$

3. Let $(-1, 3)$ be P_1 and $(-6, -4)$ be P_2.
$$m = \frac{y_2 - y_1}{x_2 - x_1} = \frac{-4 - 3}{-6 - (-1)} = \frac{-7}{-5} = \frac{7}{5}$$

5. Let $(2, 8)$ be P_1 and $(7, 2)$ be P_2.
$$m = \frac{y_2 - y_1}{x_2 - x_1} = \frac{2 - 8}{7 - 2} = \frac{-6}{5} = -\frac{6}{5}$$

7. Let $(-2, 5)$ be P_1 and $(1, -5)$ be P_2.
$$m = \frac{y_2 - y_1}{x_2 - x_1} = \frac{-5 - 5}{1 - (-2)} = \frac{-10}{3} = -\frac{10}{3}$$

9. Let $(4, -1)$ be P_1 and $(-4, -7)$ be P_2.
$$m = \frac{y_2 - y_1}{x_2 - x_1} = \frac{-7 - (-1)}{-4 - 4} = \frac{-6}{-8} = \frac{3}{4}$$

11. Let $(3, -4)$ be P_1 and $(2, -4)$ be P_2.
$$m = \frac{y_2 - y_1}{x_2 - x_1} = \frac{-4 - (-4)}{2 - 3} = \frac{0}{-1} = 0$$

13. Let $(-6, -1)$ be P_1 and $(-2, -7)$ be P_2.
$$m = \frac{y_2 - y_1}{x_2 - x_1} = \frac{-7 - (-1)}{-2 - (-6)}$$
$$m = \frac{-6}{4} = -\frac{3}{2}$$

15. Let $(-2, 4)$ be P_1 and $(-2, -6)$ be P_2.
The slope is undefined because $x_1 = x_2$.

17. Let $(-1, 10)$ be P_1 and $(-9, 2)$ be P_2.
$$m = \frac{y_2 - y_1}{x_2 - x_1} = \frac{2 - 10}{-9 - (-1)}$$
$$m = \frac{-8}{-8} = 1$$

19. Let (a, b) be P_1 and (c, d) be P_2.
$$m = \frac{y_2 - y_1}{x_2 - x_1} = \frac{d - b}{c - a}$$

21. Let $(7, 8)$ be P_1 and $(2, y)$ be P_2.
$$\frac{y - 8}{2 - 7} = \frac{4}{5}$$
$$\frac{y - 8}{-5} = \frac{4}{5}$$
$$5(y - 8) = 4(-5)$$
$$5y - 40 = -20$$
$$5y = 20$$
$$y = 4$$

Problem Set 8.3

23. Let $(-2, -4)$ be P_1 and $(x, 2)$ be P_2.

$$\frac{2-(-4)}{x-(-2)} = -\frac{3}{2}$$

$$\frac{6}{x+2} = -\frac{3}{2}$$

$$-3(x+2) = 6(2)$$

$$-3x-6 = 12$$

$$-3x = 18$$

$$x = -6$$

In Problems 25 − 32, the answers will vary but sample points are given.

25. From $(3, 2)$ move 2 units up and 3 units right to $(6, 4)$.
From $(6, 4)$ move 2 units up and 3 units right to $(9, 6)$.
From $(9, 6)$ move 2 units up and 3 units right to $(12, 8)$.

27. From $(-2, -4)$ move 1 unit up and 2 units right to $(0, -3)$.
From $(0, -3)$ move 1 unit up and 2 units right to $(2, -2)$.
From $(2, -2)$ move 1 unit up and 2 units right to $(4, -1)$.

29. From $(-3, 4)$ move 3 units down and 4 units right to $(1, 1)$.
From $(1, 1)$ move 3 units down and 4 units right to $(5, -2)$.
From $(5, -2)$ move 3 units down and 4 units right to $(9, -5)$.

31. From $(4, -5)$ move 2 units down and 1 unit right to $(5, -7)$.
From $(5, -7)$ move 2 units down and 1 unit right to $(6, -9)$.
From $(6, -9)$ move 2 units down and 1 unit right to $(7, -11)$.

33. The line falls from left to right, so slope is negative.

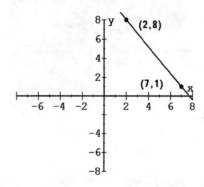

35. The line rises from left to right, so the slope is positive.

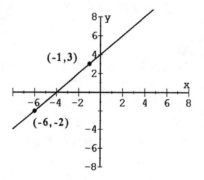

37. The line is horizontal, so the slope is zero.

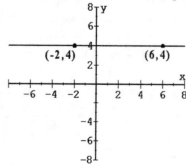

224

39. The line falls from left to right, so the slope is negative.

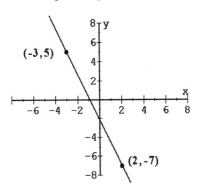

41. $3x + 2y = 6$ **Point**

$x = 0:$ $0 + 2y = 6$ $\overline{(0,\ 3)}$

            $2y = 6$ $(2,\ 0)$

             $y = 3$

$y = 0:$ $3x + 0 = 6$

             $x = 2$

$$m = \frac{0 - 3}{2 - 0} = \frac{-3}{2} = -\frac{3}{2}$$

43. $5x - 4y = 20$ **Point**

$x = 0:$ $0 - 4y = 20$ $\overline{(0,\ -5)}$

            $-4y = 20$ $(4,\ 0)$

             $y = -5$

$y = 0:$ $5x + 0 = 20$

             $5x = 20$

             $x = 4$

$$m = \frac{0 - (-5)}{4 - 0} = \frac{5}{4}$$

45. $x + 5y = 6$ **Point**

$y = 0:$ $x + 0 = 6$ $\overline{(6,\ 0)}$

             $x = 6$ $(1,\ 1)$

$y = 1:$ $x + 5 = 6$

             $x = 1$

$$m = \frac{1 - 0}{1 - 6} = \frac{1}{-5} = -\frac{1}{5}$$

47. $2x - y = -7$ **Point**

$x = 0:$ $0 - y = -7$ $\overline{(0,\ 7)}$

            $-y = -7$ $(1,\ 9)$

             $y = 7$

$x = 1:$ $2 - y = -7$

            $-y = -9$

             $y = 9$

$$m = \frac{9 - 7}{1 - 0} = \frac{2}{1} = 2$$

49. $y = 3$ **Point**

$x = 0:$ $y = 3$ $\overline{(0,\ 3)}$

$x = 2:$ $y = 3$ $(2,\ 3)$

$$m = \frac{3 - 3}{2 - 0} = \frac{0}{2} = 0$$

51. $-2x + 5y = 9$

$x = 0:$ $0 + 5y = 9$

            $5y = 9$

             $y = \dfrac{9}{5}$

$y = 0:$ $-2x + 0 = 9$

           $-2x = 9$

             $x = -\dfrac{9}{2}$

Point

$$\left(0,\ \frac{9}{5}\right)$$

$$\left(-\frac{9}{2},\ 0\right)$$

$$m = \frac{0 - \dfrac{9}{5}}{-\dfrac{9}{2} - 0} = \frac{-\dfrac{9}{5}}{-\dfrac{9}{2}}$$

$$m = \left(-\frac{9}{5}\right)\left(-\frac{2}{9}\right) = \frac{2}{5}$$

Problem Set 8.3

53. $6x - 5y = -30$

$x = 0:$

$0 - 5y = -30$
$-5y = -30$
$y = 6$

$y = 0:$

$6x + 0 = -30$
$6x = -30$
$x = -5$

Point
$(0, 6)$
$(-5, 0)$

$$m = \frac{0 - 6}{-5 - 0} = \frac{-6}{-5} = \frac{6}{5}$$

55. $y = -3x - 1$

$x = 0:$

$y = 0 - 1$
$y = -1$

$x = 1:$

$y = -3 - 1$
$y = -4$

Point
$(0, -1)$
$(1, -4)$

$$m = \frac{-4 - (-1)}{1 - 0} = \frac{-3}{1} = -3$$

57. $y = 4x$

$x = 0:$ $y = 0$

$x = 1:$ $y = 4$

Point
$(0, 0)$
$(1, 4)$

$$m = \frac{4 - 0}{1 - 0} = \frac{4}{1} = 4$$

59. $y = \frac{2}{3}x - \frac{1}{2}$

$x = 0:$

$y = 0 - \frac{1}{2}$
$y = -\frac{1}{2}$

$x = 3:$

$y = \frac{2}{3}(3) - \frac{1}{2}$
$y = 2 - \frac{1}{2}$
$y = \frac{4}{2} - \frac{1}{2}$
$y = \frac{3}{2}$

Point
$\left(0, -\frac{1}{2}\right)$
$\left(3, \frac{3}{2}\right)$

$$m = \frac{\frac{3}{2} - \left(-\frac{1}{2}\right)}{3 - 0} = \frac{\frac{4}{2}}{3} = \frac{2}{3}$$

61. The grade of the highway can be calculated as a ratio as follows:

$$\frac{\text{distance highway rises}}{\text{horizontal distance}} = \frac{135}{2640} = 0.051136...$$

The grade of the highway is 5.1%.

63. Let x represent the measure of the run of the stairs. Solve a proportion comparing the rise to the run of the stairs.

$$\frac{\text{rise}}{\text{run}} \quad \frac{3}{5} = \frac{19}{x}$$

$3x = 19(5)$
$3x = 95$
$x = 31.667$ or 32 to the nearest whole number.

The measure of the run of the stairs is 32 centimeters.

65. Let x represent the vertical drop for the sewage pipe.

$$\frac{\text{fall}}{100 \text{ feet}} \quad \frac{2\frac{1}{4}}{100} = \frac{x}{45}$$

$100x = 45\left(\frac{9}{4}\right)$
$400x = 45(9)$
$400x = 405$
$x = 1.0125$ or 1.0 to the nearest tenth.

The vertical drop must be 1.0 feet for 45 feet.

PROBLEM SET **8.4** **Writing Equations of Lines**

Problems 1 − 12 can be done by using the general approach demonstrated in Example 1 in the text or by using the point-slope form.

1. The general approach is shown for this problem. The slope for the points $(2, 3)$ and (x, y) is $\frac{2}{3}$.
$$\frac{y - 3}{x - 2} = \frac{2}{3}$$
$$2(x - 2) = 3(y - 3)$$
$$2x - 4 = 3y - 9$$
$$2x - 3y = -5$$

3. The general approach is shown for this problem. The slope for the points $(-3, -5)$ and (x, y) is $\frac{1}{2}$.
$$\frac{y - (-5)}{x - (-3)} = \frac{1}{2}$$
$$\frac{y + 5}{x + 3} = \frac{1}{2}$$
$$1(x + 3) = 2(y + 5)$$
$$x + 3 = 2y + 10$$
$$x - 2y = 7$$

5. The point-slope form is shown for this problem.
$$y - y_1 = m(x - x_1)$$
$$y - 8 = -\frac{1}{3}[x - (-4)]$$
$$3(y - 8) = -1(x + 4)$$
$$3y - 24 = -x - 4$$
$$x + 3y = 20$$

7. The point-slope form is shown for this problem.
$$y - y_1 = m(x - x_1)$$
$$y - (-7) = 0(x - 3)$$
$$y + 7 = 0$$
$$y = -7$$

9. The point-slope form is shown for this problem.
$$y - y_1 = m(x - x_1)$$
$$y - (0) = -\frac{4}{9}(x - 0)$$
$$9(y - 0) = -4(x - 0)$$
$$9y = -4x$$
$$4x + 9y = 0$$

11. The point-slope form is shown for this problem.
$$y - y_1 = m(x - x_1)$$
$$y - (-2) = 3[x - (-6)]$$
$$y + 2 = 3(x + 6)$$
$$y + 2 = 3x + 18$$
$$-3x + y = 16$$
$$3x - y = -16$$

13. Find the slope.
$$m = \frac{y_2 - y_1}{x_2 - x_1} = \frac{10 - 3}{7 - 2} = \frac{7}{5}$$
Use the slope and either point in the point-slope form.
$$y - y_1 = m(x - x_1)$$
$$y - 3 = \frac{7}{5}(x - 2)$$
$$5(y - 3) = 7(x - 2)$$
$$5y - 15 = 7x - 14$$
$$-7x + 5y = 1$$
$$7x - 5y = -1$$

15. Find the slope.
$$m = \frac{y_2 - y_1}{x_2 - x_1} = \frac{4 - (-2)}{-1 - 3} = \frac{6}{-4} = -\frac{3}{2}$$
Use the slope and either point in the point-slope form.

$$y - y_1 = m(x - x_1)$$
$$y - 4 = -\frac{3}{2}[x - (-1)]$$
$$2(y - 4) = -3(x + 1)$$
$$2y - 8 = -3x - 3$$
$$3x + 2y = 5$$

17. Find the slope.
$$m = \frac{y_2 - y_1}{x_2 - x_1} = \frac{-7 - (-2)}{-6 - (-1)} = \frac{-5}{-5} = 1$$
Use the slope and either point in the point-slope form.
$$y - y_1 = m(x - x_1)$$
$$y - (-2) = 1[x - (-1)]$$
$$y + 2 = x + 1$$
$$-x + y = -1$$
$$x - y = 1$$

19. Find the slope.
$$m = \frac{y_2 - y_1}{x_2 - x_1} = \frac{-5 - 0}{-3 - 0} = \frac{-5}{-3} = \frac{5}{3}$$
Use the slope and either point in the point-slope form.
$$y - y_1 = m(x - x_1)$$
$$y - 0 = \frac{5}{3}(x - 0)$$
$$3y = 5x$$
$$-5x + 3y = 0$$
$$5x - 3y = 0$$

21. Find the slope.
$$m = \frac{y_2 - y_1}{x_2 - x_1} = \frac{0 - 4}{7 - 0} = \frac{-4}{7} = -\frac{4}{7}$$
Use the slope and either point in the point-slope form.
$$y - y_1 = m(x - x_1)$$
$$y - 0 = -\frac{4}{7}(x - 7)$$
$$7y = -4(x - 7)$$
$$7y = -4x + 28$$
$$4x + 7y = 28$$

23. Use the slope-intercept form.
$$y = mx + b$$
$$y = \frac{3}{5}x + 2$$

25. Use the slope-intercept form.
$$y = mx + b$$
$$y = 2x - 1$$

27. Use the slope-intercept form.
$$y = mx + b$$
$$y = -\frac{1}{6}x - 4$$

29. Use the slope-intercept form.
$$y = mx + b$$
$$y = -x + \frac{5}{2}$$

31. Use the slope-intercept form.
$$y = mx + b$$
$$y = -\frac{5}{9}x - \frac{1}{2}$$

33. $y = -2x - 5$
$m = -2, b = -5$

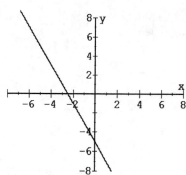

35. $3x - 5y = 15$
$$-5y = -3x + 15$$
$$y = \frac{3}{5}x - 3$$
$$m = \frac{3}{5},\ b = -3$$

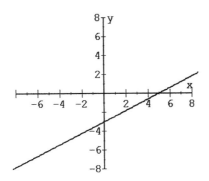

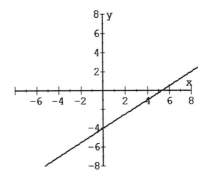

37. $-4x + 9y = 18$
$$9y = 4x + 18$$
$$y = \frac{4}{9}x + 2$$
$$m = \frac{4}{9},\ b = 2$$

41. $-2x - 11y = 11$
$$-11y = 2x + 11$$
$$y = -\frac{2}{11}x - 1$$
$$m = -\frac{2}{11},\ b = -1$$

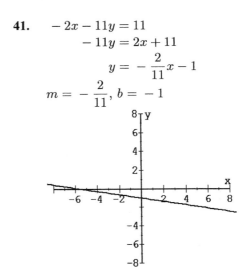

39. $-y = -\frac{3}{4}x + 4$
$$y = \frac{3}{4}x - 4$$
$$m = \frac{3}{4},\ b = -4$$

43. $9x + 7y = 0$
$$7y = -9x + 0$$
$$y = -\frac{9}{7}x + 0$$
$$m = -\frac{9}{7},\ b = 0$$

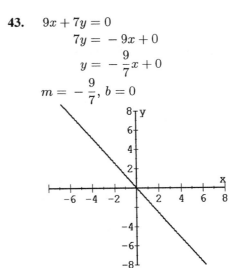

229

Problem Set 8.4

49. $5x - 2y = 6$
$-2y = -5x + 6$
$y = \dfrac{5}{2}x - 3$

$2x + 5y = 9$
$5y = -2x + 9$
$y = -\dfrac{2}{5}x + \dfrac{9}{5}$

The lines are perpendicular because the product of the slopes is
$$\left(\dfrac{5}{2}\right)\left(-\dfrac{2}{5}\right) = -1.$$

51. $4x - 3y = 12$
$-3y = -4x + 12$
$y = \dfrac{4}{3}x - 4$

$3x - 4y = 12$
$-4y = -3x + 12$
$y = \dfrac{3}{4}x - 3$

The lines intersect but are not perpendicular because the product of the slopes is
$\dfrac{4}{3} \cdot \dfrac{3}{4} = 1.$

53. $x - 3y = 7$
$-3y = -x + 7$
$y = \dfrac{1}{3}x - \dfrac{7}{3}$

$5x - 15y = 9$
$-15y = -5x + 9$
$y = \dfrac{1}{3}x - \dfrac{3}{5}$

The lines are parallel because their slopes are equal.

55. Find the slope of $2x - 3y = 6$.
$-3y = -2x + 6$
$y = \dfrac{2}{3}x - 2$
The slope is $\dfrac{2}{3}$.

Since the lines are parallel, the slopes of the two lines are equal. Use this slope and the given point in the point-slope form.
$$y - y_1 = m(x - x_1)$$
$$y - 3 = \dfrac{2}{3}(x - 4)$$
$$3(y - 3) = 2(x - 4)$$
$$3y - 9 = 2x - 8$$
$$-2x + 3y = 1$$
$$2x - 3y = -1$$

PROBLEM SET **8.5** **Solving Linear Systems by Graphing**

1.
$5x + y = 9 \qquad 3x - 2y = 4$
$5(1) + (4) = 9 \quad 3(1) - 2(4) = 4$
$9 = 9 \qquad\qquad 3 - 8 = 4$
$\qquad\qquad\qquad\qquad -5 \neq 4$

Therefore $(1, 4)$ is not a solution of the system.

3.
$x - 3y = 17 \qquad\qquad 2x + 5y = -21$
$(2) - 3(-5) = 17 \quad 2(2) + 5(-5) = -21$
$2 + 15 = 17 \qquad\qquad 4 - 25 = -21$
$17 = 17 \qquad\qquad\qquad -21 = -21$

Therefore, $(2, -5)$ is a solution of the system.

5.
$$y = 2x \qquad 3x - 4y = 5$$
$$-2 = 2(-1) \quad 3(-1) - 4(-2) = 5$$
$$-2 = -2 \qquad -3 + 8 = 5$$
$$5 = 5$$
Therefore, $(-1, -2)$ is a solution of the system.

7.
$$6x - 5y = 5 \qquad 3x + 4y = -4$$
$$6(0) - 5(-1) = 5 \quad 3(0) + 4(-1) = -4$$
$$0 + 5 = 5 \qquad 0 - 4 = -4$$
$$5 = 5 \qquad -4 = -4$$
Therefore $(0, -1)$ is a solution of the system.

9.
$$-3x - y = 4 \qquad -2x + 3y = -23$$
$$-3(4) - (-5) = 4 \quad -2(4) + 3(-5) = -23$$
$$-12 + 5 = 4 \qquad -8 - 15 = -23$$
$$-7 \neq 4 \qquad -23 = -23$$
Therefore $(4, -5)$ is not a solution of the system.

11.

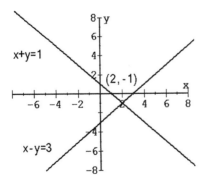

The solution set is $\{(2, -1)\}$.

13.

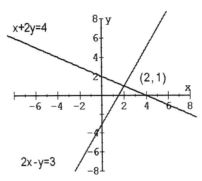

The solution set is $\{(2, 1)\}$.

15.

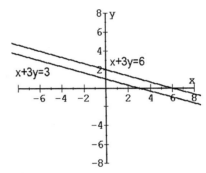

The solution is $\emptyset$.

17.

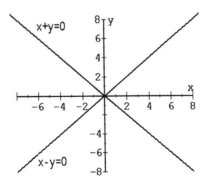

The solution set is $\{(0, 0)\}$.

231

Problem Set 8.5

19.

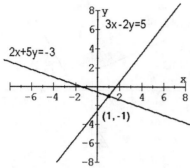

The solution set is $\{(1, -1)\}$.

21.

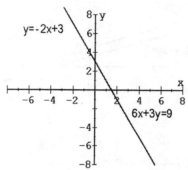

The system has infinitely many solutions.

23.

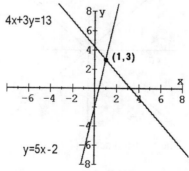

The solution set is $\{(1, 3)\}$.

25.

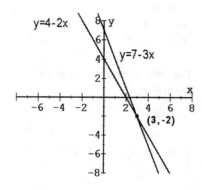

The solution set is $\{(3, -2)\}$.

27.

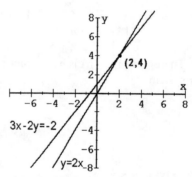

The solution set is $\{(2, 4)\}$.

29.

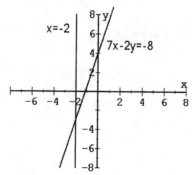

The solution set is $\{(-2, -3)\}$.

35. Use the points $(0, -2)$ and $(4, 0)$ to graph a dashed line for $x - 2y = 4$. Use $(0, 0)$ as a test point. The inequality $x - 2y < 4$ becomes $0 + 0 < 4$ which is a true statement. Therefore, $x - 2y < 4$ is satisfied by the points in the half-plane above the line $x - 2y = 4$.

Use the points $(0, 3)$ and $(1, 0)$ to graph a dashed line for $3x + y = 3$. Use $(0, 0)$ as a test point. The inequality $3x + y > 3$ becomes $0 + 0 > 3$ which is a false statement. Therefore, $3x + y > 3$ is satisfied by the points in the half-plane above the line $3x + y = 3$.

The solution set for the system is the intersection of the individual solution sets.

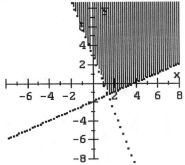

37. Use the points $(0, 4)$ and $(3, 0)$ to graph a solid line for $4x + 3y = 12$. Use $(0, 0)$ as a test point. The inequality $4x + 3y \leq 12$ becomes $0 + 0 \leq 12$ which is a true statement. Therefore, $4x + 3y \leq 12$ is satisfied by the points in the half-plane on or below the line $4x + 3y = 12$.

Use the points $(0, -4)$ and $(1, 0)$ to graph a solid line for $4x - y = 4$. Use $(0, 0)$ as a test point. The inequality $4x - y \geq 4$ becomes $0 + 0 \geq 4$ which is a false statement. Therefore, $4x - y \geq 4$ is satisfied by the points in the half-plane on or below

the line $4x - y = 4$.

The solution set for the system is the intersection of the individual solution sets.

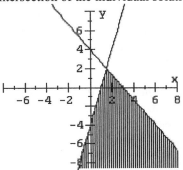

39. Use the points $(0, 0)$ and $(-1, 1)$ to graph a solid line for $y = -x$. Since $(0, 0)$ is on the line use $(3, 1)$ as a test point. The inequality $y \leq -x$ becomes $1 \leq -3$ which is a false statement. Therefore, $y \leq -x$ is satisfied by the points in the half-plane on or below the line $y = -x$.

Use the points $(0, 0)$ and $(-1, 3)$ to graph a dashed line for $y = -3x$. Since $(0, 0)$ is on the line use $(3, 1)$ as a test point. The inequality $y > -3x$ becomes $1 > -9$ which is a true statement. Therefore, $y > -3x$ is satisfied by the points in the half-plane above the line $y = -3x$.
The solution set for the system is the intersection of the individual solution sets.

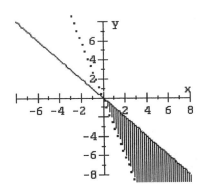

233

Problem Set 8.5

41. Use the points $(0, -2)$ and $(2, 0)$ to graph a dashed line for $y = x - 2$. Use $(0, 0)$ as a test point. The inequality $y > x - 2$ becomes $0 > 0 - 2$ which is a true statement. Therefore, $y > x - 2$ is satisfied by the points in the half-plane above the line $y = x - 2$.

Use the points $(0, 3)$ and $(-3, 0)$ to graph a dashed line for $y = x + 3$. Use $(0, 0)$ as a test point. The inequality $y < x + 3$ becomes $0 < 0 + 3$ which is a true statement. Therefore, $y < x + 3$ is satisfied by the points in the half-plane below the line $y = x + 3$.

The solution set for the system is the intersection of the individual solution sets.

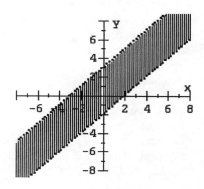

43. Use the points $(0, -2)$ and $(-4, 0)$ to graph a dashed line for $y = -\frac{1}{2}x - 2$. Use $(0, 0)$ as a test point. The inequality $y > -\frac{1}{2}x - 2$ becomes $0 > 0 - 2$ which is a true statement. Therefore, $y > -\frac{1}{2}x - 2$ is satisfied by the points in the half-plane above the line $y = -\frac{1}{2}x - 2$.

Use the points $(0, 1)$ and $(2, 0)$ to graph a dashed line for $y = -\frac{1}{2}x + 1$. Use $(0, 0)$ as a test point. The inequality $y > -\frac{1}{2}x + 1$ becomes $0 > 0 + 1$ which is a false statement. Therefore, $y > -\frac{1}{2}x + 1$ is satisfied by the points in the half-plane above the line $y = -\frac{1}{2}x + 1$.

The solution set for the system is the intersection of the individual solution sets.

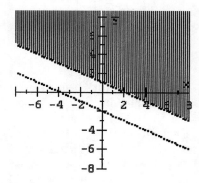

PROBLEM SET | **8.6** Elimination-by-Addition Method

1.
$$
\begin{aligned}
x + y &= 14 \\
x - y &= -2 \\
\hline
2x \phantom{{}- y} &= 12 \quad \text{Add the two equations.} \\
x &= 6
\end{aligned}
$$

Substitute 6 for x in $x + y = 14$.
$6 + y = 14$
$y = 8$
The solution set is $\{(6,\ 8)\}$.

3.
$$
\begin{aligned}
x + 4y &= -21 \\
3x - 4y &= 1 \\
\hline
-4x \phantom{{}-4y} &= -20 \quad \text{Add the two equations.} \\
x &= -5
\end{aligned}
$$

Substitute -5 for x in $x + 4y = -21$.
$-5 + 4y = -21$
$4y = -16$
$y = -4$
The solution set is $\{(-5,\ -4)\}$.

5. $y = 6 - x$ Add x to both sides. $x + y = 6$
 $x - y = -18$ Leave alone. $x - y = -18$
$$
\begin{aligned}
\hline
2x \phantom{{}- y} &= -12 \\
x &= -6
\end{aligned}
$$

Substitute -6 for x in $y = 6 - x$.
$y = 6 - (-6)$
$y = 12$
The solution set is $\{(-6,\ 12)\}$.

7. $5x + y = 23$ Multiply by 2. $10x + 2y = 46$
 $3x - 2y = 19$ Leave alone. $3x - 2y = 19$
$$
\begin{aligned}
\hline
13x \phantom{{}+ 2y} &= 65 \\
x &= 5
\end{aligned}
$$

Substitute 5 for x in $5x + y = 23$.
$5(5) + y = 23$
$25 + y = 23$
$y = -2$
The solution set is $\{(5,\ -2)\}$.

Problem Set 8.6

9.
$$x + 2y = 5$$
$$3x - 2y = 6$$
$$\overline{4x \qquad = 11} \quad \text{Add the two equations.}$$
$$x = \frac{11}{4}$$

$x + 2y = 5$ Multiply by -3. $-3x - 6y = -15$
$3x - 2y = 6$ Leave alone. $\underline{3x - 2y = \qquad 6}$
$$-8y = -9$$
$$y = \frac{9}{8}$$

The solution set is $\left\{ \left(\dfrac{11}{4}, \dfrac{9}{8} \right) \right\}$.

11.
$y = -x$ Add x to both sides. $x + y = 0$
$2x - y = -2$ Leave alone. $\underline{2x - y = -2}$
$$3x \qquad = -2$$
$$x = -\frac{2}{3}$$

Substitute $-\dfrac{2}{3}$ for y in $-x$.
$$y = -\left(-\frac{2}{3} \right) = \frac{2}{3}$$
The solution set is $\left\{ \left(-\dfrac{2}{3}, \dfrac{2}{3} \right) \right\}$.

13. $4x + 5y = 9$ Multiply by 5. $20x + 25y = \quad 45$
$5x - 6y = -50$ Multiply by -4. $\underline{-20x + 24y = 200}$
$$49y = 245$$
$$y = 5$$

Substitute 5 for y in $4x + 4y = 9$.
$$4x + 5(5) = 9$$
$$4x + 25 = 9$$
$$4x = -16$$
$$x = -4$$
The solution set is $\{(-4, 5)\}$.

15. $9x - 7y = 29$ Multiply by 5. $45x - 35y = 145$
$5x - 3y = 17$ Multiply by -9. $\underline{-45x + 27y = -153}$
$$-8y = -8$$
$$y = 1$$

Substitute 1 for y in $9x - 7y = 29$.

$$9x - 7(1) = 29$$
$$9x - 7 = 29$$
$$9x = 36$$
$$x = 4$$

The solution set is $\{(4, 1)\}$.

17. $6x + 5y = -6$ Multiply by 3. $18x + 15y = -18$
$8x - 3y = 21$ Multiply by 5. $\underline{40x - 15y = 105}$
$$58x = 87$$
$$x = \frac{87}{58} = \frac{3}{2}$$

Substitute $\frac{3}{2}$ for x in $6x + 5y = -6$.
$$6\left(\frac{3}{2}\right) + 5y = -6$$
$$9 + 5y = -6$$
$$5y = -15$$
$$y = -3$$

The solution set is $\left\{\left(\frac{3}{2}, -3\right)\right\}$.

19. $2x - 7y = -1$ Multiply by 4. $8x - 28y = -4$
$9x + 4y = -2$ Multiply by 7. $\underline{63x + 28y = -14}$
$$71x = -18$$
$$x = -\frac{18}{71}$$

$2x - 7y = -1$ Multiply by 9. $18x - 63y = -9$
$9x + 4y = -2$ Multiply by -2. $\underline{-18x - 8y = 4}$
$$-71y = -5$$
$$y = \frac{5}{71}$$

The solution set is $\left\{\left(-\frac{18}{71}, \frac{5}{71}\right)\right\}$.

21. $x + y = 750$ Multiply by -7. $-7x - 7y = -5250$
$0.07x + 0.08y = 57.5$ Multiply by 100. $\underline{7x + 8y = 5750}$
$$y = 500$$

Substitute 500 for y in $x + y = 750$.
$$x + 500 = 750$$
$$x = 250$$

The solution set is $\{(250, 500)\}$.

23. $0.09x + 0.11y = 31$ Multiply by 100. $9x + 11y = 3100$

$\qquad\qquad y = x + 100$ Subtract x from both sides. $-x + \quad y = \quad 100$

$\qquad 9x + 11y = 3100$ Leave alone. $\quad 9x + 11y = 3100$

$\qquad -x + \quad y = \quad 100$ Multiply by 9. $\underline{-9x + 9y = \quad 900}$

$\qquad\qquad\qquad\qquad\qquad\qquad\qquad\qquad\qquad 20y = 4000$

$\qquad\qquad\qquad\qquad\qquad\qquad\qquad\qquad\qquad\quad y = 200$

Substitute 200 for y in $y = x + 100$.

$200 = x + 100$

$100 = x$

The solution set is $\{(100,\ 200)\}$.

25. Let x and y represent the two numbers.

$\qquad x + y = 30$ The sum of the numbers is 30.

$\qquad \underline{x - y = 12}$ The difference of the numbers is 12.

$\quad 2x \qquad = 42$

$\qquad\quad x = 21$

Substitute 21 for x in $x + y = 30$.

$21 + y = 30$

$\qquad y = 9$

The two numbers are 21 and 9.

27. Let x represent the smaller number and y the larger number.

$\qquad y - x = 7$ Their difference is 7.

$\qquad 2y - 3x = 6$ Three times the smaller subtracted from twice the larger is 6.

$\qquad y - x = 7$ Multiply by -2. $-2y + 2x = -14$

$\qquad 2y - 3x = 6$ Leave alone. $\underline{\quad 2y - 3x = \qquad 6}$

$\qquad\qquad\qquad\qquad\qquad\qquad\qquad\qquad -x = \quad -8$

$\qquad\qquad\qquad\qquad\qquad\qquad\qquad\qquad\quad x = 8$

Substitute 8 for x in $y - x = 7$.

$y - 8 = 7$

$\qquad y = 15$

The numbers are 8 and 15.

29. Let x represent the smaller number and y the larger number.

$\qquad y = 2x$ One number is twice the other.

$\qquad 3x + 5y = 78$ The sum of three times the smaller and five times the larger is 78.

$\qquad y = 2x$ Subtract y from both sides. $2x - \quad y = \quad 0$

$\qquad 3x + 5y = 78$ Leave alone. $\qquad\quad 3x + 5y = 78$

$$2x - y = 0 \quad \text{Multiply by 5.} \quad 10x - 5y = \quad 0$$
$$3x + 5y = 78 \quad \text{Leave alone.} \quad \underline{3x + 5y = 78}$$
$$13x \qquad = 78$$
$$x = 6$$

Substitute 6 for x in $y = 2x$.
$$y = 2(6) = 12$$
The numbers are 6 and 12.

31. Let b represent the cost of one lemon and a represent the cost of one apple.
$$3b + 2a = 105 \quad \text{The cost of 3 lemons and 2 apples is \$1.05.}$$
$$2b + 3a = 120 \quad \text{The cost of 2 lemons and 3 apples is \$1.20.}$$
$$3b + 2a = 105 \quad \text{Multiply by 2.} \qquad 6b + 4b = \quad 210$$
$$2b + 3a = 120 \quad \text{Multiply by } -3. \quad \underline{-6b - 9a = -360}$$
$$-5a = -150$$
$$a = 30$$

Substitute 30 for a in $3b + 2a = 105$.
$$3b + 2(30) = 105$$
$$3b + 60 = 105$$
$$3b = 45$$
$$b = 15$$
The cost of a lemon is \$0.15 and the cost of an apple is \$0.30.

33. Let d represent the number of dimes and q the number of quarters.
$$d + q = 10 \quad \text{The number of coins is 10.}$$
$$10d + 25q = 145 \quad \text{The total value of the coins is \$1.45.}$$

$$d + q = 10 \quad \text{Multiply by } -10. \quad -10d - 10q = -100$$
$$10d + 25q = 145 \quad \text{Leave alone.} \quad \underline{10d + 25q = \quad 145}$$
$$15q = \quad 45$$
$$q = 3$$

Substitute 3 for q in $d + q = 10$.
$$d + 3 = 10$$
$$d = 7$$
He has 7 dimes and 3 quarters.

35. Let x represent the \$12 book and y represent the \$14 book.
$$x + y = 35 \quad \text{The total number of books is 35.}$$
$$12x + 14y = 462 \quad \text{The total value of the books is \$462.}$$

$$x + y = 35 \quad \text{Multiply by } -12. \quad -12x - 12y = -420$$
$$12x + 14y = 462 \quad \text{Leave alone.} \quad \underline{12x + 14y = \quad 462}$$
$$2y = \quad 42$$
$$y = 21$$

Substitute 21 for y in $x + y = 35$.
$$x + 21 = 35$$
$$x = 14$$
They bought 14 books at \$12 per book and 21 books at \$14 per book.

37. Let x represent the number of gallons of 10% salt solution and y represent the number of gallons of 15% salt solution.
$$x + y = 10 \qquad \text{The total of gallons is 10.}$$
$$0.10x + 0.15y = 0.13(10) \quad \text{The quantity of salt in the solution.}$$

$x + y = 10$	Multiply by -10.	$-10x - 10y = -100$
$0.10x + 0.15y = 1.30$	Multiply by 100.	$10x + 15y = 130$

$$5y = 30$$
$$y = 6$$

Substitute 6 for y in $x + y = 10$.
$$x + 6 = 10$$
$$x = 4$$
The quantity needed would be 4 gallons of the 10% solution and 6 gallons of the 15% solution.

39. Let x represent the investment at 10% and y represent the investment at 12%.
$$x + y = 1300 \qquad \text{The total investment was \$1300.}$$
$$0.10x + 0.12y = 146 \qquad \text{The total yearly interest was \$146.}$$

$x + y = 1300$	Multiply by -10.	$-10x - 10y = -13,000$
$0.10x + 0.12y = 146$	Multiply by 100.	$10x + 12y = 14,600$

$$2y = 1600$$
$$y = 800$$

Substitute 800 for y in $x + y = 1300$.
$$x + 800 = 1300$$
$$x = 500$$
The investments were \$500 at 10% and \$800 at 12%.

43. $\dfrac{1}{2}x - \dfrac{1}{3}y = -2$ Multiply by 12. $\qquad 6x - 4y = -24$

$\dfrac{3}{2}x + \dfrac{2}{3}y = 34$ Multiply by 6. $\qquad 9x + 4y = 204$

$$15x = 180$$
$$x = 12$$

Substitute 12 for x in $3x - 2y = -12$.
$$3(12) - 2y = -12$$
$$36 - 2y = -12$$
$$-2y = -48$$
$$y = 24$$
The solution set is $\{(12, 24)\}$.

45. $x - 4y = 6$ Multiply by -2. $-2x + 8y = -12$

   $2x - 8y = 3$ Leave alone. $\underline{ 2x - 8y = 3}$

   $0 = -9$ (Not possible)

The solution set is $\emptyset$.

PROBLEM SET | **8.7**   **Substitution Method**

1.   Substitute $2x - 1$ for y in $x + y = 14$.

$x + (2x - 1) = 14$

$\quad\quad 3x - 1 = 14$

$\quad\quad\quad 3x = 15$

$\quad\quad\quad\; x = 5$

Substitute 5 for x in $y = 2x - 1$.

$y = 2(5) - 1$

$y = 10 - 1$

$y = 9$

The solution set is $\{(5, 9)\}$.

3.   Substitute $-3x - 2$ for y in $x - y = -14$.

$x - (-3x - 2) = -14$

$\quad x + 3x + 2 = -14$

$\quad\quad\quad\; 4x = -16$

$\quad\quad\quad\; x = -4$

Substitute -4 for x in $y = -3x - 2$.

$y = -3(-4) - 2$

$y = 12 - 2$

$y = 10$

The solution set is $\{(-4, 10)\}$.

5.   Substitute $-2x + 7$ for y in $4x - 3y = -6$.

$4x - 3(-2x + 7) = -6$

$\quad 4x + 6x - 21 = -6$

$\quad\quad\; 10x - 21 = -6$

$\quad\quad\quad\quad 10x = 15$

$\quad\quad\quad\quad\; x = \dfrac{15}{10} = \dfrac{3}{2}$

Substitute $\dfrac{3}{2}$ for x in $y = -2x + 7$.

$y = -2\left(\dfrac{3}{2}\right) + 7$

$y = -3 + 7$

$y = 4$

The solution set is $\left\{\left(\dfrac{3}{2}, 4\right)\right\}$.

7.   Solve $x + y = 1$ for x.

$x + y = 1$

$\quad\; x = 1 - y$

Substitute $1 - y$ for x in $3x + 6y = 7$.

$3(1 - y) + 6y = 7$

$\quad 3 - 3y + 6y = 7$

$\quad\quad\; 3 + 3y = 7$

$\quad\quad\quad\; 3y = 4$

$\quad\quad\quad\quad y = \dfrac{4}{3}$

Substitute $\dfrac{4}{3}$ for y in $x = 1 - y$.

$x = 1 - \dfrac{4}{3}$

$x = \dfrac{3}{3} - \dfrac{4}{3} = -\dfrac{1}{3}$

The solution set is $\left\{\left(-\dfrac{1}{3}, \dfrac{4}{3}\right)\right\}$.

9.   Substitute $\frac{3}{4}y$ for x in $2x - y = 12$.

$2\left(\dfrac{3}{4}y\right) - y = 12$

$\quad \dfrac{3}{2}y - \dfrac{2}{2}y = 12$

$\quad\quad\quad \dfrac{1}{2}y = 12$

$\quad\quad\quad\quad y = 24$

Substitute 24 for y in $x = \dfrac{3}{4}y$.

$x = \dfrac{3}{4}(24) = 18$

The solution set is $\{(18, 24)\}$.

Problem Set 8.7

11. Substitute $\frac{3}{2}x$ for y in $6x - 5y = 15$.
$$6x - 5\left(\frac{3}{2}x\right) = 15$$
$$6x - \frac{15}{2}x = 15$$
$$12x - 15x = 30$$
$$-3x = 30$$
$$x = -10$$
Substitute -10 for x in $y = \frac{3}{2}x$.
$$y = \frac{3}{2}(-10) = -15$$
The solution set is $\{(-10, -15)\}$.

13. Substitute $4y - 1$ for x in $2x - 8y = 3$.
$$2(4y - 1) - 8y = 3$$
$$8y - 2 - 8y = 3$$
$$-2 = 3 \quad \text{This is not possible.}$$
The solution set is $\emptyset$.

15. Solve $7x + 2y = -2$ for y.
$$7x + 2y = -2$$
$$2y = -7x - 2$$
$$y = \frac{-7x - 2}{2}$$
Substitute $\frac{-7x - 2}{2}$ for y
in $6x + 5y = 18$.
$$6x + 5\left(\frac{-7x - 2}{2}\right) = 18$$
Multiply by 2.
$$12x + 5(-7x - 2) = 36$$
$$12x - 35x - 10 = 36$$
$$-23x - 10 = 36$$
$$-23x = 46$$
$$x = -2$$
Substitute -2 for x in $7x + 2y = -2$.
$$7(-2) + 2y = -2$$
$$-14 + 2y = -2$$
$$2y = 12$$
$$y = 6$$
The solution set is $\{(-2, 6)\}$.

17. Solve $x + 5y = -71$ for x.
$$x + 5y = -71$$
$$x = -5y - 71$$
Substitute $-5y - 71$ for x
in $8x - 3y = -9$.
$$8(-5y - 71) - 3y = -9$$
$$-40y - 568 - 3y = -9$$
$$-43y - 568 = -9$$
$$-43y = 559$$
$$y = -13$$
Substitute -13 for y in $x + 5y = -71$.
$$x + 5(-13) = -71$$
$$x - 65 = -71$$
$$x = -6$$
The solution set is $\{(-6, -13)\}$.

19. Solve $4x - 6y = 1$ for y.
$$4x - 6y = 1$$
$$-6y = -4x + 1$$
Multiply by -1.
$$6y = 4x - 1$$
$$y = \frac{4x - 1}{6}$$
Substitute $\frac{4x - 1}{6}$ for y in $2x + 3y = 4$.
$$2x + 3\left(\frac{4x - 1}{6}\right) = 4$$
Multiply by 2.
$$4x + 1(4x - 1) = 8$$
$$4x + 4x - 1 = 8$$
$$8x - 1 = 8$$
$$8x = 9$$
$$x = \frac{9}{8}$$
Substitute $\frac{9}{8}$ for x in $4x - 6y = 1$.
$$4\left(\frac{9}{8}\right) - 6y = 1$$
$$\frac{9}{2} - 6y = 1$$
$$9 - 12y = 2$$
$$-12y = -7$$
$$y = \frac{7}{12}$$
The solution set is $\left\{\left(\frac{9}{8}, \frac{7}{12}\right)\right\}$.

21. Solve $3x - 2y = 0$ for y.

$$3x - 2y = 0$$
$$-2y = -3x$$
$$y = \frac{3x}{2}$$

Substitute $\frac{3x}{2}$ for y

in $5x + 7y = 3$.

$$5x + 7\left(\frac{3x}{2}\right) = 3$$

Multiply by 2.

$$10x + 7(3x) = 6$$
$$10x + 21x = 6$$
$$31x = 6$$
$$x = \frac{6}{31}$$

Substitute $\frac{6}{31}$ for x in $3x - 2y = 0$.

$$3\left(\frac{6}{31}\right) - 2y = 0$$
$$3(6) - 62y = 0$$
$$18 - 62y = 0$$
$$-62y = -18$$
$$y = \frac{18}{62} = \frac{9}{31}$$

The solution set is $\left\{\left(\frac{6}{31}, \frac{9}{31}\right)\right\}$.

23. Substitute $x + 300$ for y
in $0.05x + 0.07y = 33$.

$$0.05x + 0.07(x + 300) = 33$$

Multiply by 100.

$$5x + 7(x + 300) = 3300$$
$$5x + 7x + 2100 = 3300$$
$$12x + 2100 = 3300$$
$$12x = 1200$$
$$x = 100$$

Substitute 100 for x in $y = x + 300$.

$$y = 100 + 300 = 400$$

The solution set is $\{(100, 400)\}$.

25. Solve $x + y = 13$ for y.

$$x + y = 13$$
$$y = 13 - x$$

Substitute $13 - x$ for y

in $0.05x + 0.1y = 1.15$.

$$0.05x + 0.1(13 - x) = 1.15$$

Multiply by 100.

$$5x + 10(13 - x) = 115$$
$$5x + 130 - 10x = 115$$
$$130 - 5x = 115$$
$$-5x = -15$$
$$x = 3$$

Substitute 3 for x in $x + y = 13$.

$$3 + y = 13$$
$$y = 10$$

The solution set is $\{(3, 10)\}$.

27. Use the elimination-by-addition method.

$5x - 4y = 14$ Multiply by 3.
$7x + 3y = -32$ Multiply by 4.

$$15x - 12y = 42$$
$$28x + 12y = -128$$
$$\overline{43x \qquad\quad = -86}$$
$$x = -2$$

Substitute -2 for x in $5x - 4y = 14$.

$$5(-2) - 4y = 14$$
$$-10 - 4y = 14$$
$$-4y = 24$$
$$y = -6$$

The solution set is $\{(-2, -6)\}$.

29. Use the substitution method.

Substitute $-x$ for y in $2x + 9y = 6$.

$$2x + 9(-x) = 6$$
$$2x - 9x = 6$$
$$-7x = 6$$
$$x = -\frac{6}{7}$$

Substitute $-\frac{6}{7}$ for x in $y = -x$.

$$y = -\left(-\frac{6}{7}\right) = \frac{6}{7}$$

The solution set is $\left\{\left(-\frac{6}{7}, \frac{6}{7}\right)\right\}$.

243

Problem Set 8.7

31. Use the elimination-by-addition method.
$x + y = 22$ Multiply by -5.
$0.6x + 0.5y = 12$ Multiply by 10.

$$-5x - 5y = -110$$
$$\underline{\;\;6x + 5y = \;\;\;\;120\;\;}$$
$$x \;\;\;\;\;= \;\;\;\;10$$

Substitute 10 for x in $x + y = 22$.
$10 + y = 22$
$\;\;\;\;\;\;\;y = 12$
The solution set is $\{(10,\ 12)\}$.

33. Use the elimination-by-addition method.
$4x - y = 0$ Multiply by 2.
$7x + 2y = 9$ Leave alone.

$$8x - 2y = 0$$
$$\underline{\;\;7x + 2y = 9\;\;}$$
$$15x \;\;\;\;\;= 9$$
$$x = \frac{9}{15} = \frac{3}{5}$$

Substitute $\frac{3}{5}$ for x in $4x - y = 0$.
$$4\left(\frac{3}{5}\right) - y = 0$$
$$\frac{12}{5} - y = 0$$
$$\frac{12}{5} = y$$
The solution set is $\left\{\left(\frac{3}{5},\ \frac{12}{5}\right)\right\}$.

35. Use the elimination-by-addition method.
$2x + y = 1$ Multiply by -3.
$6x - 7y = -57$ Leave alone.
$$-6x - 3y = -\;\;3$$
$$\underline{\;\;6x - 7y = -57\;\;}$$
$$-10y = -60$$
$$y = 6$$
Substitute 6 for y in $2x + y = 1$.
$2x + 6 = 1$
$\;\;\;\;2x = -5$
$$x = -\frac{5}{2}$$

The solution set is $\left\{\left(-\frac{5}{2},\ 6\right)\right\}$.

37. Use the elimination-by-addition method.
$6x - y = -1$ Multiply by 2.
$10x + 2y = 13$ Leave alone.

$$12x - 2y = -2$$
$$\underline{\;\;10x + 2y = \;\;\;13\;\;}$$
$$22x \;\;\;\;\;= 11$$
$$x = \frac{11}{22} = \frac{1}{2}$$
Substitute $\frac{1}{2}$ for x in $6x - y = -1$.
$$6\left(\frac{1}{2}\right) - y = -1$$
$$3 - y = -1$$
$$-y = -4$$
$$y = 4$$
The solution set is $\left\{\left(\frac{1}{2},\ 4\right)\right\}$.

39. Use the elimination-by-addition method.
$4x + 8y = 20$ Divide by -4.
$x + 2y = 5$ Leave alone.

$$-x - 2y = -5$$
$$\underline{\;\;x + 2y = \;\;\;\;5\;\;}$$
$$0 = 0 \;\;\;\;\;\;\text{(Always true.)}$$
The system has infinitely many solutions.

41. Use the substitution method.
Substitute $2y$ for x in $3x - 8y = -5$.
$3(2y) - 8y = -5$
$\;\;\;\;6y - 8y = -5$
$\;\;\;\;\;\;\;-2y = -5$
$$y = \frac{5}{2}$$
Substitute $\frac{5}{2}$ for y in $x = 2y$.
$$x = 2\left(\frac{5}{2}\right) = 5$$
The solution set is $\left\{\left(5,\ \frac{5}{2}\right)\right\}$.

43. Use the elimination-by-addition method.
$5y - 2x = -4$ Multiply by -2.
$10y = 3x + 4$ Subtract $3x$ from both sides.

$$-10y + 4x = 8$$
$$\underline{10y - 3x = 4}$$
$$x = 12$$

Substitute 12 for x in $10y = 3x + 4$.
$10y = 3(12) + 4$
$10y = 36 + 4$
$10y = 40$
$y = 4$
The solution set is $\{(12, 4)\}$.

45. Use the substitution method.
Substitute $-y - 1$ for x
in $6x - 5y = 4$.
$6(-y - 1) - 5y = 4$
$-6y - 6 - 5y = 4$
$-11y - 6 = 4$
$-11y = 10$
$$y = -\frac{10}{11}$$
Substitute $-\dfrac{10}{11}$ for y in $x = -y - 1$.
$$x = -\left(-\frac{10}{11}\right) - 1$$
$$x = \frac{10}{11} - 1$$
$$x = \frac{10}{11} - \frac{11}{11}$$
$$x = -\frac{1}{11}$$
The solution set is $\left\{\left(-\dfrac{1}{11}, -\dfrac{10}{11}\right)\right\}$.

47. Let x and y represent the two numbers.
$x + y = 46$ Their sum is 46.
$\underline{x - y = 22}$ Their difference is 22.
$2x \quad\;\; = 68$
$\qquad x = 34$
Substitute 34 for x in $x + y = 46$.
$34 + y = 46$
$\qquad y = 12$
The numbers are 12 and 34.

49. Let x represent the number of double rooms at \$28. Let y represent the number of single rooms at \$19.
$x + y = 50$ The total number of rooms is 50.
$28x + 19y = 1265$ The total revenue was \$1265.

$x + y = 50$ Multiply by -19.
$28x + 19y = 1265$ Leave alone.

$$-19x - 19y = -950$$
$$\underline{28x + 19y = 1265}$$
$$9x \qquad\quad = 315$$
$$x = 35$$
Substitute 35 for x in $x + y = 50$.
$35 + y = 50$
$\qquad y = 15$
There were 35 double rooms at \$28 and 15 single rooms at \$19.

51. Let t represent the tens digit.
Let u represent the units digit.
$t + u = 9$ Their sum is 9.
$10t + u = 9u$ The two-digit number is nine times its units digit.

$t + u = 9$ Multiply by -10.
$10t + u = 9u$ Subtract $9u$ from both sides..

$$-10t - 10u = -90$$
$$\underline{10t - 8u = 0}$$
$$-18u = -90$$
$$u = 5$$

Substitute 5 for u in $t + u = 9$.
$t + 5 = 9$
$\quad t = 4$

The tens digit is 4 and the units digit is 5.
Thus, the number is 45.

Problem Set 8.7

53. Let d represent the number of dimes.
Let q represent the number of quarters.
$10d + 25q = 1205$ The value of the
coins was $12.05.
$q = 2d + 5$ The number of
quarters is five
more than twice the
number of dimes.

Substitute $2d + 5$ for q in $10d + 25q = 1205$.
$10d + 25(2d + 5) = 1205$
$10d + 50d + 125 = 1205$
$60d + 125 = 1205$
$60d = 1080$
$d = 18$
Substitute 18 for d in $q = 2d + 5$.
$q = 2(18) + 5$
$q = 36 + 5$
$q = 41$
There were 18 dimes and 41 quarters.

55. Let t represent the tens digit.
Let u represent the units digit.
$t + u = 12$ Their sum is 12.
$t = 3u$ The tens digit is three
times the units digit.
Substitute $3u$ for t in $t + u = 12$.
$3u + u = 12$
$4u = 12$
$u = 3$
Substitute 3 for u in $t + u = 12$.
$t + 3 = 12$
$t = 9$
The tens digit is 9 and the units digit is 3.
Thus, the number is 93.

57. Let x represent the money invested at 8%.
Let y represent the money invested at 9%.
$y = x + 250$ $250 more was invested
at 9% than at 8%.
$0.08x + 0.09y = 48$ The total yearly interest
was $48.
Substitute $x + 250$ for y in
$0.08x + 0.09y = 48$.

$0.08x + 0.09(x + 250) = 48$
Multiply by 100.
$8x + 9(x + 250) = 4800$
$8x + 9x + 2250 = 4800$
$17x + 2250 = 4800$
$17x = 2550$
$x = 150$
Substitute 150 for x in $y = x + 250$.
$y = 150 + 250 = 400$
The investment was $150 at 8%
and $400 at 9%.

59. Let x represent the quantity of 30% alcohol.
Let y represent the quantity of 70% alcohol.

$x + y = 10$ The total
quantity
of solution.
$0.30x + 0.70y = 0.40(10)$ The total
quantity
of alcohol.
$x + y = 10$ Multiply by -3.
$0.30x + 0.70y = 4$ Multiply by 10.

$-3x - 3y = -30$
$\underline{3x + 7y = 40}$
$4y = 10$
$y = 2.5$
Substitute 2.5 for y in $x + y = 10$.
$x + 2.5 = 10$
$x = 7.5$
The quantity of each solution to be used
would be 7.5 liters of 30% alcohol and
2.5 liters of 70% alcohol.

65. Solve $5x - 3y = 0$ for x.
$5x - 3y = 0$
$5x = 3y$
$x = \dfrac{3}{5}y$
Substitute $\dfrac{3}{5}y$ for x in $4x + 7y = 0$.

$$4\left(\frac{3}{5}y\right) + 7y = 0$$

$$\frac{12}{5}y + 7y = 0$$

$$12y + 35y = 0$$

$$47y = 0$$

$$y = 0$$

Substitute 0 for y in $5x - 3y = 0$.

$$5x - 0 = 0$$

$$5x = 0$$

$$x = 0$$

The solution set is $\{(0, 0)\}$.

67. Substitute $-4y + 5$ for x in $2x + 8y = -1$.

$$2(-4y + 5) + 8y = -1$$

$$-8y + 10 + 8y = -1$$

$$10 = -1 \quad \text{Not possible.}$$

The solution set is $\emptyset$.

CHAPTER 8   **Review Problem Set**

1. $2x - 5y = 10$

$x = 0$	$y = 0$
$2(0) - 5y = 10$	$2x - 5(0) = 10$
$-5y = 10$	$2x = 10$
$y = -2$	$x = 5$
$(0, -2)$	$(5, 0)$

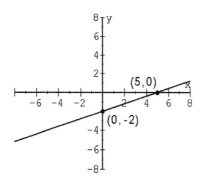

2. $y = -\dfrac{1}{3}x + 1$

$$m = -\frac{1}{3} \quad b = 1$$

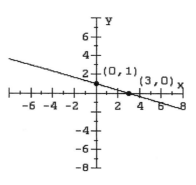

3. $y = 2x^2 + 1$

$x = 1$	$x = 2$	$x = 0$
$y = 2(1)^2 + 1$	$y = 2(2)^2 + 1$	$y = 2(0)^2 + 1$
$y = 2(1) + 1$	$y = 2(4) + 1$	$y = 2(0) + 1$
$y = 2 + 1$	$y = 8 + 1$	$y = 0 + 1$
$y = 3$	$y = 9$	$y = 1$
$(1, 3)$	$(2, 9)$	$(0, 1)$

$x = -1$	$x = -2$
$y = 2(-1)^2 + 1$	$y = 2(-2)^2 + 1$
$y = 2(1) + 1$	$y = 2(4) + 1$
$y = 2 + 1$	$y = 8 + 1$
$y = 3$	$y = 9$
$(-1, 3)$	$(-2, 9)$

247

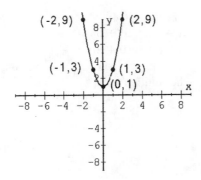

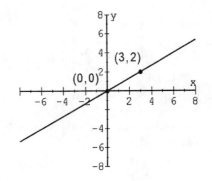

4. $y = -x^2 - 2$

$x = 0$	$x = 1$	$x = -1$
$y = -(0)^2 - 2$	$y = -(1)^2 - 2$	$y = -(-1)^2 - 2$
$y = 0 - 2$	$y = -1 - 2$	$y = -1 - 2$
$y = -2$	$y = -3$	$y = -3$
$(0, -2)$	$(1, -3)$	$(-1, -3)$

6. $y = -x^3$

$x = 0$	$x = 1$	$x = -1$
$y = -(0)^3$	$y = -(1)^3$	$y = -(-1)^3$
$y = 0$	$y = -1$	$y = 1$
$(0, 0)$	$(1, -1)$	$(-1, 1)$

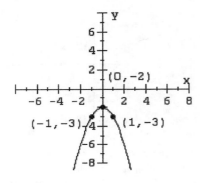

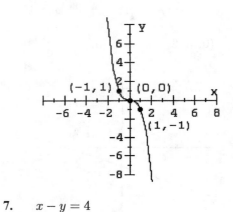

5. $2x - 3y = 0$

$x = 0$	$x = 3$
$2(0) - 3y = 0$	$2(3) - 3y = 0$
$-3y = \dfrac{0}{-3}$	$-3y = -6$
$y = 0$	$y = \dfrac{-6}{-3} = 2$
$(0, 0)$	$(3, 2)$

7. $x - y = 4$

$x = 0$	$y = 0$
$0 - y = 4$	$x - 0 = 4$
$-y = 4$	$x = 4$
$y = -4$	
$(0, -4)$	$(4, 0)$

248

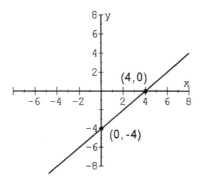

8. $x + 2y = -2$

$x = 0$	$y = 0$
$0 + 2y = -2$	$x + 2(0) = -2$
$2y = -2$	$x + 0 = -2$
$y = -1$	$x = -2$
$(0, -1)$	$(-2, 0)$

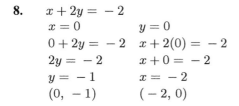

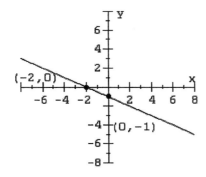

9. $y = \dfrac{2}{3}x - 1$

$x = 0$	$x = 3$
$y = \dfrac{2}{3}(0) - 1$	$y = \dfrac{2}{3}(3) - 1$
$y = 0 - 1$	$y = 2 - 1$
$y = -1$	$y = 1$
$(0, -1)$	$(3, 1)$

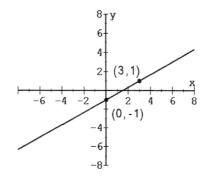

10. $y = 3x$

$x = 0$	$x = 1$	$x = -2$
$y = 3(0)$	$y = 3(1)$	$y = 3(-2)$
$y = 0$	$y = 3$	$y = -6$
$(0, 0)$	$(1, 3)$	$(-2, -6)$

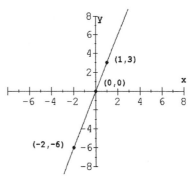

11. $2x - 5y = 10$

$$-5y = -2x + 10$$
$$y = \frac{-2}{-5}x + \frac{10}{-5}$$
$$y = \frac{2}{5}x - 2$$

$$m = \frac{2}{5} \qquad b = -2$$

Chapter 8 Review Problem Set

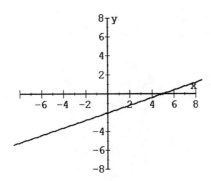

12. $y = -\dfrac{1}{3}x + 1$

$m = -\dfrac{1}{3} \quad b = 1$

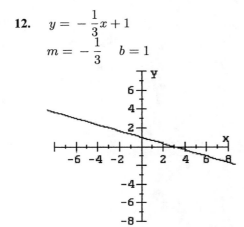

13. $x + 2y = 2$

$2y = -x + 2$

$y = \dfrac{-1}{2}x + \dfrac{2}{2}$

$y = -\dfrac{1}{2}x + 1$

$m = -\dfrac{1}{2} \quad b = 1$

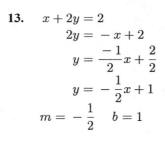

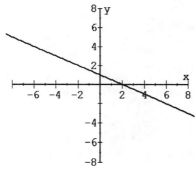

14. $3x + y = -2$

$y = -3x - 2$

$m = -3 \quad b = -2$

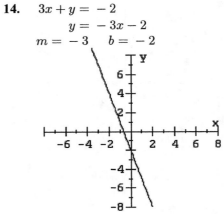

15. $2x - y = 4$

$-y = -2x + 4$

$y = 2x - 4$

$m = 2 \quad b = -4$

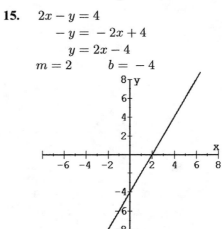

16. $3x - 4y = 12$

$-4y = -3x + 12$

$y = \dfrac{-3}{-4}x + \dfrac{12}{-4}$

$y = \dfrac{3}{4}x - 3$

$m = \dfrac{3}{4} \quad b = -3$

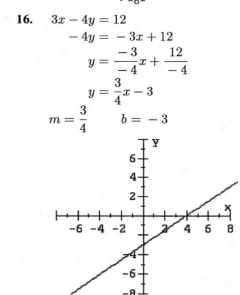

17. Let $(3, -4)$ be P_1 and $(-2, 5)$ be P_2.

$$m = \frac{y_2 - y_1}{x_2 - x_1} = \frac{5 - (-4)}{-2 - 3} = \frac{9}{-5} = -\frac{9}{5}$$

18. $5x - 6y = 30$
$$-6y = -5x + 30$$
$$y = \frac{-5}{-6}x + \frac{30}{-6}$$
$$y = \frac{5}{6}x - 5$$
$$m = \frac{5}{6}$$

19. The point-slope form is shown for this problem.
$$y - y_1 = m(x - x_1)$$
$$y - (-3) = -\frac{5}{7}(x - 2)$$
$$7(y + 3) = -5(x - 2)$$
$$7y + 21 = -5x + 10$$
$$5x + 7y = -11$$

20. $m = \dfrac{-3 - 5}{-1 - 2} = \dfrac{-8}{-3} = \dfrac{8}{3}$
$$y - y_1 = m(x - x_1)$$
$$y - 5 = \frac{8}{3}(x - 2)$$
$$3(y - 5) = 3\left[\frac{8}{3}(x - 2)\right]$$
$$3(y - 5) = 8(x - 2)$$
$$3y - 15 = 8x - 16$$
$$-15 = 8x - 3y - 16$$
$$1 = 8x - 3y$$
$$8x - 3y = 1$$

21. Use the slope-intercept form.
$$y = mx + b$$
$$y = \frac{2}{9}x - 1$$
$$9y = 2x - 9$$
$$-2x + 9y = -9$$
$$2x - 9y = 9$$

22. A line perpendicular to the x-axis will be a vertical line. The equation of a vertical line through $(2, 4)$ is $x = 2$.

23.

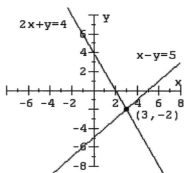

The solution set is $\{(3, -2)\}$.

24. $\left(\begin{array}{c} 2x - y = 1 \\ 3x - 2y = -5 \end{array}\right)$

Multiply equation (1) by -2 and add the result to equation (2).
$$-4x + 2y = -2$$
$$\underline{3x - 2y = -5}$$
$$-x = -7$$
$$x = 7$$
Substitute $x = 7$ into equation (1).
$$2(7) - y = 1$$
$$14 - y = 1$$
$$-y = -13$$
$$y = 13$$
The solution set is $\{(7, 13)\}$.

25. Use the substitution method.
Substitute $-3y + 1$ for x in $2x + 5y = 7$.
$$2(-3y + 1) + 5y = 7$$
$$-6y + 2 + 5y = 7$$
$$-y + 2 = 7$$
$$-y = 5$$
$$y = -5$$
Substitute -5 for y in $x = -3y + 1$.

$x = -3(-5) + 1$

$x = 15 + 1$

$x = 16$

The solution set is $\{(16, -5)\}$.

26. $\begin{pmatrix} 3x + 2y = 7 \\ 4x - 5y = 3 \end{pmatrix}$

Multiply equation (1) by -4.

$-12x - 8y = -28$ (3)

Multiply equiation (2) by 3.

$12x - 15y = 9$ (4)

Add equation (3) and equation (4).

$-12x - 8y = -28$

$\underline{12x - 15y = 9}$

$-23y = -19$

$y = \dfrac{-19}{-23} = \dfrac{19}{23}$

Substitute $y = \dfrac{19}{23}$ into equation (1).

$3x + 2\left(\dfrac{19}{23}\right) = 7$

$3x + \dfrac{38}{23} = 7$

$3x = 7 - \dfrac{38}{23}$

$3x = \dfrac{161}{23} - \dfrac{38}{23}$

$3x = \dfrac{123}{23}$

$\dfrac{1}{3}(3x) = \dfrac{1}{3}\left(\dfrac{123}{23}\right)$

$x = \dfrac{41}{23}$

The solution set is $\left\{\left(\dfrac{41}{23}, \dfrac{19}{23}\right)\right\}$.

27. Use the elimination-by-addition method.

$9x + 2y = 140$ Leave alone.

$x + 5y = 135$ Multiply by -9.

$9x + 2y = 140$

$\underline{-9x - 45y = -1215}$

$-43y = -1075$

$y = 25$

Substitute 25 for y in $x + 5y = 135$.

$x + 5(25) = 135$

$x + 125 = 135$

$x = 10$

The solution set is $\{(10, 25)\}$.

28. $\begin{pmatrix} \dfrac{1}{2}x + \dfrac{1}{4}y = -5 \\ \dfrac{2}{3}x - \dfrac{1}{2}y = 0 \end{pmatrix}$

Multiply equation (1) by 4.

$2x + y = -20$

Multiply equation (2) by 6.

$4x - 3y = 0$

The equivalent system is

$\begin{pmatrix} 2x + y = -20 \\ 4x - 3y = 0 \end{pmatrix}$ $\begin{matrix} (3) \\ (4) \end{matrix}$

Multiply equation (3) by 3 and add the result to equation (4).

$6x + 3y = -60$

$\underline{4x - 3y = 0}$

$10x = -60$

$x = -6$

Substitute $x = -6$ into equation (3).

$2(-6) + y = -20$

$-12 + y = -20$

$y = -8$

The solution set is $\{(-6, -8)\}$.

29. Use the elimination-by-addition method.

$x + y = 1000$ Multiply by -7.

$0.07x + 0.09y = 82$ Multiply by 100.

$-7x - 7y = -7000$

$\underline{7x + 9y = 8200}$

$2y = 1200$

$y = 600$

Substitute 600 for y in $x + y = 1000$.

$x + 600 = 1000$

$x = 400$

The solution set is $\{(400, 600)\}$.

30. $\begin{pmatrix} y = 5x + 2 \\ 10x - 2y = 1 \end{pmatrix}$

Substitute in equation (2) for y.

$10x - 2(5x + 2) = 1$

$10x - 10x - 4 = 1$

$-4 = 1$

Since $-4 \neq 1$ the system is inconsistent and the solution set is $\emptyset$.

31. Use the substitution method.

Substitute $3x - 2$ for y in $5x - 7y = 9$.

$5x - 7(3x - 2) = 9$

$5x - 21x + 14 = 9$

$-16x + 14 = 9$

$-16x = -5$

$x = \dfrac{5}{16}$

Substitute $\dfrac{5}{16}$ for x in $y = 3x - 2$.

$y = 3\left(\dfrac{5}{16}\right) - 2$

$y = \dfrac{15}{16} - \dfrac{32}{16}$

$y = -\dfrac{17}{16}$

The solution set is $\left\{\left(\dfrac{5}{16}, -\dfrac{17}{16}\right)\right\}$.

32. $\begin{pmatrix} 10t + u = 6u \\ t + u = 12 \end{pmatrix} \Rightarrow \begin{pmatrix} 10t - 5u = 0 \\ t + u = 12 \end{pmatrix}$ $\begin{matrix} (1) \\ (2) \end{matrix}$

Multiply equation (2) by 5 and add the result to equation (1).

$5t + 5u = 60$

$\underline{10t - 5u = 0}$

$15t = 60$

$t = 4$

Substitute $t = 4$ into equation (2).

$4 + u = 12$

$u = 8$

The solution set is $t = 4$ and $u = 8$.

33. Use the substitution method.

Substitute $2u$ for t in $10t + u - 36 = 10u + t$.

$10(2u) + u - 36 = 10u + (2u)$

$20u + u - 36 = 10u + 2u$

$21u - 36 = 12u$

$-36 = -9u$

$4 = u$

Substitute 4 for u in $t = 2u$.

$t = 2(4) = 8$

The solution set is $t = 8$ and $u = 4$.

34. $\begin{pmatrix} u = 2t + 1 \\ 10t + u + 10u + t = 100 \end{pmatrix} \Rightarrow$

$\begin{pmatrix} u = 2t + 1 \\ 11t + 11u = 110 \end{pmatrix}$

Substitute for u into equation (2).

$11t + 11(2t + 1) = 110$

$11t + 22t + 11 = 110$

$33t + 11 = 110$

$33t = 99$

$t = 3$

Substitute $t = 3$ into equation (1).

$u = 2(3) + 1$

$u = 6 + 1$

$u = 7$

The solution set is $t = 3$ and $u = 7$.

35. Use the substitution method.

Substitute $-\dfrac{2}{3}x$ for y.

in $\dfrac{1}{3}x - y = -9$.

$\dfrac{1}{3}x - \left(-\dfrac{2}{3}x\right) = -9$

$\dfrac{1}{3}x + \dfrac{2}{3}x = -9$

$x = -9$

Substitute -9 for x in $y = -\dfrac{2}{3}x$.

$y = -\dfrac{2}{3}(-9) = 6$

The solution set is $\{(-9, 6)\}$.

Chapter 8 Review Problem Set

36. $y > \dfrac{2}{3}x - 1$

Use $(0, -1)$ and $(3, 1)$ to graph a dashed line for $y = \dfrac{2}{3}x - 1$.

Use $(0, 0)$ for a test point.

$y > \dfrac{2}{3}x - 1$

$0 > \dfrac{2}{3}(0) - 1$

$0 > -1$

Since $0 > -1$ is a true statement, the solution set is the half-plane that contains $(0, 0)$.

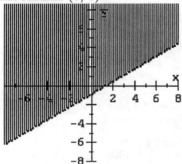

37. $x - 2y \le 4$

Use $(0, -2)$ and $(4, 0)$ to graph a solid line for $x - 2y = 4$.

Use $(0, 0)$ for a test point.

$x - 2y \le 4$

$0 - 2(0) \le 4$

$0 \le 4$

Since $0 \le 4$ is a true statement, the solution set is the half-plane that contains $(0, 0)$.

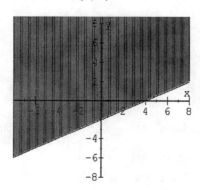

38. $y \le -2x$

Use $(0, 0)$ and $(1, -2)$ to graph a solid line for $y = -2x$.

Use $(3, 3)$ for a test point.

$y \le -2x$

$3 \le -2(3)$

$3 \le -6$

Since $3 \le -6$ is a false statement, the solution set is the half-plane that does **not** contain $(3, 3)$.

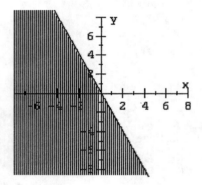

39. $3x + 2y > -6$

Use $(0, -3)$ and $(-2, 0)$ to graph a dashed line for $3x + 2y = -6$.

Use $(0, 0)$ for a test point.

$3x + 2y > -6$

$3(0) + 2(0) > -6$

$0 > -6$

Since $0 > -6$ is a true statement, the solution set is the half-plane that contains $(0, 0)$.

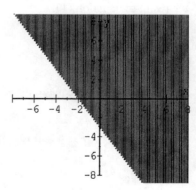

40. Let x = one number

y = other number

$$\left(\begin{matrix} x + y = 113 \\ x = 2y - 1 \end{matrix}\right)$$

Use substitution to solve.

$$x + y = 113$$
$$(2y - 1) + y = 113$$
$$3y - 1 = 113$$
$$3y = 114$$
$$y = 38$$

Substitute $y = 38$ into equation (2).

$$x = 2(38) - 1$$
$$x = 76 - 1$$
$$x = 75$$

The numbers are 75 and 38.

41. Let x represent the money invested at 9%.
Let y represent the money invested at 11%.

$$y = x + 50$$

The money invested at 11% is $50 more than the money invested at 9%.

$$0.09x + 0.11y = 55.50$$

The yearly interest was $55.50.
Substitute $x + 50$ for y in
$0.09x + 0.11y = 55.50$

$$0.09x + 0.11(x + 50) = 55.50$$

Multiply by 100.

$$9x + 11(x + 50) = 5550$$
$$9x + 11x + 550 = 5550$$
$$20x + 550 = 5550$$
$$20x = 5000$$
$$x = 250$$

Substitute 250 for x in $y = x + 50$.
$y = 250 + 50 = 300$
The money invested was $250 at 9% and $300 at 11%.

42. Let n represent the number of nickels and d the number of dimes.
$n + d = 43$

The number of coins is 43.
$5n + 10d = 340$
The total value of the coins is 340 cents.
$n + d = 43$ ⠀⠀⠀Multiply by -5.
$5n + 10d = 340$⠀⠀Leave alone.

$$-5n - 5d = -215$$
$$\underline{5n + 10d = 340}$$
$$5d = 125$$
$$d = 25$$

Substitute 25 for d in $n + d = 43$.
$$n + 25 = 43$$
$$n = 18$$
She has 18 nickels and 25 dimes.

43. Let l represent the length of the rectangle.
Let w represent the width of the rectangle.
$l = 3w + 1$
The length is one more than three times the width.
$2l + 2w = 50$
The perimeter is 50 inches.

Substitute $3w + 1$ for l in $2l + 2w = 50$.
$$2(3w + 1) + 2w = 50$$
$$6w + 2 + 2w = 50$$
$$8w + 2 = 50$$
$$8w = 48$$
$$w = 6$$
Substitute 6 for w in $l = 3w + 1$.
$$l = 3(6) + 1$$
$$l = 18 + 1$$
$$l = 19$$
The rectangle has a length of 19 inches and a width of 6 inches.

44. Let x = length of rectangle
y = width of rectangle

$$\left(\begin{matrix} y = x - 5 \\ 2x + 2y = 38 \end{matrix}\right)$$

Substitute for y in equation (2).
$$2x + 2(x - 5) = 38$$
$$2x + 2x - 10 = 38$$
$$4x - 10 = 38$$
$$4x = 48$$
$$x = 12$$
Substitute $x = 12$ into equation (1).
$$y = 12 - 5$$
$$y = 7$$
The length is 12 inches and the width is 7 inches.

45. Let $x =$ number of quarters
$y =$ number of dimes

$$\left(\begin{array}{c} x + y = 32 \\ 25x + 10y = 485 \end{array} \right)$$

Multiply equation (1) by -25 and add the resulting equation to equation (2).
$$-25x - 25y = -800$$
$$\underline{25x + 10y = \quad 485}$$
$$-15y = -315$$
$$y = 21$$
Substitute $y = 21$ in equation (1).
$$x + 21 = 32$$
$$x = 11$$
There are 11 quarters and 21 dimes.

46. Let a and b represent the angles.
$$a + b = 90$$
The angles are complementary.
$$a = 2b - 6$$
One angle is 6° less than twice the other angle.

Substitute $2b - 6$ for a in $a + b = 90$.
$$(2b - 6) + b = 90$$
$$3b - 6 = 90$$
$$3b = 96$$
$$b = 32$$
Substitute 32 for b in $a + b = 90$.
$$a + 32 = 90$$
$$a = 58$$

The two angles are 32° and 58°.

47. Let a represent the larger angle.
Let b represent the smaller angle.
$$a + b = 180$$
The angles are supplementary.
$$a = 3b - 20$$
The larger angle is 20° less than three times the smaller angle.
Substitute $3b - 20$ for a in $a + b = 180$.
$$(3b - 20) + b = 180$$
$$4b - 20 = 180$$
$$4b = 200$$
$$b = 50$$
Substitute 50 for b in $a + b = 180$.
$$a + 50 = 180$$
$$a = 130$$
The two angles are 50° and 130°.

48. Let c represent the cost of a cheeseburger and m the cost of a milkshake.
$$4c + 5m = 835$$
The cost of four cheeseburgers and five milkshakes is $8.35.
$$2m = c + 35$$
Two milkshakes cost 35 cents more than one cheeseburger.

$$4c + 5m = 835 \quad \text{Leave alone.}$$
$$-c + 2m = 35 \quad \text{Multiply by 4.}$$

$$4c + 5m = 835$$
$$\underline{-4c + 8m = 140}$$
$$13m = 975$$
$$m = 75$$
Substitute 75 for m in $2m = c + 35$.
$$2(75) = c + 35$$
$$150 = c + 35$$
$$115 = c$$
The cost of a cheeseburger is $1.15 and a milkshake is $.75

49. Let p represent the cost of a can of prune juice. Let t represent the cost of a can of tomato juice.

$3p + 2t = 385$
The cost of three prune and 2 tomato cans.
$2p + 3t = 355$
The cost of two prune and 3 tomato cans.

$3p + 2t = 385$ Multiply by -2.
$2p + 3t = 355$ Multiply by 3.

$-6p - 4t = -770$
$\underline{6p + 9t = 1065}$
$5t = 295$
$t = 59$

Substitute 59 for t in $3p + 2t = 385$.
$3p + 2(59) = 385$
$3p + 118 = 385$
$3p = 267$
$p = 89$
Prune juice costs \$0.89 per can and tomato juice costs \$0.59 per can.

CHAPTER 8 | **Test**

1. $5x + 3y = 15$
$3y = -5x + 15$
$y = \dfrac{-5}{3}x + 5$

$m = -\dfrac{5}{3} \quad b = 5$

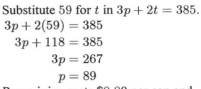

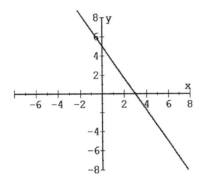

2. $-2x + y = -4$
$y = 2x - 4$
$m = 2 \qquad b = -4$

3. $y = -\dfrac{1}{2}x - 2$

$m = -\dfrac{1}{2} \quad b = -2$

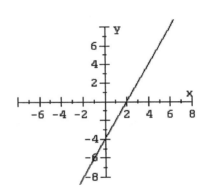

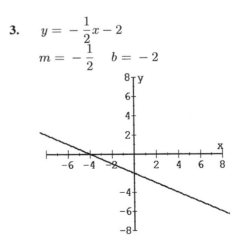

Chapter 8 Test

4. $3x + y = 0$

$y = -3x + 0$

$m = -3 \quad b = 0$

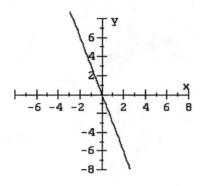

5. $\dfrac{3}{2} = \dfrac{13 - 7}{x - 4}$

$\dfrac{3}{2} = \dfrac{6}{x - 4}$

$3(x - 4) = 2(6)$

$3x - 12 = 12$

$3x = 24$

$x = 8$

6. $-\dfrac{3}{5} = \dfrac{5 - y}{6 - 1}$

$\dfrac{-3}{5} = \dfrac{5 - y}{5}$

$-3(5) = 5(5 - y)$

$-15 = 25 - 5y$

$-40 = -5y$

$8 = y$

7. Answers may vary.

$(7, 6); (11, 7)$

8. Answers may vary.

$(3, -2); (4, -5)$

9. $m = \dfrac{\text{rise}}{\text{run}} = \dfrac{85}{1850}$

$m = 0.046$

$0.046 \times 100\% = 4.6\%$

10. To find the x-intercept let $y = 0$.NEW

$y = 4x + 8$

$0 = 4x + 8$

$-4x = 8$

$x = -2$

The x-intercept is -2.

11. Use the slope-intercept form.

$y = mx + b$

$y = -\dfrac{3}{5}x + 4$

$5y = -3x + 20$

$3x + 5y = 20$

12. Use the point-slope form.

$y - y_1 = m(x - x_1)$

$y - (-2) = \dfrac{4}{9}(x - 4)$

$9(y + 2) = 4(x - 4)$

$9y + 18 = 4x - 16$

$-4x + 9y = -34$

$4x - 9y = 34$

13. Find the slope.

$m = \dfrac{y_2 - y_1}{x_2 - x_1} = \dfrac{-3 - 6}{-2 - 4} = \dfrac{-9}{-6} = \dfrac{3}{2}$

Use the slope and either point in the point-slope form.

$y - y_1 = m(x - x_1)$

$y - 6 = \dfrac{3}{2}(x - 4)$

$2(y - 6) = 3(x - 4)$

$2y - 12 = 3x - 12$

$-3x + 2y = 0$

$3x - 2y = 0$

258

14.

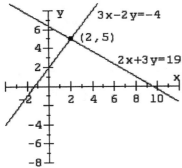

$3x - 2y = -4$

$(2, 5)$

$2x + 3y = 19$

The solution set is $\{(2, 5)\}$.

15. $x - 3y = -9$ Multiply by -4.
$4x + 7y = 40$ Leave alone.
$-4x + 12y = 36$
$\underline{4x + \ 7y = 40}$
$19y = 76$
$y = 4$
Substitute 4 for y in $x - 3y = -9$.
$x - 3(4) = -9$
$x - 12 = -9$
$x = 3$
The solution set is $\{(3, 4)\}$.

16. Solve $5x + y = -14$ for y.
$5x + y = -14$
$y = -5x - 14$
Substitute $-5x - 14$ for y in
$6x - 7y = -66$.
$6x - 7(-5x - 14) = -66$
$6x + 35x + 98 = -66$
$41x + 98 = -66$
$41x = -164$
$x = -4$
Substitute -4 for x in $y = -5x - 14$.
$y = -5(-4) - 14$
$y = 20 - 14$
$y = 6$
The solution set is $\{(-4, 6)\}$.

17. Use the elimination-by-addition method.
$2x - 7y = 26$ Multiply by 2.
$3x + 2y = -11$ Multiply by 7.

$4x - 14y = \quad 52$
$\underline{21x + 14y = -77}$
$25x \qquad\ = -25$
$x = -1$
Substitute -1 for x in $2x - 7y = 26$.
$2(-1) - 7y = 26$
$-2 - 7y = 26$
$-7y = 28$
$y = -4$
The solution set is $\{(-1, -4)\}$.

18. Use the elimination-by-addition method.
$8x + 5y = -6$ Leave alone.
$4x - y = 18$ Multiply by 5.

$8x + 5y = -6$
$\underline{20x - 5y = \ 90}$
$28x \qquad = 84$
$x = 3$
Substitute 3 for x in $4x - y = 18$.
$4(3) - y = 18$
$12 - y = 18$
$-y = 6$
$y = -6$
The solution set is $\{(3, -6)\}$.

19. $5x + 3y = 15$
$x = 0$ $y = 0$
$5(0) + 3y = 15$ $5x + 3(0) = 15$
$3y = 15$ $5x = 15$
$y = 5$ $x = 3$
$(0, 5)$ $(3, 0)$

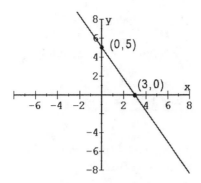

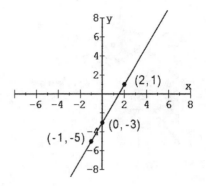

20. $y = 2x^2 - 3$

$x = 0$	$x = 1$	$x = 2$
$y = 2(0)^2 - 3$	$y = 2(1)^2 - 3$	$y = 2(2)^2 - 3$
$y = 0 + -3$	$y = 2(1) - 3$	$y = 2(4) - 3$
$y = -3$	$y = 2 - 3$	$y = 8 - 3$
$(0, -3)$	$y = -1$	$y = 5$
	$(1, -1)$	$(2, 5)$

$x = -1$	$x = -2$
$y = 2(-1)^2 - 3$	$y = 2(-2)^2 - 3$
$y = 2(1) - 3$	$y = 2(4) - 3$
$y = 2 - 3$	$y = 8 - 3$
$y = -1$	$y = 5$
$(-1, -1)$	$(-2, 5)$

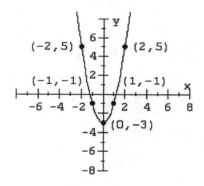

21. $y = 2x - 3$

$x = 0$	$x = 2$	$x = -1$
$y = 2(0) - 3$	$y = 2(2) - 3$	$y = 2(-1) - 3$
$y = 0 - 3$	$y = 4 - 3$	$y = -2 - 3$
$y = -3$	$y = 1$	$y = -5$
$(0, -3)$	$(2, 1)$	$(-1, -5)$

22. $y \geq 2x - 4$

Use the points $(0, -4)$ and $(2, 0)$ to graph a solid line for $y = 2x - 4$. Use $(0, 0)$ as a test point.

$y \geq 2x - 4$ becomes $0 \geq 0 - 4$ which is a true statement. Therefore, the solution set is the half-plane that contains the origin.

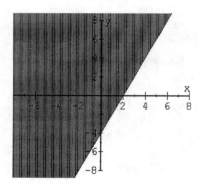

23. $x + 3y < -3$

Use the points $(0, -1)$ and $(-3, 0)$ to graph a dashed line for $x + 3y = -3$. Use $(0, 0)$ as a test point.

$x + 3y < -3$ becomes $0 + 0 < -3$ which is a false statement. Therefore, the solution set is the half-plane that does not contain the origin.

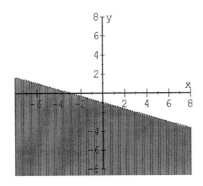

24. Let x represent the cost of a ream of paper.
Let y represent the cost of a notebook.

$3x + 4y = 19.63$
The cost of 3 reams of paper and 4 notebooks

$4x + y = 16.25$
The cost of 4 reams of paper and one notebook

$3x + 4y = 19.63$ Leave alone.
$4x + y = 16.25$ Multiply by -4.

$$3x + 4y = 19.63$$
$$\underline{-16x - 4y = -65.00}$$
$$-13x = -45.37$$
$$x = 3.49$$

Substitute 3.49 for x in $3x + 4y = 19.63$.
$$3(3.49) + 4y = 19.63$$
$$10.47 + 4y = 19.63$$
$$4y = 9.16$$
$$y = 2.29$$

A ream of paper costs \$3.49 and
a notebook costs \$2.29.

25. Let l represent the length of the rectangle.
Let w represent the width of the rectangle.

$l = 2w - 1$
The length is one less than twice the width.

$2l + 2w = 40$
The perimeter is 40 inches.

Substitute $2w - 1$ for l in $2l + 2w = 40$.
$$2(2w - 1) + 2w = 40$$
$$4w - 2 + 2w = 40$$
$$6w - 2 = 40$$
$$6w = 42$$
$$w = 7$$

Substitute 7 for w in $l = 2w - 1$.
$$l = 2(7) - 1$$
$$l = 14 - 1$$
$$l = 13$$

The rectangle has a length of 13 inches
and a width of 6 inches.

Chapter 9 Square Roots and Radicals

PROBLEM SET **9.1** **Roots and Radicals**

1. $\sqrt{49} = 7$ because $7^2 = 49$.

3. $-\sqrt{64} = -8$ because $-(8^2) = -64$.

5. $\sqrt{121} = 11$ because $11^2 = 121$.

7. $\sqrt{3600} = 60$ because $60^2 = 3600$.

9. $-\sqrt{1600} = -40$ because $-(40^2) = -1600$.

11. $\sqrt{6400} = 80$ because $80^2 = 6400$.

13. $\sqrt{324} = 18$ because $18^2 = 324$.

15. $\sqrt{\dfrac{25}{9}} = \dfrac{5}{3}$ because $\left(\dfrac{5}{3}\right)^2 = \dfrac{25}{9}$.

17. $\sqrt{0.16} = 0.4$ because $(0.4)^2 = 0.16$.

19. $\sqrt[3]{27} = 3$ because $3^3 = 27$.

21. $-\sqrt[3]{-8} = -(-2) = 2$ because $(-2)^3 = -8$.

23. $-\sqrt[3]{729} = -9$ because $-(9)^3 = -729$.

25. $\sqrt[3]{-216} = -6$ because $(-6)^3 = -216$.

For Problems 27 − 57, use a calculator and check answers in the text.

27. $\sqrt{576} = 24$

29. $\sqrt{2304} = 48$

31. $\sqrt{784} = 28$

33. $\sqrt{4225} = 65$

35. $\sqrt{3364} = 58$

37. $\sqrt[3]{3375} = 15$

39. $\sqrt[3]{9261} = 21$

41. $\sqrt{19} \approx 4.36$

43. $\sqrt{50} \approx 7.07$

45. $\sqrt{75} \approx 8.66$

47. $\sqrt{95} \approx 9.75$

49. $\sqrt{4325} \approx 66$

51. $\sqrt{1175} \approx 34$

53. $\sqrt{9501} \approx 97$

55. $\sqrt[3]{7814} \approx 20$

57. $\sqrt{1000} \approx 32$

59. $7\sqrt{2} + 14\sqrt{2} = (7 + 14)\sqrt{2} = 21\sqrt{2}$

61. $17\sqrt{7} - 9\sqrt{7} = (17 - 9)\sqrt{7} = 8\sqrt{7}$

63. $4\sqrt[3]{2} + 7\sqrt[3]{2} = 11\sqrt[3]{2}$

65. $9\sqrt[3]{7} + 2\sqrt[3]{5} - 6\sqrt[3]{7} = 3\sqrt[3]{7} + 2\sqrt[3]{5}$

67. $8\sqrt{2} - 4\sqrt{3} - 9\sqrt{2} + 6\sqrt{3} =$
$(8 - 9)\sqrt{2} + (-4 + 6)\sqrt{3} =$
$-\sqrt{2} + 2\sqrt{3}$

69. $6\sqrt{7} + 5\sqrt{10} - 8\sqrt{10} - 4\sqrt{7} - 11\sqrt{7} + \sqrt{10} =$
$(6 - 4 - 11)\sqrt{7} + (5 - 8 + 1)\sqrt{10} =$
$-9\sqrt{7} - 2\sqrt{10}$

71. $9\sqrt{3} + \sqrt{3} = 10\sqrt{3} \approx$
$10(1.732) \approx 17.3$

73. $9\sqrt{5} - 3\sqrt{5} = 6\sqrt{5} \approx$
$6(2.236) \approx 13.4$

75. $14\sqrt{2} - 15\sqrt{2} = -\sqrt{2} \approx$
$-(1.414) \approx -1.4$

77. $8\sqrt{7} - 4\sqrt{7} + 6\sqrt{7} = 10\sqrt{7} \approx$
$10(2.646) \approx 26.5$

79. $4\sqrt{3} - 2\sqrt{2} \approx$
$4(1.732) - 2(1.414) \approx$
$6.928 - 2.828 \approx 4.1$

81. $9\sqrt{6} - 3\sqrt{5} + 2\sqrt{6} - 7\sqrt{5} - \sqrt{6} =$
$10\sqrt{6} - 10\sqrt{5} \approx 10(2.449) - 10(2.236) \approx$
$24.49 - 22.36 \approx 2.1$

83. $4\sqrt{11} - 5\sqrt{11} - 7\sqrt{11} + 2\sqrt{11} - 3\sqrt{11} =$
$-9\sqrt{11} \approx -29.8$

85. $L = 2:$ $T = 2\pi\sqrt{\dfrac{L}{32}} = 2(3.14)\sqrt{\dfrac{2}{32}}$
$= 6.28(0.25) = 1.6$ seconds

$L = 3.5:$ $T = 2\pi\sqrt{\dfrac{L}{32}} = 2(3.14)\sqrt{\dfrac{3.5}{32}}$
$= 6.28(0.33) = 2.1$ seconds

$L = 4:$ $T = 2\pi\sqrt{\dfrac{L}{32}} = 2(3.14)\sqrt{\dfrac{4}{32}}$
$= 6.28(0.35) = 2.2$ seconds

87. $d = 75:$ $T = \sqrt{\dfrac{d}{16}} = \sqrt{\dfrac{75}{16}}$
$= \sqrt{4.6875} = 2.2$ seconds

$d = 125:$ $T = \sqrt{\dfrac{d}{16}} = \sqrt{\dfrac{125}{16}}$
$= \sqrt{7.8125} = 2.8$ seconds

$d = 5280:$ $T = \sqrt{\dfrac{d}{16}} = \sqrt{\dfrac{5280}{16}}$
$= \sqrt{330} = 18.2$ seconds

PROBLEM SET **9.2** **Simplifying Radicals**

1. $\sqrt{24} = \sqrt{4}\sqrt{6} = 2\sqrt{6}$

3. $\sqrt{18} = \sqrt{9}\sqrt{2} = 3\sqrt{2}$

5. $\sqrt{27} = \sqrt{9}\sqrt{3} = 3\sqrt{3}$

7. $\sqrt{40} = \sqrt{4}\sqrt{10} = 2\sqrt{10}$

9. $\sqrt[3]{-54} = \sqrt[3]{-27}\sqrt[3]{2} = -3\sqrt[3]{2}$

11. $\sqrt{80} = \sqrt{16}\sqrt{5} = 4\sqrt{5}$

13. $\sqrt{117} = \sqrt{9}\sqrt{13} = 3\sqrt{13}$

15. $4\sqrt{72} = 4\sqrt{36}\sqrt{2} = 4(6)\sqrt{2} = 24\sqrt{2}$

17. $3\sqrt[3]{40} = 3\sqrt[3]{8}\sqrt[3]{5} = 3(2)\sqrt[3]{5} = 6\sqrt[3]{5}$

19. $-5\sqrt{20} = -5\sqrt{4}\sqrt{5} =$
$-5(2)\sqrt{5} = -10\sqrt{5}$

21. $-8\sqrt{96} = -8\sqrt{16}\sqrt{6} =$
$-8(4)\sqrt{6} = -32\sqrt{6}$

23. $\dfrac{3}{2}\sqrt{8} = \dfrac{3}{2}\sqrt{4}\sqrt{2} = \dfrac{3}{2}(2)\sqrt{2} = 3\sqrt{2}$

25. $\dfrac{3}{4}\sqrt{12} = \dfrac{3}{4}\sqrt{4}\sqrt{3} = \dfrac{3}{4}(2)\sqrt{3} = \dfrac{3}{2}\sqrt{3}$

27. $-\dfrac{2}{3}\sqrt{45} = -\dfrac{2}{3}\sqrt{9}\sqrt{5} =$
$-\dfrac{2}{3}(3)\sqrt{5} = -2\sqrt{5}$

29. $-\dfrac{1}{4}\sqrt[3]{32} = -\dfrac{1}{4}\sqrt[3]{8}\sqrt[3]{4} =$
$-\dfrac{1}{4}(2)\sqrt[3]{4} = -\dfrac{1}{2}\sqrt[3]{4}$

31. $\sqrt{x^2y^3} = \sqrt{x^2y^2}\sqrt{y} = xy\sqrt{y}$

263

Problem Set 9.2

33. $\sqrt{2x^2y} = \sqrt{x^2}\sqrt{2y} = x\sqrt{2y}$

35. $\sqrt{8x^2} = \sqrt{4x^2}\sqrt{2} = 2x\sqrt{2}$

37. $\sqrt{27a^3b} = \sqrt{9a^2}\sqrt{3ab} = 3a\sqrt{3ab}$

39. $\sqrt[3]{64x^4y^2} = \sqrt[3]{64x^3}\sqrt[3]{xy^2} = 4x\sqrt[3]{xy^2}$

41. $\sqrt{63x^4y^2} = \sqrt{9x^4y^2}\sqrt{7} = 3x^2y\sqrt{7}$

43. $3\sqrt{48x^2} = 3\sqrt{16x^2}\sqrt{3} =$
$3(4x)\sqrt{3} = 12x\sqrt{3}$

45. $-6\sqrt{72x^7} = -6\sqrt{36x^6}\sqrt{2x} =$
$-6(6x^3)\sqrt{2x} = -36x^3\sqrt{2x}$

47. $\frac{2}{9}\sqrt{54xy} = \frac{2}{9}\sqrt{9}\sqrt{6xy} =$
$\frac{2}{9}(3)\sqrt{6xy} = \frac{2}{3}\sqrt{6xy}$

49. $\frac{1}{8}\sqrt[3]{250x^4} = \frac{1}{8}\sqrt[3]{125x^3}\sqrt[3]{2x} =$
$\frac{1}{8}(5x)\sqrt[3]{2x} = \frac{5}{8}x\sqrt[3]{2x}$

51. $-\frac{2}{3}\sqrt{169a^8} = -\frac{2}{3}(13a^4) = -\frac{26}{3}a^4$

53. $7\sqrt{32} + 5\sqrt{2} =$
$7\left(\sqrt{16}\sqrt{2}\right) + 5\sqrt{2} =$
$7(4)\sqrt{2} + 5\sqrt{2} =$
$28\sqrt{2} + 5\sqrt{2} = 33\sqrt{2}$

55. $4\sqrt{45} - 9\sqrt{5} =$
$4\sqrt{9}\sqrt{5} - 9\sqrt{5} =$
$4(3)\sqrt{5} - 9\sqrt{5} =$
$12\sqrt{5} - 9\sqrt{5} = 3\sqrt{5}$

57. $2\sqrt[3]{54} + 6\sqrt[3]{16} =$
$2\sqrt[3]{27}\sqrt[3]{2} + 6\sqrt[3]{8}\sqrt[3]{2} =$
$2(3)\sqrt[3]{2} + 6(2)\sqrt[3]{2} =$
$6\sqrt[3]{2} + 12\sqrt[3]{2} = 18\sqrt[3]{2}$

59. $4\sqrt{63} - 7\sqrt{28} =$
$4\sqrt{9}\sqrt{7} - 7\sqrt{4}\sqrt{7} =$
$4(3)\sqrt{7} - 7(2)\sqrt{7} =$
$12\sqrt{7} - 14\sqrt{7} = -2\sqrt{7}$

61. $5\sqrt{12} + 3\sqrt{27} - 2\sqrt{75} =$
$5\sqrt{4}\sqrt{3} + 3\sqrt{9}\sqrt{3} - 2\sqrt{25}\sqrt{3} =$
$5(2)\sqrt{3} + 3(3)\sqrt{3} - 2(5)\sqrt{3} =$
$10\sqrt{3} + 9\sqrt{3} - 10\sqrt{3} = 9\sqrt{3}$

63. $\frac{1}{2}\sqrt{20} + \frac{2}{3}\sqrt{45} - \frac{1}{4}\sqrt{80} =$
$\frac{1}{2}\sqrt{4}\sqrt{5} + \frac{2}{3}\sqrt{9}\sqrt{5} - \frac{1}{4}\sqrt{16}\sqrt{5} =$
$\frac{1}{2}(2)\sqrt{5} + \frac{2}{3}(3)\sqrt{5} - \frac{1}{4}(4)\sqrt{5} =$
$\sqrt{5} + 2\sqrt{5} - \sqrt{5} = 2\sqrt{5}$

65. $3\sqrt{8} - 5\sqrt{20} - 7\sqrt{18} - 9\sqrt{125} =$
$3\sqrt{4}\sqrt{2} - 5\sqrt{4}\sqrt{5} - 7\sqrt{9}\sqrt{2} - 9\sqrt{25}\sqrt{5} =$
$3(2)\sqrt{2} - 5(2)\sqrt{5} - 7(3)\sqrt{2} - 9(5)\sqrt{5} =$
$6\sqrt{2} - 10\sqrt{5} - 21\sqrt{2} - 45\sqrt{5} =$
$-15\sqrt{2} - 55\sqrt{5}$

69a. $\sqrt{162} = \sqrt{9}\sqrt{18} = 3\sqrt{9}\sqrt{2} = 3\cdot 3\sqrt{2} = 9\sqrt{2}$
  b. $\sqrt{279} = \sqrt{9}\sqrt{31} = 3\sqrt{31}$
  c. $\sqrt{275} = \sqrt{25}\sqrt{11} = 5\sqrt{11}$
  d. $\sqrt{212} = \sqrt{4}\sqrt{53} = 2\sqrt{53}$

71a. $3\sqrt{10} - 4\sqrt{24} + 6\sqrt{65} =$
  $3(3.162) - 4(4.899) + 6(8.062) =$
  $9.486 - 19.596 + 48.372 = 38.262$

  b. $9\sqrt{27} + 5\sqrt{37} - 3\sqrt{80} =$
  $9(5.196) + 5(6.083) - 3(8.944) =$
  $46.764 + 30.415 - 26.832 = 50.347$

  c. $12\sqrt{5} + 13\sqrt{18} + 9\sqrt{47} =$
  $12(2.236) + 13(4.243) + 9(6.856) =$
  $26.832 + 55.159 + 61.704 = 143.695$

  d. $3\sqrt{98} - 4\sqrt{83} - 7\sqrt{120} =$
  $3(9.899) - 4(9.110) - 7(10.954) =$
  $29.697 - 36.440 - 76.678 = -83.421$

264

e. $4\sqrt{170} + 2\sqrt{198} + 5\sqrt{227} =$
$4(13.038) + 2(14.071) + 5(15.067) =$
$52.152 + 28.142 + 75.335 = 155.629$

f. $-3\sqrt{256} - 6\sqrt{287} + 11\sqrt{321} =$
$-3(16) - 6(16.941) + 11(17.916) =$
$-48 - 101.646 + 197.076 = 47.430$

PROBLEM SET **9.3** **More on Simplifying Radicals**

1. $\sqrt{\dfrac{16}{25}} = \dfrac{\sqrt{16}}{\sqrt{25}} = \dfrac{4}{5}$

3. $-\sqrt{\dfrac{81}{9}} = -\sqrt{9} = -3$

5. $\sqrt{\dfrac{1}{64}} = \dfrac{\sqrt{1}}{\sqrt{64}} = \dfrac{1}{8}$

7. $\sqrt[3]{\dfrac{125}{64}} = \dfrac{\sqrt[3]{125}}{\sqrt[3]{64}} = \dfrac{5}{4}$

9. $-\sqrt{\dfrac{25}{256}} = -\dfrac{\sqrt{25}}{\sqrt{256}} = -\dfrac{5}{16}$

11. $\sqrt{\dfrac{19}{25}} = \dfrac{\sqrt{19}}{\sqrt{25}} = \dfrac{\sqrt{19}}{5}$

13. $\sqrt{\dfrac{8}{49}} = \dfrac{\sqrt{8}}{\sqrt{49}} = \dfrac{\sqrt{4}\sqrt{2}}{7} = \dfrac{2\sqrt{2}}{7}$

15. $\dfrac{\sqrt[3]{375}}{\sqrt[3]{216}} = \dfrac{\sqrt[3]{125}\sqrt[3]{3}}{\sqrt[3]{216}} = \dfrac{5\sqrt[3]{3}}{6}$

17. $\dfrac{\sqrt{12}}{\sqrt{36}} = \dfrac{\sqrt{4}\sqrt{3}}{6} = \dfrac{2\sqrt{3}}{6} = \dfrac{\sqrt{3}}{3}$

19. $\sqrt{\dfrac{3}{2}} = \dfrac{\sqrt{3}}{\sqrt{2}} \cdot \dfrac{\sqrt{2}}{\sqrt{2}} = \dfrac{\sqrt{6}}{2}$

21. $\sqrt{\dfrac{5}{8}} = \dfrac{\sqrt{5}}{\sqrt{8}} \cdot \dfrac{\sqrt{2}}{\sqrt{2}} = \dfrac{\sqrt{10}}{\sqrt{16}} = \dfrac{\sqrt{10}}{4}$

23. $\dfrac{\sqrt{56}}{\sqrt{8}} = \sqrt{\dfrac{56}{8}} = \sqrt{7}$

25. $\dfrac{\sqrt{63}}{\sqrt{7}} = \sqrt{\dfrac{63}{7}} = \sqrt{9} = 3$

27. $\dfrac{\sqrt{5}}{\sqrt{18}} = \dfrac{\sqrt{5}}{3\sqrt{2}} \cdot \dfrac{\sqrt{2}}{\sqrt{2}} = \dfrac{\sqrt{10}}{6}$

29. $\dfrac{\sqrt{4}}{\sqrt{27}} = \dfrac{2}{3\sqrt{3}} \cdot \dfrac{\sqrt{3}}{\sqrt{3}} = \dfrac{2\sqrt{3}}{9}$

31. $\sqrt{\dfrac{1}{24}} = \dfrac{\sqrt{1}}{\sqrt{24}} = \dfrac{1}{2\sqrt{6}} \cdot \dfrac{\sqrt{6}}{\sqrt{6}} = \dfrac{\sqrt{6}}{12}$

33. $\dfrac{2\sqrt{3}}{\sqrt{5}} = \dfrac{2\sqrt{3}}{\sqrt{5}} \cdot \dfrac{\sqrt{5}}{\sqrt{5}} = \dfrac{2\sqrt{15}}{5}$

35. $\dfrac{4\sqrt{2}}{3\sqrt{3}} = \dfrac{4\sqrt{2}}{3\sqrt{3}} \cdot \dfrac{\sqrt{3}}{\sqrt{3}} = \dfrac{4\sqrt{6}}{9}$

37. $\dfrac{3\sqrt{7}}{4\sqrt{12}} = \dfrac{3\sqrt{7}}{8\sqrt{3}} \cdot \dfrac{\sqrt{3}}{\sqrt{3}} = \dfrac{3\sqrt{21}}{24} = \dfrac{\sqrt{21}}{8}$

39. $\sqrt{4\dfrac{1}{9}} = \sqrt{\dfrac{37}{9}} = \dfrac{\sqrt{37}}{\sqrt{9}} = \dfrac{\sqrt{37}}{3}$

41. $\dfrac{3}{\sqrt{x}} = \dfrac{3}{\sqrt{x}} \cdot \dfrac{\sqrt{x}}{\sqrt{x}} = \dfrac{3\sqrt{x}}{x}$

43. $\dfrac{5}{\sqrt{2x}} = \dfrac{5}{\sqrt{2x}} \cdot \dfrac{\sqrt{2x}}{\sqrt{2x}} = \dfrac{5\sqrt{2x}}{2x}$

45. $\sqrt{\dfrac{3}{x}} = \dfrac{\sqrt{3}}{\sqrt{x}} = \dfrac{\sqrt{3}}{\sqrt{x}} \cdot \dfrac{\sqrt{x}}{\sqrt{x}} = \dfrac{\sqrt{3x}}{x}$

47. $\sqrt{\dfrac{12}{x^2}} = \dfrac{2\sqrt{3}}{\sqrt{x^2}} = \dfrac{2\sqrt{3}}{x}$

Problem Set 9.3

49. $\dfrac{\sqrt{2x}}{\sqrt{5y}} = \dfrac{\sqrt{2x}}{\sqrt{5y}} \cdot \dfrac{\sqrt{5y}}{\sqrt{5y}} = \dfrac{\sqrt{10xy}}{5y}$

51. $\dfrac{\sqrt{5x}}{\sqrt{27y}} = \dfrac{\sqrt{5x}}{3\sqrt{3y}} \cdot \dfrac{\sqrt{3y}}{\sqrt{3y}} = \dfrac{\sqrt{15xy}}{9y}$

53. $\dfrac{\sqrt{2x^3}}{\sqrt{8y}} = \dfrac{x\sqrt{2x}}{2\sqrt{2y}} \cdot \dfrac{\sqrt{2y}}{\sqrt{2y}} = \dfrac{x\sqrt{4xy}}{4y} =$

$\dfrac{2x\sqrt{xy}}{4y} = \dfrac{x\sqrt{xy}}{2y}$

55. $\sqrt{\dfrac{9}{x^3}} = \dfrac{\sqrt{9}}{x\sqrt{x}} \cdot \dfrac{\sqrt{x}}{\sqrt{x}} = \dfrac{3\sqrt{x}}{x^2}$

57. $\dfrac{4}{\sqrt{x^7}} = \dfrac{4}{x^3\sqrt{x}} \cdot \dfrac{\sqrt{x}}{\sqrt{x}} = \dfrac{4\sqrt{x}}{x^4}$

59. $\dfrac{3\sqrt{x}}{2\sqrt{y^3}} = \dfrac{3\sqrt{x}}{2y\sqrt{y}} \cdot \dfrac{\sqrt{y}}{\sqrt{y}} = \dfrac{3\sqrt{xy}}{2y^2}$

61. $7\sqrt{3} + \sqrt{\dfrac{1}{3}} = 7\sqrt{3} + \dfrac{\sqrt{1}}{\sqrt{3}} \cdot \dfrac{\sqrt{3}}{\sqrt{3}} =$

$7\sqrt{3} + \dfrac{\sqrt{3}}{3} = \left(7 + \dfrac{1}{3}\right)\sqrt{3} =$

$\left(\dfrac{21}{3} + \dfrac{1}{3}\right)\sqrt{3} = \dfrac{22}{3}\sqrt{3}$

63. $4\sqrt{10} - \sqrt{\dfrac{2}{5}} = 4\sqrt{10} - \dfrac{\sqrt{2}}{\sqrt{5}} \cdot \dfrac{\sqrt{5}}{\sqrt{5}} =$

$4\sqrt{10} - \dfrac{\sqrt{10}}{5} = \left(4 - \dfrac{1}{5}\right)\sqrt{10} =$

$\left(\dfrac{20}{5} - \dfrac{1}{5}\right)\sqrt{10} = \dfrac{19}{5}\sqrt{10}$

65. $-2\sqrt{5} - 5\sqrt{\dfrac{1}{5}} = -2\sqrt{5} - \dfrac{5\sqrt{1}}{\sqrt{5}} \cdot \dfrac{\sqrt{5}}{\sqrt{5}} =$

$-2\sqrt{5} - \dfrac{5\sqrt{5}}{5} = -2\sqrt{5} - \sqrt{5} =$

$-3\sqrt{5}$

67. $-3\sqrt{6} - \dfrac{5\sqrt{2}}{\sqrt{3}} = -3\sqrt{6} - \dfrac{5\sqrt{2}}{\sqrt{3}} \cdot \dfrac{\sqrt{3}}{\sqrt{3}} =$

$-3\sqrt{6} - \dfrac{5\sqrt{6}}{3} = \left(-3 - \dfrac{5}{3}\right)\sqrt{6} =$

$\left(-\dfrac{9}{3} - \dfrac{5}{3}\right)\sqrt{6} = -\dfrac{14}{3}\sqrt{6}$

69. $4\sqrt{12} + \dfrac{3}{\sqrt{3}} - 5\sqrt{27} =$

$8\sqrt{3} + \dfrac{3}{\sqrt{3}} \cdot \dfrac{\sqrt{3}}{\sqrt{3}} - 15\sqrt{3} =$

$8\sqrt{3} + \dfrac{3\sqrt{3}}{3} - 15\sqrt{3} =$

$8\sqrt{3} + \sqrt{3} - 15\sqrt{3} =$

$(8 + 1 - 15)\sqrt{3} = -6\sqrt{3}$

71. $\dfrac{9\sqrt{5}}{\sqrt{3}} - 6\sqrt{60} + \dfrac{10\sqrt{3}}{\sqrt{5}} =$

$\dfrac{9\sqrt{5}}{\sqrt{3}} \cdot \dfrac{\sqrt{3}}{\sqrt{3}} - 12\sqrt{15} + \dfrac{10\sqrt{3}}{\sqrt{5}} \cdot \dfrac{\sqrt{5}}{\sqrt{5}} =$

$\dfrac{9\sqrt{15}}{3} - 12\sqrt{15} + \dfrac{10\sqrt{15}}{5} =$

$3\sqrt{15} - 12\sqrt{15} + 2\sqrt{15} = -7\sqrt{15}$

75. $\dfrac{\sqrt[3]{7}}{\sqrt[3]{3}} = \dfrac{\sqrt[3]{7}}{\sqrt[3]{3}} \cdot \dfrac{\sqrt[3]{9}}{\sqrt[3]{9}} = \dfrac{\sqrt[3]{63}}{\sqrt[3]{27}} = \dfrac{\sqrt[3]{63}}{3}$

77. $\sqrt[3]{\dfrac{8}{25}} = \dfrac{\sqrt[3]{8}}{\sqrt[3]{25}} \cdot \dfrac{\sqrt[3]{5}}{\sqrt[3]{5}} = \dfrac{2\sqrt[3]{5}}{\sqrt[3]{125}} = \dfrac{2\sqrt[3]{5}}{5}$

79. $\sqrt[3]{\dfrac{7}{32}} = \dfrac{\sqrt[3]{7}}{\sqrt[3]{32}} \cdot \dfrac{\sqrt[3]{2}}{\sqrt[3]{2}} = \dfrac{\sqrt[3]{14}}{\sqrt[3]{64}} = \dfrac{\sqrt[3]{14}}{4}$

PROBLEM SET **9.4**   **Products and Quotients Involving Radicals**

1. $\sqrt{7}\sqrt{5} = \sqrt{35}$

3. $\sqrt{6}\sqrt{8} = \sqrt{48} = \sqrt{16}\sqrt{3} = 4\sqrt{3}$

5. $\sqrt{5}\sqrt{10} = \sqrt{50} = \sqrt{25}\sqrt{2} = 5\sqrt{2}$

7. $\sqrt[3]{9}\sqrt[3]{6} = \sqrt[3]{54} = \sqrt[3]{27}\sqrt[3]{2} = 3\sqrt[3]{2}$

9. $\sqrt{8}\sqrt{12} = \sqrt{96} = \sqrt{16}\sqrt{6} = 4\sqrt{6}$

11. $\left(3\sqrt{3}\right)\left(5\sqrt{7}\right) =$
$3\cdot5\cdot\sqrt{3}\cdot\sqrt{7} =$
$15\sqrt{21}$

13. $\left(-\sqrt[3]{6}\right)\left(5\sqrt[3]{4}\right) = -5\sqrt[3]{24} =$
$-5\sqrt[3]{8}\sqrt[3]{3} = -5(2)\sqrt[3]{3} = -10\sqrt[3]{3}$

15. $\left(3\sqrt{6}\right)\left(4\sqrt{6}\right) = 3\cdot4\cdot\sqrt{6}\cdot\sqrt{6} =$
$12\sqrt{36} = 12\cdot6 = 72$

17. $\left(5\sqrt{2}\right)\left(4\sqrt{12}\right) = 5\cdot4\cdot\sqrt{2}\cdot\sqrt{12} =$
$20\sqrt{24} = 20\sqrt{4}\sqrt{6} = 40\sqrt{6}$

19. $\left(4\sqrt{3}\right)\left(2\sqrt{15}\right) = 4\cdot2\cdot\sqrt{3}\cdot\sqrt{15} =$
$8\sqrt{45} = 8\sqrt{9}\sqrt{5} = 8\cdot3\cdot\sqrt{5} = 24\sqrt{5}$

21. $\sqrt{2}\left(\sqrt{3}+\sqrt{5}\right) = \sqrt{2}\sqrt{3}+\sqrt{2}\sqrt{5} =$
$\sqrt{6}+\sqrt{10}$

23. $\sqrt{6}\left(\sqrt{2}-5\right) = \sqrt{6}\sqrt{2}-\sqrt{6}(5) =$
$\sqrt{12}-5\sqrt{6} = \sqrt{4}\sqrt{3}-5\sqrt{6} =$
$2\sqrt{3}-5\sqrt{6}$

25. $\sqrt[3]{2}\left(\sqrt[3]{4}+\sqrt[3]{10}\right) = \sqrt[3]{8}+\sqrt[3]{20} =$
$2+\sqrt[3]{20}$

27. $\sqrt{12}\left(\sqrt{6}-\sqrt{8}\right) =$
$\sqrt{12}\sqrt{6}-\sqrt{12}\sqrt{8} =$

$\sqrt{72}-\sqrt{96} =$
$\sqrt{36}\sqrt{2}-\sqrt{16}\sqrt{6} =$
$6\sqrt{2}-4\sqrt{6}$

29. $4\sqrt{3}\left(\sqrt{2}-2\sqrt{5}\right) =$
$4\sqrt{3}\left(\sqrt{2}\right)-4\sqrt{3}\left(2\sqrt{5}\right) =$
$4\sqrt{6}-8\sqrt{15}$

31. $\left(\sqrt{2}+6\right)\left(\sqrt{2}+9\right) =$
$\sqrt{2}\left(\sqrt{2}+9\right)+6\left(\sqrt{2}+9\right) =$
$2+9\sqrt{2}+6\sqrt{2}+54 =$
$56+15\sqrt{2}$

33. $\left(\sqrt{6}-5\right)\left(\sqrt{6}+3\right) =$
$\sqrt{6}\left(\sqrt{6}+3\right)-5\left(\sqrt{6}+3\right) =$
$6+3\sqrt{6}-5\sqrt{6}-15 =$
$-9-2\sqrt{6}$

35. $\left(\sqrt{3}+\sqrt{6}\right)\left(\sqrt{6}+\sqrt{8}\right) =$
$\sqrt{3}\left(\sqrt{6}+\sqrt{8}\right)+\sqrt{6}\left(\sqrt{6}+\sqrt{8}\right) =$
$\sqrt{18}+\sqrt{24}+\sqrt{36}+\sqrt{48} =$
$\sqrt{9}\sqrt{2}+\sqrt{4}\sqrt{6}+6+\sqrt{16}\sqrt{3} =$
$3\sqrt{2}+2\sqrt{6}+6+4\sqrt{3}$

37. $\left(5+\sqrt{10}\right)\left(5-\sqrt{10}\right) =$
$5^2-\left(\sqrt{10}\right)^2 = 25-10 = 15$

39. $\left(3\sqrt{2}-\sqrt{3}\right)\left(3\sqrt{2}+\sqrt{3}\right) =$
$\left(3\sqrt{2}\right)^2-\left(\sqrt{3}\right)^2 = 9(2)-3 =$
$18-3 = 15$

41. $\left(5\sqrt{3}+2\sqrt{6}\right)\left(5\sqrt{3}-2\sqrt{6}\right) =$
$\left(5\sqrt{3}\right)^2-\left(2\sqrt{6}\right)^2 = 25(3)-4(6) =$
$75-24 = 51$

Problem Set 9.4

43. $\sqrt{xy}\sqrt{x} = \sqrt{x^2y} = \sqrt{x^2}\sqrt{y} = x\sqrt{y}$

45. $\sqrt[3]{25x^2}\sqrt[3]{5x} = \sqrt[3]{125x^3} = 5x$

47. $(4\sqrt{a})(3\sqrt{ab}) = 12\sqrt{a^2b} = 12a\sqrt{b}$

49. $\sqrt{2x}\left(\sqrt{3x} - \sqrt{6y}\right) =$
$\sqrt{2x}\left(\sqrt{3x}\right) - \sqrt{2x}\left(\sqrt{6y}\right) =$
$\sqrt{6x^2} - \sqrt{12xy} =$
$x\sqrt{6} - 2\sqrt{3xy}$

51. $(\sqrt{x}+5)(\sqrt{x}-3) =$
$\sqrt{x}(\sqrt{x}-3) + 5(\sqrt{x}-3) =$
$x - 3\sqrt{x} + 5\sqrt{x} - 15 =$
$x + 2\sqrt{x} - 15$

53. $(\sqrt{x}+7)(\sqrt{x}-7) =$
$(\sqrt{x})^2 - 7^2 = x - 49$

55. $\dfrac{3}{\sqrt{2}+4} = \dfrac{3}{\sqrt{2}+4} \cdot \dfrac{\sqrt{2}-4}{\sqrt{2}-4} =$
$\dfrac{3(\sqrt{2}-4)}{2-16} = \dfrac{3\sqrt{2}-12}{-14} = \dfrac{-3\sqrt{2}+12}{14}$

57. $\dfrac{8}{\sqrt{6}-2} = \dfrac{8}{\sqrt{6}-2} \cdot \dfrac{\sqrt{6}+2}{\sqrt{6}+2} =$
$\dfrac{8(\sqrt{6}+2)}{6-4} = \dfrac{8(\sqrt{6}+2)}{2} =$
$4(\sqrt{6}+2) = 4\sqrt{6}+8$

59. $\dfrac{2}{\sqrt{5}+\sqrt{3}} = \dfrac{2}{\sqrt{5}+\sqrt{3}} \cdot \dfrac{\sqrt{5}-\sqrt{3}}{\sqrt{5}-\sqrt{3}} =$
$\dfrac{2(\sqrt{5}-\sqrt{3})}{5-3} = \dfrac{2(\sqrt{5}-\sqrt{3})}{2} =$
$\sqrt{5} - \sqrt{3}$

61. $\dfrac{10}{2-3\sqrt{3}} = \dfrac{10}{2-3\sqrt{3}} \cdot \dfrac{2+3\sqrt{3}}{2+3\sqrt{3}} =$

$\dfrac{10\left(2+3\sqrt{3}\right)}{4-27} = \dfrac{20+30\sqrt{3}}{-23} =$
$\dfrac{-20-30\sqrt{3}}{23}$

63. $\dfrac{4}{\sqrt{x}-2} = \dfrac{4}{\sqrt{x}-2} \cdot \dfrac{\sqrt{x}+2}{\sqrt{x}+2} =$
$\dfrac{4(\sqrt{x}+2)}{x-4} = \dfrac{4\sqrt{x}+8}{x-4}$

65. $\dfrac{\sqrt{x}}{\sqrt{x}+3} = \dfrac{\sqrt{x}}{\sqrt{x}+3} \cdot \dfrac{\sqrt{x}-3}{\sqrt{x}-3} =$
$\dfrac{\sqrt{x}(\sqrt{x}-3)}{x-9} = \dfrac{x-3\sqrt{x}}{x-9}$

67. $\dfrac{\sqrt{a}+2}{\sqrt{a}-5} = \dfrac{\sqrt{a}+2}{\sqrt{a}-5} \cdot \dfrac{\sqrt{a}+5}{\sqrt{a}+5} =$
$\dfrac{(\sqrt{a}+2)(\sqrt{a}+5)}{a-25} =$
$\dfrac{a+5\sqrt{a}+2\sqrt{a}+10}{a-25} =$
$\dfrac{a+7\sqrt{a}+10}{a-25}$

69. $\dfrac{2+\sqrt{3}}{3-\sqrt{2}} = \dfrac{2+\sqrt{3}}{3-\sqrt{2}} \cdot \dfrac{3+\sqrt{2}}{3+\sqrt{2}} =$
$\dfrac{\left(2+\sqrt{3}\right)\left(3+\sqrt{2}\right)}{9-2} =$
$\dfrac{6+2\sqrt{2}+3\sqrt{3}+\sqrt{6}}{7}$

73. **55)** $\dfrac{3}{\sqrt{2}+4} \approx \dfrac{3}{1.414+4} \approx$
$\dfrac{3}{5.414} \approx 0.554;$
$\dfrac{-3\sqrt{2}+12}{14} \approx \dfrac{-3(1.414)+12}{14} =$
$\dfrac{7.758}{14} \approx 0.554$

268

56) $\dfrac{5}{\sqrt{3}+7} \approx \dfrac{5}{1.732+7} \approx$

$\dfrac{5}{8.732} \approx 0.572;$

$\dfrac{-5\sqrt{3}+35}{46} \approx \dfrac{-5(1.732)+35}{46} \approx$

$\dfrac{26.34}{46} \approx 0.573$

57) $\dfrac{8}{\sqrt{6}-2} \approx \dfrac{8}{2.449-2} \approx$

$\dfrac{8}{0.449} \approx 17.817;$

$4\sqrt{6}+8 \approx 4(2.449)+8 \approx$

$9.796+8 \approx 17.796$

58) $\dfrac{10}{3-\sqrt{7}} \approx \dfrac{10}{3-2.646} \approx$

$\dfrac{10}{-0.354} \approx -28.249;$

$15+5\sqrt{7} \approx 15+5(2.646) \approx$

$15+13.23 \approx 28.23$

59) $\dfrac{2}{\sqrt{5}+\sqrt{3}} \approx \dfrac{2}{2.236+1.732} \approx$

$\dfrac{2}{3.968} \approx 0.504;$

$\sqrt{5}-\sqrt{3} \approx 2.236-1.732 \approx$
0.504

60) $\dfrac{3}{\sqrt{6}+\sqrt{5}} \approx \dfrac{3}{2.449+2.236} \approx$

$\dfrac{3}{4.685} \approx 0.640;$
$3\sqrt{6}-3\sqrt{5} \approx$
$3(2.449)-3(2.236) \approx$
$7.347-6.708 \approx 0.639$

61) $\dfrac{10}{2-3\sqrt{3}} \approx \dfrac{10}{2-3(1.732)} \approx$

$\dfrac{10}{-3.196} \approx -3.129;$

$\dfrac{-20+30\sqrt{3}}{23} \approx \dfrac{-20-30(1.732)}{23} \approx$

$\dfrac{-71.96}{23} \approx -3.129$

62) $\dfrac{5}{3\sqrt{2}-4} \approx \dfrac{5}{3(1.414)-4} \approx$

$\dfrac{5}{0.242} \approx 20.661;$

$\dfrac{15\sqrt{2}+20}{2} \approx \dfrac{15(1.414)+20}{2} \approx$

$\dfrac{41.21}{2} \approx 20.605$

PROBLEM SET **9.5** **Solving Radical Equations**

1. $\sqrt{x}=7$
$\left(\sqrt{x}\right)^2 = 7^2$
$\qquad x = 49$
CHECK
$\sqrt{49} \overset{?}{=} 7$
$\quad 7 = 7$
The solution set is $\{49\}$.

3. $\sqrt{2x}=6$
$\left(\sqrt{2x}\right)^2 = 6^2$
$\qquad 2x = 36$
$\qquad x = 18$
CHECK
$\sqrt{2(18)} \overset{?}{=} 6$
$\quad \sqrt{36} \overset{?}{=} 6$
$\qquad 6 = 6$
The solution set is $\{18\}$.

Problem Set 9.5

5.
$$\sqrt{3x} = -6$$
$$\left(\sqrt{3x}\right)^2 = (-6)^2$$
$$3x = 36$$
$$x = 12$$
CHECK
$$\sqrt{2(12)} \overset{?}{=} -6$$
$$\sqrt{36} \overset{?}{=} -6$$
$$6 \neq -6$$
Since 12 does not check, the solution set is $\emptyset$.

7.
$$\sqrt{4x} = 3$$
$$\left(\sqrt{4x}\right)^2 = 3^2$$
$$4x = 9$$
$$x = \frac{9}{4}$$
CHECK
$$\sqrt{4\left(\frac{9}{4}\right)} \overset{?}{=} 3$$
$$\sqrt{9} \overset{?}{=} 3$$
$$3 = 3$$
The solution set is $\left\{\frac{9}{4}\right\}$.

9.
$$3\sqrt{x} = 2$$
$$\left(3\sqrt{x}\right)^2 = 2^2$$
$$9x = 4$$
$$x = \frac{4}{9}$$
CHECK
$$3\sqrt{\frac{4}{9}} \overset{?}{=} 2$$
$$3\left(\frac{2}{3}\right) \overset{?}{=} 2$$
$$2 = 2$$
The solution set is $\left\{\frac{4}{9}\right\}$.

11.
$$\sqrt{2n-3} = 5$$
$$\left(\sqrt{2n-3}\right)^2 = (5)^2$$
$$2n - 3 = 25$$
$$2n = 28$$
$$n = 14$$
CHECK
$$\sqrt{2(14)-3} \overset{?}{=} 5$$
$$\sqrt{25} \overset{?}{=} 5$$
$$5 = 5$$
The solution set is $\{14\}$.

13.
$$\sqrt{5y+2} = -1$$
$$\left(\sqrt{5y+2}\right)^2 = (-1)^2$$
$$5y + 2 = 1$$
$$5y = -1$$
$$y = -\frac{1}{5}$$
CHECK
$$\sqrt{5\left(-\frac{1}{5}\right)+2} \overset{?}{=} -1$$
$$\sqrt{1} \overset{?}{=} -1$$
$$1 \neq -1$$
Since $-\frac{1}{5}$ does not check, the solution set is $\emptyset$.

15.
$$\sqrt{6x-5} - 3 = 0$$
$$\sqrt{6x-5} = 3$$
$$\left(\sqrt{6x-5}\right)^2 = (3)^2$$
$$6x - 5 = 9$$
$$6x = 14$$
$$x = \frac{14}{6} = \frac{7}{3}$$
CHECK
$$\sqrt{6\left(\frac{7}{3}\right)-5} - 3 \overset{?}{=} 0$$
$$\sqrt{9} - 3 \overset{?}{=} 0$$
$$3 - 3 \overset{?}{=} 0$$
$$0 = 0$$
The solution set is $\left\{\frac{7}{3}\right\}$.

17.
$$5\sqrt{x} = 30$$
$$\sqrt{x} = 6$$
$$\left(\sqrt{x}\right)^2 = (6)^2$$
$$x = 36$$
CHECK
$$5\sqrt{36} \stackrel{?}{=} 30$$
$$5(6) \stackrel{?}{=} 30$$
$$30 = 30$$
The solution set is $\{36\}$.

19.
$$\sqrt{3a - 2} = \sqrt{2a + 4}$$
$$\left(\sqrt{3a - 2}\right)^2 = \left(\sqrt{2a + 4}\right)^2$$
$$3a - 2 = 2a + 4$$
$$a - 2 = 4$$
$$a = 6$$
CHECK
$$\sqrt{3(6) - 2} \stackrel{?}{=} \sqrt{2(6) + 4}$$
$$\sqrt{16} \stackrel{?}{=} \sqrt{16}$$
$$4 = 4$$
The solution set is $\{6\}$.

21.
$$\sqrt{7x - 3} = \sqrt{4x + 3}$$
$$\left(\sqrt{7x - 3}\right)^2 = \left(\sqrt{4x + 3}\right)^2$$
$$7x - 3 = 4x + 3$$
$$3x - 3 = 3$$
$$3x = 6$$
$$x = 2$$
CHECK
$$\sqrt{7(2) - 3} \stackrel{?}{=} \sqrt{4(2) + 3}$$
$$\sqrt{11} = \sqrt{11}$$
The solution set is $\{2\}$.

23.
$$2\sqrt{y + 1} = 5$$
$$\left(2\sqrt{y + 1}\right)^2 = (5)^2$$
$$4(y + 1) = 25$$
$$4y + 4 = 25$$
$$4y = 21$$
$$y = \frac{21}{4}$$

CHECK
$$2\sqrt{\frac{21}{4} + 1} \stackrel{?}{=} 5$$
$$2\sqrt{\frac{25}{4}} \stackrel{?}{=} 5$$
$$2\left(\frac{5}{2}\right) \stackrel{?}{=} 5$$
$$5 = 5$$
The solution set is $\left\{\dfrac{21}{4}\right\}$.

25.
$$\sqrt{x + 3} = x + 3$$
$$\left(\sqrt{x + 3}\right)^2 = (x + 3)^2$$
$$x + 3 = x^2 + 6x + 9$$
$$0 = x^2 + 5x + 6$$
$$0 = (x + 2)(x + 3)$$
$$x + 2 = 0 \quad \text{or} \quad x + 3 = 0$$
$$x = -2 \quad \text{or} \quad x = -3$$
CHECK
$$\sqrt{-2 + 3} \stackrel{?}{=} -2 + 3$$
$$\sqrt{1} = 1$$

$$\sqrt{-3 + 3} \stackrel{?}{=} -3 + 3$$
$$\sqrt{0} = 0$$
The solution set is $\{-3, \ -2\}$.

27.
$$\sqrt{-2x + 28} = x - 2$$
$$\left(\sqrt{-2x + 28}\right)^2 = (x - 2)^2$$
$$-2x + 28 = x^2 - 4x + 4$$
$$0 = x^2 - 2x - 24$$
$$0 = (x + 4)(x - 6)$$
$$x + 4 = 0 \quad \text{or} \quad x - 6 = 0$$
$$x = -4 \quad \text{or} \quad x = 6$$
CHECK
$$\sqrt{-2(-4) + 28} \stackrel{?}{=} -4 - 2$$
$$\sqrt{36} \stackrel{?}{=} -6$$
$$6 \neq -6$$
$$\sqrt{-2(6) + 28} \stackrel{?}{=} 6 - 2$$
$$\sqrt{16} \stackrel{?}{=} 4$$
$$4 = 4$$
The solution set is $\{6\}$.

271

Problem Set 9.5

29.
$$\sqrt{3n-4} = \sqrt{n}$$
$$\left(\sqrt{3n-4}\right)^2 = \left(\sqrt{n}\right)^2$$
$$3n - 4 = n$$
$$-4 = -2n$$
$$2 = n$$
CHECK
$$\sqrt{3(2)-4} \overset{?}{=} \sqrt{2}$$
$$\sqrt{2} = \sqrt{2}$$
The solution set is $\{2\}$.

31.
$$\sqrt{3x} = x - 6$$
$$\left(\sqrt{3x}\right)^2 = (x-6)^2$$
$$3x = x^2 - 12x + 36$$
$$0 = x^2 - 15x + 36$$
$$0 = (x - 12)(x - 3)$$
$$x - 12 = 0 \quad \text{or} \quad x - 3 = 0$$
$$x = 12 \quad \text{or} \quad x = 3$$
CHECK
$$\sqrt{3(12)} \overset{?}{=} 12 - 6$$
$$\sqrt{36} \overset{?}{=} 6$$
$$6 = 6$$

$$\sqrt{3(3)} \overset{?}{=} 3 - 6$$
$$\sqrt{9} \overset{?}{=} -3$$
$$3 \neq -3$$
The solution set is $\{12\}$.

33.
$$4\sqrt{x} + 5 = x$$
$$4\sqrt{x} = x - 5$$
$$\left(4\sqrt{x}\right)^2 = (x-5)^2$$
$$16x = x^2 - 10x + 25$$
$$0 = x^2 - 26x + 25$$
$$0 = (x - 25)(x - 1)$$
$$x - 25 = 0 \quad \text{or} \quad x - 1 = 0$$
$$x = 25 \quad \text{or} \quad x = 1$$

CHECK
$$4\sqrt{25} + 5 \overset{?}{=} 25$$
$$4(5) + 5 \overset{?}{=} 25$$
$$25 = 25$$

$$4\sqrt{1} + 5 \overset{?}{=} 1$$
$$9 \neq 1$$
The solution set is $\{25\}$.

35.
$$\sqrt{x^2 + 27} = x + 3$$
$$\left(\sqrt{x^2 + 27}\right)^2 = (x+3)^2$$
$$x^2 + 27 = x^2 + 6x + 9$$
$$18 = 6x$$
$$3 = x$$
CHECK
$$\sqrt{(3)^2 + 27} \overset{?}{=} 3 + 3$$
$$\sqrt{36} \overset{?}{=} 6$$
$$6 = 6$$
The solution set is $\{3\}$.

37.
$$\sqrt{x^2 + 2x + 3} = x + 2$$
$$\left(\sqrt{x^2 + 2x + 3}\right)^2 = (x+2)^2$$
$$x^2 + 2x + 3 = x^2 + 4x + 4$$
$$-2x = 1$$
$$x = -\frac{1}{2}$$
CHECK
$$\sqrt{\left(-\frac{1}{2}\right)^2 + 2\left(-\frac{1}{2}\right) + 3} \overset{?}{=} -\frac{1}{2} + 2$$
$$\sqrt{\frac{1}{4} - 1 + 3} \overset{?}{=} \frac{3}{2}$$
$$\sqrt{\frac{9}{4}} \overset{?}{=} \frac{3}{2}$$
$$\frac{3}{2} = \frac{3}{2}$$
The solution set is $\left\{-\frac{1}{2}\right\}$.

39. $\sqrt{8x} - 2 = x$

$\sqrt{8x} = x + 2$

$\left(\sqrt{8x}\right)^2 = (x+2)^2$

$8x = x^2 + 4x + 4$

$0 = x^2 - 4x + 4$

$0 = (x-2)(x-2)$

$x - 2 = 0$ or $x - 2 = 0$

$x = 2$ or $x = 2$

CHECK

$\sqrt{8(2)} - 2 \stackrel{?}{=} 2$

$\sqrt{16} - 2 \stackrel{?}{=} 2$

$4 - 2 \stackrel{?}{=} 2$

$2 = 2$

The solution set is $\{2\}$.

41. $S = 40: D = \dfrac{S^2}{30f} = \dfrac{40^2}{30(0.95)} =$

$\dfrac{1600}{28.5} = 56$ to the nearest foot.

The car will skid approximately 56 feet.

$S = 55: D = \dfrac{S^2}{30f} = \dfrac{55^2}{30(0.95)} =$

$\dfrac{3025}{28.5} = 106$ to the nearest foot.

The car will skid approximately 106 feet.

$S = 65: D = \dfrac{S^2}{30f} = \dfrac{65^2}{30(0.95)} =$

$\dfrac{4225}{28.5} = 148$ to the nearest foot.

The car will skid approximately 148 feet.

43. $T = 2: L = \dfrac{32T^2}{4\pi^2} = \dfrac{32(2)^2}{4(3.14)^2} =$

$\dfrac{128}{39.4384} = 3.2$ to the nearest tenth.

The length of the pendulum would
be approximately 3.2 feet.

$T = 2.5: L = \dfrac{32T^2}{4\pi^2} = \dfrac{32(2.5)^2}{4(3.14)^2} =$

$\dfrac{200}{39.4384} = 5.1$ to the nearest tenth.

The length of the pendulum would
be approximately 5.1 feet.

$T = 3: L = \dfrac{32T^2}{4\pi^2} = \dfrac{32(3)^2}{4(3.14)^2} =$

$\dfrac{288}{39.4384} = 7.3$ to the nearest tenth.

The length of the pendulum would
be approximately 7.3 feet.

47. $\sqrt{x-2} = \sqrt{x+7} - 1$

Square both sides.

$\left(\sqrt{x-2}\right)^2 = \left(\sqrt{x+7}-1\right)^2$

$x - 2 = (x+7) - 2\sqrt{x+7} + 1$

Combine like terms.

$x - 2 = x + 8 - 2\sqrt{x+7}$

Subtract $(x+8)$ from both sides.

$-10 = -2\sqrt{x+7}$

Square both sides.

$(-10)^2 = \left(-2\sqrt{x+7}\right)^2$

$100 = 4(x+7)$

$100 = 4x + 28$

$72 = 4x$

$18 = x$

CHECK

$\sqrt{18-2} \stackrel{?}{=} \sqrt{18+7} - 1$

$\sqrt{16} \stackrel{?}{=} \sqrt{25} - 1$

$4 = 4$

The solution set is $\{18\}$.

Problem Set 9.5

49. $\sqrt{2n+1} - \sqrt{n-3} = 2$

Add $\sqrt{n-3}$ to both sides.
$$\sqrt{2n+1} = 2 + \sqrt{n-3}$$
Square both sides.
$$\left(\sqrt{2n+1}\right)^2 = \left(2 + \sqrt{n-3}\right)^2$$
$$2n+1 = 4 + 4\sqrt{n-3} + (n-3)$$
Combine like terms.
$$2n+1 = n+1 + 4\sqrt{n-3}$$
Subtract $(n+1)$ from both sides.
$$n = 4\sqrt{n-3}$$
Square both sides.
$$(n)^2 = \left(4\sqrt{n-3}\right)^2$$
$$n^2 = 16(n-3)$$
$$n^2 = 16n - 48$$
$$n^2 - 16n + 48 = 0$$
$$(n-4)(n-12) = 0$$
$$n-4 = 0 \quad \text{or} \quad n-12 = 0$$
$$n = 4 \quad \text{or} \quad n = 12$$

CHECK
$$\sqrt{2(4)+1} - \sqrt{4-3} \overset{?}{=} 2$$
$$\sqrt{9} - \sqrt{1} \overset{?}{=} 2$$
$$3 - 1 = 2$$

$$\sqrt{2(12)+1} - \sqrt{12-3} \overset{?}{=} 2$$
$$\sqrt{25} - \sqrt{9} \overset{?}{=} 2$$
$$5 - 3 = 2$$
The solution set is $\{4,\ 12\}$.

51. $\sqrt[3]{x+7} = 4$
$$\left(\sqrt[3]{x+7}\right)^3 = 4^3$$
$$x+7 = 64$$
$$x = 57$$

CHECK
$$\sqrt[3]{57+7} \overset{?}{=} 4$$
$$\sqrt[3]{64} \overset{?}{=} 4$$
$$4 = 4$$
The solution set is $\{57\}$.

53. $\sqrt[3]{2x-4} = 6$
$$\left(\sqrt[3]{2x-4}\right)^3 = 6^3$$
$$2x-4 = 216$$
$$2x = 220$$
$$x = 110$$

CHECK
$$\sqrt[3]{2(110)-4} \overset{?}{=} 6$$
$$\sqrt[3]{220-4} \overset{?}{=} 6$$
$$\sqrt[3]{216} \overset{?}{=} 6$$
$$6 = 6$$
The solution set is $\{110\}$.

55. $\sqrt[3]{4x-8} = 10$
$$\left(\sqrt[3]{4x-8}\right)^3 = 10^3$$
$$4x-8 = 1000$$
$$4x = 1008$$
$$x = 252$$

CHECK
$$\sqrt[3]{4(252)-8} \overset{?}{=} 10$$
$$\sqrt[3]{1008-8} \overset{?}{=} 10$$
$$\sqrt[3]{1000} \overset{?}{=} 10$$
$$10 = 10$$
The solution set is $\{252\}$.

CHAPTER 9 **Review Problem Set**

1. $\sqrt{64} = 8$ because $8^2 = 64$.

2. $-\sqrt{49} = -7$ because $-(7)^2 = -49$.

3. $\sqrt{1600} = 40$ because $40^2 = 1600$.

4. $\sqrt{\dfrac{81}{25}} = \dfrac{9}{5}$ because $\left(\dfrac{9}{5}\right)^2 = \dfrac{81}{25}$.

5. $-\sqrt{\dfrac{4}{9}} = -\dfrac{2}{3}$ because $\left(\dfrac{2}{3}\right)^2 = \dfrac{4}{9}$.

6. $\sqrt{\dfrac{49}{36}} = \dfrac{7}{6}$ because $\left(\dfrac{7}{6}\right)^2 = \dfrac{49}{36}$.

7. $\sqrt{20} = \sqrt{4}\sqrt{5} = 2\sqrt{5}$

8. $\sqrt{32} = \sqrt{16}\sqrt{2} = 4\sqrt{2}$

9. $5\sqrt{8} = 5\sqrt{4}\sqrt{2} = 5 \cdot 2\sqrt{2} = 10\sqrt{2}$

10. $\sqrt{80} = \sqrt{16}\sqrt{5} = 4\sqrt{5}$

11. $2\sqrt[3]{-125} = 2(-5) = -10$

12. $\dfrac{\sqrt[3]{40}}{\sqrt[3]{8}} = \sqrt[3]{\dfrac{40}{8}} = \sqrt[3]{5}$

13. $\dfrac{\sqrt{36}}{\sqrt{7}} = \dfrac{6}{\sqrt{7}} \cdot \dfrac{\sqrt{7}}{\sqrt{7}} = \dfrac{6\sqrt{7}}{7}$

14. $\sqrt{\dfrac{7}{8}} = \dfrac{\sqrt{7}}{2\sqrt{2}} = \dfrac{\sqrt{7}}{2\sqrt{2}} \cdot \dfrac{\sqrt{2}}{\sqrt{2}} = \dfrac{\sqrt{14}}{4}$

15. $\sqrt{\dfrac{8}{24}} = \sqrt{\dfrac{1}{3}} = \dfrac{\sqrt{1}}{\sqrt{3}} \cdot \dfrac{\sqrt{3}}{\sqrt{3}} = \dfrac{\sqrt{3}}{3}$

16. $\dfrac{3\sqrt{2}}{\sqrt{5}} = \dfrac{3\sqrt{2}}{\sqrt{5}} \cdot \dfrac{\sqrt{5}}{\sqrt{5}} = \dfrac{3\sqrt{10}}{5}$

17. $\dfrac{4\sqrt{3}}{\sqrt{12}} = \dfrac{4\sqrt{3}}{\sqrt{4}\sqrt{3}} = \dfrac{4}{\sqrt{4}} = \dfrac{4}{2} = 2$

18. $\dfrac{5\sqrt{2}}{2\sqrt{3}} = \dfrac{5\sqrt{2}}{2\sqrt{3}} \cdot \dfrac{\sqrt{3}}{\sqrt{3}} = \dfrac{5\sqrt{6}}{6}$

19. $\dfrac{-3\sqrt{2}}{\sqrt{27}} = \dfrac{-3\sqrt{2}}{3\sqrt{3}} \cdot \dfrac{\sqrt{3}}{\sqrt{3}} = \dfrac{-3\sqrt{6}}{3\sqrt{9}} =$
     $-\dfrac{\sqrt{6}}{\sqrt{9}} = -\dfrac{\sqrt{6}}{3}$

20. $\dfrac{4\sqrt{6}}{3\sqrt{12}} = \dfrac{4\sqrt{1}}{3\sqrt{2}} \cdot \dfrac{\sqrt{2}}{\sqrt{2}} = \dfrac{4\sqrt{2}}{6} = \dfrac{2\sqrt{2}}{3}$

21. $\sqrt{27} = \sqrt{9}\sqrt{3} = 3\sqrt{3} \approx 3(1.73) \approx 5.2$

22. $\dfrac{2}{\sqrt{3}} = \dfrac{2}{\sqrt{3}} \cdot \dfrac{\sqrt{3}}{\sqrt{3}} = \dfrac{2\sqrt{3}}{3} \approx \dfrac{2(1.73)}{3} \approx 1.2$

23. $3\sqrt{12} + \sqrt{48} = 3\sqrt{4}\sqrt{3} + \sqrt{16}\sqrt{3} =$
     $3 \cdot 2\sqrt{3} + 4\sqrt{3} = 6\sqrt{3} + 4\sqrt{3} =$
     $10\sqrt{3} \approx 10(1.73) \approx 17.3$

24. $2\sqrt{27} - 2\sqrt{75} = 2\sqrt{9}\sqrt{3} - 2\sqrt{25}\sqrt{3} =$
     $2 \cdot 3\sqrt{3} - 2 \cdot 5\sqrt{3} = 6\sqrt{3} - 10\sqrt{3} =$
     $-4\sqrt{3} = -4(1.73) = -6.9$

25. $\sqrt{12a^2b^3} = \sqrt{4a^2b^2}\sqrt{3b} = 2ab\sqrt{3b}$

26. $\sqrt{50xy^4} = \sqrt{25y^4}\sqrt{2x} = 5y^2\sqrt{2x}$

27. $\sqrt{48x^3y^2} = \sqrt{16x^2y^2}\sqrt{3x} = 4xy\sqrt{3x}$

28. $\sqrt[3]{125a^2b} = \sqrt[3]{125}\sqrt[3]{a^2b} = 5\sqrt[3]{a^2b}$

29. $\dfrac{4}{3}\sqrt{27xy^2} = \dfrac{4}{3}\sqrt{9y^2}\sqrt{3x} =$
     $\dfrac{4}{3}(3y)\sqrt{3x} = 4y\sqrt{3x}$

Chapter 9 Review Problem Set

30. $\dfrac{3}{4}\left(\sqrt[3]{24x^3}\right) = \dfrac{3\left(\sqrt[3]{8x^3}\sqrt[3]{3}\right)}{4} =$

$\dfrac{3\left(2x\sqrt[3]{3}\right)}{4} = \dfrac{6x\sqrt[3]{3}}{4} = \dfrac{3x\sqrt[3]{3}}{2}$

31. $\dfrac{\sqrt{2x}}{\sqrt{5y}} = \dfrac{\sqrt{2x}}{\sqrt{5y}}\cdot\dfrac{\sqrt{5y}}{\sqrt{5y}} = \dfrac{\sqrt{10xy}}{5y}$

32. $\dfrac{\sqrt{72x}}{\sqrt{16y}} = \dfrac{6\sqrt{2x}}{4\sqrt{y}}\cdot\dfrac{\sqrt{y}}{\sqrt{y}} =$

$\dfrac{6\sqrt{2xy}}{4y} = \dfrac{3\sqrt{2xy}}{2y}$

33. $\sqrt{\dfrac{4}{x}} = \dfrac{\sqrt{4}}{\sqrt{x}}\cdot\dfrac{\sqrt{x}}{\sqrt{x}} = \dfrac{\sqrt{4x}}{x} = \dfrac{2\sqrt{x}}{x}$

34. $\sqrt{\dfrac{2x^3}{9}} = \dfrac{\sqrt{2x^3}}{\sqrt{9}} = \dfrac{\sqrt{x^2}\sqrt{2x}}{3} = \dfrac{x\sqrt{2x}}{3}$

35. $\dfrac{3\sqrt{x}}{4\sqrt{y^3}} = \dfrac{3\sqrt{x}}{4y\sqrt{y}}\cdot\dfrac{\sqrt{y}}{\sqrt{y}} = \dfrac{3\sqrt{xy}}{4y^2}$

36. $\dfrac{-2\sqrt{x^2y}}{5\sqrt{xy}} = \dfrac{-2\sqrt{x}}{5}$

37. $\left(\sqrt{6}\right)\left(\sqrt{12}\right) = \sqrt{72} = \sqrt{36}\sqrt{2} = 6\sqrt{2}$

38. $\left(2\sqrt{3}\right)\left(3\sqrt{6}\right) = 6\sqrt{18} = 6\sqrt{9}\sqrt{2} =$
$6\cdot3\sqrt{2} = 18\sqrt{2}$

39. $\left(-5\sqrt{8}\right)\left(2\sqrt{2}\right) = -10\sqrt{16} =$
$-10(4) = -40$

40. $\left(2\sqrt[3]{7}\right)\left(5\sqrt[3]{4}\right) = 10\sqrt[3]{28}$

41. $\sqrt[3]{2}\left(\sqrt[3]{3} + \sqrt[3]{4}\right) = \sqrt[3]{6} + \sqrt[3]{8} = \sqrt[3]{6} + 2$

42. $3\sqrt{5}\left(\sqrt{8} - 2\sqrt{12}\right) =$
$3\sqrt{5}\sqrt{8} - 3\sqrt{5}\left(2\sqrt{12}\right) =$
$3\sqrt{40} - 6\sqrt{60} =$
$3\sqrt{4}\sqrt{10} - 6\sqrt{4}\sqrt{15} =$
$6\sqrt{10} - 12\sqrt{15}$

43. $\left(\sqrt{3} + \sqrt{5}\right)\left(\sqrt{3} + \sqrt{7}\right) =$
$\sqrt{3}\left(\sqrt{3} + \sqrt{7}\right) + \sqrt{5}\left(\sqrt{3} + \sqrt{7}\right) =$
$\sqrt{9} + \sqrt{21} + \sqrt{15} + \sqrt{35} =$
$3 + \sqrt{21} + \sqrt{15} + \sqrt{35}$

44. $\left(2\sqrt{3} + 3\sqrt{2}\right)\left(\sqrt{3} - 5\sqrt{2}\right) =$
$2\sqrt{3}\left(\sqrt{3} - 5\sqrt{2}\right) + 3\sqrt{2}\left(\sqrt{3} - 5\sqrt{2}\right) =$
$2\cdot3 - 10\sqrt{6} + 3\sqrt{6} - 15\cdot2 =$
$6 - 7\sqrt{6} - 30 = -24 - 7\sqrt{6}$

45. $\left(\sqrt{6} + 2\sqrt{7}\right)\left(3\sqrt{6} - \sqrt{7}\right) =$
$\sqrt{6}\left(3\sqrt{6} - \sqrt{7}\right) + 2\sqrt{7}\left(3\sqrt{6} - \sqrt{7}\right) =$
$3\sqrt{36} - \sqrt{42} + 6\sqrt{42} - 2\sqrt{49} =$
$18 - \sqrt{42} + 6\sqrt{42} - 14 = 4 + 5\sqrt{42}$

46. $\left(3 + 2\sqrt{5}\right)\left(4 - 3\sqrt{5}\right) =$
$3\left(4 - 3\sqrt{5}\right) + 2\sqrt{5}\left(4 - 3\sqrt{5}\right) =$
$12 - 9\sqrt{5} + 8\sqrt{5} - 6\cdot5 =$
$12 - \sqrt{5} - 30 = -18 - \sqrt{5}$

47. $\dfrac{5}{\sqrt{7} - \sqrt{5}} = \dfrac{5}{\sqrt{7} - \sqrt{5}}\cdot\dfrac{\sqrt{7} + \sqrt{5}}{\sqrt{7} + \sqrt{5}} =$
$\dfrac{5\left(\sqrt{7} + \sqrt{5}\right)}{7 - 5} = \dfrac{5\sqrt{7} + 5\sqrt{5}}{2}$

48. $\dfrac{\sqrt{6}}{\sqrt{3} - \sqrt{2}} = \dfrac{\sqrt{6}}{\sqrt{3} - \sqrt{2}}\cdot\dfrac{\sqrt{3} + \sqrt{2}}{\sqrt{3} + \sqrt{2}} =$
$\dfrac{\sqrt{6}\left(\sqrt{3} + \sqrt{2}\right)}{3 - 2} = \dfrac{\sqrt{18} + \sqrt{12}}{1} =$
$3\sqrt{2} + 2\sqrt{3}$

49. $\dfrac{2}{3\sqrt{2} - \sqrt{6}} = \dfrac{2}{3\sqrt{2} - \sqrt{6}} \cdot \dfrac{3\sqrt{2} + \sqrt{6}}{3\sqrt{2} + \sqrt{6}} =$

$\dfrac{2\left(3\sqrt{2} + \sqrt{6}\right)}{18 - 6} = \dfrac{2\left(3\sqrt{2} + \sqrt{6}\right)}{12} =$

$\dfrac{3\sqrt{2} + \sqrt{6}}{6}$

50. $\dfrac{\sqrt{6}}{3\sqrt{7} + 2\sqrt{10}} =$

$\dfrac{\sqrt{6}}{3\sqrt{7} + 2\sqrt{10}} \cdot \dfrac{3\sqrt{7} - 2\sqrt{10}}{3\sqrt{7} - 2\sqrt{10}} =$

$\dfrac{\sqrt{6}\left(3\sqrt{7} - 2\sqrt{10}\right)}{63 - 40} =$

$\dfrac{3\sqrt{42} - 2\sqrt{60}}{23} = \dfrac{3\sqrt{42} - 4\sqrt{15}}{23}$

51. $2\sqrt{50} + 3\sqrt{72} - 5\sqrt{8} =$
$2\sqrt{25}\sqrt{2} + 3\sqrt{36}\sqrt{2} - 5\sqrt{4}\sqrt{2} =$
$10\sqrt{2} + 18\sqrt{2} - 10\sqrt{2} = 18\sqrt{2}$

52. $\sqrt{8x} - 3\sqrt{18x} = \sqrt{4}\sqrt{2x} - 3\sqrt{9}\sqrt{2x} =$
$2\sqrt{2x} - 9\sqrt{2x} = -7\sqrt{2x}$

53. $9\sqrt[3]{2} - 5\sqrt[3]{16} = 9\sqrt[3]{2} - 5\sqrt[3]{8}\sqrt[3]{2} =$
$9\sqrt[3]{2} - 5(2)\sqrt[3]{2} = 9\sqrt[3]{2} - 10\sqrt[3]{2} =$
$-\sqrt[3]{2}$

54. $3\sqrt{10} + \sqrt{\dfrac{2}{5}} = 3\sqrt{10} + \dfrac{\sqrt{2}}{\sqrt{5}} \cdot \dfrac{\sqrt{5}}{\sqrt{5}} =$

$3\sqrt{10} + \dfrac{\sqrt{10}}{5} = \left(\dfrac{15}{5} + \dfrac{1}{5}\right)\sqrt{10} =$

$\dfrac{16}{5}\sqrt{10}$

55. $4\sqrt{20} - \dfrac{3}{\sqrt{5}} + \sqrt{45} =$

$4\sqrt{4}\sqrt{5} - \dfrac{3}{\sqrt{5}} \cdot \dfrac{\sqrt{5}}{\sqrt{5}} + \sqrt{9}\sqrt{5} =$

$8\sqrt{5} - \dfrac{3\sqrt{5}}{5} + 3\sqrt{5} =$

$\left(\dfrac{40}{5} - \dfrac{3}{5} + \dfrac{15}{5}\right)\sqrt{5} = \dfrac{52}{5}\sqrt{5}$

56. $\sqrt{\dfrac{2}{3}} - 2\sqrt{54} = \dfrac{\sqrt{2}}{\sqrt{3}} \cdot \dfrac{\sqrt{3}}{\sqrt{3}} - 2\sqrt{9}\sqrt{6} =$

$\dfrac{\sqrt{6}}{3} - 6\sqrt{6} = \left(\dfrac{1}{3} - \dfrac{18}{3}\right)\sqrt{6} =$

$-\dfrac{17}{3}\sqrt{6}$

57. $\sqrt{5x + 6} = 6$

$\left(\sqrt{5x + 6}\right)^2 = (6)^2$

$5x + 6 = 36$

$5x = 30$

$x = 6$

CHECK

$\sqrt{5(6) + 6} \overset{?}{=} 6$

$\sqrt{30 + 6} \overset{?}{=} 6$

$\sqrt{36} \overset{?}{=} 6$

$6 = 6$

The solution set is $\{6\}$.

58. $\sqrt{6x + 1} = \sqrt{3x + 13}$

$\left(\sqrt{6x + 1}\right)^2 = \left(\sqrt{3x + 13}\right)^2$

$6x + 1 = 3x + 13$

$3x = 12$

$x = 4$

CHECK

$\sqrt{6(4) + 1} \overset{?}{=} \sqrt{3(4) + 13}$

$\sqrt{24 + 1} \overset{?}{=} \sqrt{12 + 13}$

$\sqrt{25} \overset{?}{=} \sqrt{25}$

$5 = 5$

The solution set is $\{4\}$.

Chapter 9 Review Problem Set

59.
$$3\sqrt{n} = n$$
$$\left(3\sqrt{n}\right)^2 = n^2$$
$$9n = n^2$$
$$0 = n^2 - 9n$$
$$0 = n(n-9)$$
$$n = 0 \quad \text{or} \quad n - 9 = 0$$
$$n = 0 \quad \text{or} \quad n = 9$$

CHECK
$$3\sqrt{0} \overset{?}{=} 0$$
$$0 = 0$$

$$3\sqrt{9} \overset{?}{=} 9$$
$$3(3) = 9$$
The solution set is $\{0, 9\}$.

60.
$$\sqrt{y+5} = y + 5$$
$$\left(\sqrt{y+5}\right)^2 = (y+5)^2$$
$$y + 5 = y^2 + 10y + 25$$
$$0 = y^2 + 9y + 20$$
$$0 = (y+4)(y+5)$$
$$y + 4 = 0 \quad \text{or} \quad y + 5 = 0$$
$$y = -4 \quad \text{or} \quad y = -5$$

CHECK
$$\sqrt{-4+5} \overset{?}{=} -4 + 5$$
$$\sqrt{1} = 1$$

$$\sqrt{-5+5} \overset{?}{=} -5 + 5$$
$$\sqrt{0} = 0$$
The solution set is $\{-5, \ -4\}$.

61.
$$\sqrt{-3a+10} = a - 2$$
$$\left(\sqrt{-3a+10}\right)^2 = (a-2)^2$$
$$-3a + 10 = a^2 - 4a + 4$$
$$0 = a^2 - a - 6$$
$$0 = (a+2)(a-3)$$
$$a + 2 = 0 \quad \text{or} \quad a - 3 = 0$$
$$a = -2 \quad \text{or} \quad a = 3$$

CHECK
$$\sqrt{-3(-2)+10} \overset{?}{=} -2 - 2$$
$$\sqrt{16} \overset{?}{=} -4$$
$$4 \neq -4$$

$$\sqrt{-3(3)+10} \overset{?}{=} 3 - 2$$
$$\sqrt{1} \overset{?}{=} 1$$
$$1 = 1$$
The solution set is $\{3\}$.

62.
$$3 - \sqrt{2x-1} = 2$$
$$-\sqrt{2x-1} = -1$$
$$\left(-\sqrt{2x-1}\right)^2 = (-1)^2$$
$$2x - 1 = 1$$
$$2x = 2$$
$$x = 1$$

CHECK
$$3 - \sqrt{2(1)-1} \overset{?}{=} 2$$
$$3 - \sqrt{1} \overset{?}{=} 2$$
$$3 - 1 = 2$$
$$2 = 2$$
The solution set is $\{1\}$.

63. $\sqrt{2116} = 46$

64. $\sqrt{4356} = 66$

65. $\sqrt{5184} = 72$

66. $\sqrt{690} = 26$ to the nearest whole number.

67. $\sqrt{2185} = 47$ to the nearest whole number.

68. $\sqrt{5500} = 74$ to the nearest whole number.

CHAPTER 9 Test

1. $-\sqrt{\dfrac{64}{49}} = -\dfrac{8}{7}$ because

$-\left(\dfrac{8}{7}\right)^2 = -\dfrac{64}{49}.$

2. $\sqrt{0.0025} = 0.05$ because

$(0.05)^2 = 0.0025.$

3. $\sqrt{8} = \sqrt{4}\sqrt{2} = 2\sqrt{2} \approx$

$2(1.41) \approx 2.8$

4. $-\sqrt{32} = -\sqrt{16}\sqrt{2} =$

$-4\sqrt{2} \approx -4(1.41) \approx -5.6$

5. $\dfrac{3}{\sqrt{2}} = \dfrac{3}{\sqrt{2}} \cdot \dfrac{\sqrt{2}}{\sqrt{2}} = \dfrac{3\sqrt{2}}{2} \approx$

$\dfrac{3(1.41)}{2} \approx \dfrac{4.23}{2} \approx 2.115 \approx 2.1$

6. $\sqrt{45} = \sqrt{9}\sqrt{5} = 3\sqrt{5}$

7. $-4\sqrt[3]{54} = -4\sqrt[3]{27}\sqrt[3]{2} =$

$-4(3)\sqrt[3]{2} = -12\sqrt[3]{2}$

8. $\dfrac{2\sqrt{3}}{3\sqrt{6}} = \dfrac{2\sqrt{1}}{3\sqrt{2}} \cdot \dfrac{\sqrt{2}}{\sqrt{2}} = \dfrac{2\sqrt{2}}{6} = \dfrac{\sqrt{2}}{3}$

9. $\sqrt{\dfrac{25}{2}} = \dfrac{5}{\sqrt{2}} \cdot \dfrac{\sqrt{2}}{\sqrt{2}} = \dfrac{5\sqrt{2}}{2}$

10. $\dfrac{\sqrt{24}}{\sqrt{36}} = \dfrac{\sqrt{4}\sqrt{6}}{6} = \dfrac{2\sqrt{6}}{6} = \dfrac{\sqrt{6}}{3}$

11. $\sqrt{\dfrac{5}{8}} = \dfrac{\sqrt{5}}{2\sqrt{2}} \cdot \dfrac{\sqrt{2}}{\sqrt{2}} = \dfrac{\sqrt{10}}{4}$

12. $\sqrt[3]{-250x^4y^3} = \sqrt[3]{-125x^3y^3}\sqrt[3]{2x} =$

$-5xy\sqrt[3]{2x}$

13. $\dfrac{\sqrt{3x}}{\sqrt{5y}} = \dfrac{\sqrt{3x}}{\sqrt{5y}} \cdot \dfrac{\sqrt{5y}}{\sqrt{5y}} = \dfrac{\sqrt{15xy}}{5y}$

14. $\dfrac{3}{4}\sqrt{48x^3y^2} = \dfrac{3}{4}\sqrt{16x^2y^2}\sqrt{3x} =$

$\dfrac{3}{4}(4xy)\sqrt{3x} = 3xy\sqrt{3x}$

15. $\left(\sqrt{8}\right)\left(\sqrt{12}\right) = \sqrt{96} = \sqrt{16}\sqrt{6} = 4\sqrt{6}$

16. $\left(6\sqrt[3]{5}\right)\left(4\sqrt[3]{2}\right) = 24\sqrt[3]{10}$

17. $\sqrt{6}\left(2\sqrt{12} - 3\sqrt{8}\right) = 2\sqrt{72} - 3\sqrt{48} =$

$2\sqrt{36}\sqrt{2} - 3\sqrt{16}\sqrt{3} = 12\sqrt{2} - 12\sqrt{3}$

18. $\left(2\sqrt{5} + \sqrt{3}\right)\left(\sqrt{5} - 3\sqrt{3}\right) =$

$2\sqrt{5}\left(\sqrt{5} - 3\sqrt{3}\right) + \sqrt{3}\left(\sqrt{5} - 3\sqrt{3}\right) =$

$2 \cdot 5 - 6\sqrt{15} + \sqrt{15} - 3 \cdot 3 =$

$10 - 5\sqrt{15} - 9 = 1 - 5\sqrt{15}$

19. $\dfrac{\sqrt{6}}{\sqrt{12} + \sqrt{2}} =$

$\dfrac{\sqrt{6}}{\sqrt{12} + \sqrt{2}} \cdot \dfrac{\sqrt{12} - \sqrt{2}}{\sqrt{12} - \sqrt{2}} =$

$\dfrac{\sqrt{72} - \sqrt{12}}{12 - 2} = \dfrac{6\sqrt{2} - 2\sqrt{3}}{10} =$

$\dfrac{2\left(3\sqrt{2} - \sqrt{3}\right)}{10} = \dfrac{3\sqrt{2} - \sqrt{3}}{5}$

Chapter 9 Test

20. $2\sqrt{24} - 4\sqrt{54} + 3\sqrt{96} =$
$2\sqrt{4}\sqrt{6} - 4\sqrt{9}\sqrt{6} + 3\sqrt{16}\sqrt{6} =$
$4\sqrt{6} - 12\sqrt{6} + 12\sqrt{6} = 4\sqrt{6}$

21. $\sqrt{500} = 22$ to the nearest whole number.

22. $\sqrt{3x+1} = 4$
$\left(\sqrt{3x+1}\right)^2 = (4)^2$
$3x + 1 = 16$
$3x = 15$
$x = 5$
CHECK
$\sqrt{3(5)+1} \overset{?}{=} 4$
$\sqrt{16} \overset{?}{=} 4$
$4 = 4$
The solution set is $\{5\}$.

23. $\sqrt{2x-5} = -4$
$\left(\sqrt{2x-5}\right)^2 = (-4)^2$
$2x - 5 = 16$
$2x = 21$
$x = \dfrac{21}{2}$
CHECK
$\sqrt{2\left(\dfrac{21}{2}\right)-5} \overset{?}{=} -4$
$\sqrt{21-5} \overset{?}{=} -4$
$\sqrt{16} \overset{?}{=} -4$
$4 \neq -4$
The solution set is $\emptyset$.

24. $\sqrt{n-3} = 3 - n$
$\left(\sqrt{n-3}\right)^2 = (3-n)^2$
$n - 3 = 9 - 6n + n^2$
$0 = n^2 - 7n + 12$
$0 = (n-3)(n-4)$
$n - 3 = 0 \quad \text{or} \quad n - 4 = 0$
$n = 3 \quad \text{or} \quad\quad n = 4$
CHECK
$\sqrt{3-3} \overset{?}{=} 3 - 3$
$\sqrt{0} = 0$

$\sqrt{4-3} \overset{?}{=} 3 - 4$
$\sqrt{1} \overset{?}{=} -1$
$1 \neq -1$
The solution set is $\{3\}$.

25. $\sqrt{3x+6} = x + 2$
$\left(\sqrt{3x+6}\right)^2 = (x+2)^2$
$3x + 6 = x^2 + 4x + 4$
$0 = x^2 + x - 2$
$0 = (x+2)(x-1)$
$x + 2 = 0 \quad\quad \text{or} \quad x - 1 = 0$
$x = -2 \quad \text{or} \quad\quad x = 1$
CHECK
$\sqrt{3(-2)+6} \overset{?}{=} -2 + 2$
$\sqrt{0} \overset{?}{=} 0$
$0 = 0$

$\sqrt{3(1)+6} \overset{?}{=} 1 + 2$
$\sqrt{9} \overset{?}{=} 3$
$3 = 3$
The solution set is $\{-2, 1\}$.

CHAPTERS 1-9 **Cumulative Review**

1. $-2^6 = -64$

2. $\left(\dfrac{1}{4}\right)^{-3} = \dfrac{1^{-3}}{4^{-3}} = \dfrac{4^3}{1^3} = 64$

3. $\left(\dfrac{1}{3} - \dfrac{1}{4}\right)^{-2} = \left(\dfrac{4}{12} - \dfrac{3}{12}\right)^{-2} =$
$\left(\dfrac{1}{12}\right)^{-2} = 12^2 = 144$

4. $-\sqrt{64} = -8$

5. $\sqrt{\dfrac{4}{9}} = \dfrac{2}{3}$

6. $3^0 + 3^{-1} + 3^{-2} =$
$1 + \dfrac{1}{3} + \dfrac{1}{3^2} = \dfrac{9}{9} + \dfrac{3}{9} + \dfrac{1}{9} = \dfrac{13}{9}$

7. $3(2x - 1) - 4(2x + 3) - (x + 6) =$
$6x - 3 - 8x - 12 - x - 6 =$
$-3x - 21 =$
$-3(-4) - 21$, when $x = -4$
$= 12 - 21 = -9$

8. $(3x^2 - 4x - 6) - (3x^2 + 3x + 1) =$
$3x^2 - 4x - 6 - 3x^2 - 3x - 1 =$
$-7x - 7 =$
$-7(6) - 7$, when $x = 6$
$= -42 - 7 = -49$

9. $2(a - b) - 3(2a + b) + 2(a - 3b) =$
$2a - 2b - 6a - 3b + 2a - 6b =$
$-2a - 11b =$
$-2(-2) - 11(3)$, when $a = -2$, $b = 3$
$= 4 - 33 = -29$

10. $x^2 - 2xy + y^2 = (x - y)^2 =$
$[5 - (-2)]^2$, when $x = 5$, $y = -2$

$= (5 + 2)^2 = 7^2 = 49$

11. $\dfrac{3}{4x} + \dfrac{5}{2x} - \dfrac{7}{x} =$
$\dfrac{3}{4x} + \dfrac{5}{2x}\left(\dfrac{2}{2}\right) - \dfrac{7}{x}\left(\dfrac{4}{4}\right) =$
$\dfrac{3 + 10 - 28}{4x} = -\dfrac{15}{4x}$

12. $\dfrac{3}{x - 2} - \dfrac{4}{x + 3} =$
$\dfrac{3}{x - 2}\left(\dfrac{x + 3}{x + 3}\right) - \dfrac{4}{x + 3}\left(\dfrac{x - 2}{x - 2}\right) =$
$\dfrac{3(x + 3) - 4(x - 2)}{(x - 2)(x + 3)} =$
$\dfrac{3x + 9 - 4x + 8}{(x - 2)(x + 3)} = \dfrac{-x + 17}{(x - 2)(x + 3)}$

13. $\dfrac{3x}{7y} \div \dfrac{6x}{35y^2} = \dfrac{\cancel{3}x}{7\cancel{y}} \cdot \dfrac{\overset{5}{\cancel{35}}\;\overset{y}{\cancel{y^2}}}{\underset{2}{\cancel{6}\cancel{x}}} = \dfrac{5y}{2}$

14. $\dfrac{x - 2}{x^2 + x - 6} \cdot \dfrac{x^2 + 6x + 9}{x^2 - x - 12} =$
$\dfrac{\cancel{x - 2}}{\cancel{(x + 3)}\,\cancel{(x - 2)}} \cdot \dfrac{\cancel{(x + 3)}\,(x + 3)}{(x - 4)\cancel{(x + 3)}} =$
$\dfrac{1}{x - 4}$

15. $\dfrac{7}{x^2 + 3x - 18} - \dfrac{8}{x - 3} =$
$\dfrac{7}{(x + 6)(x - 3)} - \dfrac{8}{x - 3}\left(\dfrac{x + 6}{x + 6}\right) =$
$\dfrac{7 - 8(x + 6)}{(x + 6)(x - 3)} = \dfrac{7 - 8x - 48}{(x + 6)(x - 3)} =$
$\dfrac{-8x - 41}{(x + 6)(x - 3)}$

16. $(-3xy)(-4y^2)(5x^3y) =$
$(-3)(-4)(5)(x^{1+3})(y^{1+2+1}) =$
$60x^4y^4$

17. $(-4x^{-5})(2x^3) = -8x^{-2} = \dfrac{-8}{x^2}$

18. $\dfrac{-12a^{-2}b^3}{4a^{-5}b^4} = -3a^3b^{-1} = \dfrac{-3a^3}{b}$

19. $(3n^4)^{-1} = 3^{-1}n^{-4} = \dfrac{1}{3n^4}$

20. $(9x-2)(3x+4) =$
$9x(3x) + (36-6)x - 2(4) =$
$27x^2 + 30x - 8$

21. $(-x-1)(5x+7) =$
$-x(5x) + (-7-5)x - 1(7) =$
$-5x^2 - 12x - 7$

22. $(3x+1)(2x^2-x-4) =$
$3x(2x^2-x-4) + 1(2x^2-x-4) =$
$6x^3 - 3x^2 - 12x + 2x^2 - x - 4 =$
$6x^3 - x^2 - 13x - 4$

23. $\dfrac{15x^6y^8 - 20x^3y^5}{5x^3y^2} =$

$\dfrac{15x^6y^8}{5x^3y^2} - \dfrac{20x^3y^5}{5x^3y^2} =$

$3x^3y^6 - 4y^3$

24.
$$\begin{array}{r} 2x^2 - 2x - 3 \\ 5x+1\overline{\smash{\big)}10x^3 - 8x^2 - 17x - 3} \\ \underline{10x^3 + 2x^2} \\ -10x^2 - 17x - 3 \\ \underline{-10x^2 - 2x} \\ -15x - 3 \\ \underline{-15x - 3} \end{array}$$

25. $\dfrac{\dfrac{1}{x} - \dfrac{1}{y}}{\dfrac{1}{xy}} = \dfrac{xy\left(\dfrac{1}{x} - \dfrac{1}{y}\right)}{xy\left(\dfrac{1}{xy}\right)} =$

$\dfrac{xy\left(\dfrac{1}{x}\right) - xy\left(\dfrac{1}{y}\right)}{1} = y - x$

26. $\dfrac{2}{1500} = \dfrac{x}{3500}$
$2(3500) = 1500x$
$7000 = 1500x$
$4.67 = x$
4.67 gallons of paint will be needed.

27. $\qquad 18 = x(72)$
$\qquad \dfrac{18}{72} = x$
$\qquad 0.25 = x$
$0.25 \times 100\% = 25\%$

28. $V = \dfrac{1}{3}Bh$, $V = 432$ and $h = 12$
$432 = \dfrac{1}{3}(B)(12)$
$432 = 4B$
$108 = B$

29. $P = 2L + 2W$, $L = 25$ and $W = 40$
$P = 2(25) + 2(40)$
$P = 50 + 80$
$P = 130$
130 feet of fence is needed.

30. Surface Area $= 4\pi r^2$
Surface Area $= 4(3.14)(5)^2$
Surface Area $= 4(3.14)(25)$
Surface Area $= 314$
The surface area is 314 square inches.

31 a. $85000 = (8.5)(10^4)$
   b. $0.0009 = (9)(10^{-4})$
   c. $0.00000104 = (1.04)(10^{-6})$
   d. $53000000 = (5.3)(10^7)$

32. $12x^3 + 14x^2 - 40x =$
$2x(6x^2 + 7x - 20) =$
$2x(6x^2 + 15x - 8x - 20) =$
$2x[3x(2x + 5) - 4(2x + 5)] =$
$2x(2x + 5)(3x - 4)$

33. $12x^2 - 27 =$
$3(4x^2 - 9) =$
$3(2x + 3)(2x - 3)$

34. $xy + 3x - 2y - 6 =$
$x(y + 3) - 2(y + 3) =$
$(y + 3)(x - 2)$

35. $30 + 19x - 5x^2 =$
$30 + 25x - 6x - 5x^2 =$
$5(6 + 5x) - x(6 + 5x) =$
$(6 + 5x)(5 - x)$

36. $4x^4 - 4 =$
$4(x^4 - 1) =$
$4(x^2 + 1)(x^2 - 1) =$
$4(x^2 + 1)(x + 1)(x - 1)$

37. $21x^2 + 22x - 8 =$
$21x^2 - 6x + 28x - 8 =$
$3x(7x - 2) + 4(7x - 2) =$
$(7x - 2)(3x + 4)$

38. $4\sqrt{28} = 4\left(\sqrt{4}\sqrt{7}\right) = 4\left(2\sqrt{7}\right) = 8\sqrt{7}$

39. $-\sqrt{45} = -\sqrt{9}\sqrt{5} = -3\sqrt{5}$

40. $\sqrt{\dfrac{36}{5}} = \dfrac{\sqrt{36}}{\sqrt{5}} = \dfrac{6}{\sqrt{5}} \cdot \dfrac{\sqrt{5}}{\sqrt{5}} = \dfrac{6\sqrt{5}}{5}$

41. $\dfrac{5\sqrt{8}}{6\sqrt{12}} = \dfrac{5\left(2\sqrt{2}\right)}{6\left(2\sqrt{3}\right)} =$

$\dfrac{5\sqrt{2}}{6\sqrt{3}} \cdot \dfrac{\sqrt{3}}{\sqrt{3}} = \dfrac{5\sqrt{6}}{18}$

42. $\sqrt{72xy^5} = \sqrt{36y^4}\sqrt{2xy} = 6y^2\sqrt{2xy}$

43. $\dfrac{-2\sqrt{ab^2}}{5\sqrt{b}} = \dfrac{-2\sqrt{b}\sqrt{ab}}{5\sqrt{b}} = \dfrac{-2\sqrt{ab}}{5}$

44. $\left(3\sqrt{8}\right)\left(4\sqrt{2}\right) = 12\sqrt{16} = 12 \cdot 4 = 48$

45. $6\sqrt{2}\left(9\sqrt{8} - 3\sqrt{12}\right) =$
$6\sqrt{2}\left(9\sqrt{8}\right) - 6\sqrt{2}\left(3\sqrt{12}\right) =$
$54\sqrt{16} - 18\sqrt{24} =$
$54 \cdot 4 - 18 \cdot 2\sqrt{6} = 216 - 36\sqrt{6}$

46. $\left(3\sqrt{2} - \sqrt{7}\right)\left(3\sqrt{2} + \sqrt{7}\right) =$
$\left(3\sqrt{2}\right)^2 - \left(\sqrt{7}\right)^2 =$
$9 \cdot 2 - 7 = 18 - 7 = 11$

47. $\dfrac{4}{\sqrt{3} + \sqrt{2}} =$

$\dfrac{4}{\sqrt{3} + \sqrt{2}} \cdot \dfrac{\sqrt{3} - \sqrt{2}}{\sqrt{3} - \sqrt{2}} =$

$\dfrac{4\left(\sqrt{3} - \sqrt{2}\right)}{1} = 4\sqrt{3} - 4\sqrt{2}$

48. $\dfrac{-6}{3\sqrt{5} - \sqrt{6}} =$

$\dfrac{-6}{3\sqrt{5} - \sqrt{6}} \cdot \dfrac{3\sqrt{5} + \sqrt{6}}{3\sqrt{5} + \sqrt{6}} =$

$\dfrac{-6\left(3\sqrt{5} + \sqrt{6}\right)}{45 - 6} =$

$\dfrac{-6\left(3\sqrt{5} + 6\right)}{39} =$

$$\frac{-2\left(3\sqrt{5}+\sqrt{6}\right)}{13} =$$

$$\frac{-6\sqrt{5}-2\sqrt{6}}{13} = -\frac{6\sqrt{5}+2\sqrt{6}}{13}$$

49. $3\sqrt{50} - 7\sqrt{72} + 4\sqrt{98} =$
$3\sqrt{25}\sqrt{2} - 7\sqrt{36}\sqrt{2} + 4\sqrt{49}\sqrt{2} =$
$15\sqrt{2} - 42\sqrt{2} + 28\sqrt{2} = \sqrt{2}$

50. $\frac{2}{3}\sqrt{20} - \frac{3}{4}\sqrt{45} + \sqrt{80} =$

$\frac{2\sqrt{4}\sqrt{5}}{3} - \frac{3\sqrt{9}\sqrt{5}}{4} + \sqrt{16}\sqrt{5} =$

$\frac{4\sqrt{5}}{3}\left(\frac{4}{4}\right) - \frac{9\sqrt{5}}{4}\left(\frac{3}{3}\right) + \frac{4\sqrt{5}}{1}\left(\frac{12}{12}\right) =$

$\frac{16\sqrt{5} - 27\sqrt{5} + 48\sqrt{5}}{12} = \frac{37\sqrt{5}}{12}$

51. $3x - 6y = -6$

$x = 0:$ $0 - 6y = -6$
                   $y = 1$

$y = 0:$ $3x - 0 = -6$
                   $x = -2$

$x = 2:$ $6 - 6y = -6$
                 $-6y = -12$
                   $y = 2$

Point
$(0, 1)$
$(-2, 0)$
$(2, 2)$

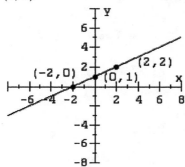

52. $y = \frac{1}{3}x + 4$

Let $x = 0$ $x = 3$

$y = \frac{1}{3}(0) + 4$ $y = \frac{1}{3}(3) + 4$

$y = 4$ $y = 1 + 4$

$(0, 4)$ $y = 5$
                   $(3, 5)$

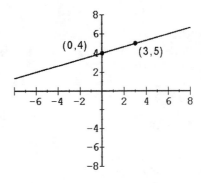

53. $y = -\frac{2}{5}x + 3$

Let $x = 0$ $x = 5$

$y = -\frac{2}{5}(0) + 3$ $y = -\frac{2}{5}(5) + 3$

$y = 3$ $y = -2 + 3$

$(0, 3)$ $y = 1$
                   $(5, 1)$

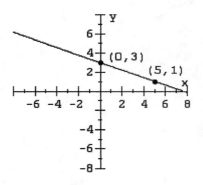

284

54. $y - 2x = 0$
$y = 2x$

Let $x = 0$ $x = 1$
$y = 2(0)$ $y = 2(1)$
$y = 0$ $y = 2$
$(0, 0)$ $(1, 2)$

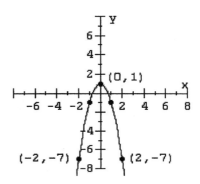

55. $y = -2x^2 + 1$

$x = -2:$ $y = -2(-2)^2 + 1$
             $y = -2(4) + 1$
             $y = -7$

$x = -1:$ $y = -2(-1)^2 + 1$
             $y = -2(1) + 1$
             $y = -1$

$x = 0:$ $y = -2(0)^2 + 1$
             $y = -2(0) + 1$
             $y = 1$

$x = 1:$ $y = -2(1)^2 + 1$
             $y = -2(1) + 1$
             $y = -1$

$x = 2:$ $y = -2(2)^2 + 1$
             $y = -2(4) + 1$
             $y = -7$

Point
$(-2, -7)$
$(-1, -1)$
$(0, 1)$
$(1, -1)$
$(2, -7)$

56. $y = -2x^3$

$x = -2:$ $y = -2(-2)^3$
             $y = -2(-8)$
             $y = 16$

$x = -1:$ $y = -2(-1)^3$
             $y = -2(-1)$
             $y = 2$

$x = 0:$ $y = -2(0)^3$
             $y = -2(0)$
             $y = 0$

$x = 1:$ $y = -2(1)^3$
             $y = -2(1)$
             $y = -2$

$x = 2:$ $y = -2(2)^3$
             $y = -2(8)$
             $y = -16$

Point
$(-2, 16)$
$(-1, 2)$
$(0, 0)$
$(1, -2)$
$(2, -16)$

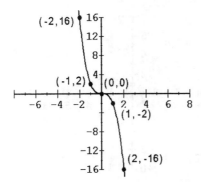

57. $y = -x$

$x = -1:$ $y = -(-1) = 1$

$x = 0:$ $y = 0$

$x = 2:$ $y = -(2) = -2$

$$\underline{\text{Point}}$$
$(-1, 1)$
$(0, 0)$
$(2, -2)$

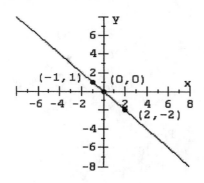

58. Use $(0, -6)$ and $(3, 0)$ to graph a solid line for $y = 2x - 6$. Use $(0, 0)$ as a test point. The given inequality $y \geq 2x - 6$ becomes $0 \geq 0 - 6$ which is a true statement. Therefore, the solution set is the half-plane that contains the origin.

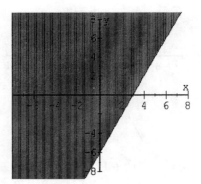

59. Use $(0, 3)$ and $(-2, 0)$ to graph a dashed line for $3x - 2y = -6$. Use $(0, 0)$ as a test point. The given inequality $3x - 2y < -6$ becomes $0 - 0 < -6$ which is a false statement. Therefore, the solution set is the half-plane that does NOT contain the origin.

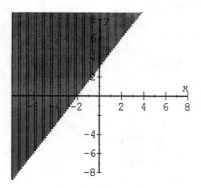

60. Use $(0, -2)$ and $(-4, 0)$ to graph a dashed line for $-2x - 4y = 8$. Use $(0, 0)$ as a test point. The given inequality $-2x - 4y > 8$ becomes $0 - 0 > 8$ which is a false statement. Therefore, the solution set is the half-plane that does NOT contain the origin.

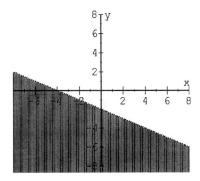

61. Let $P_1 = (-3, 6)$ and $P_2 = (2, -4)$.

$$m = \frac{y_2 - y_1}{x_2 - x_1} = \frac{-4 - 6}{2 - (-3)} = \frac{-10}{5} = -2$$

62. $4x - 7y = 12$
$$-7y = -4x + 12$$
$$y = \frac{4}{7}x - \frac{12}{7}$$
The slope is $\frac{4}{7}$.

63. Use the point-slope form with $m = \frac{2}{3}$ and point $= (7, 2)$.
$$y - y_1 = m(x - x_1)$$
$$y - 2 = \frac{2}{3}(x - 7)$$
$$3y - 6 = 2(x - 7)$$
$$3y - 6 = 2x - 14$$
$$-2x + 3y = -8 \quad \text{or} \quad 2x - 3y = 8$$

64. Find the slope.
Let $P_1 = (-4,)$ and $P_2 = (-1, -3)$.
$$m = \frac{y_2 - y_1}{x_2 - x_1} = \frac{-3 - 1}{-1 - (-4)} = \frac{-4}{3} = -\frac{4}{3}$$
Use the slope and either point in the point-slope form.
$$y - y_1 = m(x - x_1)$$
$$y - 1 = -\frac{4}{3}[x - (-4)]$$
$$3y - 3 = -4(x + 4)$$
$$3y - 3 = -4x - 16$$
$$4x + 3y = -13$$

65. Use the slope-intercept form with $m = -\frac{1}{4}$ and $b = -3$.
$$y = mx + b$$
$$y = -\frac{1}{4}x - 3$$
$$4y = -x - 12$$
$$x + 4y = -12$$

66. A line perpendicular to the x-axis would be a vertical line. Since it contains the point $(4, -5)$, the line would be $x = 4$.

67. $y = 3x - 5$
$3x + 4y = -5$
Substitute $3x - 5$ for y in $3x + 4y = -5$.
$$3x + 4(3x - 5) = -5$$
$$3x + 12x - 20 = -5$$
$$15x - 20 = -5$$
$$15x = 15$$
$$x = 1$$
Substitute 1 for x in $y = 3x - 5$.
$$y = 3(1) - 5 = 3 - 5 = -2$$
The solution set is $\{(1, -2)\}$.

68. $4x - 3y = -20$ Multiply by 5.

$3x + 5y = 14$ Multiply by 3.

$$20x - 15y = -100$$
$$\underline{9x + 15y = 42}$$
$$29x = -58$$
$$x = -2$$

Substitute -2 for x in $3x + 5y = 14$.

$$3(-2) + 5y = 14$$
$$-6 + 5y = 14$$
$$5y = 20$$
$$y = 4$$

The solution set is $\{(-2, 4)\}$.

69. $\dfrac{1}{2}x - \dfrac{2}{3}y = -11$ Multiply by 6.

$\dfrac{1}{3}x + \dfrac{5}{6}y = 8$ Multiply by 6.

$3x - 4y = -66$ Multiply by 5.

$2x + 5y = 48$ Multiply by 4.

$$15x - 20y = -330$$
$$\underline{8x + 20y = 192}$$
$$23x = -138$$
$$x = -6$$

Substitute -6 for x in $3x - 4y = -66$.

$$3(-6) - 4y = -66$$
$$-18 - 4y = -66$$
$$-4y = -48$$
$$y = 12$$

The solution set is $\{(-6, 12)\}$.

70. $2x + 7y = 22$ Multiply by -2.

$4x - 5y = -13$ Leave alone.

$$-4x - 14y = -44$$
$$\underline{4x - 5y = -13}$$
$$-19y = -57$$
$$y = 3$$

Substitute 3 for y in $2x + 7y = 22$.

$$2x + 7(3) = 22$$
$$2x + 21 = 22$$
$$2x = 1$$
$$x = \frac{1}{2}$$

The solution set is $\left\{ \left(\dfrac{1}{2}, 3 \right) \right\}$.

71. $-2(n - 1) + 4(2n - 3) = 4(n + 6)$

$$-2n + 2 + 8n - 12 = 4n + 24$$
$$6n - 10 = 4n + 24$$
$$2n - 10 = 24$$
$$2n = 34$$
$$n = 17$$

The solution set is $\{17\}$.

72. $\dfrac{4}{x - 1} = \dfrac{-1}{x + 6}, \ x \neq -6 \text{ or } x \neq 1$

Cross products are equal.

$$4(x + 6) = -1(x - 1)$$
$$4x + 24 = -x + 1$$
$$5x + 24 = 1$$
$$5x = -23$$
$$x = -\frac{23}{5}$$

The solution set is $\left\{ -\dfrac{23}{5} \right\}$.

73. $\dfrac{t - 1}{3} - \dfrac{t + 2}{4} = -\dfrac{5}{12}$

Multiply both sides by 12.

$$4(t - 1) - 3(t + 2) = -5$$
$$4t - 4 - 3t - 6 = -5$$
$$t - 10 = -5$$
$$t = 5$$

The solution set is $\{5\}$.

74.
$$-7 - 2n - 6n = 7n - 5n + 12$$
$$-7 - 8n = 2n + 12$$
$$-7 - 10n = 12$$
$$-10n = 19$$
$$n = -\frac{19}{10}$$
The solution set is $\left\{ -\dfrac{19}{10} \right\}$.

75.
$$\frac{n-5}{2} = 3 - \frac{n+4}{5}$$
Multiply both sides by 10.
$$5(n-5) = 10(3) - 2(n+4)$$
$$5n - 25 = 30 - 2n - 8$$
$$5n - 25 = 22 - 2n$$
$$7n - 25 = 22$$
$$7n = 47$$
$$n = \frac{47}{7}$$
The solution set is $\left\{ \dfrac{47}{7} \right\}$.

76.
$$0.11x + 0.14(x + 400) = 181$$
Multiply by 100.
$$11x + 14(x + 400) = 18,100$$
$$11x + 14x + 5600 = 18,100$$
$$25x + 5600 = 18,100$$
$$25x = 12,500$$
$$x = 500$$
The solution set is $\{500\}$.

77.
$$\frac{x}{60-x} = 7 + \frac{4}{60-x}, \ x \neq 60$$
Multiply by $(60 - x)$.
$$(60-x)\left(\frac{x}{60-x}\right) = (60-x)\left(7 + \frac{4}{60-x}\right)$$
$$x = 7(60-x) + 4$$
$$x = 420 - 7x + 4$$
$$8x = 424$$
$$x = 53$$
The solution set is $\{53\}$.

78.
$$1 + \frac{x+1}{2x} = \frac{3}{4}, \ x \neq 0$$
Multiply both sides by $4x$.
$$4x + 2(x+1) = 3x$$
$$4x + 2x + 2 = 3x$$
$$6x + 2 = 3x$$
$$2 = -3x$$
$$-\frac{2}{3} = x$$
The solution set is $\left\{ -\dfrac{2}{3} \right\}$.

79.
$$x^2 + 4x - 12 = 0$$
$$(x+6)(x-2) = 0$$
$$x + 6 = 0 \quad \text{or} \quad x - 2 = 0$$
$$x = -6 \quad \text{or} \quad x = 2$$
The solution set is $\{-6, 2\}$.

80.
$$2x^2 - 8 = 0$$
$$2(x^2 - 4) = 0$$
$$2(x+2)(x-2) = 0$$
$$x + 2 = 0 \quad \text{or} \quad x - 2 = 0$$
$$x = -2 \quad \text{or} \quad x = 2$$
The solution set is $\{-2, 2\}$.

81.
$$\sqrt{3x - 6} = 9$$
$$\left(\sqrt{3x-6}\right)^2 = 9^2$$
$$3x - 6 = 81$$
$$3x = 87$$
$$x = 29$$

CHECK
$$\sqrt{3(29) - 6} \stackrel{?}{=} 9$$
$$\sqrt{87 - 6} \stackrel{?}{=} 9$$
$$\sqrt{81} \stackrel{?}{=} 9$$
$$9 = 9$$
The solution set is $\{29\}$.

Chapters 1-9 Cumulative Review

82. $\sqrt{3n} - 2 = 7$

$\qquad \sqrt{3n} = 9$

$\qquad \left(\sqrt{3n}\right)^2 = 9^2$

$\qquad 3n = 81$

$\qquad n = 27$

CHECK

$\sqrt{3(27)} - 2 \overset{?}{=} 7$

$\qquad \sqrt{81} - 2 \overset{?}{=} 7$

$\qquad 9 - 2 \overset{?}{=} 7$

$\qquad 7 = 7$

The solution set is $\{27\}$.

83. $-3n - 4 \le 11$

$\qquad -3n \le 15$

Dividing by -3 reverses the inequality.

$\qquad n \ge -5$

The solution set is $\{n | n \ge -5\}$

or $[-5, \infty)$.

84. $-5 > 3n - 4 - 7n$

$\qquad -5 > -4n - 4$

$\qquad -1 > -4n$

Dividing by -4 reverses the inequality.

$\qquad \dfrac{1}{4} < n$

The solution set is $\left\{ n | n > \dfrac{1}{4} \right\}$

or $\left(\dfrac{1}{4}, \infty \right)$.

85. $2(x - 2) + 3(x + 4) > 6$

$\qquad 2x - 4 + 3x + 12 > 6$

$\qquad 5x + 8 > 6$

$\qquad 5x > -2$

$\qquad x > -\dfrac{2}{5}$

The solution set is $\left\{ x | x > -\dfrac{2}{5} \right\}$

or $\left(-\dfrac{2}{5}, \infty \right)$.

86. $\dfrac{1}{2}n - \dfrac{2}{3}n < -1$

Multiply both sides by 6.

$\qquad 3n - 4n < -6$

$\qquad -n < -6$

Multiplying by -1 reverses the inequality.

$\qquad n > 6$

The solution set is $\{n | n > 6\}$ or $(6, \infty)$.

87. $\dfrac{x + 1}{2} + \dfrac{x - 2}{6} < \dfrac{3}{8}$

$24\left(\dfrac{x + 1}{2} + \dfrac{x - 2}{6} \right) < 24\left(\dfrac{3}{8} \right)$

$12(x + 1) + 4(x - 2) < 3(3)$

$\qquad 12x + 12 + 4x - 8 < 9$

$\qquad 16x + 4 < 9$

$\qquad 16x < 5$

$\qquad x < \dfrac{5}{16}$

The solution set is $\left\{ x | x < \dfrac{5}{16} \right\}$

or $\left(-\infty, \dfrac{5}{16} \right)$.

88. $\dfrac{x - 3}{7} - \dfrac{x - 2}{4} \le \dfrac{9}{14}$

Multiply both sides by 28.

$4(x - 3) - 7(x - 2) \le 2(9)$

$\qquad 4x - 12 - 7x + 14 \le 18$

$\qquad -3x + 2 \le 18$

$\qquad -3x \le 16$

$\qquad x \ge -\dfrac{16}{3}$

The solution set is $\left\{ x | x \ge -\dfrac{16}{3} \right\}$

or $\left[-\dfrac{16}{5}, \infty \right)$.

89. Let a represent the measure of the smaller angle, then $180 - a$ represents the larger angle.

$$180 - a = 2a - 15$$
$$180 - 3a = -15$$
$$-3a = -195$$
$$a = 65$$

If $a = 65$, then $180 - a = 115$. The measures of the supplementary angles are $65°$ and $115°$.

90. Let n represent the smaller number, then $3n - 2$ represents the larger number. Their sum is 50.

$$n + (3n - 2) = 50$$
$$4n - 2 = 50$$
$$4n = 52$$
$$n = 13$$

If $n = 13$, then $3n - 2 = 3(13) - 2 = 37$. The numbers are 13 and 37.

91. Let x represent the first odd number. Then $x + 2$ represents the second odd number.

$$x^2 + (x + 2)^2 = 130$$
$$x^2 + x^2 + 4x + 4 = 130$$
$$2x^2 + 4x - 126 = 0$$
$$2\left(x^2 + 2x - 63\right) = 0$$
$$2(x + 9)(x - 7) = 0$$
$$x + 9 = 0 \quad \text{or} \quad x - 7 = 0$$
$$x = -9 \quad \text{or} \quad x = 7$$

Since -9 is not a whole number, the consecutive whole numbers are 7 and 9.

92. Let n represent the number of nickels, then $2n + 1$ is the number of dimes, and $3n + 4$ is the number of quarters.

$$n + (2n + 1) + (3n + 4) = 47$$
$$6n + 5 = 47$$
$$6n = 42$$
$$n = 7 \text{ nickels}$$

If $n = 7$, then $2n + 1 = 2(7) + 1 = 14 + 1 = 15$ dimes.
If $n = 7$, then $3n + 4 = 3(7) + 4 =$

$21 + 4 = 25$ quarters.
There were 7 nickels, 15 dimes, and 25 quarters.

93. Let t represent the taxes on the $\$90,000$ home. The problem can be solved as a proportion.

$$\frac{\text{taxes}}{\text{value}} \quad \frac{1050}{70,000} = \frac{t}{90,000}$$

Cross products are equal.
$$70,000t = 90,000(1050)$$

Divide both sides by $10,000$.
$$7t = 9(1050)$$
$$7t = 9450$$
$$t = 1350$$

The taxes would be $\$1350$.

94. Let s represent the selling price. Profit is 60% of cost.

Selling price $=$ Cost $+$ Profit
$$s = 30 + (60\%)(30)$$
$$s = 30 + 0.6(30)$$
$$s = 30 + 18$$
$$s = 48$$

The selling price would be $\$48$.

95. Let t represent Polly's time, then $t + 1$ represents Rosa's time. A diagram of the problem would be as follows:

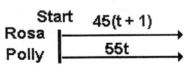

Distances are equal.

The two distances are the same.
$$55t = 45(t + 1)$$
$$55t = 45t + 45$$
$$10t = 45$$
$$t = \frac{45}{10} = 4\frac{1}{2}$$

It would take Polly $4\frac{1}{2}$ hours to overtake Rosa.

Chapters 1-9 Cumulative Review

96. Let x represent the amount of pure acid to be added. Then $100 + x$ represents the amount of final solution.

$$\left(\begin{array}{c}\text{pure}\\\text{acid}\\\text{in 10\%}\\\text{solution}\end{array}\right) + \left(\begin{array}{c}\text{pure}\\\text{acid}\\\text{to be}\\\text{added}\end{array}\right) = \left(\begin{array}{c}\text{pure}\\\text{acid}\\\text{in final}\\\text{solution}\end{array}\right)$$

$$(10\%)(100) + x = (20\%)(100 + x)$$
$$0.10(100) + x = 0.20(100 + x)$$
$$10[0.10(100) + x] = 10[0.20(100 + x)]$$
$$1(100) + 10x = 2(100 + x)$$
$$100 + 10x = 200 + 2x$$
$$100 + 8x = 200$$
$$8x = 100$$
$$x = 12.5$$

We must add 12.5 milliliters of pure acid.

97. Let x represent the score on the fourth algebra test.

$$\frac{85 + 90 + 86 + x}{4} \geq 88$$
$$85 + 90 + 86 + x \geq 352$$
$$261 + x \geq 352$$
$$x \geq 91$$

The score on the last test needs to be 91 or better.

98. Let w represent the number of games they must win of the remaining 20 games. They will play a total of $70 + 72 + 20 = 162$ games. To win more than 50% of their games they would need to win more than 81 games.

$$w + 70 > 81$$
$$w > 11$$

They must win more than 11 of the remaining games.

99. Let t represent the time working together. The sum of the individual rates equals the rate working together.

$$\text{Seth's rate} + \text{Butch's rate} = \text{Rate together}$$
$$\frac{1}{20} + \frac{1}{30} = \frac{1}{t}$$
$$60t\left(\frac{1}{20} + \frac{1}{30}\right) = 60t\left(\frac{1}{t}\right)$$
$$3t + 2t = 60$$
$$5t = 60$$
$$t = 12$$

It would take 12 minutes to complete the job working together.

Chapter 10 Quadratic Equations

PROBLEM SET **10.1** Quadratic Equations

1. $x^2 + 15x = 0$
$x(x + 15) = 0$
$x = 0$ or $x + 15 = 0$
$x = 0$ or $x = -15$
The solution set is $\{-15, 0\}$.

3. $n^2 = 12n$
$n^2 - 12n = 0$
$n(n - 12) = 0$
$n = 0$ or $n - 12 = 0$
$n = 0$ or $n = 12$
The solution set is $\{0, 12\}$.

5. $3y^2 = 15y$
$y^2 = 5y$
$y^2 - 5y = 0$
$y(y - 5) = 0$
$y = 0$ or $y - 5 = 0$
$y = 0$ or $y = 5$
The solution set is $\{0, 5\}$.

7. $x^2 - 9x + 8 = 0$
$(x - 8)(x - 1) = 0$
$x - 8 = 0$ or $x - 1 = 0$
$x = 8$ or $x = 1$
The solution set is $\{1, 8\}$.

9. $x^2 - 5x - 14 = 0$
$(x - 7)(x + 2) = 0$
$x - 7 = 0$ or $x + 2 = 0$
$x = 7$ or $x = -2$
The solution set is $\{-2, 7\}$.

11. $n^2 + 5n - 6 = 0$
$(n + 6)(n - 1) = 0$

$n + 6 = 0$ or $n - 1 = 0$
$n = -6$ or $n = 1$
The solution set is $\{-6, 1\}$.

13. $6y^2 + 7y - 5 = 0$
$(3y + 5)(2y - 1) = 0$
$3y + 5 = 0$ or $2y - 1 = 0$
$3y = -5$ or $2y = 1$
$y = -\dfrac{5}{3}$ or $y = \dfrac{1}{2}$
The solution set is $\left\{-\dfrac{5}{3}, \dfrac{1}{2}\right\}$.

15. $30x^2 - 37x + 10 = 0$
$(5x - 2)(6x - 5) = 0$
$5x - 2 = 0$ or $6x - 5 = 0$
$5x = 2$ or $6x = 5$
$x = \dfrac{2}{5}$ or $x = \dfrac{5}{6}$
The solution set is $\left\{\dfrac{2}{5}, \dfrac{5}{6}\right\}$.

17. $4x^2 - 4x + 1 = 0$
$(2x - 1)(2x - 1) = 0$
$2x - 1 = 0$ or $2x - 1 = 0$
$2x = 1$ or $2x = 1$
$x = \dfrac{1}{2}$ or $x = \dfrac{1}{2}$
The solution set is $\left\{\dfrac{1}{2}\right\}$.

19. $x^2 = 64$
$x = \pm\sqrt{64}$
$x = \pm 8$
The solution set is $\{-8, 8\}$.

293

Problem Set 10.1

21. $x^2 = \dfrac{25}{9}$

$x = \pm\sqrt{\dfrac{25}{9}}$

$x = \pm\dfrac{5}{3}$

The solution set is $\left\{-\dfrac{5}{3}, \dfrac{5}{3}\right\}$.

23. $4x^2 = 64$

$x^2 = 16$

$x = \pm\sqrt{16}$

$x = \pm 4$

The solution set is $\{-4, 4\}$.

25. $n^2 = 14$

$n = \pm\sqrt{14}$

The solution set is $\left\{-\sqrt{14}, \sqrt{14}\right\}$.

27. $n^2 + 16 = 0$

$n^2 = -16$

There are no real number solutions because n^2 will always be nonnegative.

29. $y^2 = 32$

$y = \pm\sqrt{32}$

$y = \pm\sqrt{16}\sqrt{2}$

$y = \pm 4\sqrt{2}$

The solution set is $\left\{-4\sqrt{2}, 4\sqrt{2}\right\}$.

31. $3x^2 - 54 = 0$

$x^2 - 18 = 0$

$x^2 = 18$

$x = \pm\sqrt{18}$

$x = \pm\sqrt{9}\sqrt{2}$

$x = \pm 3\sqrt{2}$

The solution set is $\left\{-3\sqrt{2}, 3\sqrt{2}\right\}$.

33. $2x^2 = 9$

$x^2 = \dfrac{9}{2}$

$x = \pm\sqrt{\dfrac{9}{2}}$

$x = \pm\dfrac{\sqrt{9}}{\sqrt{2}}$

$x = \pm\dfrac{3}{\sqrt{2}} \cdot \dfrac{\sqrt{2}}{\sqrt{2}}$

$x = \pm\dfrac{3\sqrt{2}}{2}$

The solution set is $\left\{-\dfrac{3\sqrt{2}}{2}, \dfrac{3\sqrt{2}}{2}\right\}$.

35. $8n^2 = 25$

$n^2 = \dfrac{25}{8}$

$n = \pm\sqrt{\dfrac{25}{8}}$

$n = \pm\dfrac{\sqrt{25}}{\sqrt{8}}$

$n = \pm\dfrac{5}{2\sqrt{2}} \cdot \dfrac{\sqrt{2}}{\sqrt{2}}$

$n = \pm\dfrac{5\sqrt{2}}{4}$

The solution set is $\left\{-\dfrac{5\sqrt{2}}{4}, \dfrac{5\sqrt{2}}{4}\right\}$.

37. $(x-1)^2 = 4$

$x - 1 = \pm\sqrt{4}$

$x - 1 = \pm 2$

$x - 1 = 2$ or $x - 1 = -2$

$x = 3$ or $x = -1$

The solution set is $\{-1, 3\}$.

39. $(x+3)^2 = 25$

$x + 3 = \pm\sqrt{25}$

$x + 3 = \pm 5$

$x + 3 = 5$ or $x + 3 = -5$

$x = 2$ or $x = -8$

The solution set is $\{-8, 2\}$.

41. $(3x-2)^2 = 49$

$3x-2 = \pm\sqrt{49}$

$3x-2 = \pm 7$

$3x-2 = 7$ or $3x-2 = -7$

$3x = 9$ or $3x = -5$

$x = 3$ or $x = -\dfrac{5}{3}$

The solution set is $\left\{-\dfrac{5}{3}, 3\right\}$.

43. $(x+6)^2 = 5$

$x+6 = \pm\sqrt{5}$

$x+6 = \sqrt{5}$ or $x+6 = -\sqrt{5}$

$x = -6+\sqrt{5}$ or $x = -6-\sqrt{5}$

The solution set is

$\left\{-6+\sqrt{5},\ -6-\sqrt{5}\right\}$.

45. $(n-1)^2 = 8$

$n-1 = \pm\sqrt{8} = \pm 2\sqrt{2}$

$n-1 = 2\sqrt{2}$ or $n-1 = -2\sqrt{2}$

$n = 1+2\sqrt{2}$ or $n = 1-2\sqrt{2}$

The solution set is

$\left\{1+2\sqrt{2},\ 1-2\sqrt{2}\right\}$.

47. $(2n+3)^2 = 20$

$2n+3 = \pm\sqrt{20} = \pm 2\sqrt{5}$

$2n+3 = 2\sqrt{5}$ or $2n+3 = -2\sqrt{5}$

$2n = -3+2\sqrt{5}$ or $2n = -3-2\sqrt{5}$

$n = \dfrac{-3+2\sqrt{5}}{2}$ or $n = \dfrac{-3-2\sqrt{5}}{2}$

The solution set is

$\left\{\dfrac{-3-2\sqrt{5}}{2}, \dfrac{-3+2\sqrt{5}}{2}\right\}$.

49. $(4x-1)^2 = -2$

The equation has no real number solutions because $(4x-1)^2$ will always be nonnegative.

51. $(3x-5)^2 - 40 = 0$

$(3x-5)^2 = 40$

$3x-5 = \pm\sqrt{40} = \pm 2\sqrt{10}$

$3x-5 = 2\sqrt{10}$ or $3x-5 = -2\sqrt{10}$

$3x = 5+2\sqrt{10}$ or $3x = 5-2\sqrt{10}$

$x = \dfrac{5+2\sqrt{10}}{3}$ or $x = \dfrac{5-2\sqrt{10}}{3}$

The solution set is

$\left\{\dfrac{5-2\sqrt{10}}{3}, \dfrac{5+2\sqrt{10}}{3}\right\}$.

53. $2(7x-1)^2 + 5 = 37$

$2(7x-1)^2 = 32$

$(7x-1)^2 = 16$

$7x-1 = \pm\sqrt{16} = \pm 4$

$7x-1 = 4$ or $7x-1 = -4$

$7x = 5$ or $7x = -3$

$x = \dfrac{5}{7}$ or $x = \dfrac{-3}{7}$

The solution set is $\left\{-\dfrac{3}{7}, \dfrac{5}{7}\right\}$.

55. $2(x+8)^2 - 9 = 91$

$2(x+8)^2 = 100$

$(x+8)^2 = 50$

$x+8 = \pm\sqrt{50} = \pm 5\sqrt{2}$

$x+8 = 5\sqrt{2}$ or $x+8 = -5\sqrt{2}$

$x = -8+5\sqrt{2}$ or $x = -8-5\sqrt{2}$

The solution set is

$\left\{-8-5\sqrt{2},\ -8+5\sqrt{2}\right\}$.

57. $c^2 = a^2 + b^2$

$c^2 = 1^2 + 7^2 = 1+49$

$c^2 = 50$

$c = \sqrt{50} = 5\sqrt{2}$ inches

Problem Set 10.1

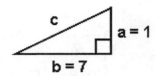

59.
$$c^2 = a^2 + b^2$$
$$8^2 = a^2 + 6^2$$
$$64 = a^2 + 36$$
$$28 = a^2$$
$$a = \sqrt{28} = 2\sqrt{7} \text{ meters}$$

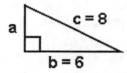

61.
$$a^2 + b^2 = c^2$$
$$10^2 + b^2 = 12^2$$
$$100 + b^2 = 144$$
$$b^2 = 44$$
$$b = \sqrt{44} = 2\sqrt{11} \text{ feet}$$

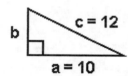

63. If $c = 8$ inches, then
$$a = \frac{1}{2}c = \frac{1}{2}(8) = 4 \text{ inches.}$$

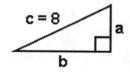

$$c^2 = a^2 + b^2$$
$$8^2 = 4^2 + b^2$$
$$64 = 16 + b^2$$
$$48 = b^2$$
$$b = \sqrt{48} = 4\sqrt{3} \text{ inches}$$

65. If $a = 6$ feet, then
$$c = 2a = 2(6) = 12 \text{ feet.}$$

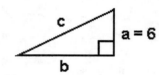

$$c^2 = a^2 + b^2$$
$$12^2 = 6^2 + b^2$$
$$144 = 36 + b^2$$
$$108 = b^2$$
$$b = \sqrt{108} = 6\sqrt{3} \text{ feet}$$

67. Substitute $2a$ for c in the Pythagorean Theorem.

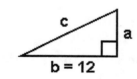

$$c^2 = a^2 + b^2$$
$$(2a)^2 = a^2 + 12^2$$
$$4a^2 = a^2 + 144$$
$$3a^2 = 144$$
$$a^2 = 48$$
$$a = \sqrt{48} = 4\sqrt{3} \text{ meters}$$
$$c = 2a = 2\left(4\sqrt{3}\right) = 8\sqrt{3} \text{ meters}$$

69. If $b = 10$ inches, then $a = 10$ inches.

$c^2 = a^2 + b^2$
$c^2 = 10^2 + 10^2$
$c^2 = 100 + 100$
$c^2 = 200$
$c = \sqrt{200} = 10\sqrt{2}$ inches

71. Let $b = a$ in the Pythagorean Theorem.

$a^2 + b^2 = c^2$
$a^2 + a^2 = 9^2$
$2a^2 = 81$
$a^2 = \dfrac{81}{2}$
$a = b = \sqrt{\dfrac{81}{2}} = \dfrac{\sqrt{81}}{\sqrt{2}}$
$a = b = \dfrac{9}{\sqrt{2}} \cdot \dfrac{\sqrt{2}}{\sqrt{2}} = \dfrac{9\sqrt{2}}{2}$ meters

73. Using the Pythagorean Theorem, let c represent the length of the ladder which is 18 feet, then let a represent the height of the windowsill above the ground which is 16 feet.

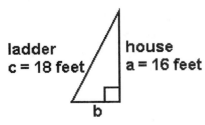

$a^2 + b^2 = c^2$
$16^2 + b^2 = 18^2$
$256 + b^2 = 324$
$b^2 = 68$
$b = \sqrt{68} = 8.2$ to nearest tenth
The ladder is approximately 8.2 feet from the foundation of the house.

75. Let c represent the length of the diagonal of the rectangle.

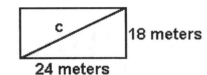

$c^2 = a^2 + b^2$
$c^2 = 18^2 + 24^2$
$c^2 = 324 + 576$
$c^2 = 900$
$c = \sqrt{900} = 30$
The length of the diagonal is 30 meters.

77. Let x represent a side of the square Parking Lot.

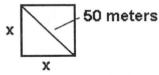

$a^2 + b^2 = c^2$
$x^2 + x^2 = 50^2$
$2x^2 = 2500$
$x^2 = 1250$
$x = \sqrt{1250} = 35$ to the nearest whole number.

The length of a side of the lot is approximately 35 meters.

Problem Set 10.1

81. Let a represent the altitude of the triangle. The hypotenuse of the triangle, formed by the altitude, one-half of the base and one side of the equilateral triangle, is $c = 6$ centimeters. Let b represent one-half of the base.

$$a^2 + b^2 = c^2$$
$$a^2 + 3^2 = 6^2$$
$$a^2 + 9 = 36$$
$$a^2 = 27$$
$$a = \sqrt{27} = 5.2 \text{ to the nearest tenth}$$

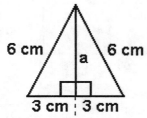

The altitude would measure approximately 5.2 centimeters.

83. Let x represent the length of a diagonal of the bottom of the rectangle. Let y represent the length of a diagonal from the lower corner to the diagonally opposite upper corner. First, find the value of x by substituting 8 and 6 for the legs.

$$x^2 = 8^2 + 6^2$$
$$x^2 = 64 + 36 = 100$$
$$x = \sqrt{100} = 10$$

Next, find the value of y by substituting 10 for one leg and 4 for the other leg.

$$y^2 = (10)^2 + 4^2$$
$$y^2 = 100 + 16 = 116$$
$$y = \sqrt{116} = 10.8 \text{ to the nearest tenth}$$

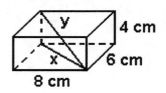

The length of the diagonal is approximately 10.8 centimeters.

PROBLEM SET **10.2** **Completing the Square**

1.
$$x^2 + 8x - 1 = 0$$
$$x^2 + 8x = 1$$
$$x^2 + 8x + 16 = 1 + 16$$
$$(x + 4)^2 = 17$$
$$x + 4 = \pm\sqrt{17}$$
$$x + 4 = \sqrt{17} \quad \text{or} \quad x + 4 = -\sqrt{17}$$
$$x = -4 + \sqrt{17} \quad \text{or} \quad x = -4 - \sqrt{17}$$
The solution set is
$$\left\{-4 - \sqrt{17},\ -4 + \sqrt{17}\right\}.$$

3.
$$x^2 + 10x + 2 = 0$$
$$x^2 + 10x = -2$$
$$x^2 + 10x + 25 = -2 + 25$$
$$(x + 5)^2 = 23$$
$$x + 5 = \pm\sqrt{23}$$
$$x + 5 = \sqrt{23} \quad \text{or} \quad x + 5 = -\sqrt{23}$$
$$x = -5 + \sqrt{23} \quad \text{or} \quad x = -5 - \sqrt{23}$$
The solution set is
$$\left\{-5 - \sqrt{23},\ -5 + \sqrt{23}\right\}.$$

5.
$$x^2 - 4x - 4 = 0$$
$$x^2 - 4x = 4$$
$$x^2 - 4x + 4 = 4 + 4$$
$$(x-2)^2 = 8$$
$$x - 2 = \pm\sqrt{8} = \pm 2\sqrt{2}$$
$$x - 2 = 2\sqrt{2} \quad \text{or} \quad x - 2 = -2\sqrt{2}$$
$$x = 2 + 2\sqrt{2} \quad \text{or} \quad x = 2 - 2\sqrt{2}$$
The solution set is $\left\{ 2 - 2\sqrt{2},\, 2 + 2\sqrt{2} \right\}$.

7.
$$x^2 + 6x + 12 = 0$$
$$x^2 + 6x = -12$$
$$x^2 + 6x + 9 = -12 + 9$$
$$(x+3)^2 = -3$$
The right side is negative.
There are no real number solutions because $(x+3)^2$ will always be nonnegative.

9.
$$n^2 + 2n = 17$$
$$n^2 + 2n + 1 = 17 + 1$$
$$(n+1)^2 = 18$$
$$n + 1 = \pm\sqrt{18} = \pm 3\sqrt{2}$$
$$n+1 = 3\sqrt{2} \quad \text{or} \quad n+1 = -3\sqrt{2}$$
$$n = -1 + 3\sqrt{2} \quad \text{or} \quad n = -1 - 3\sqrt{2}$$
The solution set is
$$\left\{ -1 - 3\sqrt{2},\, -1 + 3\sqrt{2} \right\}.$$

11.
$$x^2 + x - 3 = 0$$
$$x^2 + x = 3$$
$$x^2 + x + \frac{1}{4} = 3 + \frac{1}{4}$$
$$\left(x + \frac{1}{2}\right)^2 = \frac{13}{4}$$
$$x + \frac{1}{2} = \pm\sqrt{\frac{13}{4}} = \pm\frac{\sqrt{13}}{2}$$
$$x + \frac{1}{2} = \frac{\sqrt{13}}{2} \quad \text{or} \quad x + \frac{1}{2} = -\frac{\sqrt{13}}{2}$$
$$x = -\frac{1}{2} + \frac{\sqrt{13}}{2} \quad \text{or} \quad x = -\frac{1}{2} - \frac{\sqrt{13}}{2}$$
$$x = \frac{-1 + \sqrt{13}}{2} \quad \text{or} \quad x = \frac{-1 - \sqrt{13}}{2}$$

The solution set is $\left\{ \dfrac{-1 - \sqrt{13}}{2},\, \dfrac{-1 + \sqrt{13}}{2} \right\}$.

13.
$$a^2 - 5a = 2$$
$$a^2 - 5a + \frac{25}{4} = 2 + \frac{25}{4}$$
$$\left(a - \frac{5}{2}\right)^2 = \frac{33}{4}$$
$$a - \frac{5}{2} = \pm\sqrt{\frac{33}{4}} = \pm\frac{\sqrt{33}}{2}$$
$$a - \frac{5}{2} = \frac{\sqrt{33}}{2} \quad \text{or} \quad a - \frac{5}{2} = -\frac{\sqrt{33}}{2}$$
$$a = \frac{5}{2} + \frac{\sqrt{33}}{2} \quad \text{or} \quad a = \frac{5}{2} - \frac{\sqrt{33}}{2}$$
$$a = \frac{5 + \sqrt{33}}{2} \quad \text{or} \quad a = \frac{5 - \sqrt{33}}{2}$$
The solution set is $\left\{ \dfrac{5 - \sqrt{33}}{2},\, \dfrac{5 + \sqrt{33}}{2} \right\}$.

15.
$$2x^2 + 8x - 3 = 0$$
$$2x^2 + 8x = 3$$
$$x^2 + 4x = \frac{3}{2}$$
$$x^2 + 4x + 4 = \frac{3}{2} + 4$$
$$(x+2)^2 = \frac{11}{2}$$
$$x + 2 = \pm\sqrt{\frac{11}{2}} = \pm\frac{\sqrt{22}}{2}$$
$$x+2 = \frac{\sqrt{22}}{2} \quad \text{or} \quad x+2 = -\frac{\sqrt{22}}{2}$$
$$x = -2 + \frac{\sqrt{22}}{2} \quad \text{or} \quad x = -2 - \frac{\sqrt{22}}{2}$$
$$x = \frac{-4 + \sqrt{22}}{2} \quad \text{or} \quad x = \frac{-4 - \sqrt{22}}{2}$$
The solution set is $\left\{ \dfrac{-4 - \sqrt{22}}{2},\, \dfrac{-4 + \sqrt{22}}{2} \right\}$.

Problem Set 10.2

17. $3x^2 + 12x - 2 = 0$

$$3x^2 + 12x = 2$$

$$x^2 + 4x = \frac{2}{3}$$

$$x^2 + 4x + 4 = \frac{2}{3} + 4$$

$$(x+2)^2 = \frac{14}{3}$$

$$x + 2 = \pm\sqrt{\frac{14}{3}} = \pm\frac{\sqrt{42}}{3}$$

$$x+2 = \frac{\sqrt{42}}{3} \quad \text{or} \quad x+2 = -\frac{\sqrt{42}}{3}$$

$$x = -2 + \frac{\sqrt{42}}{3} \quad \text{or} \quad x = -2 - \frac{\sqrt{42}}{3}$$

$$x = \frac{-6 + \sqrt{42}}{3} \quad \text{or} \quad x = \frac{-6 - \sqrt{42}}{3}$$

The solution set is

$$\left\{ \frac{-6 - \sqrt{42}}{3}, \frac{-6 + \sqrt{42}}{3} \right\}.$$

19. $2t^2 - 4t + 1 = 0$

$$2t^2 - 4t = -1$$

$$t^2 - 2t = -\frac{1}{2}$$

$$t^2 - 2t + 1 = -\frac{1}{2} + 1$$

$$(t-1)^2 = \frac{1}{2}$$

$$t - 1 = \pm\sqrt{\frac{1}{2}} = \pm\frac{\sqrt{2}}{2}$$

$$t - 1 = \frac{\sqrt{2}}{2} \quad \text{or} \quad t - 1 = -\frac{\sqrt{2}}{2}$$

$$t = 1 + \frac{\sqrt{2}}{2} \quad \text{or} \quad t = 1 - \frac{\sqrt{2}}{2}$$

$$t = \frac{2 + \sqrt{2}}{2} \quad \text{or} \quad t = \frac{2 - \sqrt{2}}{2}$$

The solution set is

$$\left\{ \frac{2 - \sqrt{2}}{2}, \frac{2 + \sqrt{2}}{2} \right\}.$$

21. $5n^2 + 10n + 6 = 0$

$$5n^2 + 10n = -6$$

$$n^2 + 2n = -\frac{6}{5}$$

$$n^2 + 2n + 1 = -\frac{6}{5} + 1$$

$$(n+1)^2 = -\frac{1}{5}$$

The right side is negative.
There are no real number solutions
because $(n+1)^2$ will always be
nonnegative.

23. $-n^2 + 9n = 4$

$$n^2 - 9n = -4$$

$$n^2 - 9n + \frac{81}{4} = -4 + \frac{81}{4}$$

$$\left(n - \frac{9}{2}\right)^2 = \frac{65}{4}$$

$$n - \frac{9}{2} = \pm\sqrt{\frac{65}{4}} = \pm\frac{\sqrt{65}}{2}$$

$$n - \frac{9}{2} = \frac{\sqrt{65}}{2} \quad \text{or} \quad n - \frac{9}{2} = -\frac{\sqrt{65}}{2}$$

$$n = \frac{9}{2} + \frac{\sqrt{65}}{2} \quad \text{or} \quad n = \frac{9}{2} - \frac{\sqrt{65}}{2}$$

$$n = \frac{9 + \sqrt{65}}{2} \quad \text{or} \quad n = \frac{9 - \sqrt{65}}{2}$$

The solution set is

$$\left\{ \frac{9 - \sqrt{65}}{2}, \frac{9 + \sqrt{65}}{2} \right\}.$$

25. $2x^2 + 3x - 1 = 0$

$$2x^2 + 3x = 1$$

$$x^2 + \frac{3}{2}x = \frac{1}{2}$$

$$x^2 + \frac{3}{2}x + \frac{9}{16} = \frac{1}{2} + \frac{9}{16}$$

$$\left(x + \frac{3}{4}\right)^2 = \frac{17}{16}$$

$$x + \frac{3}{4} = \pm\sqrt{\frac{17}{16}} = \pm\frac{\sqrt{17}}{4}$$

$x + \dfrac{3}{4} = \dfrac{\sqrt{17}}{4}$ or $x + \dfrac{3}{4} = -\dfrac{\sqrt{17}}{4}$

$x = -\dfrac{3}{4} + \dfrac{\sqrt{17}}{4}$ or $x = -\dfrac{3}{4} - \dfrac{\sqrt{17}}{4}$

$x = \dfrac{-3 + \sqrt{17}}{4}$ or $x = \dfrac{-3 - \sqrt{17}}{4}$

The solution set is
$$\left\{ \dfrac{-3 - \sqrt{17}}{4}, \dfrac{-3 + \sqrt{17}}{4} \right\}.$$

27. $3x^2 + 2x - 2 = 0$

$3x^2 + 2x = 2$

$x^2 + \dfrac{2}{3}x = \dfrac{2}{3}$

$x^2 + \dfrac{2}{3}x + \dfrac{1}{9} = \dfrac{2}{3} + \dfrac{1}{9}$

$\left(x + \dfrac{1}{3} \right)^2 = \dfrac{7}{9}$

$x + \dfrac{1}{3} = \pm \sqrt{\dfrac{7}{9}} = \pm \dfrac{\sqrt{7}}{3}$

$x + \dfrac{1}{3} = \dfrac{\sqrt{7}}{3}$ or $x + \dfrac{1}{3} = -\dfrac{\sqrt{7}}{3}$

$x = -\dfrac{1}{3} + \dfrac{\sqrt{7}}{3}$ or $x = -\dfrac{1}{3} - \dfrac{\sqrt{7}}{3}$

$x = \dfrac{-1 + \sqrt{7}}{3}$ or $x = \dfrac{-1 - \sqrt{7}}{3}$

The solution set is
$$\left\{ \dfrac{-1 - \sqrt{7}}{3}, \dfrac{-1 + \sqrt{7}}{3} \right\}.$$

29. $n(n + 2) = 168$

$n^2 + 2n = 168$

$n^2 + 2n + 1 = 168 + 1$

$(n + 1)^2 = 169$

$n + 1 = \pm \sqrt{169} = \pm 13$

$n + 1 = 13$ or $n + 1 = -13$

$n = -1 + 13$ or $n = -1 - 13$

$n = 12$ or $n = -14$

The solution set is $\{-14, 12\}$.

31. $n(n - 4) = 165$

$n^2 - 4n = 165$

$n^2 - 4n + 4 = 165 + 4$

$(n - 2)^2 = 169$

$n - 2 = \pm \sqrt{169} = \pm 13$

$n - 2 = 13$ or $n - 2 = -13$

$n = 2 + 13$ or $n = 2 - 13$

$n = 15$ or $n = -11$

The solution set is $\{-11, 15\}$.

33 a. $x^2 + 4x - 12 = 0$

$(x + 6)(x - 2) = 0$

$x + 6 = 0$ or $x - 2 = 0$

$x = -6$ or $x = 2$

The solution set is $\{-6, 2\}$.

b. $x^2 + 4x - 12 = 0$

$x^2 + 4x = 12$

$x^2 + 4x + 4 = 12 + 4$

$(x + 2)^2 = 16$

$x + 2 = \pm \sqrt{16} = \pm 4$

$x + 2 = 4$ or $x + 2 = -4$

$x = -2 + 4$ or $x = -2 - 4$

$x = 2$ or $x = -6$

The solution set is $\{-6, 2\}$.

35 a. $x^2 + 12x + 27 = 0$

$(x + 9)(x + 3) = 0$

$x + 9 = 0$ or $x + 3 = 0$

$x = -9$ or $x = -3$

The solution set is $\{-9, -3\}$.

b. $x^2 + 12x + 27 = 0$

$x^2 + 12x = -27$

$x^2 + 12x + 36 = -27 + 36$

$(x + 6)^2 = 9$

$x + 6 = \pm \sqrt{9} = \pm 3$

$x + 6 = 3$ or $x + 6 = -3$

$x = -6 + 3$ or $x = -6 - 3$

$x = -3$ or $x = -9$

The solution set is $\{-9, -3\}$.

Problem Set 10.2

37 a. $n^2 - 3n - 40 = 0$
$(n + 5)(n - 8) = 0$
$n + 5 = 0 \quad$ or $\quad n - 8 = 0$
$\quad n = -5 \quad$ or $\quad n = 8$
The solution set is $\{-5, 8\}$.

b. $n^2 - 3n - 40 = 0$
$n^2 - 3n = 40$
$n^2 - 3n + \dfrac{9}{4} = 40 + \dfrac{9}{4}$
$\left(n - \dfrac{3}{2}\right)^2 = \dfrac{169}{4}$
$n - \dfrac{3}{2} = \pm\sqrt{\dfrac{169}{4}} = \pm\dfrac{13}{2}$
$n - \dfrac{3}{2} = \dfrac{13}{2} \quad$ or $\quad n - \dfrac{3}{2} = -\dfrac{13}{2}$
$n = \dfrac{3}{2} + \dfrac{13}{2} \quad$ or $\quad n = \dfrac{3}{2} - \dfrac{13}{2}$
$n = \dfrac{16}{2} = 8 \quad$ or $\quad n = \dfrac{-10}{2} = -5$
The solution set is $\{-5, 8\}$.

39a. $2n^2 - 9n + 4 = 0$
$(2n - 1)(n - 4) = 0$
$2n - 1 = 0 \quad$ or $\quad n - 4 = 0$
$\quad 2n = 1 \quad$ or $\quad n = 4$
$\quad n = \dfrac{1}{2} \quad$ or $\quad n = 4$
The solution set is $\left\{\dfrac{1}{2}, 4\right\}$.

b. $2n^2 - 9n + 4 = 0$
$2n^2 - 9n = -4$
$n^2 - \dfrac{9}{2}n = -2$
$n^2 - \dfrac{9}{2}n + \dfrac{81}{16} = -2 + \dfrac{81}{16}$
$\left(n - \dfrac{9}{4}\right)^2 = \dfrac{49}{16}$
$n - \dfrac{9}{4} = \pm\sqrt{\dfrac{49}{16}} = \pm\dfrac{7}{4}$

$n - \dfrac{9}{4} = \dfrac{7}{4} \quad$ or $\quad n - \dfrac{9}{4} = -\dfrac{7}{4}$
$n = \dfrac{9}{4} + \dfrac{7}{4} \quad$ or $\quad n = \dfrac{9}{4} - \dfrac{7}{4}$
$n = \dfrac{16}{4} = 4 \quad$ or $\quad n = \dfrac{2}{4} = \dfrac{1}{2}$
The solution set is $\left\{\dfrac{1}{2}, 4\right\}$.

41 a. $4n^2 + 4n - 15 = 0$
$(2n - 3)(2n + 5) = 0$
$2n - 3 = 0 \quad$ or $\quad 2n + 5 = 0$
$\quad 2n = 3 \quad$ or $\quad 2n = -5$
$\quad n = \dfrac{3}{2} \quad$ or $\quad n = -\dfrac{5}{2}$
The solution set is $\left\{-\dfrac{5}{2}, \dfrac{3}{2}\right\}$.

b. $4n^2 + 4n - 15 = 0$
$4n^2 + 4n = 15$
$n^2 + n = \dfrac{15}{4}$
$n^2 + n + \dfrac{1}{4} = \dfrac{15}{4} + \dfrac{1}{4}$
$\left(n + \dfrac{1}{2}\right)^2 = \dfrac{16}{4} = 4$
$n + \dfrac{1}{2} = \pm\sqrt{4} = \pm 2$
$n + \dfrac{1}{2} = 2 \quad$ or $\quad n + \dfrac{1}{2} = -2$
$n = -\dfrac{1}{2} + 2 \quad$ or $\quad n = -\dfrac{1}{2} - 2$
$n = \dfrac{3}{2} \quad$ or $\quad n = -\dfrac{5}{2}$
The solution set is $\left\{-\dfrac{5}{2}, \dfrac{3}{2}\right\}$.

45.

$$ax^2 + bx + c = 0$$

$$ax^2 + bx = -c$$

$$x^2 + \frac{b}{a}x = -\frac{c}{a}$$

$$x^2 + \frac{b}{a}x + \frac{b^2}{4a^2} = -\frac{c}{a} + \frac{b^2}{4a^2}$$

$$\left(x + \frac{b}{2a}\right)^2 = \frac{-4ac + b^2}{4a^2} = \frac{b^2 - 4ac}{4a^2}$$

$$x + \frac{b}{2a} = \pm\sqrt{\frac{b^2 - 4ac}{4a^2}} = \pm\frac{\sqrt{b^2 - 4ac}}{2a}$$

$$x + \frac{b}{2a} = \frac{\sqrt{b^2 - 4ac}}{2a}$$

$$x = -\frac{b}{2a} + \frac{\sqrt{b^2 - 4ac}}{2a}$$

$$x = \frac{-b + \sqrt{b^2 - 4ac}}{2a}$$

or

$$x + \frac{b}{2a} = -\frac{\sqrt{b^2 - 4ac}}{2a}$$

$$x = -\frac{b}{2a} - \frac{\sqrt{b^2 - 4ac}}{2a}$$

$$x = \frac{-b - \sqrt{b^2 - 4ac}}{2a}$$

The solution set is

$$\left\{\frac{-b - \sqrt{b^2 - 4ac}}{2a}, \frac{-b + \sqrt{b^2 - 4ac}}{2a}\right\}.$$

PROBLEM SET | **10.3** **The Quadratic Formula**

1. $x^2 - 5x - 6 = 0$

$$x = \frac{-(-5) \pm \sqrt{(-5)^2 - 4(1)(-6)}}{2(1)}$$

$$x = \frac{5 \pm \sqrt{49}}{2}$$

$$x = \frac{5 \pm 7}{2}$$

$$x = \frac{5 - 7}{2} \quad \text{or} \quad x = \frac{5 + 7}{2}$$

$$x = \frac{-2}{2} = -1 \quad \text{or} \quad x = \frac{12}{2} = 6$$

The solution set is $\{-1, 6\}$.

3. $x^2 + 5x = 36$

$$x^2 + 5x - 36 = 0$$

$$x = \frac{-(5) \pm \sqrt{(5)^2 - 4(1)(-36)}}{2(1)}$$

$$x = \frac{-5 \pm \sqrt{169}}{2}$$

$$x = \frac{-5 \pm 13}{2}$$

$$x = \frac{-5 - 13}{2} \quad \text{or} \quad x = \frac{-5 + 13}{2}$$

$$x = \frac{-18}{2} = -9 \quad \text{or} \quad x = \frac{8}{2} = 4$$

The solution set is $\{-9, 4\}$.

5. $n^2 - 2n - 5 = 0$

$$n = \frac{-(-2) \pm \sqrt{(-2)^2 - 4(1)(-5)}}{2(1)}$$

$$n = \frac{2 \pm \sqrt{24}}{2}$$

$$n = \frac{2 \pm 2\sqrt{6}}{2}$$

$$n = 1 \pm \sqrt{6}$$

The solution set is $\left\{1 - \sqrt{6}, 1 + \sqrt{6}\right\}$.

7. $a^2 - 5a - 2 = 0$

$$a = \frac{-(-5) \pm \sqrt{(-5)^2 - 4(1)(-2)}}{2(1)}$$

$$a = \frac{5 \pm \sqrt{33}}{2}$$

The solution set is $\left\{\frac{5 - \sqrt{33}}{2}, \frac{5 + \sqrt{33}}{2}\right\}.$

Problem Set 10.3

9. $x^2 - 2x + 6 = 0$

$$x = \frac{-(-2) \pm \sqrt{(-2)^2 - 4(1)(6)}}{2(1)}$$

$$x = \frac{2 \pm \sqrt{-20}}{2}$$

Since $\sqrt{-20}$ is not a real number, this equation has no real number solutions.

11. $y^2 + 4y + 2 = 0$

$$y = \frac{-(4) \pm \sqrt{(4)^2 - 4(1)(2)}}{2(1)}$$

$$y = \frac{-4 \pm \sqrt{8}}{2}$$

$$y = \frac{-4 \pm 2\sqrt{2}}{2}$$

$$y = -2 \pm \sqrt{2}$$

The solution set is
$$\left\{ -2 - \sqrt{2},\ -2 + \sqrt{2} \right\}.$$

13. $x^2 - 6x = 0$ (Note: $c = 0$)

$$x = \frac{-(-6) \pm \sqrt{(-6)^2 - 4(1)(0)}}{2(1)}$$

$$x = \frac{6 \pm \sqrt{36}}{2}$$

$$x = \frac{6 \pm 6}{2}$$

$$x = \frac{6 - 6}{2} \quad \text{or} \quad x = \frac{6 + 6}{2}$$

$$x = \frac{0}{2} = 0 \quad \text{or} \quad x = \frac{12}{2} = 6$$

The solution set is $\{0, 6\}$.

15. $2x^2 = 7x$

$2x^2 - 7x = 0$ (Note: $c = 0$)

$$x = \frac{-(-7) \pm \sqrt{(-7)^2 - 4(2)(0)}}{2(2)}$$

$$x = \frac{7 \pm \sqrt{49}}{4}$$

$$x = \frac{7 \pm 7}{4}$$

$$x = \frac{7 - 7}{4} \quad \text{or} \quad x = \frac{7 + 7}{4}$$

$$x = \frac{0}{4} = 0 \quad \text{or} \quad x = \frac{14}{4} = \frac{7}{2}$$

The solution set is $\left\{ 0, \dfrac{7}{2} \right\}$.

17. $n^2 - 34n + 288 = 0$

$$n = \frac{-(-34) \pm \sqrt{(-34)^2 - 4(1)(288)}}{2(1)}$$

$$n = \frac{34 \pm \sqrt{4}}{2}$$

$$n = \frac{34 \pm 2}{2}$$

$$n = \frac{34 - 2}{2} \quad \text{or} \quad n = \frac{34 + 2}{2}$$

$$n = \frac{32}{2} = 16 \quad \text{or} \quad n = \frac{36}{2} = 18$$

The solution set is $\{16, 18\}$.

19. $x^2 + 2x - 80 = 0$

$$x = \frac{-(2) \pm \sqrt{(2)^2 - 4(1)(-80)}}{2(1)}$$

$$x = \frac{-2 \pm \sqrt{324}}{2}$$

$$x = \frac{-2 \pm 18}{2}$$

$$x = \frac{-2 - 18}{2} \quad \text{or} \quad x = \frac{-2 + 18}{2}$$

$$x = \frac{-20}{2} = -10 \quad \text{or} \quad x = \frac{16}{2} = 8$$

The solution set is $\{-10, 8\}$.

21. $t^2 + 4t + 4 = 0$

$$t = \frac{-(4) \pm \sqrt{(4)^2 - 4(1)(4)}}{2(1)}$$

$$t = \frac{-4 \pm \sqrt{0}}{2}$$

$$t = \frac{-4 \pm 0}{2}$$

$$t = \frac{-4}{2} = -2$$

The solution set is $\{-2\}$.

23. $6x^2 + x - 2 = 0$

$$x = \frac{-(1) \pm \sqrt{(1)^2 - 4(6)(-2)}}{2(6)}$$

$$x = \frac{-1 \pm \sqrt{49}}{12}$$

$$x = \frac{-1 \pm 7}{12}$$

$$x = \frac{-1 - 7}{12} \quad \text{or} \quad x = \frac{-1 + 7}{12}$$

$$x = \frac{-8}{12} = -\frac{2}{3} \quad \text{or} \quad x = \frac{6}{12} = \frac{1}{2}$$

The solution set is $\left\{-\frac{2}{3}, \frac{1}{2}\right\}$.

25. $5x^2 + 3x - 2 = 0$

$$x = \frac{-(3) \pm \sqrt{(3)^2 - 4(5)(-2)}}{2(5)}$$

$$x = \frac{-3 \pm \sqrt{49}}{10}$$

$$x = \frac{-3 \pm 7}{10}$$

$$x = \frac{-3 - 7}{10} \quad \text{or} \quad x = \frac{-3 + 7}{10}$$

$$x = \frac{-10}{10} = -1 \quad \text{or} \quad x = \frac{4}{10} = \frac{2}{5}$$

The solution set is $\left\{-1, \frac{2}{5}\right\}$.

27. $12x^2 + 19x = -5$

$12x^2 + 19x + 5 = 0$

$$x = \frac{-(19) \pm \sqrt{(19)^2 - 4(12)(5)}}{2(12)}$$

$$x = \frac{-19 \pm \sqrt{121}}{24}$$

$$x = \frac{-19 \pm 11}{24}$$

$$x = \frac{-19 - 11}{24} \quad \text{or} \quad x = \frac{-19 + 11}{24}$$

$$x = \frac{-30}{24} = -\frac{5}{4} \quad \text{or} \quad x = \frac{-8}{24} = -\frac{1}{3}$$

The solution set is $\left\{-\frac{5}{4}, -\frac{1}{3}\right\}$.

29. $2x^2 + 5x - 6 = 0$

$$x = \frac{-(5) \pm \sqrt{(5)^2 - 4(2)(-6)}}{2(2)}$$

$$x = \frac{-5 \pm \sqrt{73}}{4}$$

The solution set is

$$\left\{\frac{-5 - \sqrt{73}}{4}, \frac{-5 + \sqrt{73}}{4}\right\}.$$

31. $3x^2 + 4x - 1 = 0$

$$x = \frac{-(4) \pm \sqrt{(4)^2 - 4(3)(-1)}}{2(3)}$$

$$x = \frac{-4 \pm \sqrt{28}}{6}$$

$$x = \frac{-4 \pm 2\sqrt{7}}{6}$$

$$x = \frac{-2 \pm \sqrt{7}}{3}$$

The solution set is

$$\left\{\frac{-2 - \sqrt{7}}{3}, \frac{-2 + \sqrt{7}}{3}\right\}.$$

Problem Set 10.3

33. $16x^2 + 24x + 9 = 0$

$$x = \frac{-(24) \pm \sqrt{(24)^2 - 4(16)(9)}}{2(16)}$$

$$x = \frac{-24 \pm \sqrt{0}}{32}$$

$$x = \frac{-24 \pm 0}{32} = -\frac{3}{4}$$

The solution set is $\left\{ -\frac{3}{4} \right\}$.

35. $4n^2 + 8n - 1 = 0$

$$n = \frac{-(8) \pm \sqrt{(8)^2 - 4(4)(-1)}}{2(4)}$$

$$n = \frac{-8 \pm \sqrt{80}}{8}$$

$$n = \frac{-8 \pm 4\sqrt{5}}{8}$$

$$n = \frac{-2 \pm \sqrt{5}}{2}$$

The solution set is $\left\{ \frac{-2-\sqrt{5}}{2}, \frac{-2+\sqrt{5}}{2} \right\}$.

37. $6n^2 + 9n + 1 = 0$

$$n = \frac{-(9) \pm \sqrt{(9)^2 - 4(6)(1)}}{2(6)}$$

$$n = \frac{-9 \pm \sqrt{57}}{12}$$

The solution set is

$$\left\{ \frac{-9-\sqrt{57}}{12}, \frac{-9+\sqrt{57}}{12} \right\}.$$

39. $2y^2 - y - 4 = 0$

$$y = \frac{-(-1) \pm \sqrt{(-1)^2 - 4(2)(-4)}}{2(2)}$$

$$y = \frac{1 \pm \sqrt{33}}{4}$$

The solution set is $\left\{ \frac{1-\sqrt{33}}{4}, \frac{1+\sqrt{33}}{4} \right\}$.

41. $4t^2 + 5t + 3 = 0$

$$t = \frac{-(5) \pm \sqrt{(5)^2 - 4(4)(3)}}{2(4)}$$

$$t = \frac{-5 \pm \sqrt{-23}}{8}$$

Since $\sqrt{-23}$ is not a real number, this equation has no real number solutions.

43. $7x^2 + 5x - 4 = 0$

$$x = \frac{-(5) \pm \sqrt{(5)^2 - 4(7)(-4)}}{2(7)}$$

$$x = \frac{-5 \pm \sqrt{137}}{14}$$

The solution set is

$$\left\{ \frac{-5-\sqrt{137}}{14}, \frac{-5+\sqrt{137}}{14} \right\}.$$

45. $7 = 3x^2 - x$

$0 = 3x^2 - x - 7$

$$x = \frac{-(-1) \pm \sqrt{(-1)^2 - 4(3)(-7)}}{2(3)}$$

$$x = \frac{1 \pm \sqrt{85}}{6}$$

The solution set is $\left\{ \frac{1-\sqrt{85}}{6}, \frac{1+\sqrt{85}}{6} \right\}$.

47. $n^2 + 23n = -126$

$n^2 + 23n + 126 = 0$

$$n = \frac{-(23) \pm \sqrt{(23)^2 - 4(1)(126)}}{2(1)}$$

$$n = \frac{-23 \pm \sqrt{25}}{2}$$

$$n = \frac{-23 \pm 5}{2}$$

$$n = \frac{-23 - 5}{2} \quad \text{or} \quad n = \frac{-23+5}{2}$$

$$n = \frac{-28}{2} = -14 \quad \text{or} \quad n = \frac{-18}{2} = -9$$

The solution set is $\{-14, -9\}$.

306

53. $x^2 - 5x - 19 = 0$

$$x = \frac{-(-5) \pm \sqrt{(-5)^2 - 4(1)(-19)}}{2(1)}$$

$$x = \frac{5 \pm \sqrt{101}}{2}$$

$$x = \frac{5 \pm 10.0499}{2}$$

$$x = \frac{5 - 10.0499}{2} \quad \text{or} \quad x = \frac{5 + 10.0499}{2}$$

$$x = \frac{-5.0499}{2} \quad \text{or} \quad x = \frac{15.0499}{2}$$

$$x = -2.52 \quad \text{or} \quad x = 7.52$$

The solution set is $\{-2.52, 7.52\}$.

55. $x^2 + 6x - 17 = 0$

$$x = \frac{-(6) \pm \sqrt{(6)^2 - 4(1)(-17)}}{2(1)}$$

$$x = \frac{-6 \pm \sqrt{104}}{2}$$

$$x = \frac{-6 \pm 10.1980}{2}$$

$$x = \frac{-6 - 10.1980}{2} \quad \text{or} \quad x = \frac{-6 + 10.1980}{2}$$

$$x = \frac{-16.1980}{2} \quad \text{or} \quad x = \frac{4.1980}{2}$$

$$x = -8.10 \quad \text{or} \quad x = 2.10$$

The solution set is $\{-8.10, 2.10\}$.

57. $3x^2 + 7x - 13 = 0$

$$x = \frac{-(7) \pm \sqrt{(7)^2 - 4(3)(-13)}}{2(3)}$$

$$x = \frac{-7 \pm \sqrt{205}}{6}$$

$$x = \frac{-7 \pm 14.3178}{6}$$

$$x = \frac{-7 - 14.3178}{6} \quad \text{or} \quad x = \frac{-7+14.3178}{6}$$

$$x = \frac{-21.3178}{6} \quad \text{or} \quad x = \frac{7.3178}{6}$$

$$x = -3.55 \quad \text{or} \quad x = 1.22$$

The solution set is $\{-3.55, 1.22\}$.

59. $4x^2 - 9x - 19 = 0$

$$x = \frac{-(-9) \pm \sqrt{(-9)^2 - 4(4)(-19)}}{2(4)}$$

$$x = \frac{9 \pm \sqrt{385}}{8}$$

$$x = \frac{9 \pm 19.6214}{8}$$

$$x = \frac{9 - 19.6214}{8} \quad \text{or} \quad x = \frac{9 + 19.6214}{8}$$

$$x = \frac{-10.6214}{8} \quad \text{or} \quad x = \frac{28.6214}{8}$$

$$x = -1.33 \quad \text{or} \quad x = 3.58$$

The solution set is $\{-1.33, 3.58\}$.

61. $-5x^2 + x + 21 = 0$

$5x^2 - x - 21 = 0$

$$x = \frac{-(-1) \pm \sqrt{(-1)^2 - 4(5)(-21)}}{2(5)}$$

$$x = \frac{1 \pm \sqrt{421}}{10}$$

$$x = \frac{1 \pm 20.5183}{10}$$

$$x = \frac{1 - 20.5183}{10} \quad \text{or} \quad x = \frac{1 + 20.5183}{10}$$

$$x = \frac{-19.5183}{10} \quad \text{or} \quad x = \frac{21.5183}{10}$$

$$x = -1.95 \quad \text{or} \quad x = 2.15$$

The solution set is $\{-1.95, 2.15\}$.

Problem Set 10.4

PROBLEM SET **10.4**   **Solving Quadratic Equations - Which Method?**

For the problems in this set, the method chosen to solve the problem is not the only possible method. If a different method is used, the solution sets obtained should agree.

1.
$$x^2 + 4x = 45$$
$$x^2 + 4x - 45 = 0$$
$$(x + 9)(x - 5) = 0$$
$$x + 9 = 0 \quad \text{or} \quad x - 5 = 0$$
$$x = -9 \quad \text{or} \quad x = 5$$
The solution set is $\{-9, 5\}$.

3.
$$(5n + 6)^2 = 49$$
$$5n + 6 = \pm\sqrt{49} = \pm 7$$
$$5n + 6 = -7 \quad \text{or} \quad 5n + 6 = 7$$
$$5n = -13 \quad \text{or} \quad 5n = 1$$
$$n = -\frac{13}{5} \quad \text{or} \quad n = \frac{1}{5}$$
The solution set is $\left\{-\frac{13}{5}, \frac{1}{5}\right\}$.

5.
$$t^2 - t - 2 = 0$$
$$(t - 2)(t + 1) = 0$$
$$t - 2 = 0 \quad \text{or} \quad t + 1 = 0$$
$$t = 2 \quad \text{or} \quad t = -1$$
The solution set is $\{-1, 2\}$.

7.
$$8x = 3x^2$$
$$0 = 3x^2 - 8x$$
$$0 = x(3x - 8)$$
$$x = 0 \quad \text{or} \quad 3x - 8 = 0$$
$$x = 0 \quad \text{or} \quad 3x = 8$$
$$x = 0 \quad \text{or} \quad x = \frac{8}{3}$$
The solution set is $\left\{0, \frac{8}{3}\right\}$.

9.
$$9x^2 - 6x + 1 = 0$$
$$(3x - 1)^2 = 0$$
$$3x - 1 = 0$$
$$3x = 1$$
$$x = \frac{1}{3}$$
The solution set is $\left\{\frac{1}{3}\right\}$.

11.
$$5n^2 = \sqrt{8}n$$
$$5n^2 - \sqrt{8}n = 0$$
$$n\left(5n - \sqrt{8}\right) = 0$$
$$n = 0 \quad \text{or} \quad 5n - \sqrt{8} = 0$$
$$n = 0 \quad \text{or} \quad 5n = \sqrt{8} = 2\sqrt{2}$$
$$n = 0 \quad \text{or} \quad n = \frac{2\sqrt{2}}{5}$$
The solution set is $\left\{0, \frac{2\sqrt{2}}{5}\right\}$.

13.
$$n^2 - 14n = 19$$
$$n^2 - 14n + 49 = 19 + 49$$
$$(n - 7)^2 = 68$$
$$n - 7 = \pm\sqrt{68} = \pm 2\sqrt{17}$$
$$n - 7 = -2\sqrt{17} \quad \text{or} \quad n - 7 = 2\sqrt{17}$$
$$n = 7 - 2\sqrt{17} \quad \text{or} \quad n = 7 + 2\sqrt{17}$$
The solution set is
$$\left\{7 - 2\sqrt{17}, 7 + 2\sqrt{17}\right\}.$$

15.
$$5x^2 - 2x - 7 = 0$$
$$(5x - 7)(x + 1) = 0$$
$$5x - 7 = 0 \quad \text{or} \quad x + 1 = 0$$
$$5x = 7 \quad \text{or} \quad x = -1$$
$$x = \frac{7}{5} \quad \text{or} \quad x = -1$$
The solution set is $\left\{-1, \frac{7}{5}\right\}$.

17. $15x^2 + 28x + 5 = 0$
$(5x + 1)(3x + 5) = 0$
$5x + 1 = 0 \quad$ or $\quad 3x + 5 = 0$
$\qquad 5x = -1 \quad$ or $\qquad 3x = -5$
$\qquad x = -\dfrac{1}{5} \quad$ or $\qquad x = -\dfrac{5}{3}$
The solution set is $\left\{ -\dfrac{5}{3}, \ -\dfrac{1}{5} \right\}$.

19. $x^2 - \sqrt{8}x - 7 = 0$
$x^2 - 2\sqrt{2}x = 7$
$x^2 - 2\sqrt{2}x + 2 = 7 + 2$
$\left(x - \sqrt{2} \right)^2 = 9$
$x - \sqrt{2} = \pm\sqrt{9} = \pm 3$
$x = \sqrt{2} \pm 3$
The solution set is $\left\{ \sqrt{2} - 3, \ \sqrt{2} + 3 \right\}$.

21. $y^2 + 5y = 84$
$y^2 + 5y - 84 = 0$
$(y + 12)(y - 7) = 0$
$y + 12 = 0 \quad$ or $\quad y - 7 = 0$
$\qquad y = -12 \quad$ or $\qquad y = 7$
The solution set is $\{ -12, \ 7 \}$.

23. $2n = 3 + \dfrac{3}{n}$
Multiply both sides by n.
$2n^2 = 3n + 3$
$2n^2 - 3n - 3 = 0$
$n = \dfrac{-(-3) \pm \sqrt{(-3)^2 - 4(2)(-3)}}{2(2)}$
$n = \dfrac{3 \pm \sqrt{33}}{4}$
The solution set is $\left\{ \dfrac{3 - \sqrt{33}}{4}, \ \dfrac{3 + \sqrt{33}}{4} \right\}$.

25. $3x^2 - 9x - 12 = 0$
$x^2 - 3x - 4 = 0$
$(x - 4)(x + 1) = 0$
$x - 4 = 0 \quad$ or $\quad x + 1 = 0$
$\qquad x = 4 \quad$ or $\qquad x = -1$
The solution set is $\{ -1, \ 4 \}$.

27. $2x^2 - 3x + 7 = 0$
$x = \dfrac{-(-3) \pm \sqrt{(-3)^2 - 4(2)(7)}}{2(2)}$
$x = \dfrac{3 \pm \sqrt{-47}}{4}$
Since $\sqrt{-47}$ is not a real number, this equation has no real number solutions.

29. $n(n - 46) = -480$
$n^2 - 46n = -480$
$n^2 - 46n + 480 = 0$
$n = \dfrac{-(-46) \pm \sqrt{(-46)^2 - 4(1)(480)}}{2(1)}$
$n = \dfrac{46 \pm \sqrt{196}}{2}$
$n = \dfrac{46 \pm 14}{2}$
$n = \dfrac{46 - 14}{2} \quad$ or $\quad n = \dfrac{46 + 14}{2}$
$n = \dfrac{32}{2} = 16 \quad$ or $\quad n = \dfrac{60}{2} = 30$
The solution set is $\{16, \ 30\}$.

31. $n - \dfrac{3}{n} = -1$
Multiply both sides by n.
$n^2 - 3 = -n$
$n^2 + n - 3 = 0$
$n = \dfrac{-(1) \pm \sqrt{(1)^2 - 4(1)(-3)}}{2(1)}$
$n = \dfrac{-1 \pm \sqrt{13}}{2}$
The solution set is
$\left\{ \dfrac{-1 - \sqrt{13}}{2}, \ \dfrac{-1 + \sqrt{13}}{2} \right\}$.

Problem Set 10.4

33.
$$x + \frac{1}{x} = \frac{25}{12}$$
Multiply both sides by $12x$.
$$12x^2 + 12 = 25x$$
$$12x^2 - 25x + 12 = 0$$
$$(3x - 4)(4x - 3) = 0$$
$$3x - 4 = 0 \quad \text{or} \quad 4x - 3 = 0$$
$$3x = 4 \quad \text{or} \quad 4x = 3$$
$$x = \frac{4}{3} \quad \text{or} \quad x = \frac{3}{4}$$
The solution set is $\left\{ \frac{3}{4}, \frac{4}{3} \right\}$.

35.
$$t^2 + 12t + 36 = 49$$
$$(t + 6)^2 = 49$$
$$t + 6 = \pm \sqrt{49} = \pm 7$$
$$t + 6 = -7 \quad \text{or} \quad t + 6 = 7$$
$$t = -13 \quad \text{or} \quad t = 1$$
The solution set is $\{ -13, 1 \}$.

37.
$$x^2 - 28x + 187 = 0$$
$$x = \frac{-(-28) \pm \sqrt{(-28)^2 - 4(1)(187)}}{2(1)}$$
$$x = \frac{28 \pm \sqrt{36}}{2}$$
$$x = \frac{28 \pm 6}{2}$$
$$x = \frac{28 - 6}{2} \quad \text{or} \quad x = \frac{28 + 6}{2}$$
$$x = \frac{22}{2} = 11 \quad \text{or} \quad x = \frac{34}{2} = 17$$
The solution set is $\{11, 17\}$.

39.
$$\frac{x^2}{3} - x = -\frac{1}{2}$$
Multiply both sides by 6.
$$2x^2 - 6x = -3$$
$$2x^2 - 6x + 3 = 0$$
$$x = \frac{-(-6) \pm \sqrt{(-6)^2 - 4(2)(3)}}{2(2)}$$
$$x = \frac{6 \pm \sqrt{12}}{4} = \frac{6 \pm 2\sqrt{3}}{4}$$
$$x = \frac{3 \pm \sqrt{3}}{2}$$
The solution set is $\left\{ \frac{3 - \sqrt{3}}{2}, \frac{3 + \sqrt{3}}{2} \right\}$.

41.
$$\frac{2}{x + 2} - \frac{1}{x} = 3$$
Multiply both sides by $x(x + 2)$.
$$2x - (x + 2) = 3x(x + 2)$$
$$x - 2 = 3x^2 + 6x$$
$$0 = 3x^2 + 5x + 2$$
$$0 = (3x + 2)(x + 1)$$
$$3x + 2 = 0 \quad \text{or} \quad x + 1 = 0$$
$$3x = -2 \quad \text{or} \quad x = -1$$
$$x = -\frac{2}{3} \quad \text{or} \quad x = -1$$
The solution set is $\left\{ -1, -\frac{2}{3} \right\}$.

43.
$$\frac{2}{3n - 1} = \frac{n + 2}{6}$$
Cross products are equal.
$$(3n - 1)(n + 2) = 2(6)$$
$$3n^2 + 5n - 2 = 12$$
$$3n^2 + 5n - 14 = 0$$
$$n = \frac{-(5) \pm \sqrt{(5)^2 - 4(3)(-14)}}{2(3)}$$
$$n = \frac{-5 \pm \sqrt{193}}{6}$$
The solution set is
$$\left\{ \frac{-5 - \sqrt{193}}{6}, \frac{-5 + \sqrt{193}}{6} \right\}.$$

310

45.

$(n-2)(n+4) = 7$
$n^2 + 2n - 8 = 7$
$n^2 + 2n - 15 = 0$
$(n-3)(n+5) = 0$

$n - 3 = 0$ or $n + 5 = 0$
$n = 3$ or $n = -5$
The solution set is $\{-5, 3\}$.

PROBLEM SET **10.5** Solving Problems Using Quadratic Equations

1. Let n and $n + 1$ represent the consecutive whole numbers.

$$n(n+1) = 306$$
$$n^2 + n = 306$$
$$n^2 + n - 306 = 0$$
$$(n+18)(n-17) = 0$$
$$n + 18 = 0 \quad \text{or} \quad n - 17 = 0$$
$$n = -18 \quad \text{or} \quad n = 17$$

Discard the negative number since whole numbers are wanted. The numbers are 17 and $17 + 1 = 18$.

3. Let x and y represent the two positive integers.

$x + y = 44$ Their sum is 44.
$xy = 475$ Their product is 475.

Solve $x + y = 44$ for y.
$x + y = 44$
$ y = 44 - x$
Substitute $44 - x$ for y in $xy = 475$.
$x(44 - x) = 475$
$44x - x^2 = 475$
$ 0 = x^2 - 44x + 475$
$ 0 = (x - 25)(x - 19)$
$x - 25 = 0 \quad \text{or} \quad x - 19 = 0$
$ x = 25 \quad \text{or} \quad x = 19$
If $x = 25$, then $y = 44 - 25 = 19$.
If $x = 19$, then $y = 44 - 19 = 25$.
The numbers would be 25 and 19.

5. Let x and y represent the numbers.
$x + y = 6$ Their sum is 6.
$xy = 4$ Their product is 4.

Solve $x + y = 6$ for y.
$x + y = 6$
$ y = 6 - x$
Substitute $6 - x$ for y in $xy = 4$.
$x(6 - x) = 4$
$6x - x^2 = 4$
$ 0 = x^2 - 6x + 4$
$$x = \frac{-(-6) \pm \sqrt{(-6)^2 - 4(1)(4)}}{2(1)}$$
$$x = \frac{6 \pm \sqrt{20}}{2}$$
$$x = \frac{6 \pm 2\sqrt{5}}{2}$$
$$x = 3 \pm \sqrt{5}$$
If $x = 3 + \sqrt{5}$, then $y = 6 - \left(3 + \sqrt{5}\right)$
$$= 3 - \sqrt{5}.$$
If $x = 3 - \sqrt{5}$, then $y = 6 - \left(3 - \sqrt{5}\right)$
$$= 3 + \sqrt{5}.$$
The numbers would be $3 - \sqrt{5}$ and $3 + \sqrt{5}$.

7. Let x represent the number, then
$\dfrac{1}{x}$ represents its reciprocal.

$$x + \frac{1}{x} = \frac{3\sqrt{2}}{2}$$

Their sum is $\frac{3\sqrt{2}}{2}$.

$$x^2 + 1 = \frac{3\sqrt{2}}{2}x$$

Multiplied both sides by x.

$$x^2 - \frac{3\sqrt{2}}{2}x = -1$$

$$x^2 - \frac{3\sqrt{2}}{2}x + \frac{18}{16} = -1 + \frac{18}{16}$$

$$\left(x - \frac{3\sqrt{2}}{4}\right)^2 = \frac{2}{16}$$

$$x - \frac{3\sqrt{2}}{4} = \pm\sqrt{\frac{2}{16}} = \pm\frac{\sqrt{2}}{4}$$

$$x = \frac{3\sqrt{2}}{4} \pm \frac{\sqrt{2}}{4}$$

$$x = \frac{3\sqrt{2} \pm \sqrt{2}}{4}$$

$$x = \frac{3\sqrt{2} + \sqrt{2}}{4} \quad \text{or} \quad x = \frac{3\sqrt{2} - \sqrt{2}}{4}$$

$$x = \frac{4\sqrt{2}}{4} = \sqrt{2} \quad \text{or} \quad x = \frac{2\sqrt{2}}{4} = \frac{\sqrt{2}}{2}$$

The number could be $\sqrt{2}$ or $\frac{\sqrt{2}}{2}$.

9. Let n, $n + 2$ and $n + 4$ represent the three consecutive even whole numbers.

$$n^2 + (n + 2)^2 + (n + 4)^2 = 596$$

The sum of their squares is 596.

$$n^2 + n^2 + 4n + 4 + n^2 + 8n + 16 = 596$$
$$3n^2 + 12n + 20 = 596$$
$$3n^2 + 12n - 576 = 0$$
$$n^2 + 4n - 192 = 0$$
$$(n + 16)(n - 12) = 0$$
$$n + 16 = 0 \quad \text{or} \quad n - 12 = 0$$
$$n = -16 \quad \text{or} \quad n = 12$$

Since whole numbers are needed, the negative number is discarded. Thus, the numbers are 12, $12 + 2 = 14$, and $12 + 4 = 16$.

11. Let x represent the number, then $\frac{1}{2}x$ represents one-half of the number.

$$x^2 + \left(\frac{1}{2}x\right)^2 = 80$$

The sum of their squares is 80.

$$x^2 + \frac{1}{4}x^2 = 80$$

Multiply both sides by 4.

$$4x^2 + x^2 = 320$$
$$5x^2 = 320$$
$$x^2 = 64$$
$$x = \pm\sqrt{64} = \pm 8$$

The number could be either -8 or 8.

13. Let w represent the width of the rectangle, then $2w - 4$ represents the length of the rectangle.

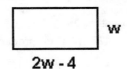

2w - 4

$$w(2w - 4) = 96$$
$$2w^2 - 4w = 96$$
$$2w^2 - 4w - 96 = 0$$
$$w^2 - 2w - 48 = 0$$
$$(w - 8)(w + 6) = 0$$
$$w - 8 = 0 \quad \text{or} \quad w + 6 = 0$$
$$w = 8 \quad \text{or} \quad w = -6$$

The negative solution must be discarded. Thus, the rectangle is 8 meters by $2(8) - 4 = 12$ meters.

15. Let l represent the length and w represent the width of the rectangle.

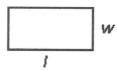

$2l + 2w = 80$ The perimeter is 80 centimeters.
$lw = 375$ The area is 375 square centimeters.

Solve $2l + 2w = 80$ for l.
$2l + 2w = 80$
$l + w = 40$
$l = 40 - w$
Substitute $40 - w$ for l in $lw = 375$.
$w(40 - w) = 375$
$40w - w^2 = 375$
$0 = w^2 - 40w + 375$
$0 = (w - 25)(w - 15)$
$w - 25 = 0$ or $w - 15 = 0$
$w = 25$ or $w = 15$
If $w = 25$, then $l = 40 - 25 = 15$.
If $w = 15$, then $l = 40 - 15 = 25$.
Thus, the rectangle is 15 centimeters by 25 centimeters.

17. Let w represent the width and $\frac{26}{9}w$ the length of the tennis court.

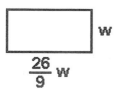

$w\left(\frac{26}{9}w\right) = 2106$
$\frac{26}{9}w^2 = 2106$
$26w^2 = 18{,}954$
$w^2 = 729$
$w = \pm\sqrt{729} = \pm 27$
Discard the negative solution.

If $w = 27$, then $l = \frac{26}{9}(27) = 78$.
The tennis court would be 27 feet by 78 feet.

19. Let s represent the number of seats per row, then $s - 5$ represents the number of rows.

$s(s - 5) = 300$ The product of the seats per row and the number of rows is 300.

$s^2 - 5s = 300$
$s^2 - 5s - 300 = 0$
$(s - 20)(s + 15) = 0$
$s - 20 = 0$ or $s + 15 = 0$
$s = 20$ or $s = -15$
Discard the negative solution.
There are 20 seats per row and
$20 - 5 = 15$ rows in the auditorium.

21. Let l represent the length and w represent the width of the original rectangle. Then $l + 3$ and $w + 3$ represent the lenght and width, respectively, of the larger rectangle.

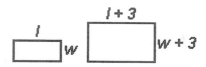

$lw = 63$
The original area is 63 square feet.
$(l + 3)(w + 3) = 63 + 57 = 120$
If the length and width are each increased by 3, the area is increased by 57.

Solve $lw = 63$ for l.
$lw = 63$
$l = \frac{63}{w}$
Substitute $\frac{63}{w}$ for l in the second equation.

313

$$\left(\frac{63}{w} + 3\right)(w + 3) = 120$$

$$63 + \frac{189}{w} + 3w + 9 = 120$$

$$63w + 189 + 3w^2 + 9w = 120w$$

$$3w^2 + 72w + 189 = 120w$$

$$3w^2 - 48w + 189 = 0$$

$$w^2 - 16w + 63 = 0$$

$$(w - 9)(w - 7) = 0$$

$$w - 9 = 0 \quad \text{or} \quad w - 7 = 0$$

$$w = 9 \quad \text{or} \quad w = 7$$

If $w = 9$, then $l = \dfrac{63}{9} = 7$.

If $w = 7$, then $l = \dfrac{63}{7} = 9$.

The original rectangle is 7 feet by 9 feet.

23. Let a and b represent the length of the legs of the right triangle.

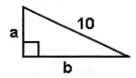

$a + b = 14$	The sum of the lengths is 14 inches.
$a^2 + b^2 = 10^2$	The Pythagorean Theorem using $c = 10$ inches.

Solve $a + b = 14$ for a.

$$a + b = 14$$
$$a = 14 - b$$

Substitute $14 - b$ for a in $a^2 + b^2 = 100$.

$$(14 - b)^2 + b^2 = 100$$
$$196 - 28b + b^2 + b^2 = 100$$
$$196 - 28b + 2b^2 = 100$$
$$2b^2 - 28b + 96 = 0$$
$$b^2 - 14b + 48 = 0$$
$$(b - 8)(b - 6) = 0$$
$$b - 8 = 0 \quad \text{or} \quad b - 6 = 0$$
$$b = 8 \quad \text{or} \quad b = 6$$

If $b = 8$, then $a = 14 - 8 = 6$.

If $b = 6$, then $a = 14 - 6 = 8$.

The lengths of the legs would be 6 inches and 8 inches.

25. Let x represent the uniform width of the frame. Then $(5 + 2x)$ and $(7 + 2x)$ represent the dimensions of the picture and frame together. The area is 80 square inches.

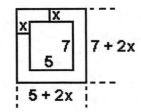

$$(5 + 2x)(7 + 2x) = 80$$
$$35 + 24x + 4x^2 = 80$$
$$4x^2 + 24x - 45 = 0$$
$$(2x + 15)(2x - 3) = 0$$
$$2x + 15 = 0 \quad \text{or} \quad 2x - 3 = 0$$
$$2x = -15 \quad \text{or} \quad 2x = 3$$
$$x = -\frac{15}{2} \quad \text{or} \quad x = \frac{3}{2} = 1\frac{1}{2}$$

Discard the negative solution. The frame width would be $1\dfrac{1}{2}$ inches.

27. Let s represent the number of students. Let c represent the cost per student of the trip.

$cs = 3000$	The cost of the trip was \$3000.
$(c - 25)(s + 10) = 3000$	With ten more students, the cost was \$25 less per student.

Solve $cs = 3000$ for c.

$$cs = 3000$$
$$c = \frac{3000}{s}$$

Substitute $\dfrac{3000}{s}$ for c in the second equation.

314

$$\left(\frac{3000}{s} - 25\right)(s + 10) = 3000$$

$$3000 + \frac{30,000}{s} - 25s - 250 = 3000$$

Multiply both sides by s.

$$3000s + 30,000 - 25s^2 - 250s = 3000s$$
$$-25s^2 + 2750s + 30,000 = 3000s$$
$$-25s^2 - 250s + 30,000 = 0$$

Divide by -25.

$$s^2 + 10s - 1200 = 0$$
$$(s - 30)(s + 40) = 0$$
$$s - 30 = 0 \quad \text{or} \quad s + 40 = 0$$
$$s = 30 \quad \text{or} \qquad s = -40$$

Discard the negative solution.
There were 30 students on the trip.

29. Let x represent the length of a side of the first square. Let y represent the length of a side of the second square. Then $4x$ represents the length of the wire used for the first square and $4y$ represents the length of the wire used for the second square.

$4x + 4y = 56$ The sum of the lengths of the two wires is 56 inches.

$x^2 + y^2 = 100$ The sum of the areas of the two squares is 100 square inches.

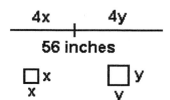

Solve $4x + 4y = 56$ for y.

$$4x + 4y = 56$$
$$x + y = 14$$
$$y = 14 - x$$

Substitute $14 - x$ for y in $x^2 + y^2 = 100$.

$$x^2 + (14 - x)^2 = 100$$
$$x^2 + 196 - 28x + x^2 = 100$$
$$2x^2 - 28x + 196 = 100$$
$$2x^2 - 28x + 96 = 0$$
$$x^2 - 14x + 48 = 0$$
$$(x - 6)(x - 8) = 0$$
$$x - 6 = 0 \quad \text{or} \quad x - 8 = 0$$
$$x = 6 \quad \text{or} \qquad x = 8$$

If $x = 6$, then $y = 14 - 6 = 8$.
If $x = 8$, then $y = 14 - 8 = 6$.
Thus, a side of one square is 6 inches for a wire length of $4(6) = 24$ inches and a side of the other square is 8 inches for a wire length of $4(8) = 32$ inches.

31. Let r represent her rate for the first 36 miles, then $r - 4$ represents the rate for the last 14 miles. Let t represent her time for the first 36 miles, then $3 - t$ represents the time for the last 14 miles.

$rt = 36$ first part

$(r - 4)(3 - t) = 14$ second part .

Solve $rt = 36$ for t.

$$rt = 36$$
$$t = \frac{36}{r}$$

Substitute $\frac{36}{r}$ for t in the second equation.

$$(r - 4)\left(3 - \frac{36}{r}\right) = 14$$
$$3r - 36 - 12 + \frac{144}{r} = 14$$

Multiply both sides by r.

$$3r^2 - 36r - 12r + 144 = 14r$$
$$3r^2 - 48r + 144 = 14r$$
$$3r^2 - 62r + 144 = 0$$
$$(3r - 8)(r - 18) = 0$$
$$3r - 8 = 0 \quad \text{or} \quad r - 18 = 0$$
$$3r = 8 \qquad \text{or} \qquad\qquad r = 18$$
$$r = \frac{8}{3} \quad \text{or} \qquad\qquad r = 18$$

If $r = \frac{8}{3}$, then $r - 4 = \frac{8}{3} - 4 = -\frac{4}{3}$ which is not reasonable.
If $r = 18$, then $r - 4 = 18 - 4 = 14$. The rate for the first 36 miles was 18 miles per hour.

Chapter 10 Review Problem Set

1. $(2x + 7)^2 = 25$

$2x + 7 = \pm\sqrt{25} = \pm 5$

$2x + 7 = 5 \quad$ or $\quad 2x + 7 = -5$

$2x = -2 \quad$ or $\qquad 2x = -12$

$x = -1 \quad$ or $\qquad x = -6$

The solution set is $\{-6, \ -1\}$.

2. $x^2 + 8x = -3$

$x^2 + 8x + 16 = -3 + 16$

$(x + 4)^2 = 13$

$x + 4 = \pm\sqrt{13}$

$x + 4 = -\sqrt{13} \quad$ or $\quad x + 4 = \sqrt{13}$

$x = -4 - \sqrt{13} \quad$ or $\qquad x = -4 + \sqrt{13}$

The solution set is

$\left\{-4 - \sqrt{13}, \ -4 + \sqrt{13}\right\}$.

3. $21x^2 - 13x + 2 = 0$

$(3x - 1)(7x - 2) = 0$

$3x - 1 = 0 \quad$ or $\quad 7x - 2 = 0$

$3x = 1 \quad$ or $\qquad 7x = 2$

$x = \dfrac{1}{3} \quad$ or $\qquad x = \dfrac{2}{7}$

The solution set is $\left\{\dfrac{2}{7}, \dfrac{1}{3}\right\}$.

4. $x^2 = 17x$

$x^2 - 17x = 0$

$x(x - 17) = 0$

$x = 0 \quad$ or $\quad x - 17 = 0$

$x = 0 \quad$ or $\qquad x = 17$

The solution set is $\{0, 17\}$.

5. $n - \dfrac{4}{n} = -3$

$n^2 - 4 = -3n$

$n^2 + 3n - 4 = 0$

$(n + 4)(n - 1) = 0$

$n + 4 = 0 \quad$ or $\quad n - 1 = 0$

$n = -4 \quad$ or $\qquad n = 1$

The solution set is $\{-4, 1\}$.

6. $n^2 - 26n + 165 = 0$

$n = \dfrac{-(-26) \pm \sqrt{(-26)^2 - 4(1)(165)}}{2(1)}$

$n = \dfrac{26 \pm \sqrt{16}}{2}$

$n = \dfrac{26 \pm 4}{2} = 13 \pm 2$

$n = 13 - 2 = 11 \quad$ or $\quad n = 13 + 2 = 15$

The solution set is $\{11, 15\}$.

7. $3a^2 + 7a - 1 = 0$

$a = \dfrac{-(7) \pm \sqrt{(7)^2 - 4(3)(-1)}}{2(3)}$

$a = \dfrac{-7 \pm \sqrt{61}}{6}$

The solution set is

$\left\{\dfrac{-7 - \sqrt{61}}{6}, \dfrac{-7 + \sqrt{61}}{6}\right\}$.

8. $4x^2 - 4x + 1 = 0$

$(2x - 1)^2 = 0$

$2x - 1 = 0$

$2x = 1$

$x = \dfrac{1}{2}$

The solution set is $\left\{\dfrac{1}{2}\right\}$.

9. $5x^2 + 6x + 7 = 0$

$$x = \frac{-(6) \pm \sqrt{(6)^2 - 4(5)(7)}}{2(5)}$$

$$x = \frac{-6 \pm \sqrt{-104}}{10}$$

Since $\sqrt{-104}$ is not a real number, this equation has no real number solutions.

10. $3x^2 + 18x + 15 = 0$
$x^2 + 6x + 5 = 0$
$(x + 5)(x + 1) = 0$
$x + 5 = 0$ or $x + 1 = 0$
$x = -5$ or $x = -1$
The solution set is $\{-5, -1\}$.

11. $3(x - 2)^2 - 2 = 4$
$3(x - 2)^2 = 6$
$(x - 2)^2 = 2$
$x - 2 = \pm\sqrt{2}$
$x = 2 \pm \sqrt{2}$
The solution set is
$\left\{2 - \sqrt{2}, 2 + \sqrt{2}\right\}$.

12. $x^2 + 4x - 14 = 0$

$$x = \frac{-(4) \pm \sqrt{(4)^2 - 4(1)(-14)}}{2(1)}$$

$$x = \frac{-4 \pm \sqrt{72}}{2}$$

$$x = \frac{-4 \pm 6\sqrt{2}}{2} = -2 \pm 3\sqrt{2}$$

The solution set is
$\left\{-2 - 3\sqrt{2}, -2 + 3\sqrt{2}\right\}$.

13. $y^2 = 45$
$y = \pm\sqrt{45}$
$y = \pm 3\sqrt{5}$
The solution set is $\left\{-3\sqrt{5}, 3\sqrt{5}\right\}$.

14. $x(x - 6) = 27$
$x^2 - 6x = 27$
$x^2 - 6x - 27 = 0$
$(x - 9)(x + 3) = 0$
$x - 9 = 0$ or $x + 3 = 0$
$x = 9$ or $x = -3$
The solution set is $\{-3, 9\}$.

15. $x^2 = x$
$x^2 - x = 0$
$x(x - 1) = 0$
$x = 0$ or $x - 1 = 0$
$x = 0$ or $x = 1$
The solution set is $\{0, 1\}$.

16. $n^2 - 4n - 3 = 6$
$n^2 - 4n - 9 = 0$

$$n = \frac{-(-4) \pm \sqrt{(-4)^2 - 4(1)(-9)}}{2(1)}$$

$$n = \frac{4 \pm \sqrt{52}}{2}$$

$$n = \frac{4 \pm 2\sqrt{13}}{2} = 2 \pm \sqrt{13}$$

The solution set is $\left\{2 - \sqrt{13}, 2 + \sqrt{13}\right\}$.

17. $n^2 - 44n + 480 = 0$
$(n - 24)(n - 20) = 0$
$n - 24 = 0$ or $n - 20 = 0$
$n = 24$ or $n = 20$
The solution set is $\{20, 24\}$.

Chapter 10 Review Problem Set

18.
$$\frac{x^2}{4} = x + 1$$
$$x^2 = 4x + 4$$
$$x^2 - 4x - 4 = 0$$
$$x = \frac{-(-4) \pm \sqrt{(-4)^2 - 4(1)(-4)}}{2(1)}$$
$$x = \frac{4 \pm \sqrt{32}}{2}$$
$$x = \frac{4 \pm 4\sqrt{2}}{2} = 2 \pm 2\sqrt{2}$$
The solution set is $\left\{ 2 - 2\sqrt{2},\ 2 + 2\sqrt{2} \right\}$.

19.
$$\frac{5x - 2}{3} = \frac{2}{x + 1}$$
Cross products are equal.
$$(5x - 2)(x + 1) = 2(3)$$
$$5x^2 + 3x - 2 = 6$$
$$5x^2 + 3x - 8 = 0$$
$$(5x + 8)(x - 1) = 0$$
$$5x + 8 = 0 \quad \text{or} \quad x - 1 = 0$$
$$5x = -8 \quad \text{or} \quad x = 1$$
$$x = -\frac{8}{5} \quad \text{or} \quad x = 1$$
The solution set is $\left\{ -\frac{8}{5},\ 1 \right\}$.

20.
$$\frac{-1}{3x - 1} = \frac{2x + 1}{-2}$$
Cross products are equal.
$$(3x - 1)(2x + 1) = -1(-2)$$
$$6x^2 + x - 1 = 2$$
$$6x^2 + x - 3 = 0$$
$$x = \frac{-(1) \pm \sqrt{(1)^2 - 4(6)(-3)}}{2(6)}$$
$$x = \frac{-1 \pm \sqrt{73}}{12}$$
The solution set is
$$\left\{ \frac{-1 - \sqrt{73}}{12},\ \frac{-1 + \sqrt{73}}{12} \right\}.$$

21.
$$\frac{5}{x - 3} + \frac{4}{x} = 6$$
Multiply both sides by $x(x - 3)$.
$$5x + 4(x - 3) = 6x(x - 3)$$
$$5x + 4x - 12 = 6x^2 - 18x$$
$$9x - 12 = 6x^2 - 18x$$
$$0 = 6x^2 - 27x + 12$$
$$0 = 2x^2 - 9x + 4$$
$$0 = (2x - 1)(x - 4)$$
$$2x - 1 = 0 \quad \text{or} \quad x - 4 = 0$$
$$2x = 1 \quad \text{or} \quad x = 4$$
$$x = \frac{1}{2} \quad \text{or} \quad x = 4$$
The solution set is $\left\{ \frac{1}{2},\ 4 \right\}$.

22.
$$\frac{1}{x + 2} - \frac{2}{x} = 3$$
Multiply both sides by $x(x + 2)$.
$$x - 2(x + 2) = 3x(x + 2)$$
$$x - 2x - 4 = 3x^2 + 6x$$
$$-x - 4 = 3x^2 + 6x$$
$$0 = 3x^2 + 7x + 4$$
$$0 = (3x + 4)(x + 1)$$
$$3x + 4 = 0 \quad \text{or} \quad x + 1 = 0$$
$$3x = -4 \quad \text{or} \quad x = -1$$
$$x = -\frac{4}{3} \quad \text{or} \quad x = -1$$
The solution set is $\left\{ -\frac{4}{3},\ -1 \right\}$.

23. Let l represent the length and w represent the width of the rectangle.

$2l + 2w = 42$ The perimeter is 42 inches.

$lw = 108$ The area is 108 square inches.

Solve $2l + 2w = 42$ for l.
$$2l + 2w = 42$$
$$l + w = 21$$
$$l = 21 - w$$
Substitute $21 - w$ for l in $lw = 108$.
$$w(21 - w) = 108$$
$$21w - w^2 = 108$$
$$0 = w^2 - 21w + 108$$
$$0 = (w - 9)(w - 12)$$
$$w - 9 = 0 \quad \text{or} \quad w - 12 = 0$$
$$w = 9 \quad \text{or} \quad w = 12$$
If $w = 9$, then $l = 21 - 9 = 12$.
If $w = 12$, then $l = 21 - 12 = 9$.
Thus, the rectangle is 9 inches by 12 inches.

24. Let n and $n + 1$ represent the two consecutive whole numbers.
$$n(n + 1) = 342 \quad \text{Their product is 342.}$$
$$n^2 + n = 342$$
$$n^2 + n - 342 = 0$$
$$(n + 19)(n - 18) = 0$$
$$n + 19 = 0 \quad \text{or} \quad n - 18 = 0$$
$$n = -19 \quad \text{or} \quad n = 18$$
Since whole numbers must be positive the negative solution will not satisfy the condition. Thus, the numbers are 18 and $18 + 1 = 19$.

25. Let n, $n + 2$ and $n + 4$ represent the three consecutive whole odd numbers.
$$n^2 + (n + 2)^2 + (n + 4)^2 = 251$$
The sum of their squares is 251.
$$n^2 + n^2 + 4n + 4 + n^2 + 8n + 16 = 251$$
$$3n^2 + 12n + 20 = 251$$
$$3n^2 + 12n - 231 = 0$$
$$n^2 + 4n - 77 = 0$$
$$(n + 11)(n - 7) = 0$$
$$n + 11 = 0 \quad \text{or} \quad n - 7 = 0$$
$$n = -11 \quad \text{or} \quad n = 7$$
Discard the negative solution.
If $n = 7$, $n + 2 = 7 + 2 = 9$,
and $n + 4 = 7 + 4 = 11$.
Thus, the numbers are 7, 9, and 11.

26. Let x represent a side of the smaller square, then $3x$ represents a side of the larger square.

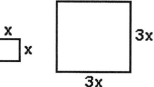

$$x^2 + (3x)^2 = 50 \quad \text{The combined area is}$$
50 square meters.
$$x^2 + 9x^2 = 50$$
$$10x^2 = 50$$
$$x^2 = 5$$
$$x = \pm \sqrt{5}$$
Discard the negative solutions. The lengths of the sides are $\sqrt{5}$ meters and $3\sqrt{5}$ meters.

27. Let a represent the length of one leg of the triangle, then $a - 2$ represents the length of the other leg.

$$a^2 + (a - 2)^2 = \left(2\sqrt{13}\right)^2$$
Use the Pythagorean Theorem.
$$a^2 + a^2 - 4a + 4 = 4(13)$$
$$2a^2 - 4a + 4 = 52$$
$$2a^2 - 4a - 48 = 0$$
$$a^2 - 2a - 24 = 0$$
$$(a - 6)(a + 4) = 0$$
$$a - 6 = 0 \quad \text{or} \quad a + 4 = 0$$
$$a = 6 \quad \text{or} \quad a = -4$$
Discard the negative solution. The lengths of the legs are 6 yards and $6 - 2 = 4$ yards.

28. Let n represent the number of shares purchased. Let v represent the purchase price per share. The $n - 20$ represents the number of shares sold and $v + 8$ represents the selling price per share.

$nv = 720$ The original purchase was \$720.

$(n - 20)(v + 8) = 800$

The sale price was \$720 + \$80 = \$800

Solve $nv = 720$ for v.

$$v = \frac{720}{n}$$

Substitute $\dfrac{720}{n}$ for v in the second equation.

$$(n - 20)\left(\frac{720}{n} + 8\right) = 800$$

$$720 + 8n - \frac{14,400}{n} - 160 = 800$$

Multiply both sides by n.

$$720n + 8n^2 - 14,400 - 160n = 800n$$

$$8n^2 + 560n - 14,400 = 800n$$

$$8n^2 - 240n - 14,400 = 0$$

$$n^2 - 30n - 1800 = 0$$

$$(n - 60)(n + 30) = 0$$

$n - 60 = 0$ or $n + 30 = 0$

$n = 60$ or $n = -30$

Discard the negative solution.

If $n = 60$, then $v = \dfrac{720}{60} = 12$.

The number of shares sold was $60 - 20 = 40$ shares at $12 + 8 = \$20$ per share.

29. Let x represent the width of the strip to be added. Then $(40 + x)$ and $(60 + x)$ represent the dimensions of the parking lot after the addition. The original area is 2400 square meters.

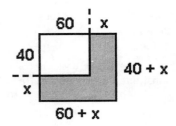

$$(40 + x)(60 + x) = 2400 + 1100$$

$$2400 + 100x + x^2 = 3500$$

$$x^2 + 100x - 1100 = 0$$

$$(x + 110)(x - 10) = 0$$

$x + 110 = 0$ or $x - 10 = 0$

$\qquad x = -110$ or $\qquad x = 10$

Discard the negative solution. The width of the strip would be 10 meters.

30. Let r represent Jean's rate, then $r - 3$ represents Jay's rate. Let t represent Jean's time, then $t - 2$ represents Jay's time.

$rt = 336$ \qquad Jean's trip was 336 miles.

$(r - 3)(t - 2) = 225$ Jay's trip was 225 miles.

Solve $rt = 336$ for t.

$$t = \frac{336}{r}$$

Substitute $\dfrac{336}{r}$ for t in the second equation.

$$(r - 3)\left(\frac{336}{r} - 2\right) = 225$$

$$336 - 2r - \frac{1008}{r} + 6 = 225$$

Multiply both sides by r.

$$336r - 2r^2 - 1008 + 6r = 225r$$

$$-2r^2 + 342r - 1008 = 225r$$

$$-2r^2 + 117r - 1008 = 0$$

$$2r^2 - 117r + 1008 = 0$$

$$r = \frac{-(-117) \pm \sqrt{(-117)^2 - 4(2)(1008)}}{2(2)}$$

$$r = \frac{117 \pm \sqrt{5625}}{4}$$

$$r = \frac{117 \pm 75}{4}$$

$$r = \frac{117 - 75}{4} \quad \text{or} \quad r = \frac{117 + 75}{4}$$

$$r = \frac{42}{4} = \frac{21}{2} \quad \text{or} \quad r = \frac{192}{4} = 48$$

If $r = \frac{21}{2} = 10\frac{1}{2}$, then $r - 3 = 7\frac{1}{2}$.

If $r = 48$, then $r - 3 = 45$.

If Jean's rate is $10\frac{1}{2}$ miles per hour, then

Jay's rate is $7\frac{1}{2}$ miles per hour. Or, if

Jean's rate is 48 miles per hour, then Jay's

rate is 45 miles per hour.

31. Let a represent the length of the two equal legs.

$a^2 + a^2 = 12^2$
Use the Pythagorean Theorem
$$2a^2 = 144$$
$$a^2 = 72$$
$$a = \pm\sqrt{72} = \pm 6\sqrt{2}$$
Discard the negative solution. The length of each leg is $6\sqrt{2}$ inches.

CHAPTER 10 Test

1. Let $a = 4$ and $b = 6$ in the Pythagorean Theorem.

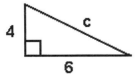

32. Let c represent the length of the hypotenuse.
Then $\frac{1}{2}c$ represents the length of the other leg.

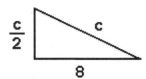

$$8^2 + \left(\frac{1}{2}c\right)^2 = c^2$$
The Pythagorean Theorem
$$64 + \frac{1}{4}c^2 = c^2$$
$$256 + c^2 = 4c^2$$
$$-3c^2 = -256$$
$$c^2 = \frac{256}{3}$$
$$c = \pm\sqrt{\frac{256}{3}} = \pm\frac{16}{\sqrt{3}} = \pm\frac{16\sqrt{3}}{3}$$
Discard the negative solution. The length
of the hypotenuse is $\dfrac{16\sqrt{3}}{3}$ centimeters.

$$c^2 = a^2 + b^2$$
$$c^2 = 4^2 + 6^2$$
$$c^2 = 16 + 36$$
$$c^2 = 52$$
$$c = \sqrt{52} = 2\sqrt{13} \text{ inches}$$

2. Let $c = 14$ and $a = 5$ in the Pythagorean Theorem.

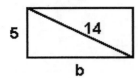

$$c^2 = a^2 + b^2$$
$$14^2 = 5^2 + b^2$$
$$196 = 25 + b^2$$
$$171 = b^2$$
$$b = \sqrt{171} = 13 \text{ to the nearest whole number.}$$

The length of the rectangle is approximately 13 meters.

3. Let x represent a side of the square. Use the Pythagorean Theorem.

$$x^2 + x^2 = 10^2$$
$$2x^2 = 100$$
$$x^2 = 50$$
$$x = \sqrt{50} = 7 \text{ to the nearest whole number.}$$

The length of a side of the square would be approximately 7 inches.

4. Let a represent the side opposite the 60° angle. The hypotenuse is 8 centimeters which is twice the side opposite the 30° angle.

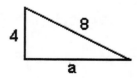

$$a^2 + 4^2 = 8^2$$
$$a^2 + 16 = 64$$
$$a^2 = 48$$
$$a = \sqrt{48} = 4\sqrt{3}$$

The length of a side opposite the 60° angle is $4\sqrt{3}$ centimeters.

5. $(3x + 2)^2 = 49$

$$3x + 2 = \pm\sqrt{49} = \pm 7$$
$$3x + 2 = -7 \quad \text{or} \quad 3x + 2 = 7$$
$$3x = -9 \quad \text{or} \quad 3x = 5$$
$$x = -3 \quad \text{or} \quad x = \frac{5}{3}$$

The solution set is $\left\{ -3, \dfrac{5}{3} \right\}$.

6. $4x^2 = 64$

$$x^2 = 16$$
$$x = \pm\sqrt{16} = \pm 4$$

The solution set is $\{ -4, 4 \}$.

7. $8x^2 - 10x + 3 = 0$

$$(2x - 1)(4x - 3) = 0$$
$$2x - 1 = 0 \quad \text{or} \quad 4x - 3 = 0$$
$$2x = 1 \quad \text{or} \quad 4x = 3$$
$$x = \frac{1}{2} \quad \text{or} \quad x = \frac{3}{4}$$

The solution set is $\left\{ \dfrac{1}{2}, \dfrac{3}{4} \right\}$.

8. $x^2 - 3x - 5 = 0$

$$x = \frac{-(-3) \pm \sqrt{(-3)^2 - 4(1)(-5)}}{2(1)}$$
$$x = \frac{3 \pm \sqrt{29}}{2}$$

The solution set is
$$\left\{ \frac{3 - \sqrt{29}}{2}, \frac{3 + \sqrt{29}}{2} \right\}.$$

9.
$$n^2 + 2n = 9$$
$$n^2 + 2n + 1 = 9 + 1$$
$$(n + 1)^2 = 10$$
$$n + 1 = \pm\sqrt{10}$$
$$n + 1 = -\sqrt{10} \quad \text{or} \quad n + 1 = \sqrt{10}$$
$$n = -1 - \sqrt{10} \quad \text{or} \quad n = -1 + \sqrt{10}$$
The solution set is
$$\left\{ -1 - \sqrt{10}, \ -1 + \sqrt{10} \right\}.$$

10.
$$(2x - 1)^2 = -16$$
$$2x - 1 = \pm\sqrt{-16}$$
Since $\sqrt{-16}$ is not a real number, this equation has no real number solutions.

11.
$$y^2 + 10y = 24$$
$$y^2 + 10y - 24 = 0$$
$$(y + 12)(y - 2) = 0$$
$$y + 12 = 0 \quad \text{or} \quad y - 2 = 0$$
$$y = -12 \quad \text{or} \quad y = 2$$
The solution set is $\{ -12, \ 2 \}$.

12.
$$2x^2 - 3x - 4 = 0$$
$$x = \frac{-(-3) \pm \sqrt{(-3)^2 - 4(2)(-4)}}{2(2)}$$
$$x = \frac{3 \pm \sqrt{41}}{4}$$
The solution set is
$$\left\{ \frac{3 - \sqrt{41}}{4}, \ \frac{3 + \sqrt{41}}{4} \right\}.$$

13.
$$\frac{x - 2}{3} = \frac{4}{x + 1}$$
Cross products are equal.
$$(x - 2)(x + 1) = 3(4)$$
$$x^2 - x - 2 = 12$$
$$x^2 - x - 14 = 0$$
$$x = \frac{-(-1) \pm \sqrt{(-1)^2 - 4(1)(-14)}}{2(1)}$$
$$x = \frac{1 \pm \sqrt{57}}{2}$$
The solution set is
$$\left\{ \frac{1 - \sqrt{57}}{2}, \ \frac{1 + \sqrt{57}}{2} \right\}.$$

14.
$$\frac{2}{x - 1} + \frac{1}{x} = \frac{5}{2}$$
Multiply both sides by $2x(x - 1)$.
$$4x + 2(x - 1) = 5x(x - 1)$$
$$4x + 2x - 2 = 5x^2 - 5x$$
$$6x - 2 = 5x^2 - 5x$$
$$0 = 5x^2 - 11x + 2$$
$$0 = (5x - 1)(x - 2)$$
$$5x - 1 = 0 \quad \text{or} \quad x - 2 = 0$$
$$5x = 1 \quad \text{or} \quad x = 2$$
$$x = \frac{1}{5} \quad \text{or} \quad x = 2$$
The solution set is $\left\{ \dfrac{1}{5}, \ 2 \right\}$.

15.
$$n(n - 28) = -195$$
$$n^2 - 28n = -195$$
$$n^2 - 28n + 195 = 0$$
$$(n - 15)(n - 13) = 0$$
$$n - 15 = 0 \quad \text{or} \quad n - 13 = 0$$
$$n = 15 \quad \text{or} \quad n = 13$$
The solution set is $\{13, \ 15\}$.

Chapter 10 Test

16.
$$n + \frac{3}{n} = \frac{19}{4}$$
Multiply both sides by $4n$.
$$4n^2 + 12 = 19n$$
$$4n^2 - 19n + 12 = 0$$
$$(4n - 3)(n - 4) = 0$$
$$4n - 3 = 0 \quad \text{or} \quad n - 4 = 0$$
$$4n = 3 \quad \text{or} \quad n = 4$$
$$n = \frac{3}{4} \quad \text{or} \quad n = 4$$
The solution set is $\left\{\frac{3}{4}, 4\right\}$.

17.
$$(2x + 1)(3x - 2) = -2$$
$$6x^2 - x - 2 = -2$$
$$6x^2 - x = 0$$
$$x(6x - 1) = 0$$
$$x = 0 \quad \text{or} \quad 6x - 1 = 0$$
$$x = 0 \quad \text{or} \quad 6x = 1$$
$$x = 0 \quad \text{or} \quad x = \frac{1}{6}$$
The solution set is $\left\{0, \frac{1}{6}\right\}$.

18.
$$(7x + 2)^2 - 4 = 21$$
$$(7x + 2)^2 = 25$$
$$7x + 2 = \pm\sqrt{25} = \pm 5$$
$$7x + 2 = -5 \quad \text{or} \quad 7x + 2 = 5$$
$$7x = -7 \quad \text{or} \quad 7x = 3$$
$$x = -1 \quad \text{or} \quad x = \frac{3}{7}$$
The solution set is $\left\{-1, \frac{3}{7}\right\}$.

19.
$$(4x - 1)^2 = 27$$
$$4x - 1 = \pm\sqrt{27} = \pm 3\sqrt{3}$$
$$4x - 1 = -3\sqrt{3} \quad \text{or} \quad 4x - 1 = 3\sqrt{3}$$
$$4x = 1 - 3\sqrt{3} \quad \text{or} \quad 4x = 1 + 3\sqrt{3}$$
$$x = \frac{1 - 3\sqrt{3}}{4} \quad \text{or} \quad x = \frac{1 + 3\sqrt{3}}{4}$$

The solution set is

$$\left\{\frac{1 - 3\sqrt{3}}{4}, \frac{1 + 3\sqrt{3}}{4}\right\}.$$

20. $n^2 - 5n + 7 = 0$
$$n = \frac{-(-5) \pm \sqrt{(-5)^2 - 4(1)(7)}}{2(1)}$$
$$n = \frac{5 \pm \sqrt{-3}}{2}$$
Since $\sqrt{-3}$ is not a real number, this equation has no real number solutions.

21. Let r represent the number of rows, then $2r - 1$ represents the number of seats per row.

$$r(2r - 1) = 120 \quad \begin{array}{l} \text{The product of the} \\ \text{seats per row and the} \\ \text{number of rows is 120.} \end{array}$$

$$2r^2 - r = 120$$
$$2r^2 - r - 120 = 0$$
$$(2r + 15)(r - 8) = 0$$
$$2r + 15 = 0 \quad \text{or} \quad r - 8 = 0$$
$$2r = -15 \quad \text{or} \quad r = 8$$
$$r = -\frac{15}{2} \quad \text{or} \quad r = 8$$
Discard the negative solution. There are $2r - 1 = 2(8) - 1 = 15$ seats per row.

22. Let r represent Stan's rate, then $r + 2$ represents Abu's rate. Let t represent Stan's time, then $t - 2$ represents Abu's time.
$$rt = 72 \qquad \text{Stan's trip was 72 miles.}$$
$$(r + 2)(t - 2) = 56 \quad \text{Abu's trip was 56 miles.}$$

Solve $rt = 72$ for t.
$$t = \frac{72}{r}$$
Substitute $\frac{72}{r}$ for t in the second equation.

$$(r+2)\left(\frac{72}{r}-2\right)=56$$

$$72-2r+\frac{144}{r}-4=56$$

Multiply both sides by r.

$$72r-2r^2+144-4r=56r$$
$$-2r^2+68r+144=56r$$
$$-2r^2+12r+144=0$$
$$r^2-6r-72=0$$
$$(r-12)(r+6)=0$$
$$r-12=0 \quad \text{or} \quad r+6=0$$
$$r=12 \quad \text{or} \quad r=-6$$

If $r=12$, then $r+2=14$.
Abu's rate is 14 miles per hour.

23. Let n and $n+2$ represent the two consecutive odd whole numbers.

$$n(n+2)=255 \quad \text{Their product is 255.}$$
$$n^2+2n=255$$
$$n^2+2n-255=0$$
$$(n+17)(n-15)=0$$
$$n+17=0 \quad \text{or} \quad n-15=0$$
$$n=-17 \quad \text{or} \quad n=15$$

If $n=15$, $n+2=15+2=17$.
Discard the negative solution.
The numbers are 15 and 17.

24. Let x represent a side of the smaller square, then $2x+1$ represents a side of the larger square.

$$x^2+(2x+1)^2=97$$

The combined area is 97 square feet.

$$x^2+4x^2+4x+1=97$$
$$5x^2+4x+1=97$$
$$5x^2+4x-96=0$$
$$(5x+24)(x-4)=0$$
$$5x+24=0 \quad \text{or} \quad x-4=0$$
$$5x=-24 \quad \text{or} \quad x=4$$
$$x=-\frac{24}{5} \quad \text{or} \quad x=4$$

Discard the negative solution. The lengths of the sides of the larger square are $2(4)+1=9$ feet.

25. Let n represent the number of shares of stock she purchased. Then $n-4$ represents the number of shares she sold. Let x represent the original price of a single share, then $x+2$ represents the selling price per share.

$$nx=160 \qquad \text{The original purchase.}$$
$$(x+2)(n-4)=160 \quad \text{The sale of the stock.}$$

Solve $nx=160$ for x.

$$nx=160$$
$$x=\frac{160}{n}$$

Substitute $\frac{160}{n}$ for x in the second equation.

$$\left(\frac{160}{n}+2\right)(n-4)=160$$
$$160-\frac{640}{n}+2n-8=160$$

Multiply both sides by n.

$$160n-640+2n^2-8n=160n$$
$$2n^2+152n-640=160n$$
$$2n^2-8n-640=0$$
$$n^2-4n-320=0$$
$$(n+16)(n-20)=0$$
$$n+16=0 \quad \text{or} \quad n-20=0$$
$$n=-16 \quad \text{or} \quad n=20$$

Discard the negative solution. The original purchase was 20 shares.

Chapter 11 Additional Topics

PROBLEM SET **11.1** Equations and Inequalities Involving Absolute Value

1. $|x| = 4$ is equivalent to $x = -4$ or $x = 4$.
The solution set is $\{-4, 4\}$.

3. $|x| < 1$ means that x must be less than one unit away from zero. Therefore, $|x| < 1$ is equivalent to $x > -1$ and $x < 1$. The solution set is $\{x | x > -1 \text{ and } x < 1\}$ or $(-1, 1)$.

5. $|x| \geq 2$ means that x must be equal to or more than two unit away from zero. Therefore, $|x| \geq 2$ is equivalent to $x \leq -2$ or $x \geq 2$. The solution set is $\{x | x \leq -2 \text{ or } x \geq 2\}$ or $(-\infty, -2] \cup [2, \infty)$.

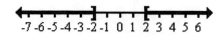

7. $|x + 2| = 1$ is equivalent to $x + 2 = -1$ or $x + 2 = 1$.
$$x + 2 = -1 \quad \text{or} \quad x + 2 = 1$$
$$x = -3 \quad \text{or} \quad x = -1$$
The solution set is $\{-3, -1\}$.

9. $|x - 1| = 2$ is equivalent to $x - 1 = -2$ or $x - 1 = 2$.
$$x - 1 = -2 \quad \text{or} \quad x - 1 = 2$$
$$x = -1 \quad \text{or} \quad x = 3$$
The solution set is $\{-1, 3\}$.

11. $|x - 2| \leq 2$ is equivalent to $x - 2 \geq -2$ and $x - 2 \leq 2$.
$$x - 2 \geq -2 \quad \text{and} \quad x - 2 \leq 2$$
$$x \geq 0 \quad \text{and} \quad x \leq 4$$
The solution set is $\{x | x \geq 0 \text{ and } x \leq 4\}$ or $[0, 4]$.

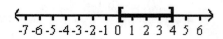

13. $|x + 1| > 3$ is equivalent to $x + 1 < -3$ or $x + 1 > 3$.
$$x + 1 < -3 \quad \text{or} \quad x + 1 > 3$$
$$x < -4 \quad \text{or} \quad x > 2$$
The solution set is $\{x | x < -4 \text{ or } x > 2\}$ or $(-\infty, -4) \cup (2, \infty)$.

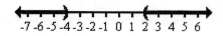

15. $|2x + 1| = 3$ is equivalent to $2x + 1 = -3$ or $2x + 1 = 3$.
$$2x + 1 = -3 \quad \text{or} \quad 2x + 1 = 3$$
$$2x = -4 \quad \text{or} \quad 2x = 2$$
$$x = -2 \quad \text{or} \quad x = 1$$
The solution set is $\{-2, 1\}$.

17. $|5x - 2| = 4$ is equivalent to
$5x - 2 = -4$ or $5x - 2 = 4$.
$$5x - 2 = -4 \quad \text{or} \quad 5x - 2 = 4$$
$$5x = -2 \quad \text{or} \quad 5x = 6$$
$$x = -\frac{2}{5} \quad \text{or} \quad x = \frac{6}{5}$$
The solution set is $\left\{ -\dfrac{2}{5}, \dfrac{6}{5} \right\}$.

19. $|2x - 3| \geq 1$ is equivalent to
$2x - 3 \leq -1$ or $2x - 3 \geq 1$.
$$2x - 3 \leq -1 \quad \text{or} \quad 2x - 3 \geq 1$$
$$2x \leq 2 \quad \text{or} \quad 2x \geq 4$$
$$x \leq 1 \quad \text{or} \quad x \geq 2$$
The solution set is $\{x | x \leq 1 \text{ or } x \geq 2\}$
or $(-\infty, 1] \cup [2, \infty)$.

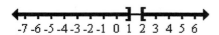

21. $|4x + 3| < 2$ is equivalent to
$4x + 3 > -2$ and $4x + 3 < 2$.
$$4x + 3 > -2 \quad \text{and} \quad 4x + 3 < 2$$
$$4x > -5 \quad \text{and} \quad 4x < -1$$
$$x > -\frac{5}{4} \quad \text{and} \quad x < -\frac{1}{4}$$
The solution set is
$\left\{ x | x > -\dfrac{5}{4} \text{ and } x < -\dfrac{1}{4} \right\}$ or $\left(-\dfrac{5}{4}, -\dfrac{1}{4} \right)$.

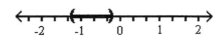

23. $|3x + 6| = 0$ is equivalent to $3x + 6 = 0$.
$$3x + 6 = 0$$
$$3x = -6$$
$$x = -2$$
The solution set is $\{-2\}$.

25. $|3x - 2| > 0$ is equivalent to $3x - 2 \neq 0$.
$$3x - 2 \neq 0$$
$$3x \neq 2$$
$$x \neq \frac{2}{3}$$
The solution set is
$\left\{ x | x \neq \dfrac{2}{3} \right\}$ or $\left(-\infty, \dfrac{2}{3} \right) \cup \left(\dfrac{2}{3}, \infty \right)$.

27. $|3x - 1| = 17$ is equivalent to
$3x - 1 = -17$ or $3x - 1 = 17$.
$$3x - 1 = -17 \quad \text{or} \quad 3x - 1 = 17$$
$$3x = -16 \quad \text{or} \quad 3x = 18$$
$$x = -\frac{16}{3} \quad \text{or} \quad x = 6$$
The solution set is $\left\{ -\dfrac{16}{3}, 6 \right\}$.

29. $|2x + 1| > 9$ is equivalent to
$2x + 1 < -9$ or $2x + 1 > 9$.
$$2x + 1 < -9 \quad \text{or} \quad 2x + 1 > 9$$
$$2x < -10 \quad \text{or} \quad 2x > 8$$
$$x < -5 \quad \text{or} \quad x > 4$$
The solution set is
$\{x | x < -5 \text{ or } x > 4\}$ or
$(-\infty, -5) \cup (4, \infty)$.

31. $|3x - 5| < 19$ is equivalent to
$3x - 5 > -19$ and $3x - 5 < 19$.
$$3x - 5 > -19 \quad \text{and} \quad 3x - 5 < 19$$
$$3x > -14 \quad \text{and} \quad 3x < 24$$
$$x > -\frac{14}{3} \quad \text{and} \quad x < 8$$
The solution set is
$\left\{ x | x > -\dfrac{14}{3} \text{ and } x < 8 \right\}$
or $\left(-\dfrac{14}{3}, 8 \right)$.

Problem Set 11.1

33. $|-3x - 1| = 17$ is equivalent to
$-3x - 1 = -17$ or $-3x - 1 = 17$.
$$-3x - 1 = -17 \quad \text{or} \quad -3x - 1 = 17$$
$$-3x = -16 \quad \text{or} \quad -3x = 18$$
$$x = \frac{16}{3} \quad \text{or} \quad x = -6$$
The solution set is $\left\{ -6, \frac{16}{3} \right\}$.

35. $|4x - 7| \leq 31$ is equivalent to
$4x - 7 \geq -31$ and $4x - 7 \leq 31$.
$$4x - 7 \geq -31 \quad \text{and} \quad 4x - 7 \leq 31$$
$$4x \geq -24 \quad \text{and} \quad 4x \leq 38$$
$$x \geq -6 \quad \text{and} \quad x \leq \frac{38}{4} = \frac{19}{2}$$
The solution set is
$\left\{ x | x \geq -6 \text{ and } x \leq \frac{19}{2} \right\}$ or $\left[-6, \frac{19}{2} \right]$.

37. $|5x + 3| \geq 18$ is equivalent to
$5x + 3 \leq -18$ or $5x + 3 \geq 18$.
$$5x + 3 \leq -18 \quad \text{or} \quad 5x + 3 \geq 18$$
$$5x \leq -21 \quad \text{or} \quad 5x \geq 15$$
$$x \leq -\frac{21}{5} \quad \text{or} \quad x \geq 3$$
The solution set is $\left\{ x | x \leq -\frac{21}{5} \text{ or } x \geq 3 \right\}$
or $\left(-\infty, -\frac{21}{5} \right] \cup [3, \infty)$.

39. $|-x - 2| < 4$ is equivalent to
$-x - 2 > -4$ and $-x - 2 < 4$.
$$-x - 2 > -4 \quad \text{and} \quad -x - 2 < 4$$
$$-x > -2 \quad \text{and} \quad -x < 6$$
$$x < 2 \quad \text{and} \quad x > -6$$
The solution set is
$\{ x | x > -6 \text{ and } x < 2 \}$ or $(-6, 2)$.

41. $|-2x + 1| > 6$ is equivalent to
$-2x + 1 < -6$ or $-2x + 1 > 6$.
$$-2x + 1 < -6 \quad \text{or} \quad -2x + 1 > 6$$
$$-2x < -7 \quad \text{or} \quad -2x > 5$$
$$x > \frac{7}{2} \quad \text{or} \quad x < -\frac{5}{2}$$

The solution set is $\left\{ x | x < -\frac{5}{2} \text{ or } x > \frac{7}{2} \right\}$
or $\left(-\infty, -\frac{5}{2} \right) \cup \left(\frac{7}{2}, \infty \right)$.

43. $|7x| = 0$ has only one solution since absolute value is defined to be the distance between the number and zero on a number line. In this case the distance is zero. The solution set is $\{0\}$.

45. $|x - 6| > -4$ will be true for all real numbers since the absolute value of $x - 6$, regardless of what number is substituted for x, will always be greater than -4. The solution set is $\{ x | x \text{ is a real number} \}$ or $(-\infty, \infty)$.

47. $|x + 4| < -7$ has no solutions since we cannot obtain an absolute value less than -7. The solution set is $\emptyset$.

49. $|x + 6| \leq 0$ has only one solution since absolute value is defined to be the distance between the number and zero on a number line. In this case the distance is zero and the x would be -6. The solution set is $\{ -6 \}$.

53. $-2 < x - 6 < 8$
Add 6 to the left side, middle, and right side.
$$4 < x < 14$$
The solution set is $\{ x | 4 < x < 14 \}$ or $(4, 14)$.

55. $1 \leq 2x + 3 \leq 11$
Subtract 3 from the left side, middle, and right side.
$$-2 \leq 2x \leq 8$$
Divide through by 2.
$$-1 \leq x \leq 4$$
The solution set is $\{ x | -1 \leq x \leq 4 \}$ or $[-1, 4]$.

57. $-4 < \dfrac{x-1}{3} < 2$

Multiply through by 3.

$-12 < x - 1 < 6$

Add 1 to the left side, middle, and right side.

$-11 < \quad x \quad < 7$

The solution set is $\{x|-11 < x < 7\}$ or $(-11, 7)$.

59. $-3 < x + 4 < 3$

Subtract 4 from the left side, middle, and right side.

$-7 < \quad x \quad < -1$

The solution set is $\{x|-7 < x < -1\}$ or $(-7, -1)$.

61. $-7 < 2x - 5 < 7$

Add 5 to the left side, middle, and right side.

$-2 < \quad 2x \quad < 12$

Divide through by 2.

$-1 < \quad x \quad < 6$

The solution set is $\{x|-1 < x < 6\}$ or $(-1, 6)$.

PROBLEM SET $\;$ **11.2** $\;$ **3X3 Systems of Equations**

1. $\begin{pmatrix} 3x + y + 2z = 6 \\ 6y + 5z = -4 \\ -4z = 8 \end{pmatrix}$

Solve equation (3).

$-4z = 8$

$z = -2$

Substitute $z = -2$ into equation (2).

$6y + 5(-2) = -4$

$6y - 10 = -4$

$6y = 6$

$y = 1$

Substitute $z = -2$ and $y = 1$ into equation (1).

$3x + (1) + 2(-2) = 6$

$3x + 1 - 4 = 6$

$3x - 3 = 6$

$3x = 9$

$x = 3$

The solution set is $\{(3, 1, -2)\}$.

3. $\begin{pmatrix} x + 2y - z = 1 \\ y + 2z = 11 \\ 2y - z = 2 \end{pmatrix}$

Multiply equation (2) by -2 and add the result to equation (3).

$\begin{aligned} -2y - 4z &= -22 \\ 2y - z &= 2 \\ \hline -5z &= -20 \\ z &= 4 \end{aligned}$

Substitute $z = 4$ into equation (2).

$y + 2(4) = 11$

$y + 8 = 11$

$y = 3$

Substitute $y = 3$ and $z = 4$ into equation (1).

$x + 2(3) - (4) = 1$

$x + 6 - 4 = 1$

$x + 2 = 1$

$x = -1$

The solution set is $\{(-1, 3, 4)\}$.

5. $\begin{pmatrix} 4x + 3y - 2z = 9 \\ 2x + y = 7 \\ 3x - 2y = 21 \end{pmatrix}$

Multiply equation (2) by 2 and add the result to equation (3).

$\begin{aligned} 4x + 2y &= 14 \\ 3x - 2y &= 21 \\ \hline 7x &= 35 \\ x &= 5 \end{aligned}$

329

Problem Set 11.2

Substitute $x = 5$ into equation (2).

$2(5) + y = 7$

$10 + y = 7$

$y = -3$

Substitute $x = 5$ and $y = -3$ into equation (1).

$4(5) + 3(-3) - 2z = 9$

$20 - 9 - 2z = 9$

$11 - 2z = 9$

$-2z = -2$

$z = 1$

The solution set is $\{(5, -3, 1)\}$.

7. $\begin{pmatrix} x + 2y - 3z = -11 \\ 2x - y + 2z = 3 \\ 4x + 3y + z = 6 \end{pmatrix}$

Multiply equation (1) by -2 and add the result to equation (2).

$-2x - 4y + 6z = 22$

$\underline{2x - y + 2z = 3}$

$-5y + 8z = 25$ (4)

Multiply equation (1) by -4 and add the result to equation (3).

$-4x - 8y + 12z = 44$

$\underline{4x + 3y + z = 6}$

$-5y + 13z = 50$ (5)

Multiply equation (4) by -1 and add the result to equation (5).

$5y - 8z = -25$

$\underline{-5y + 13z = 50}$

$5z = 25$

$z = 5$

Substitute $z = 5$ into equation (4).

$-5y + 8(5) = 25$

$-5y + 40 = 25$

$-5y = -15$

$y = 3$

Substitute $y = 3$ and $z = 5$ into equation (1).

$x + 2(3) - 3(5) = -11$

$x + 6 - 15 = -11$

$x - 9 = -11$

$x = -2$

The solution set is $\{(-2, 3, 5)\}$.

9. $\begin{pmatrix} 4x - 3y + z = 14 \\ 2x + y - 3z = 16 \\ 3x - 4y + 2z = 9 \end{pmatrix}$

Multiply equation (1) by 3 and add the result to equation (2).

$12x - 9y + 3z = 42$

$\underline{2x + y - 3z = 16}$

$14x - 8y = 58$ (4)

Multiply equation (1) by -2 and add the result to equation (3).

$-8x + 6y - 2z = -28$

$\underline{3x - 4y + 2z = 9}$

$-5x + 2y = -19$ (5)

Multiply equation (5) by 4 and add the result to equation (4).

$-20x + 8y = -76$

$\underline{14x - 8y = 58}$

$-6x = -18$

$x = 3$

Substitute $x = 3$ into equation (5).

$-5(3) + 2y = -19$

$-15 + 2y = -19$

$2y = -4$

$y = -2$

Substitute $x = 3$ and $y = -2$ into equation (1).

$4(3) - 3(-2) + z = 14$

$12 + 6 + z = 14$

$18 + z = 14$

$z = -4$

The solution set is $\{(3, -2, -4)\}$.

11. $\begin{pmatrix} 2x + y + 4z = 5 \\ 5x - 2y + z = -10 \\ 3x + 3y - 2z = 4 \end{pmatrix}$

Multiply equation (1) by 2 and add the result to equation (2).

$4x + 2y + 8z = 10$

$\underline{5x - 2y + z = -10}$

$9x + 9z = 0$ (4)

Multiply equation (1) by -3 and add the result to equation (3).

$-6x - 3y - 12z = -15$

$\underline{3x + 3y - 2z = \quad 4}$

$-3x \qquad - 14z = -11 \quad (5)$

Multiply equation (5) by 3 and add the result to equation (4).

$-9x - 42z = -33$

$\underline{9x + \quad 9z = \quad 0}$

$-33z = -33$

$z = 1$

Substitute $z = 1$ into equation (4).

$9x + 9(1) = 0$

$9x + 9 = 0$

$9x = -9$

$x = -1$

Substitute $x = -1$ and $z = 1$ into equation (1).

$2(-1) + y + 4(1) = 5$

$-2 + y + 4 = 5$

$y + 2 = 5$

$y = 3$

The solution set is $\{(-1, 3, 1)\}$.

13. $\begin{pmatrix} x + 3y - 4z = 11 \\ 3x - y + 2z = 5 \\ 2x + 5y - z = 8 \end{pmatrix}$

Multiply equation (1) by -3 and add the result to equation (2).

$-3x - 9y + 12z = -33$

$\underline{3x - \quad y + \quad 2z = \quad 5}$

$-10y + 14z = -28 \quad (4)$

Multiply equation (1) by -2 and add the result to equation (3).

$-2x - 6y + 8z = -22$

$\underline{2x + 5y - \quad z = \quad 8}$

$-y + 7z = -14 \quad (5)$

Multiply equation (5) by -10 and add the result to equation (4).

$-10y - 70z = \quad 140$

$\underline{10y + 14z = -28}$

$-56z = 112$

$z = -2$

Substitute $z = -2$ into equation (4).

$-10y + 14(-2) = -28$

$-10y - 28 = -28$

$-10y = 0$

$y = \dfrac{0}{-10} = 0$

Substitute $y = 0$ and $z = -2$ into equation (1).

$x + 3(0) - 4(-2) = 11$

$x + 0 + 8 = 11$

$x + 8 = 11$

$x = 3$

The solution set is $\{(3, 0, -2)\}$.

15. $\begin{pmatrix} 3x + y - 2z = 3 \\ 2x - 3y + 4z = -2 \\ 4x + z = 6 \end{pmatrix}$

Multiply equation (1) by 3 and add the result to equation (2).

$9x + 3y - 6z = \quad 9$

$\underline{2x - 3y + 4z = -2}$

$11x \qquad - 2z = \quad 7 \quad (4)$

Multiply equation (3) by 2 and add the result to equation (4).

$8x + 2z = 12$

$\underline{11x - 2z = \quad 7}$

$19x \qquad = 19$

$x = 1$

Substitute $x = 1$ into equation (3).

$4(1) + z = 6$

$4 + z = 6$

$z = 2$

Substitute $x = 1$ and $z = 2$ into equation (1).

$3(1) + y - 2(2) = 3$

$3 + y - 4 = 3$

$y - 1 = 3$

$y = 4$

The solution set is $\{(1, 4, 2)\}$.

331

Problem Set 11.2

17. Let $x =$ number of quarters
$\quad\quad y =$ number of dimes
$\quad\quad z =$ number of nickels

$x + y + z = 20$
$y + z = x \quad\quad\quad\quad \Rightarrow \quad -x + y + z = 0$
$25x + 10y + 5z = 340$

$$\begin{pmatrix} x + y + z = 20 \\ -x + y + z = 0 \\ 25x + 10y + 5z = 340 \end{pmatrix}$$

Add equation (1) and equation (2).
$$\begin{array}{r} x + y + z = 20 \\ -x + y + z = 0 \\ \hline 2y + 2z = 20 \quad (4) \end{array}$$
Multiply equation (1) by -25 and add
the result to equation (3).
$$\begin{array}{r} -25x - 25y - 25z = -500 \\ 25x + 10y + 5z = 340 \\ \hline -15y - 20z = -160 \quad (5) \end{array}$$
Multiply equation (4) by 10 and add
the result to equation (5).
$$\begin{array}{r} 20y + 20z = 200 \\ -15y - 20z = -160 \\ \hline 5y = 40 \\ y = 8 \end{array}$$

Substitute $y = 8$ into equation (4).
$2(8) + 2z = 20$
$\quad 16 + 2z = 20$
$\quad\quad\quad 2z = 4$
$\quad\quad\quadz = 2$
Substitute $y = 8$ and $z = 2$ into
equation (1).
$x + 8 + 2 = 20$
$\quad x + 10 = 20$
$\quad\quad\quad x = 10$
There are 10 quarters, 8 dimes, and 2 nickels.

19. Let $a =$ measure of $\angle A$
$\quad\quad b =$ measure of $\angle B$
$\quad\quad c =$ measure of $\angle C$

$a = 5b \quad\quad\quad\quad \Rightarrow \quad a - 5b = 0$
$b + c = a - 60 \quad\quad \Rightarrow \quad -a + b + c = -60$
$a + b + c = 180$

$$\begin{pmatrix} a + b + c = 180 \\ -a + b + c = -60 \\ a - 5b = 0 \end{pmatrix}$$

Multiply equation (1) by -1 and add
the result to equation (2).
$$\begin{array}{r} -a - b - c = -180 \\ -a + b + c = -60 \\ \hline -2a = -240 \\ a = 120 \end{array}$$
Substitute $a = 120$ into equation (3).
$120 - 5b = 0$
$\quad\quad -5b = -120$
$\quad\quad\quadb = 24$
Substitute $a = 120$ and $b = 24$ into
equation (1).
$120 + 24 + c = 180$
$\quad\quad 144 + c = 180$
$\quad\quad\quad\quad\quad c = 36$
The measure of $\angle A = 120°$,
the measure of $\angle B = 24°$, and
the measure of $\angle C = 36°$.

21. Let $x =$ wages per hour of plumber
$\quad\quad y =$ wages per hour of apprentice
$\quad\quad z =$ wages per hour of laborer

$x + y + z = 80$
$x = y + z + 20 \quad \Rightarrow \quad x - y - z = 20$
$x = 5z \quad\quad\quad\quad\quad \Rightarrow \quad x - 5z = 0$

$$\begin{pmatrix} x + y + z = 80 \\ x - y - z = 20 \\ x - 5z = 0 \end{pmatrix}$$

Add equation (1) and equation (2).
$$\begin{array}{r} x + y + z = 80 \\ x - y - z = 20 \\ \hline 2x = 100 \\ x = 50 \end{array}$$

332

Substitute $x = 50$ into equation (3).

$50 - 5z = 0$

$-5z = -50$

$z = 10$

Substitute $x = 50$ and $z = 10$ into equation (1).

$50 + y + 10 = 80$

$y + 60 = 80$

$y = 20$

The plumber's wages are \$50 per hour, the apprentice's wages are \$20 per hour, and the laborer's wages are \$10 per hour.

23. Let $x =$ price per pound of peaches

$y =$ price per pound of cherries

$z =$ price per pound of pears

$$\begin{pmatrix} 2x + y + 3z = 5.64 \\ x + 2y + 2z = 4.65 \\ 2x + 4y + z = 7.23 \end{pmatrix}$$

Multiply equation (1) by -2 and add the result to equation (2).

$-4x - 2y - 6z = -11.28$

$\underline{x + 2y + 2z = \quad 4.65}$

$-3x \qquad - 4z = - \ 6.63 \quad (4)$

Multiply equation (1) by -4 and add the result to equation (3).

$-8x - 4y - 12z = -22.56$

$\underline{2x + 4y + \quad z = \quad 7.23}$

$-6x \qquad - 11z = -15.33 \quad (5)$

Multiply equation (4) by -2 and add the result to equation (5).

$6x + 8z = \quad 13.26$

$\underline{-6x - 11z = -15.33}$

$-3z = - \ 2.07$

$z = 0.69$

Substitute $z = 0.69$ into equation (4).

$-3x - 4(0.69) = -6.63$

$-3x - 2.76 = -6.63$

$-3x = -3.87$

$x = 1.29$

Substitute $x = 1.29$ and $z = 0.69$ into equation (1).

$2(1.29) + y + 3(0.69) = 5.64$

$2.58 + y + 2.07 = 5.64$

$y + 4.65 = 5.64$

$y = 0.99$

Peaches cost \$1.29 per pound, cherries cost \$0.99 per pound, and pears cost \$0.69 per pound.

25. Let $x =$ cost of helmet

$y =$ cost of jacket

$z =$ cost of gloves

$x + y + z = 650$

$y = x + 100 \qquad \Rightarrow \qquad -x + y = 100$

$x + z = y - 50 \qquad \Rightarrow \qquad x - y + z = -50$

$$\begin{pmatrix} x + y + z = 650 \\ -x + y = 100 \\ x - y + z = -50 \end{pmatrix}$$

Multiply equation (1) by -1 and add the result to equation (3).

$-x - y - z = -650$

$\underline{x - y + z = - \quad 50}$

$-2y \qquad = -700$

$y = 350$

Substitute $y = 350$ into equation (2).

$-x + 350 = 100$

$-x = -250$

$x = 250$

Substitute $x = 250$ and $y = 350$ into equation (1).

$250 + 350 + z = 650$

$600 + z = 650$

$z = 50$

The helmet cost \$250, the jacket cost \$350, and the gloves cost \$50.

Problem Set 11.2

Further Investigations

29. $\begin{pmatrix} x + 2y - 3z = 1 \\ 2x - y + 2z = 3 \\ 3x + y - z = 4 \end{pmatrix}$

Multiply equation (1) by -2 and add the result to equation (2).
$$-2x - 4y + 6z = -2$$
$$\underline{2x - y + 2z = 3}$$
$$-5y + 8z = 1 \qquad (4)$$
Multiply equation (1) by -3 and add the result to equation (3).
$$-3x - 6y + 9z = -3$$
$$\underline{3x + y - z = 4}$$
$$-5y + 8z = 1 \qquad (5)$$
Multiply equation (4) by -1 and add the result to equation (5).
$$5y - 8z = -1$$
$$\underline{-5y + 8z = 1}$$
$$0 = 0$$
Since $0 = 0$ is a true statement, the system is dependent and has an infinite number of solutions.

31. $\begin{pmatrix} 3x - y + 2z = -1 \\ 2x + 3y + z = 8 \\ 8x + y + 5z = 4 \end{pmatrix}$

Multiply equation (1) by 3 and add the result to equation (2).
$$9x - 3y + 6z = -3$$
$$\underline{2x + 3y + z = 8}$$
$$11x + 7z = 5 \qquad (4)$$
Add equation (1) and equation (3).
$$3x - y + 2z = -1$$
$$\underline{8x + y + 5z = 4}$$
$$11x + 7z = 3 \qquad (5)$$
Multiply equation (4) by -1 and add the result to equation (5).
$$-11x - 7z = -5$$
$$\underline{11x + 7z = 3}$$
$$0 = -2$$
Since $0 \neq -2$, the system is inconsistent and the solution set is $\emptyset$.

PROBLEM SET | **11.3** **Factional Exponents**

1. $\sqrt{81} = 9$ because $9^2 = 81$.

3. $-\sqrt{100} = -10$ because $-(10^2) = -100$.

5. $\sqrt[3]{125} = 5$ because $(5)^3 = 125$

7. $\sqrt[3]{-64} = -4$ because $(-4)^3 = -64$.

9. $\dfrac{\sqrt[3]{64}}{\sqrt{49}} = \dfrac{4}{7}$ because $\dfrac{4^3}{7^2} = \dfrac{64}{49}$.

11. $\sqrt[4]{81} = 3$ because $3^4 = 81$.

13. $\sqrt[5]{-243} = -3$ because $(-3)^5 = -243$.

15. $64^{\frac{1}{2}} = \sqrt{64} = 8$

17. $64^{\frac{2}{3}} = \left(\sqrt[3]{64}\right)^2 = 4^2 = 16$

19. $(-64)^{\frac{2}{3}} = \left(\sqrt[3]{-64}\right)^2 = (-4)^2 = 16$

334

21. $4^{\frac{5}{2}} = \left(\sqrt{4}\right)^5 = 2^5 = 32$

23. $32^{-\frac{1}{5}} = \left(\sqrt[5]{32}\right)^{-1} = 2^{-1} = \dfrac{1}{2}$

25. $-27^{\frac{1}{3}} = -\sqrt[3]{27} = -3$

27. $16^{-\frac{3}{4}} = \left(\sqrt[4]{16}\right)^{-3} = 2^{-3} = \dfrac{1}{2^3} = \dfrac{1}{8}$

29. $\left(\dfrac{2}{3}\right)^{-3} = \dfrac{1}{\left(\dfrac{2}{3}\right)^3} = \dfrac{1}{\dfrac{8}{27}} = \dfrac{27}{8}$

31. $\left(\dfrac{16}{64}\right)^{-\frac{1}{2}} = \left(\dfrac{1}{4}\right)^{-\frac{1}{2}} = 4^{\frac{1}{2}} = \sqrt{4} = 2$

33. $125^{\frac{4}{3}} = \left(\sqrt[3]{125}\right)^4 = 5^4 = 625$

35. $-16^{\frac{5}{4}} = -\left(\sqrt[4]{16}\right)^5 = -(2)^5 = -32$

37. $\left(\dfrac{1}{32}\right)^{\frac{3}{5}} = \left(\sqrt[5]{\dfrac{1}{32}}\right)^3 = \left(\dfrac{1}{2}\right)^3 = \dfrac{1}{8}$

39. $2^{\frac{1}{3}} \cdot 2^{\frac{2}{3}} = 2^{\frac{1}{3}+\frac{2}{3}} = 2^{\frac{3}{3}} = 2^1 = 2$

41. $3^{\frac{4}{3}} \cdot 3^{\frac{5}{3}} = 3^{\frac{4}{3}+\frac{5}{3}} = 3^{\frac{9}{3}} = 3^3 = 27$

43. $\dfrac{2^{\frac{1}{2}}}{2^{\frac{1}{2}}} = 1$

45. $\dfrac{3^{-\frac{2}{3}}}{3^{\frac{1}{3}}} = 3^{-\frac{1}{3}-\frac{2}{3}} = 3^{-\frac{3}{3}} = 3^{-1} = \dfrac{1}{3}$

47. $\dfrac{2^{\frac{9}{4}}}{2^{\frac{1}{4}}} = 2^{\frac{9}{4}-\frac{1}{4}} = 2^{\frac{8}{4}} = 2^2 = 4$

49. $\dfrac{7^{\frac{4}{3}}}{7^{-\frac{2}{3}}} = 7^{\frac{4}{3}-\left(-\frac{2}{3}\right)} = 7^{\frac{4}{3}+\frac{2}{3}} = 7^{\frac{6}{3}} = 7^2 = 49$

51. $x^{\frac{1}{2}} \cdot x^{\frac{1}{4}} = x^{\frac{1}{2}+\frac{1}{4}} = x^{\frac{2}{4}+\frac{1}{4}} = x^{\frac{3}{4}}$

53. $a^{\frac{2}{3}} \cdot a^{\frac{3}{4}} = a^{\frac{2}{3}+\frac{3}{4}} = a^{\frac{8}{12}+\frac{9}{12}} = a^{\frac{17}{12}}$

55. $\left(3x^{\frac{1}{4}}\right)\left(5x^{\frac{1}{3}}\right) = 3 \cdot 5 \cdot x^{\frac{1}{4}} \cdot x^{\frac{1}{3}} =$
$15x^{\frac{1}{4}+\frac{1}{3}} = 15x^{\frac{3}{12}+\frac{4}{12}} = 15x^{\frac{7}{12}}$

57. $\left(4x^{\frac{2}{3}}\right)\left(6x^{\frac{1}{4}}\right) = 4 \cdot 6 \cdot x^{\frac{2}{3}} \cdot x^{\frac{1}{4}} =$
$24x^{\frac{2}{3}+\frac{1}{4}} = 24x^{\frac{8}{12}+\frac{3}{12}} = 24x^{\frac{11}{12}}$

59. $\left(2y^{\frac{2}{3}}\right)\left(y^{-\frac{1}{4}}\right) = 2 \cdot y^{\frac{2}{3}} \cdot y^{-\frac{1}{4}} =$
$2y^{\frac{2}{3}-\frac{1}{4}} = 2y^{\frac{8}{12}-\frac{3}{12}} = 2y^{\frac{5}{12}}$

61. $\left(5n^{\frac{3}{4}}\right)\left(2n^{-\frac{1}{2}}\right) = 5 \cdot 2 \cdot n^{\frac{3}{4}} \cdot n^{-\frac{1}{2}} =$
$10n^{\frac{3}{4}-\frac{1}{2}} = 10n^{\frac{3}{4}-\frac{2}{4}} = 10n^{\frac{1}{4}}$

63. $\left(2x^{\frac{1}{3}}\right)\left(x^{-\frac{1}{2}}\right) = 2 \cdot x^{\frac{1}{3}} \cdot x^{-\frac{1}{2}} =$
$2x^{\frac{1}{3}-\frac{1}{2}} = 2x^{\frac{2}{6}-\frac{3}{6}} = 2x^{-\frac{1}{6}} = \dfrac{2}{x^{\frac{1}{6}}}$

65. $\left(5x^{\frac{1}{2}}y\right)^2 = (5)^2\left(x^{\frac{1}{2}}\right)^2(y)^2 = 25xy^2$

67. $\left(4x^{\frac{1}{4}}y^{\frac{1}{2}}\right)^3 = (4)^3\left(x^{\frac{1}{4}}\right)^3\left(y^{\frac{1}{2}}\right)^3 = 64x^{\frac{3}{4}}y^{\frac{3}{2}}$

69. $(8x^6y^3)^{\frac{1}{3}} = (8)^{\frac{1}{3}}(x^6)^{\frac{1}{3}}(y^3)^{\frac{1}{3}} =$
$\sqrt[3]{8}x^{\frac{6}{3}}y^{\frac{3}{3}} = 2x^2y$

71. $\dfrac{24x^{\frac{3}{5}}}{6x^{\frac{1}{3}}} = \dfrac{24}{6}x^{\frac{3}{5}-\frac{1}{3}} = 4x^{\frac{9}{15}-\frac{5}{15}} = 4x^{\frac{4}{15}}$

Problem Set 11.3

73. $\dfrac{48b^{\frac{1}{3}}}{12b^{\frac{3}{4}}} = \dfrac{48}{12}b^{\frac{1}{3}-\frac{3}{4}} = 4b^{\frac{4}{12}-\frac{9}{12}} = 4b^{-\frac{5}{12}} = \dfrac{4}{b^{\frac{5}{12}}}$

75. $\dfrac{27n^{-\frac{1}{3}}}{9n^{-\frac{1}{3}}} = \dfrac{27}{9}(1) = 3$

77. $\left[\dfrac{3x^{\frac{1}{3}}}{2x^{\frac{1}{2}}}\right]^2 = \dfrac{(3)^2\left(x^{\frac{1}{3}}\right)^2}{(2)^2\left(x^{\frac{1}{2}}\right)^2} = \dfrac{9x^{\frac{2}{3}}}{4x^{\frac{2}{2}}} =$

$\dfrac{9}{4}x^{\frac{2}{3}-1} = \dfrac{9}{4}x^{-\frac{1}{3}} = \dfrac{9}{4x^{\frac{1}{3}}}$

79. $\left[\dfrac{5x^{\frac{1}{2}}}{6y^{\frac{1}{3}}}\right]^3 = \dfrac{(5)^3\left(x^{\frac{1}{2}}\right)^3}{(6)^3\left(y^{\frac{1}{3}}\right)^3} = \dfrac{125x^{\frac{3}{2}}}{216y^{\frac{3}{3}}} = \dfrac{125x^{\frac{3}{2}}}{216y}$

83 a. $\sqrt[3]{21952} = 28$

b. $\sqrt[3]{42875} = 35$

c. $\sqrt[4]{83521} = 17$

d. $\sqrt[4]{3111696} = 42$

85 a. $5^{\frac{3}{2}} \approx 11.18$

b. $8^{\frac{4}{5}} \approx 5.28$

c. $17^{\frac{2}{5}} \approx 3.11$

d. $19^{\frac{5}{2}} \approx 1573.56$

e. $12^{\frac{3}{4}} \approx 6.45$

f. $14^{\frac{2}{3}} \approx 5.81$

PROBLEM SET **11.4** **Complex Numbers**

1. $\sqrt{-64} = \sqrt{-1}\,\sqrt{64} = i\sqrt{64} = 8i$

3. $\sqrt{-\dfrac{25}{9}} = \sqrt{-1}\,\sqrt{\dfrac{25}{9}} = i\sqrt{\dfrac{25}{9}} = \dfrac{5}{3}i$

5. $\sqrt{-11} = \sqrt{-1}\,\sqrt{11} = i\sqrt{11}$

7. $\sqrt{-50} = \sqrt{-1}\,\sqrt{50} = i\sqrt{25}\sqrt{2} = 5i\sqrt{2}$

9. $\sqrt{-48} = \sqrt{-1}\,\sqrt{48} = i\sqrt{16}\sqrt{3} = 4i\sqrt{3}$

11. $\sqrt{-54} = \sqrt{-1}\,\sqrt{54} = i\sqrt{9}\sqrt{6} = 3i\sqrt{6}$

13. $(3+8i)+(5+9i) =$
$(3+5)+(8+9)i =$
$8+17i$

15. $(7-6i)+(3-4i) =$
$(7+3)+(-6-4)i =$
$10-10i$

17. $(10+4i)-(6+2i) =$
$(10-6)+(4-2)i =$
$4+2i$

19. $(5+2i)-(7+8i) =$
$(5-7)+(2-8)i =$
$-2-6i$

21. $(-2-i)-(3-4i) =$
$(-2-3)+(-1+4)i =$
$-5+3i$

23. $(-4-7i)+(-8-9i) =$
$(-4-8)+(-7-9)i =$
$-12-16i$

25. $(0-6i)+(-10+2i) =$
$(0-10)+(-6+2)i =$
$-10-4i$

27. $(-9+7i)-(-8-5i) =$
$(-9+8)+(7+5)i =$
$-1+12i$

29. $(-10-4i)-(10+4i) =$
$(-10-10)+(-4-4)i =$
$-20-8i$

31. $\left(\dfrac{1}{2} + \dfrac{2}{3}i\right) + \left(\dfrac{1}{3} - \dfrac{1}{4}i\right) =$

$\left(\dfrac{1}{2} + \dfrac{1}{3}\right) + \left(\dfrac{2}{3} - \dfrac{1}{4}\right)i =$

$\left(\dfrac{3}{6} + \dfrac{2}{6}\right) + \left(\dfrac{8}{12} - \dfrac{3}{12}\right)i =$

$\dfrac{5}{6} + \dfrac{5}{12}i$

33. $\left(\dfrac{3}{5} - \dfrac{1}{4}i\right) - \left(\dfrac{2}{3} - \dfrac{5}{6}i\right) =$

$\left(\dfrac{3}{5} - \dfrac{2}{3}\right) + \left(-\dfrac{1}{4} + \dfrac{5}{6}\right)i =$

$\left(\dfrac{9}{15} - \dfrac{10}{15}\right) + \left(-\dfrac{3}{12} + \dfrac{10}{12}\right)i =$

$-\dfrac{1}{15} + \dfrac{7}{12}i$

35. $(7i)(8i) = 56i^2 =$
$56(-1) = -56 + 0i$

37. $2i(6 + 3i) = 12i + 6i^2 =$
$12i + 6(-1) = 12i - 6 =$
$-6 + 12i$

39. $-4i(-5 - 6i) = 20i + 24i^2 =$
$20i + 24(-1) = -24 + 20i$

41. $(2 + 3i)(5 + 4i) =$
$10 + 8i + 15i + 12i^2 =$

$10 + 23i + 12(-1) = -2 + 23i$

43. $(7 - 3i)(8 + i) =$
$56 + 7i - 24i - 3i^2 =$
$56 - 17i - 3(-1) =$
$59 - 17i$

45. $(-2 - 3i)(6 - 3i) =$
$-12 + 6i - 18i + 9i^2 =$
$-12 - 12i + 9(-1) =$
$-21 - 12i$

47. $(-1 - 4i)(-2 - 7i) =$
$2 + 7i + 8i + 28i^2 =$
$2 + 15i + 28(-1) =$
$-26 + 15i$

49. $(4 + 5i)^2 =$
$(4 + 5i)(4 + 5i) =$
$16 + 20i + 20i + 25i^2 =$
$16 + 40i + 25(-1) =$
$-9 + 40i$

51. $(5 - 6i)(5 + 6i) =$
$25 - 30i + 30i - 36i^2 =$
$25 - 36(-1) =$
$61 + 0i$

53. $(-2 + i)(-2 - i) =$
$4 + 2i - 2i - i^2 =$
$4 - (-1) =$
$5 + 0i$

PROBLEM SET | **11.5** **Quadratic Equations: Complex Solutions**

1. $x^2 = -64$

$x = \pm\sqrt{-64}$

$x = \pm i\sqrt{64} = \pm 8i$

The solution set is $\{-8i, 8i\}$.

3. $(x - 2)^2 = -1$

$x - 2 = \pm\sqrt{-1} = \pm i$

$x - 2 = -i$ or $x - 2 = i$

$x = 2 - i$ or $x = 2 + i$

The solution set is $\{2 - i, 2 + i\}$.

5. $(x + 5)^2 = -13$

$x + 5 = \pm\sqrt{-13} = \pm i\sqrt{13}$

$x + 5 = -i\sqrt{13}$ or $x + 5 = i\sqrt{13}$

$x = -5 - i\sqrt{13}$ or $x = -5 + i\sqrt{13}$

The solution set is

$\left\{-5 - i\sqrt{13}, \ -5 + i\sqrt{13}\right\}.$

Problem Set 11.5

7. $(x-3)^2 = -18$

$x - 3 = \pm\sqrt{-18} = \pm 3i\sqrt{2}$

$x - 3 = -3i\sqrt{2}$ or $x - 3 = 3i\sqrt{2}$

$x = 3 - 3i\sqrt{2}$ or $x = 3 + 3i\sqrt{2}$

The solution set is

$\left\{3 - 3i\sqrt{2},\ 3 + 3i\sqrt{2}\right\}.$

9. $(5x-1)^2 = 9$

$5x - 1 = \pm\sqrt{9} = \pm 3$

$5x - 1 = -3$ or $5x - 1 = 3$

$5x = -2$ or $5x = 4$

$x = -\dfrac{2}{5}$ or $x = \dfrac{4}{5}$

The solution set is $\left\{-\dfrac{2}{5},\ \dfrac{4}{5}\right\}.$

11. $a^2 - 3a - 4 = 0$

$(a+1)(a-4) = 0$

$a + 1 = 0$ or $a - 4 = 0$

$a = -1$ or $a = 4$

The solution set is $\{-1,\ 4\}.$

13. $t^2 + 6t = -12$

$t^2 + 6t + 9 = -12 + 9$

$(t+3)^2 = -3$

$t + 3 = \pm\sqrt{-3} = \pm i\sqrt{3}$

$t + 3 = -i\sqrt{3}$ or $t + 3 = i\sqrt{3}$

$t = -3 - i\sqrt{3}$ or $t = -3 + i\sqrt{3}$

The solution set is

$\left\{-3 - i\sqrt{3},\ -3 + i\sqrt{3}\right\}.$

15. $n^2 - 6n + 13 = 0$

$n = \dfrac{-(-6) \pm \sqrt{(-6)^2 - 4(1)(13)}}{2(1)}$

$n = \dfrac{6 \pm \sqrt{-16}}{2}$

$n = \dfrac{6 \pm 4i}{2} = 3 \pm 2i$

The solution set is $\{3 - 2i,\ 3 + 2i\}.$

17. $x^2 - 4x + 20 = 0$

$x = \dfrac{-(-4) \pm \sqrt{(-4)^2 - 4(1)(20)}}{2(1)}$

$x = \dfrac{4 \pm \sqrt{-64}}{2}$

$x = \dfrac{4 \pm 8i}{2} = 2 \pm 4i$

The solution set is $\{2 - 4i,\ 2 + 4i\}.$

19. $3x^2 - 2x + 1 = 0$

$x = \dfrac{-(-2) \pm \sqrt{(-2)^2 - 4(3)(1)}}{2(3)}$

$x = \dfrac{2 \pm \sqrt{-8}}{6}$

$x = \dfrac{2 \pm 2i\sqrt{2}}{6} = \dfrac{1 \pm i\sqrt{2}}{3}$

The solution set is

$\left\{\dfrac{1 - i\sqrt{2}}{3},\ \dfrac{1 + i\sqrt{2}}{3}\right\}.$

21. $2x^2 - 3x - 5 = 0$

$(2x - 5)(x + 1) = 0$

$2x - 5 = 0$ or $x + 1 = 0$

$2x = 5$ or $x = -1$

$x = \dfrac{5}{2}$ or $x = -1$

The solution set is $\left\{-1,\ \dfrac{5}{2}\right\}.$

23. $y^2 - 2y = -19$

$y^2 - 2y + 1 = -19 + 1$

$(y-1)^2 = -18$

$y - 1 = \pm\sqrt{-18} = \pm 3i\sqrt{2}$

$y - 1 = -3i\sqrt{2}$ or $y - 1 = 3i\sqrt{2}$

$y = 1 - 3i\sqrt{2}$ or $y = 1 + 3i\sqrt{2}$

The solution set is

$\left\{1 - 3i\sqrt{2},\ 1 + 3i\sqrt{2}\right\}.$

25. $x^2 - 4x + 7 = 0$

$$x = \frac{-(-4) \pm \sqrt{(-4)^2 - 4(1)(7)}}{2(1)}$$

$$x = \frac{4 \pm \sqrt{-12}}{2}$$

$$x = \frac{4 \pm 2i\sqrt{3}}{2} = 2 \pm i\sqrt{3}$$

The solution set is $\left\{ 2 - i\sqrt{3}, \, 2 + i\sqrt{3} \right\}$.

27. $4x^2 - x + 2 = 0$

$$x = \frac{-(-1) \pm \sqrt{(-1)^2 - 4(4)(2)}}{2(4)}$$

$$x = \frac{1 \pm \sqrt{-31}}{8}$$

$$x = \frac{1 \pm i\sqrt{31}}{8}$$

The solution set is
$$\left\{ \frac{1 - i\sqrt{31}}{8}, \, \frac{1 + i\sqrt{31}}{8} \right\}.$$

29. $6x^2 + 2x + 1 = 0$

$$x = \frac{-(2) \pm \sqrt{(2)^2 - 4(6)(1)}}{2(6)}$$

$$x = \frac{-2 \pm \sqrt{-20}}{12}$$

$$x = \frac{-2 \pm 2i\sqrt{5}}{12} = \frac{-1 \pm i\sqrt{5}}{6}$$

The solution set is
$$\left\{ \frac{-1 - i\sqrt{5}}{6}, \, \frac{-1 + i\sqrt{5}}{6} \right\}.$$

PROBLEM SET **11.6** **Pie, Bar, and Line Graphs**

1. Kayaks   8%
Sailboats   6%
Together   14%
14% of the boat rentals were from kayaks or sailboats.

3. Jon Boats were 20%.
20% of 2400 = .20(2400) = 480
Jon Boats were rented 480 times.

5. Sailboats   6%
Ski Boats   28%
Together   34%
The rentals not from Sailboats or Ski Boats are 100% − 34% = 66%.

7. Physics at 8% is the least popular.

9. Chemistry   13%
Physics   8%
Together   21%
21% of the students chose chemistry or physics.

11. Oceanography   22%
Astronomy   16%
Together   38%
The percent of students who did not choose oceanography or astronomy is 100% − 38% = 62%.

13. Space Center   1500
Water Park   500
Difference   1000
1000 more people preferred the space center to the water park.

15. Beach   2000
Golf Course   500
Difference   1500
The difference is 1500 people.

17. January and February

19. February and March

Problem Set 11.6

21.

Bank	8.2%
Credit Union	7.8%
Difference	0.4%

The difference in the interest rates is 0.4%.

23.

Friday	120
Saturday	180
Sunday	90
Together	390

$390 in tips were earned for Friday, Saturday, and Sunday.

25. Monday or Wednesday

27.

Sunday	$90
Monday	40
Tuesday	50
Wednesday	40
Thursday	60
Friday	120
Saturday	180
Together	$580

$580 was earned in tips for the week.

29.

High-Tech Fund	16%
Utility Fund	14%
Difference	2%

The difference in annual total return is 2%.

31.

Year 1997	15%
Year 1996	12%
Difference	3%

The change in the annual total return is 3%.

33. 1997 and 1998

35 a. Average $= \dfrac{16 + 17 + 20 + 5 + 4 + 10}{6}$

$= \dfrac{72}{6} = 12$

The average annual total return is 12%.

b. Average $= \dfrac{14 + 12 + 15 + 11 + 9 + 11}{6}$

$= \dfrac{72}{6} = 12$

The average annual total return is 12%.

c. Neither, the average is the same.

39.

Stocks	8000
Mutual Funds	14000
Bonds	6000
Annuities	5000
Gold	3000
Total	36000

Stocks $\dfrac{8000}{36000} \times 360° = 80°$

Mutual Funds $\dfrac{14000}{36000} \times 360° = 140°$

Bonds $\dfrac{6000}{36000} \times 360° = 60°$

Annuities $\dfrac{5000}{36000} \times 360° = 50°$

Gold $\dfrac{3000}{36000} \times 360° = 30°$

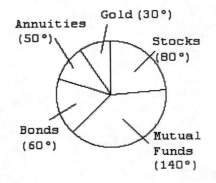

340

PROBLEM SET **11.7** Relations and Functions

1. Domain: $\{4, 6, 8, 10\}$
Range: $\{7, 11, 20, 28\}$
It is a function.

3. Domain: $\{-2, -1, 0, 1\}$
Range: $\{1, 2, 3, 4\}$
It is a function.

5. Domain: $\{4, 9\}$
Range: $\{-3, -2, 2, 3\}$
It is not a function.

7. Domain: $\{3, 4, 5, 6\}$
Range: $\{15\}$
It is a function.

9. Domain: $\{Carol\}$
Range: $\{22400, 23700, 25200\}$
It is not a function.

11. Domain: $\{-6\}$
Range: $\{1, 2, 3, 4\}$
It is not a function.

13. Domain: $\{-2, -1, 0, 1, 2\}$
Range: $\{0, 1, 4\}$
It is a function.

15. The domain is all real numbers.

17. The denominator can not equal zero.
So set $x + 8 = 0$
$$x = -8$$
Therefore -8 is excluded from the domain.
The domain is all real numbers except -8.

19. The domain is all real numbers.

21. The denominator can not equal zero.
So set $2x - 10 = 0$
$$2x = 10$$
$$x = 5$$
Therefore 5 is excluded from the domain.
The domain is all real numbers except 5.

23. The denominator can not equal zero.
So set $5x - 8 = 0$
$$5x = 8$$
$$x = \frac{8}{5}$$
Therefore $\frac{8}{5}$ is excluded from the domain.
The domain is all real numbers except $\frac{8}{5}$.

25. Since the denominator is a constant, it can not equal zero. The domain is all real numbers.

27. The denominator can not equal zero.
So set $x = 0$
Therefore 0 is excluded from the domain.
The domain is all real numbers except 0.

29. The domain is all real numbers.

31.
$$f(x) = 3x + 4$$
$$f(0) = 3(0) + 4 = 4$$
$$f(1) = 3(1) + 4 = 3 + 4 = 7$$
$$f(-1) = 3(-1) + 4 = -3 + 4 = 1$$
$$f(6) = 3(6) + 4 = 18 + 4 = 22$$

33.
$$f(x) = -5x - 1$$
$$f(3) = -5(3) - 1$$
$$= -15 - 1 = -16$$
$$f(-4) = -5(-4) - 1$$
$$= 20 - 1 = 19$$
$$f(-5) = -5(-5) - 1$$
$$= 25 - 1 = 24$$
$$f(t) = -5(t) - 1 = -5t - 1$$

Problem Set 11.7

35. $g(x) = \dfrac{2}{3}x + \dfrac{3}{4}$

$$g(3) = \dfrac{2}{3}(3) + \dfrac{3}{4} = 2 + \dfrac{3}{4}$$
$$= \dfrac{8}{4} + \dfrac{3}{4} = \dfrac{11}{4}$$
$$g\left(\dfrac{1}{2}\right) = \dfrac{2}{3}\left(\dfrac{1}{2}\right) + \dfrac{3}{4} = \dfrac{1}{3} + \dfrac{3}{4}$$
$$= \dfrac{4}{12} + \dfrac{9}{12} = \dfrac{13}{12}$$
$$g\left(-\dfrac{1}{3}\right) = \dfrac{2}{3}\left(-\dfrac{1}{3}\right) + \dfrac{3}{4} = -\dfrac{2}{9} + \dfrac{3}{4}$$
$$= -\dfrac{8}{36} + \dfrac{27}{36} = \dfrac{19}{36}$$
$$g(-2) = \dfrac{2}{3}(-2) + \dfrac{3}{4} = -\dfrac{4}{3} + \dfrac{3}{4}$$
$$= -\dfrac{16}{12} + \dfrac{9}{12} = -\dfrac{7}{12}$$

37. $f(x) = x^2 - 4$
$$f(2) = (2)^2 - 4 = 4 - 4 = 0$$
$$f(-2) = (-2)^2 - 4 = 4 - 4 = 0$$
$$f(7) = (7)^2 - 4 = 49 - 4 = 45$$
$$f(0) = (0)^2 - 4 = 0 - 4 = -4$$

39. $f(x) = -x^2 + 1$
$$f(-1) = -(-1)^2 + 1$$
$$= -1 + 1 = 0$$
$$f(2) = -(2)^2 + 1$$
$$= -4 + 1 = -3$$
$$f(-2) = -(-2)^2 + 1$$
$$= -4 + 1 = -3$$
$$f(-3) = -(-3)^2 + 1$$
$$= -9 + 1 = -8$$

41. $f(x) = 4x + 3$
$$f(5) = 4(5) + 3 = 23$$
$$f(-6) = 4(-6) + 3 = -21$$

$$g(x) = x^2 - 2x$$
$$g(-1) = (-1)^2 - 2(-1) = 3$$
$$g(4) = (4)^2 - 2(4) = 8$$

43. $f(x) = 3x^2 - x + 4$
$$f(-1) = 3(-1)^2 - (-1) + 4 = 8$$
$$f(4) = 3(4)^2 - (4) + 4 = 48$$

$$g(x) = -3x + 5$$
$$g(-1) = -3(-1) + 5 = 8$$
$$g(4) = -3(4) + 5 = -7$$

PROBLEM SET **11.8** **Applications of Functions**

1. $A(s) = s^2$
$$A(3) = 3^2 = 9$$
$$A(17) = 17^2 = 289$$
$$A(8.5) = 8.5^2 = 72.25$$
$$A(20.75) = 20.75^2 = 430.5625$$
$$A(11.25) = 11.25^2 = 126.5625$$

3. $C(n) = 12n + 44500$
$$C(35000) = 12(35000) + 44500$$
$$= 420,000 + 44,500$$
$$= 464,500$$
The cost is $464,500.

5. $s(c) = 1.5c$
$$s(4.50) = 1.5(4.50) = 6.75$$
$$s(6.75) = 1.5(6.75) = 10.13$$
$$s(9.00) = 1.5(9.00) = 13.50$$
$$s(16.40) = 1.5(16.40) = 24.60$$

7. $f(n) = .75n + 15$
$$f(20) = .75(20) + 15 = \$30.00$$
$$f(0) = .75(0) + 15 = \$15.00$$
$$f(16) = .75(16) + 15 = \$27.00$$

9. $f(x) = 0$ when $x < 2$
$g(x) = 0.30x$ when $x \geq 2$

Since $x = 8$ which is greater than 2 use
$g(x) = 0.30x$ to determine the charge.
$g(8) = 0.30(8) = 2.40$.
The charge would be $2.40.

Since $x = 3$ which is greater than 2 use
$g(x) = 0.30x$ to determine the charge.
$g(3) = 0.30(3) = 0.90$
The charge would be $0.90.

Since $x = 1$ which is less than 2
use $f(x) = 0$ to determine the charge.
$f(1) = 0$
There would be no charge or the charge
equals zero.

Since $x = 12$ which is greater than 2
use $g(x) = 0.30x$ to determine the charge.
$g(12) = 0.30(12) = 3.60$
The charge would be $3.60.

11. $f(h) = 10.50h$ when $x \leq 40$
$g(h) = 15.75h - 210$ when $x > 40$

Since $x = 35$, which is less than 40,
use $f(h) = 10.50h$ to determine his pay.
$f(35) = 10.50(35) = 367.50$
His pay is $367.50.

Since $x = 40$, which is less than or equal
to 40, use $f(h) = 10.50h$ to determine his
pay. $f(40) = 10.50(40) = 420$
His pay is $420.

Since $x = 50$, which is more than 40, use
$g(h) = 15.75h - 210$ to determine his pay.
$g(50) = 15.75(50) - 210$
$g(50) = 787.50 - 210$
$g(50) = 577.50$
His pay is $577.50.

Since $x = 20$, which is less than 40,
use $f(h) = 10.50h$ to determine his pay.
$f(20) = 10.50(20) = 210$
His pay is $210.

13. $f(n) = 48.50 + 10(n - 1)$
$f(2) = 48.50 + 10(2 - 1)$
$f(2) = 48.50 + 10(1)$
$f(2) = 48.50 + 10$
$f(2) = 58.50$

$f(3) = 48.50 + 10(3 - 1)$
$f(3) = 48.50 + 10(2)$
$f(3) = 48.50 + 20$
$f(3) = 68.50$

$f(1) = 48.50 + 10(1 - 1)$
$f(1) = 48.50 + 10(0)$
$f(1) = 48.50 + 0$
$f(1) = 48.50$

$f(4) = 48.50 + 10(4 - 1)$
$f(4) = 48.50 + 10(3)$
$f(4) = 48.50 + 30$
$f(4) = 78.50$

15. $f(x) = \dfrac{x}{10}$

$x = 0 \quad f(0) = \dfrac{0}{10} = 0$

$x = 1 \quad f(1) = \dfrac{1}{10}$

$x = 2 \quad f(2) = \dfrac{2}{10} = \dfrac{1}{5}$

$x = 3 \quad f(3) = \dfrac{3}{10}$

$x = 4 \quad f(4) = \dfrac{4}{10} = \dfrac{2}{5}$

x	0	1	2	3	4
$f(x)$	0	$\dfrac{1}{10}$	$\dfrac{1}{5}$	$\dfrac{3}{10}$	$\dfrac{2}{5}$

17. $f(n) = 0.0003n$

$n = 10,000$
$f(10000) = 0.0003(10,000) = 3$

$n = 15,000$
$f(15000) = 0.0003(15,000) = 4.5$

$n = 20,000$
$f(20000) = 0.0003(20,000) = 6$

$n = 25,000$
$f(25000) = 0.0003(25,000) = 7.5$

$n = 30,000$
$f(30000) = 0.0003(30,000) = 9$

x in gallons, $f(x)$ in pounds

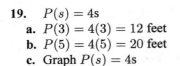

x	10000	15000	20000	25000	30000
$f(x)$	3	4.5	6	7.5	9

19. $P(s) = 4s$
 a. $P(3) = 4(3) = 12$ feet
 b. $P(5) = 4(5) = 20$ feet
 c. Graph $P(s) = 4s$

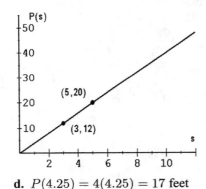

 d. $P(4.25) = 4(4.25) = 17$ feet

21. $V(t) = 32500 - 1950t$

 a. $V(6) = 32500 - 1950(6)$
    $V(6) = 32500 - 11700$
    $V(6) = 20800$

 b. $V(9) = 32500 - 1950(9)$
    $V(9) = 32500 - 17550$
    $V(9) = 14950$

 c. Graph $V(t) = 32500 - 1950t$

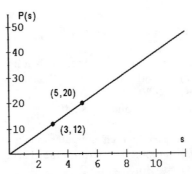

 d. $V(10) = 32500 - 1950(10)$
    $V(10) = 32500 - 19500$
    $V(10) = 13000$
    The value after 10 years will be \$13000.

 e. answers vary

 f. $0 = 32500 - 1950t$
    $-32500 = -1950t$
    $\dfrac{-32500}{-1950} = t$
    $16.7 = t$
    The value will be zero in 16.7 years.

23 a. $f(t) = \dfrac{5}{9}(t - 32)$

    $f(50) = \dfrac{5}{9}(50 - 32) = \dfrac{5}{9}(18) = 10$

    $f(41) = \dfrac{5}{9}(41 - 32) = \dfrac{5}{9}(9) = 5$

    $f(-4) = \dfrac{5}{9}(-4 - 32)$

    $= \dfrac{5}{9}(-36) = -20$

    $f(212) = \dfrac{5}{9}(212 - 32)$

    $= \dfrac{5}{9}(180) = 100$

    $f(95) = \dfrac{5}{9}(95 - 32) = \dfrac{5}{9}(63) = 35$

    $f(77) = \dfrac{5}{9}(77 - 32) = \dfrac{5}{9}(45) = 25$

    $f(59) = \dfrac{5}{9}(59 - 32) = \dfrac{5}{9}(27) = 15$

t	50	41	-4	212	95	77	59
$f(t)$	10	5	-20	100	35	25	15

b. Graph $f(t) = \dfrac{5}{9}(t - 32)$.

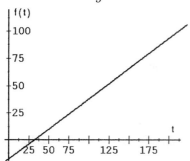

c. $f(20) = \dfrac{5}{9}(20 - 32)$

$f(20) = \dfrac{5}{9}(-12)$

$f(20) = -6.7$

CHAPTER 11   **Review Problem Set**

1.   Ankle   8%
         Hip   15%
         Knee   32%
         _____
         Together   55%

2.   Hip   15%
         Neck   4%
         Back   25%
         _____
         Together   44%

3.   Knee   32%
         Elbow   16%
         _____
         Together   48%
         The percent not seen for knee or elbow is
         $100\% - 48\% = 52\%$.

4.   Television 35
         Studying   10
         _____
         Difference 25
         There were 25 more students watching
         television than studying.

5.   Shopping 20 student
         Workout 20 student

6.   Dinner Out, Television, Movie, Shopping,
         Workout, Sports, Study

7.   1996

8.   Band   6500
         Sports   5000
         _____
         Difference 1500
         The largest difference is $1500.

9.   Average =
         $\dfrac{6500 + 5500 + 6000 + 5000 + 5500 + 6000}{6}$
         $= 5750$

         The average over the six years is $5750.

10. Average =
         $\dfrac{5000 + 5500 + 5000 + 5500 + 6000 + 6500}{6}$
         $= 5583$

         The average over the six years is $5583.

11. $|3x - 5| = 7$
         $3x - 5 = 7 \quad$ or $\quad 3x - 5 = -7$
         $3x = 12 \quad$ or $\quad 3x = -2$
         $x = 4 \quad$ or $\quad x = -\dfrac{2}{3}$
         The solution set is $\left\{ -\dfrac{2}{3}, 4 \right\}$.

345

12. $|x - 4| < 1$
$$-1 < x - 4 < 1$$
$$3 < \quad x \quad < 5$$
The solution set is $\{x | 3 < x < 5\}$
or $(3, 5)$.

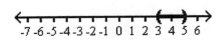

13. $|2x - 1| \geq 3$
$$2x - 1 \leq -3 \quad \text{or} \quad 2x - 1 \geq 3$$
$$2x \leq -2 \quad \text{or} \quad 2x \geq 4$$
$$x \leq -1 \quad \text{or} \quad x \geq 2$$
The solution set is $\{x | x \leq -1 \text{ or } x \geq 2\}$
or $(-\infty, -1] \cup [2, \infty)$.

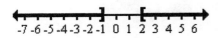

14. $|3x - 2| \leq 4$
$$-4 \leq 3x - 2 \leq 4$$
$$-2 \leq \quad 3x \quad \leq 6$$
$$-\frac{2}{3} \leq \quad x \quad \leq 2$$
The solution set is $\left\{ x | -\frac{2}{3} \leq x \leq 2 \right\}$
or $\left[-\frac{2}{3}, 2 \right]$.

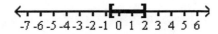

15. $|2x - 1| = 9$
$$2x - 1 = 9 \quad \text{or} \quad 2x - 1 = -9$$
$$2x = 10 \quad \text{or} \quad 2x = -8$$
$$x = 5 \quad \text{or} \quad x = -4$$
The solution set is $\{-4, 5\}$.

16. $|5x - 2| \geq 6$
$$5x - 2 \leq -6 \quad \text{or} \quad 5x - 2 \geq 6$$
$$5x \leq -4 \quad \text{or} \quad 5x \geq 8$$
$$x \leq -\frac{4}{5} \quad \text{or} \quad x \geq \frac{8}{5}$$
The solution set is $\left\{ x | x \leq -\frac{4}{5} \text{ or } x \geq \frac{8}{5} \right\}$
or $\left(-\infty, -\frac{4}{5} \right] \cup \left[\frac{8}{5}, \infty \right)$.

17. $\sqrt{\dfrac{64}{36}} = \dfrac{8}{6} = \dfrac{4}{3}$ because $\left(\dfrac{2 \cdot 4}{2 \cdot 3} \right)^2 = \dfrac{64}{36}$

18. $-\sqrt{1} = -1$

19. $\sqrt[3]{\dfrac{27}{64}} = \dfrac{3}{4}$ because $\left(\dfrac{3}{4} \right)^3 = \dfrac{27}{64}$

20. $\sqrt[3]{-125} = -5$

21. $\sqrt[4]{\dfrac{81}{16}} = \dfrac{3}{2}$ because $\left(\dfrac{3}{2} \right)^4 = \dfrac{81}{16}$

22. $25^{\frac{3}{2}} = \left(\sqrt{25} \right)^3 = 5^3 = 125$

23. $8^{\frac{5}{3}} = \left(\sqrt[3]{8} \right)^5 = 2^5 = 32$

24. $(-8)^{\frac{5}{3}} = \left(\sqrt[3]{-8} \right)^5 = (-2)^5 = -32$

25. $4^{-2} = \dfrac{1}{4^2} = \dfrac{1}{16}$

26. $4^{-\frac{1}{2}} = \dfrac{1}{4^{\frac{1}{2}}} = \dfrac{1}{\sqrt{4}} = \dfrac{1}{2}$

27. $32^{-\frac{2}{5}} = \left(\sqrt[5]{32}\right)^{-2} = 2^{-2} = \frac{1}{2^2} = \frac{1}{4}$

28. $\left(\frac{2}{3}\right)^{-1} = \frac{3}{2}$

29. $2^{\frac{7}{4}} \cdot 2^{\frac{5}{4}} = 2^{\frac{7}{4}+\frac{5}{4}} = 2^{\frac{12}{4}} = 2^3 = 8$

30. $3^{\frac{1}{3}} \cdot 3^{\frac{5}{3}} = 3^{\frac{1}{3}+\frac{5}{3}} = 3^{\frac{6}{3}} = 3^2 = 9$

31. $\dfrac{3^{\frac{1}{3}}}{3^{\frac{4}{3}}} = 3^{\frac{1}{3}-\frac{4}{3}} = 3^{-\frac{3}{3}} = 3^{-1} = \frac{1}{3}$

32. $x^{\frac{5}{6}} \cdot x^{\frac{5}{6}} = x^{\frac{5}{6}+\frac{5}{6}} = x^{\frac{10}{6}} = x^{\frac{5}{3}}$

33. $\left(3x^{\frac{1}{4}}\right)\left(2x^{\frac{3}{5}}\right) = 3 \cdot 2 \cdot x^{\frac{1}{4}} \cdot x^{\frac{3}{5}} =$
$6x^{\frac{1}{4}+\frac{3}{5}} = 6x^{\frac{5}{20}+\frac{12}{20}} = 6x^{\frac{17}{20}}$

34. $\left(9a^{\frac{1}{2}}\right)\left(4a^{-\frac{1}{3}}\right) = 9 \cdot 4 \cdot a^{\frac{1}{2}} \cdot a^{-\frac{1}{3}} =$
$36a^{\frac{3}{6}-\frac{2}{6}} = 36a^{\frac{1}{6}}$

35. $\left(3x^{\frac{1}{3}}y^{\frac{2}{3}}\right)^3 = (3)^3\left(x^{\frac{1}{3}}\right)^3\left(y^{\frac{2}{3}}\right)^3 =$
$27x^{\frac{3}{3}}y^{\frac{6}{3}} = 27xy^2$

36. $(25x^4y^6)^{\frac{1}{2}} = (25)^{\frac{1}{2}}(x^4)^{\frac{1}{2}}(y^6)^{\frac{1}{2}} =$
$\sqrt{25}x^{\frac{4}{2}}y^{\frac{6}{2}} = 5x^2y^3$

37. $\dfrac{39n^{\frac{3}{5}}}{3n^{\frac{1}{4}}} = \frac{39}{3}n^{\frac{3}{5}-\frac{1}{4}} = 13n^{\frac{12}{20}-\frac{5}{20}} = 13n^{\frac{7}{20}}$

38. $\dfrac{64n^{\frac{5}{8}}}{16n^{\frac{7}{8}}} = \frac{64}{16}n^{\frac{5}{8}-\frac{7}{8}} = 4n^{-\frac{2}{8}} = 4n^{-\frac{1}{4}} = \dfrac{4}{n^{\frac{1}{4}}}$

39. $\left[\dfrac{6x^{\frac{2}{7}}}{3x^{-\frac{5}{7}}}\right]^3 = \left[\frac{6}{3}x^{\frac{2}{7}-\left(-\frac{5}{7}\right)}\right]^3 =$
$\left(2x^{\frac{7}{7}}\right)^3 = (2x)^3 = 8x^3$

40. $\begin{pmatrix} x + 3y - z = 1 \\ 2x - y + z = 3 \\ 3x + y + 2z = 12 \end{pmatrix}$

Add equation (1) and equation (2).
$$\begin{aligned} x + 3y - z &= 1 \\ 2x - y + z &= 3 \\ \hline 3x + 2y \phantom{{}+z} &= 4 \quad (4) \end{aligned}$$

Multiply equation (1) by 2 and add the result to equation (3).
$$\begin{aligned} 2x + 6y - 2z &= 2 \\ 3x + y + 2z &= 12 \\ \hline 5x + 7y \phantom{{}+2z} &= 14 \quad (5) \end{aligned}$$

Multiply equation (4) by 5 and multiply equation (5) by -3. Then add the resulting equations.
$$\begin{aligned} 15x + 10y &= 20 \\ -15x - 21y &= -42 \\ \hline -11y &= -22 \\ y &= 2 \end{aligned}$$
Substitute for $y = 2$ into equation (4).
$$3x + 2(2) = 4$$
$$3x + 4 = 4$$
$$3x = 0$$
$$x = 0$$
Substitute $x = 0$ and $y = 2$ into equation (1).
$$x + 3y - z = 1$$
$$0 + 3(2) - z = 1$$
$$0 + 6 - z = 1$$
$$6 - z = 1$$
$$-z = -5$$
$$z = 5$$
The solution set is $\{(0, 2, 5)\}$.

Chapter 11 Review Problem Set

41. $\begin{pmatrix} 2x + 3y - z = 4 \\ x + 2y + z = 7 \\ 3x + y + 2z = 13 \end{pmatrix}$

Add equation (1) and equation (2).

$2x + 3y - z = 4$

$\underline{x + 2y + z = 7}$

$3x + 5y = 11 \quad (4)$

Multiply equation (1) by 2 and add the result to equation (3).

$4x + 6y - 2z = 8$

$\underline{3x + y + 2z = 13}$

$7x + 7y = 21 \quad (5)$

Multiply equation (4) by -7 and multiply equation (5) by 3. Then add the resulting equations.

$-21x - 35y = -77$

$\underline{21x + 21y = 63}$

$ -14y = -14$

$ y = 1$

Substitute for $y = 1$ into equation (4).

$3x + 5(1) = 11$

$3x + 5 = 11$

$3x = 6$

$x = 2$

Substitute $x = 2$ and $y = 1$ into equation (1).

$2x + 3y - z = 4$

$2(2) + 3(1) - z = 4$

$4 + 3 - z = 4$

$7 - z = 4$

$-z = -3$

$z = 3$

The solution set is $\{(2, 1, 3)\}$.

42. $(5 - 7i) + (-4 + 9i) =$
$(5 - 4) + (-7 + 9)i =$
$1 + 2i$

43. $(-3 + 2i) + (-4 - 7i) =$
$(-3 - 4) + (2 - 7)i =$
$-7 - 5i$

44. $(6 - 9i) - (4 - 5i) =$
$(6 - 4) + (-9 + 5)i =$
$2 - 4i$

45. $(-5 + 3i) - (-8 + 7i) =$
$(-5 + 8) + (3 - 7)i =$
$3 - 4i$

46. $(7 - 2i) - (6 - 4i) + (-2 + i) =$
$(7 - 6 - 2) + (-2 + 4 + 1)i =$
$-1 + 3i$

47. $(-4 + i) - (-4 - i) - (6 - 8i) =$
$(-4 + 4 - 6) + (1 + 1 + 8)i =$
$-6 + 10i$

48. $(2 + 5i)(3 + 8i) =$
$6 + 16i + 15i + 40i^2 =$
$6 + 31i + 40(-1) =$
$-34 + 31i$

49. $(4 - 3i)(1 - 2i) =$
$4 - 8i - 3i + 6i^2 =$
$4 - 11i + 6(-1) =$
$-2 - 11i$

50. $(-1 + i)(-2 + 6i) =$
$2 - 6i - 2i + 6i^2 =$
$2 - 8i + 6(-1) =$
$-4 - 8i$

51. $(-3 - 3i)(7 + 8i) =$
$-21 - 24i - 21i - 24i^2 =$
$-21 - 45i - 24(-1) =$
$3 - 45i$

52. $(2 + 9i)(2 - 9i) =$
$4 - 18i + 18i - 81i^2 =$
$4 - 81(-1) = 85$

53. $(-3 + 7i)(-3 - 7i) =$
$9 + 21i - 21i - 49i^2 =$
$9 - 49(-1) = 58$

54. $(-3 - 8i)(3 + 8i) =$
$-9 - 24i - 24i - 64i^2 =$
$-9 - 48i - 64(-1) =$

55. $(6 + 9i)(-1 - i) =$
$$-6 - 6i - 9i - 9i^2 =$$
$$-6 - 15i - 9(-1) = 3 - 15i$$

(Above 55, at top: $55 - 48i$)

56. $(x - 6)^2 = -25$
$$x - 6 = \pm\sqrt{-25} = \pm 5i$$
$$x - 6 = -5i \quad \text{or} \quad x - 6 = 5i$$
$$x = 6 - 5i \quad \text{or} \quad x = 6 + 5i$$
The solution set is $\{6 - 5i, \ 6 + 5i\}$.

57. $n^2 + 2n = -7$
$$n^2 + 2n + 1 = -7 + 1$$
$$(n + 1)^2 = -6$$
$$n + 1 = \pm\sqrt{-6} = \pm i\sqrt{6}$$
$$n + 1 = -i\sqrt{6} \quad \text{or} \quad n + 1 = i\sqrt{6}$$
$$n = -1 - i\sqrt{6} \quad \text{or} \quad n = -1 + i\sqrt{6}$$
The solution set is
$$\left\{-1 - i\sqrt{6}, \ -1 + i\sqrt{6}\right\}.$$

58. $x^2 - 2x + 17 = 0$
$$x = \frac{-(-2) \pm \sqrt{(-2)^2 - 4(1)(17)}}{2(1)}$$
$$x = \frac{2 \pm \sqrt{-64}}{2}$$
$$x = \frac{2 \pm 8i}{2} = 1 \pm 4i$$
The solution set is $\{1 - 4i, \ 1 + 4i\}$.

59. $x^2 - x + 7 = 0$
$$x = \frac{-(-1) \pm \sqrt{(-1)^2 - 4(1)(7)}}{2(1)}$$
$$x = \frac{1 \pm \sqrt{-27}}{2} = \frac{1 \pm 3i\sqrt{3}}{2}$$
The solution set is
$$\left\{\frac{1 - 3i\sqrt{3}}{2}, \ \frac{1 + 3i\sqrt{3}}{2}\right\}.$$

60. $2x^2 - x + 3 = 0$
$$x = \frac{-(-1) \pm \sqrt{(-1)^2 - 4(2)(3)}}{2(2)}$$
$$x = \frac{1 \pm \sqrt{-23}}{4}$$
$$x = \frac{1 \pm i\sqrt{23}}{4}$$
The solution set is
$$\left\{\frac{1 - i\sqrt{23}}{4}, \ \frac{1 + i\sqrt{23}}{4}\right\}.$$

61. $6x^2 - 11x + 3 = 0$
$$(3x - 1)(2x - 3) = 0$$
$$3x - 1 = 0 \quad \text{or} \quad 2x - 3 = 0$$
$$3x = 1 \quad \text{or} \quad 2x = 3$$
$$x = \frac{1}{3} \quad \text{or} \quad x = \frac{3}{2}$$
The solution set is $\left\{\dfrac{1}{3}, \dfrac{3}{2}\right\}$.

62. $-x^2 + 5x - 7 = 0$
$$x^2 - 5x + 7 = 0$$
$$x = \frac{-(-5) \pm \sqrt{(-5)^2 - 4(1)(7)}}{2(1)}$$
$$x = \frac{5 \pm \sqrt{-3}}{2}$$
$$x = \frac{5 \pm i\sqrt{3}}{2}$$
The solution set is
$$\left\{\frac{5 - i\sqrt{3}}{2}, \ \frac{5 + i\sqrt{3}}{2}\right\}.$$

Chapter 11 Review Problem Set

63.
$$-2x^2 - 3x - 6 = 0$$
$$2x^2 + 3x + 6 = 0$$
$$x = \frac{-(3) \pm \sqrt{(3)^2 - 4(2)(6)}}{2(2)}$$
$$x = \frac{-3 \pm \sqrt{-39}}{4}$$
$$x = \frac{-3 \pm i\sqrt{39}}{4}$$
The solution set is
$$\left\{ \frac{-3 - i\sqrt{39}}{4}, \frac{-3 + i\sqrt{39}}{4} \right\}.$$

64.
$$3x^2 + x + 5 = 0$$
$$x = \frac{-(1) \pm \sqrt{(1)^2 - 4(3)(5)}}{2(3)}$$
$$x = \frac{-1 \pm \sqrt{-59}}{6}$$
$$x = \frac{-1 \pm i\sqrt{59}}{6}$$
The solution set is
$$\left\{ \frac{-1 - i\sqrt{59}}{6}, \frac{-1 + i\sqrt{59}}{6} \right\}.$$

65.
$$x(4x + 1) = -3$$
$$4x^2 + x = -3$$
$$4x^2 + x + 3 = 0$$
$$x = \frac{-(1) \pm \sqrt{(1)^2 - 4(4)(3)}}{2(4)}$$
$$x = \frac{-1 \pm \sqrt{-47}}{8}$$
$$x = \frac{-1 \pm i\sqrt{47}}{8}$$
The solution set is
$$\left\{ \frac{-1 - i\sqrt{47}}{8}, \frac{-1 + i\sqrt{47}}{8} \right\}.$$

66. Domain {red, blue, green}
Range $\left\{ \dfrac{1}{8}, \dfrac{1}{4}, \dfrac{5}{8} \right\}$
It is a function.

67. Domain {3, 4, 5}
Range {5, 7, 9}
It is a function.

68. Domain {1, 2}
Range {−16, −8, 8, 16}
It is not a function since (1, 8)
and (1, −8) have 1 assigned to
two different range components.

69. Domain {2, 3, 4, 5}
Range {10}
It is a function.

70. Domain {−2, −1, 0, 1, 2}
Range {0, 1, 4}
It is a function.

71. Domain {1, 2, 3}
Range {4, 8, 10, 15}
It is not a function.

72. The denominator can not equal zero.
So set $x - 6 = 0$
$$x = 6$$
Therefore 6 is excluded from the domain.
The domain is all real numbers except 6.

73. The domain is all real numbers.

74. The domain is all real numbers.

75. The denominator can not equal zero.
So set $x + 4 = 0$
$$x = -4$$
Therefore −4 is excluded from the domain.
The domain is all real numbers except −4.

76. The denominator can not equal zero.
So set $2x - 1 = 0$
$$2x = 1$$
$$x = \frac{1}{2}$$
Therefore $\dfrac{1}{2}$ is excluded from the domain.
The domain is all real numbers except $\dfrac{1}{2}$.

77. The denominator can not equal zero.
So set $3x + 1 = 0$
$$3x = -1$$
$$x = -\frac{1}{3}$$
Therefore $-\frac{1}{3}$ is excluded from the domain.

The domain is all real numbers except $-\frac{1}{3}$.

78. $f(x) = 3x - 2$

$f(-4) = 3(-4) - 2 \quad f(0) = 3(0) - 2$
$f(-4) = -12 - 2 \quad f(0) = 0 - 2$
$f(-4) = -14 \quad\quad f(0) = -2$

$f(5) = 3(5) - 2 \quad f(a) = 3(a) - 2$
$f(5) = 15 - 2 \quad\quad f(0) = 3a - 2$
$f(5) = 13$

79. $f(x) = \dfrac{6}{x - 4}$

$f(-4) = \dfrac{6}{-4 - 4} = \dfrac{6}{-8} = -\dfrac{3}{4}$

$f(0) = \dfrac{6}{0 - 4} = \dfrac{6}{-4} = -\dfrac{3}{2}$

$f(1) = \dfrac{6}{1 - 4} = \dfrac{6}{-3} = -2$

$f(2) = \dfrac{6}{2 - 4} = \dfrac{6}{-2} = -3$

80. $f(x) = \dfrac{x}{2x + 1}$

$f(-3) = \dfrac{-3}{2(-3) + 1} = \dfrac{-3}{-6 + 1} = \dfrac{-3}{-5} = \dfrac{3}{5}$

$f(0) = \dfrac{0}{2(0) + 1} = \dfrac{0}{0 + 1} = \dfrac{0}{1} = 0$

$f(2) = \dfrac{2}{2(2) + 1} = \dfrac{2}{4 + 1} = \dfrac{2}{5}$

$f(3) = \dfrac{3}{2(3) + 1} = \dfrac{3}{6 + 1} = \dfrac{3}{7}$

81. $f(x) = x^2 + 4x - 3$
$f(-1) = (-1)^2 + 4(-1) - 3$
$\quad\quad = 1 - 4 - 3 = -6$
$f(0) = (0)^2 + 4(0) - 3$
$\quad\quad = 0 + 0 - 3 = -3$
$f(1) = (1)^2 + 4(1) - 3$
$\quad\quad = 1 + 4 - 3 = 2$
$f(2) = (2)^2 + 4(2) - 3$
$\quad\quad = 4 + 8 - 3 = 9$

82 a. $f(2) = 0.20 + 0.30(2)$
$\quad\quad = 0.20 + 0.60 = 0.80$

b. $f(5) = 0.20 + 0.30(5)$
$\quad\quad = 0.20 + 1.50 = 1.70$

c.

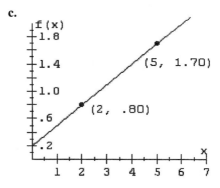

d. $f(4) = 0.20 + 0.30(4)$
$\quad\quad = 0.20 + 1.20 = 1.40$
The charge for 4 ounces will be $1.40.

83 a. $f(x) = 12 + 0.2x$
$f(50) = 12 + 0.2(50)$
$f(50) = 12 + 10$
$f(50) = 22$

b. $f(100) = 12 + 0.2(100)$
$f(100) = 12 + 20$
$f(100) = 32$

Chapter 11 Review Problem Set

c.

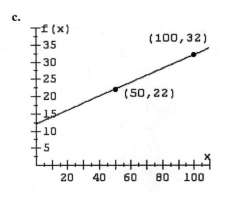

d. $f(80) = 12 + 0.2(80)$
$f(80) = 12 + 16$
$f(80) = 28$

CHAPTER 11 Test

1. Men − Armageddon
Women − Titanic

2.
English Patient	30
Saving Private Ryan	15
Difference	15

15 more women chose English Patient as a favorite instead of Saving Private Ryan.

3.
Armageddon	40
Titanic	25
Difference	15

15 more men chose Armageddon as a favorite instead of Titanic.

4.
Women	50
Men	25
Difference	25

25 more women than men chose Titanic as a favorite.

5. $(3+i)(2-5i) =$
$6 - 15i + 2i - 5i^2 =$
$6 - 13i - 5(-1) =$
$6 - 13i + 5 = 11 - 13i$

6. $\sqrt{-75} = \sqrt{-1}\sqrt{75} = i\sqrt{25}\sqrt{3} = 5i\sqrt{3}$

7. $36^{\frac{3}{2}} = \left(\sqrt{36}\right)^3 = 6^3 = 216$

8. $\left(\dfrac{2}{3}\right)^{-3} = \dfrac{2^{-3}}{3^{-3}} = \dfrac{3^3}{2^3} = \dfrac{27}{8}$

9. $\left(2x^{\frac{1}{4}}\right)\left(5x^{\frac{2}{3}}\right) = 10x^{\frac{1}{4}+\frac{2}{3}} = 10x^{\frac{11}{12}}$

10. $\dfrac{30n^{\frac{1}{2}}}{6n^{\frac{2}{5}}} = \dfrac{30}{6}n^{\frac{1}{2}-\frac{2}{5}} = 5n^{\frac{5}{10}-\frac{4}{10}} = 5n^{\frac{1}{10}}$

11. $(x-2)^2 = -16$
$x - 2 = \pm\sqrt{-16}$
$x - 2 = \pm 4i$
$x = 2 \pm 4i$
The solution set is $\{2 - 4i, \ 2 + 4i\}$.

12. $x^2 - 2x + 3 = 0$
$x = \dfrac{-(-2) \pm \sqrt{(-2)^2 - 4(1)(3)}}{2(1)}$
$x = \dfrac{2 \pm \sqrt{-8}}{2}$
$x = \dfrac{2 \pm 2i\sqrt{2}}{2} = 1 \pm i\sqrt{2}$
The solution set is $\left\{1 - i\sqrt{2}, \ 1 + i\sqrt{2}\right\}$.

13.
$$x^2 + 6x = -21$$
$$x^2 + 6x + 9 = -21 + 9$$
$$(x + 3)^2 = -12$$
$$x + 3 = \pm\sqrt{-12}$$
$$x + 3 = \pm 2i\sqrt{3}$$
$$x = -3 \pm 2i\sqrt{3}$$
The solution set is
$$\left\{-3 - 2i\sqrt{3},\ -3 + 2i\sqrt{3}\right\}.$$

14. $x^2 - 3x + 5 = 0$
$$x = \frac{-(-3) \pm \sqrt{(-3)^2 - 4(1)(5)}}{2(1)}$$
$$x = \frac{3 \pm \sqrt{-11}}{2}$$
$$x = \frac{3 \pm i\sqrt{11}}{2}$$
The solution set is
$$\left\{\frac{3 - i\sqrt{11}}{2}, \frac{3 + i\sqrt{11}}{2}\right\}.$$

15. $|x - 2| = 6$
$$x - 2 = -6 \quad \text{or} \quad x - 2 = 6$$
$$x = -4 \quad \text{or} \quad x = 8$$
The solution set is $\{-4, 8\}$.

16. $|4x + 5| = 2$ is equivalent to
$4x + 5 = -2$ or $4x + 5 = 2$.
$$4x + 5 = -2 \quad \text{or} \quad 4x + 5 = 2$$
$$4x = -7 \quad \text{or} \quad 4x = -3$$
$$x = -\frac{7}{4} \quad \text{or} \quad x = -\frac{3}{4}$$
The solution set is $\left\{-\frac{7}{4},\ -\frac{3}{4}\right\}.$

17. $|3x - 1| = -4$
No Solution. Absolute value
can not equal a negative number.
The solution set is $\emptyset$.

18. $2x^2 - x + 1 = 0$
$$x = \frac{-(-1) \pm \sqrt{(-1)^2 - 4(2)(1)}}{2(2)}$$
$$x = \frac{1 \pm \sqrt{-7}}{4}$$
$$x = \frac{1 \pm i\sqrt{7}}{4}$$
The solution set is
$$\left\{\frac{1 - i\sqrt{7}}{4}, \frac{1 + i\sqrt{7}}{4}\right\}.$$

19.
$$3x^2 + 5x - 28 = 0$$
$$(3x - 7)(x + 4) = 0$$
$$3x - 7 = 0 \quad \text{or} \quad x + 4 = 0$$
$$3x = 7 \quad \text{or} \quad x = -4$$
$$x = \frac{7}{3} \quad \text{or} \quad x = -4$$
The solution set is $\left\{-4, \frac{7}{3}\right\}.$

20. $|x + 3| \geq 2$ is equivalent to
$x + 3 \leq -2$ or $x + 3 \geq 2$.
$$x + 3 \leq -2 \quad \text{or} \quad x + 3 \geq 2$$
$$x \leq -5 \quad \text{or} \quad x \geq -1$$

The solution set is $\{x | x \leq -5 \text{ or } x \geq -1\}$
or $(-\infty, -5] \cup [-1, \infty)$.

21. $|2x - 1| < 7$
$$-7 < 2x - 1 < 7$$
$$-6 < 2x < 8$$
$$-3 < x < 4$$
The solution set is $\{x | -3 < x < 4\}$
or $(-3, 4)$.

22. The denominator can not equal zero.
So set $2x - 7 = 0$
$$2x = 7$$
$$x = \frac{7}{2}$$
Therefore $\frac{7}{2}$ is excluded from the domain.

The domain is all real numbers except $\frac{7}{2}$.

Chapter 11 Test

23. $f(x) = 5x - 6$
$f(2) = 5(2) - 6 = 10 - 6 = 4$
$f(0) = 5(0) - 6 = 0 - 6 = -6$
$f(-3) = 5(-3) - 6 = -15 - 6 = -21$
$f(4) = 5(4) - 6 = 20 - 6 = 14$

24. $P(x) = 20 + 0.10x$
$P(4270) = 20 + 0.10(4270)$
$P(4270) = 20 + 427$
$P(4270) = 447$

25. $f(x) = 0.07x \qquad x \leq 200,000$
$g(x) = 0.05x + 4000 \quad x > 200,000$
Since $165,000$ is less than $200,000$ use

$f(x) = 0.07x$ to determine the
commission.
$f(165000) = 0.07(165000) = 11550$
The commision for a selling price of
$\$165,000$ would be $\$11,550$.

Since $245,000$ is more than $165,000$ use
$g(x) = 0.05x + 4000$ to determine the
commision.

$g(245000) = 0.05(245000) + 400$
$g(245000) = 12250 + 4000$
$g(245000) = 16250$
The commision for a selling price of
$\$245,000$ would be $\$16,250$.

1.   Let x = the number
$$17 + x = 33$$
$$17 + x - 17 = 33 - 17$$
$$x = 16$$

The number is 16.

3.   Let x = the number
$$4x + 3 = 23$$
$$4x = 23 - 3$$
$$4x = 20$$
$$x = \frac{20}{4} = 5$$
The number is 5.

5.   Let x = the number
$$15 - 6x = 3$$
$$15 - 6x - 15 = 3 - 15$$
$$-6x = -12$$
$$x = \frac{-12}{-6} = 2$$
The number is 2.

7.   Let n = 1st consecutive integer
then $n + 1$ = the next consecutive integer

$$n + n + 1 = -83$$
$$2n + 1 = -83$$
$$2n + 1 - 1 = -83 - 1$$
$$2n = -84$$
$$n = \frac{-84}{2} = -42$$
The consecutive integers
are -42 and -41.

9.   Let x = 1st consecutive odd integer
then $x + 2$ = the 2nd consecutive odd integer
and $x + 4$ = the 3rd consecutive odd integer

$$x + x + 2 + x + 4 = 147$$
$$3x + 6 = 147$$
$$3x + 6 - 6 = 147 - 6$$
$$3x = 141$$
$$x = \frac{141}{3} = 47$$
The consecutive odd integers
are 47, 49 and 51.

11.   Let x = the number
$$3x - 2 = 4x + 1$$
$$3x - 2 + 2 = 4x + 1 + 2$$
$$3x = 4x + 3$$
$$3x - 4x = 4x + 3 - 4x$$
$$-x = 3$$
$$x = -3$$
The number is -3.

13.   Let n = 1st consecutive whole number
then $n + 1$ = the 2nd consecutive
whole number
and $n + 2$ = the 3rd consecutive
whole number

$$(n + n + 1) - (n + 2) = 48$$
$$n + n + 1 - n - 2 = 48$$
$$n - 1 = 48$$
$$n - 1 + 1 = 48 + 1$$
$$n = 49$$
The consecutive whole numbers
are 49, 50 and 51.

15.   Let x = the number
$$\frac{1}{2}x + \frac{3}{4}x = 40$$
$$\frac{2}{4}x + \frac{3}{4}x = 40$$
$$\frac{5}{4}x = 40$$
$$\left(\frac{4}{5}\right)\left(\frac{5}{4}x\right) = \frac{4}{5}(40)$$
$$x = 32$$
The number is 32.

17. Let $x =$ the number
$$\frac{3}{4}x - \left(\frac{1}{8}x + \frac{1}{6}x\right) = 33$$
$$\frac{3}{4}x - \frac{1}{8}x - \frac{1}{6}x = 33$$
$$24\left(\frac{3}{4}x - \frac{1}{8}x - \frac{1}{6}x\right) = 24(33)$$
$$24\left(\frac{3}{4}x\right) - 24\left(\frac{1}{8}x\right) - 24\left(\frac{1}{6}x\right) = 792$$
$$18x - 3x - 4x = 792$$
$$11x = 792$$
$$x = \frac{792}{11} = 72$$

The number is 72.

19. Let $x =$ one number
then $5x - 2 =$ the other number

$$x(5x - 2) = 24$$
$$5x^2 - 2x = 24$$
$$5x^2 - 2x - 24 = 0$$
$$(5x - 12)(x + 2) = 0$$
$$5x - 12 = 0 \quad \text{or} \quad x + 2 = 0$$
$$5x = 12 \qquad \text{or} \quad x = -2$$
$$x = \frac{12}{5} \qquad \text{or} \quad x = -2$$
When $x = \frac{12}{5}$, then $5x - 2 =$
$$5\left(\frac{12}{5}\right) - 2 = 12 - 2 = 10.$$

When $x = -2$, then $5x - 2 =$
$$5(-2) - 1 = -10 - 2 = -12.$$

The numbers are $\frac{12}{5}$ and 10 or -2 and -12.

21. Let n = 1st consecutive odd whole number
n + 2 = 2nd consecutive odd whole number

$$(n)^2 + (n + 2)^2 = 34$$
$$n^2 + n^2 + 4n + 4 = 34$$
$$2n^2 + 4n - 30 = 0$$
$$2(n^2 + 2n - 15) = 0$$
$$n^2 + 2n - 15 = 0$$
$$(n + 5)(n - 3) = 0$$
$$n + 5 = 0 \quad \text{or} \quad n - 3 = 0$$
$$n = -5 \quad \text{or} \quad n = 3$$

When $n = -5$, the number is not a whole number.
When $n = 3$, then the numbers are 3 and 5.

23. Let $x =$ larger number
then $x - 47 =$ the smaller number
$$\frac{x}{x - 47} = 6 + \frac{2}{x - 47}$$
$$(x - 47)\left[\frac{x}{x - 47}\right] = (x - 47)\left[6 + \frac{2}{x - 47}\right]$$
$$x = (x - 47)(6) + (x\text{-}47)\left(\frac{2}{x\text{-}47}\right)$$
$$x = (x - 47)(6) + 2$$
$$x = 6x - 282 + 2$$
$$x = 6x - 280$$
$$-5x = -280$$
$$x = \frac{-280}{-5} = 56$$

The numbers are 56 and $56 - 47 = 9$.

25. Let $x =$ the number to be subtracted
$$\frac{28 - x}{37 - x} = \frac{8}{11}$$
$$11(37 - x)\left(\frac{28 - x}{37 - x}\right) = 11(37 - x)\left(\frac{8}{11}\right)$$
$$11(28 - x) = 8(37 - x)$$
$$308 - 11x = 296 - 8x$$
$$12 - 11x = -8x$$
$$12 = 3x$$
$$4 = x$$
The number to be subtracted is 4.

27. Let $x = $ the number

then $\dfrac{1}{x} = $ the reciprocal of the number

$$x - \frac{1}{x} = -\frac{7}{12}$$

$$\frac{x^2}{x} - \frac{1}{x} = -\frac{7}{12}$$

$$12x\left(\frac{x^2}{x} - \frac{1}{x}\right) = 12x\left(-\frac{7}{12}\right)$$

$$12x\left(\frac{x^2}{x}\right) - 12x\left(\frac{1}{x}\right) = -7x$$

$$12x^2 - 12 = -7x$$

$$12x^2 + 7x - 12 = 0$$

$$(3x + 4)(4x - 3) = 0$$

$3x + 4 = 0 \quad$ or $\quad 4x - 3 = 0$

$3x = -4 \quad$ or $\quad 4x = 3$

$x = -\dfrac{4}{3} \quad$ or $\quad x = \dfrac{3}{4}$

The number is $-\dfrac{4}{3}$ or $\dfrac{3}{4}$.

29. Let $x = $ the number

then $\dfrac{1}{x} = $ the reciprocal of the number

$$\frac{1}{x} = \frac{5}{6} + x$$

$$6x\left(\frac{1}{x}\right) = 6x\left(\frac{5}{6} + x\right)$$

$$6x\left(\frac{1}{x}\right) = 6x\left(\frac{5}{6}\right) + 6x(x)$$

$$6 = 5x + 6x^2$$

$$0 = 6x^2 + 5x - 6$$

$$0 = (3x - 2)(2x + 3)$$

$3x - 2 = 0 \quad$ or $\quad 2x + 3 = 0$

$3x = 2 \quad\quad$ or $\quad 2x = -3$

$x = \dfrac{2}{3} \quad\quad$ or $\quad x = -\dfrac{3}{2}$

The number is $-\dfrac{3}{2}$ or $\dfrac{2}{3}$.

31. Let $x = $ larger number

then $x - 27 = $ the smaller number

$$x - \left[3(x - 27)\right] = -1$$

$$x - (3x - 81) = -1$$

$$x - 3x + 81 = -1$$

$$-2x + 81 = -1$$

$$-2x = -82$$

$$x = \frac{-82}{-2} = 41$$

The numbers are 41 and $41 - 27 = 14$.

33. Let $x = $ the number

and $99 - x = $ the other number

$$x - (99 - x) = 35$$

$$x - 99 + x = 35$$

$$2x - 99 = 35$$

$$2x = 134$$

$$x = 67$$

The number is 67 and the other number is 32.

35. Let $x = $ one number

and $y = $ the other number

$$x + y = 10$$

$$xy = 22$$

$$y = 10 - x$$

$$x(10 - x) = 22$$

$$10x - x^2 = 22$$

$$-x^2 + 10x - 22 = 0$$

$$x^2 - 10x + 22 = 0$$

$$x = \frac{-(-10) \pm \sqrt{(-10)^2 - 4(1)(22)}}{2(1)}$$

$$x = \frac{10 \pm \sqrt{100 - 88}}{2}$$

$$x = \frac{10 \pm \sqrt{12}}{2}$$

$$x = \frac{10 \pm 2\sqrt{3}}{2}$$

$$x = \frac{2(5 \pm \sqrt{3})}{2} = 5 \pm \sqrt{3}$$

The two numbers are $5 - \sqrt{3}$ and $5 + \sqrt{3}$.

37. Let x = the number

then $9 - x$ = the other number

and $\dfrac{1}{x}$ = the reciprocal of the number

and $\dfrac{1}{9 - x}$ = the reciprocal of the

other number

$$\frac{1}{x} + \frac{1}{9 - x} = \frac{1}{2}$$

$$2x(9 - x)\left(\frac{1}{x}\right) + 2x(9 - x)\left(\frac{1}{9 - x}\right) = 2x(9 - x)\left(\frac{1}{2}\right)$$

$$2(9 - x) + 2x = x(9 - x)$$

$$18 - 2x + 2x = 9x - x^2$$

$$18 = 9x - x^2$$

$$x^2 - 9x + 18 = 0$$

$$(x - 3)(x - 6) = 0$$

$$x - 3 = 0 \quad \text{or} \quad x - 6 = 0$$

$$x = 3 \qquad \text{or} \quad x = 6$$

The numbers are 3 and 6.

39. Let x = Nancy's present age

then $x + 14$ = Nancy's age fourteen

years from now

$$x + 14 = 37$$

$$x = 23$$

Nancy's present age is 23 years.

41.

	Age Now	Age in 12 years
Annilee	$\dfrac{2}{3}x$	$\dfrac{2}{3}x + 12$
Jessie	x	$x + 12$

$$\frac{2}{3}x + 12 + x + 12 = 54$$

$$\frac{2}{3}x + \frac{3}{3}x + 24 = 54$$

$$\frac{5}{3}x = 30$$

$$\frac{3}{5}\left(\frac{5}{3}x\right) = \frac{3}{5}(30)$$

$$x = 18$$

Jessie is 18 years old and
Annilee is 12 years old.

43. Let x = the number

$$3x - 4 > 9$$

$$3x > 13$$

$$x > \frac{13}{3}$$

The solution set is $\left(\dfrac{13}{3}, \infty\right)$.

45. Let x = the number

$$6x - 1 < 2x + 11$$

$$6x < 2x + 12$$

$$4x < 12$$

$$x < 3$$

The solution set is $(-\infty, 3)$.

47. Let x = score on 5th exam

$$\frac{89 + 92 + 96 + 98 + x}{5} \geq 95$$

$$\frac{375 + x}{5} \geq 95$$

$$5\left(\frac{375 + x}{5}\right) \geq 5(95)$$

$$375 + x \geq 475$$

$$x \geq 100$$

On the 5th exam Olga must
score 100 or better.

49. Let x = score on 3rd game

$$\frac{148 + 166 + x}{3} \geq 160$$

$$\frac{314 + x}{3} \geq 160$$

$$3\left(\frac{314 + x}{3}\right) \geq 3(160)$$

$$314 + x \geq 480$$

$$x \geq 166$$

On the 3rd game Gabrielle must
bowl 166 or better.

51. Let x = the length of the paper

$$8 = 1 + \frac{1}{2}x$$

$$7 = \frac{1}{2}x$$

$$2(7) = 2\left(\frac{1}{2}x\right)$$

$$14 = x$$

The length is 14 inches.

53. Let x = measure of one angle of a triangle
then $180 - x$ = measure of the supplement
of the angle

$$x - (180 - x) = 56$$
$$x - 180 + x = 56$$
$$2x - 180 = 56$$
$$2x = 236$$
$$x = 118$$

The measure of the angle is $118°$ and
its supplement mesures $62°$.

55. Let x = measure of an angle
then $90 - x$ = measure of its complement
and $180 - x$ = measure of its supplement

$$90 - x + \frac{1}{2}\left(180 - x\right) = 90$$

$$90 - x + 90 - \frac{1}{2}x = 90$$

$$180 - \frac{2}{2}x - \frac{1}{2}x = 90$$

$$180 - \frac{3}{2}x = 90$$

$$-\frac{3}{2}x = -90$$

$$-\frac{2}{3}\left(-\frac{3}{2}x\right) = -\frac{2}{3}(-90)$$

$$x = 60$$

The measure of the angle is $60°$.

57. Let x = measure of an angle
then $90 - x$ = measure of its complement
and $180 - x$ = measure of its supplement

$$180 - x = 4(90 - x) - 15$$
$$180 - x = 360 - 4x - 15$$
$$180 - x = 345 - 4x$$
$$-x = 165 - 4x$$
$$3x = 165$$
$$x = 55$$

The measure of the angle is $55°$.

59. Let x = the length of the rectangle
then $x - 7$ = the width of the rectangle

$$P = 2L + 2W$$
$$70 = 2(x) + 2(x - 7)$$
$$70 = 2x + 2x - 14$$
$$70 = 4x - 14$$
$$84 = 4x$$
$$21 = x$$

The length is 21 centimeters and the
width is $21 - 7 = 14$ centimeters.

61. Let x = the side of the square
then $x + 4$ = the side of the
equilateral triangle

Perimeter of square $= 4x$.
Perimeter of equilateral triangle $= 3(x + 4)$.

$$4x = 3(x + 4)$$
$$4x = 3x + 12$$
$$x = 12$$

The side of the square is 12inches.

63. Let x = the width of the garden
then $3 + x$ = the length of the garden

$$A = lw$$
$$40 = (3 + x)x$$
$$40 = 3x + x^2$$
$$0 = x^2 + 3x - 40$$
$$0 = (x + 8)(x - 5)$$
$$x + 8 = 0 \quad \text{or} \quad x - 5 = 0$$
$$x = -8 \quad \text{or} \quad x = 5$$
Discard $x = -8$.

The width is 5 meters and the
length is 8 meters.

65. Let a = length of one leg
then $21 - a$ = length of other leg

$$a^2 + b^2 = c^2$$
$$a^2 + (21 - a)^2 = (15)^2$$
$$a^2 + 441 - 42a + a^2 = 225$$
$$2a^2 - 42a + 216 = 0$$
$$2(a^2 - 21a + 108) = 0$$
$$a^2 - 21a + 108 = 0$$
$$(a - 12)(a - 9) = 0$$
$$a - 12 = 0 \quad \text{or} \quad a - 9 = 0$$
$$a = 12 \quad\quad \text{or} \quad a = 9$$

One leg is 12 inches and the
other is 9 inches.

67. Let x = width of sidewalk

$$(20 + 2x)(12 + 2x) = 68 + 240$$
$$240 + 64x + 4x^2 = 308$$
$$4x^2 + 64x - 68 = 0$$
$$4(x^2 + 16x - 17) = 0$$
$$x^2 + 16x - 17 = 0$$
$$(x + 17)(x - 1) = 0$$
$$x + 17 = 0 \quad \text{or} \quad x - 1 = 0$$
$$x = -17 \quad \text{or} \quad x = 1$$

Discard negative solution.

The width of the sidewalk is 1 meter.

69. Let x = the width of the box
then $2 + x$ = the length of the box

$$V = LWH$$
$$70 = (x - 4)(2 + x - 4)(2)$$
$$70 = (x - 4)(x - 2)(2)$$
$$35 = (x - 4)(x - 2)$$
$$35 = x^2 - 6x + 8$$
$$0 = x^2 - 6x - 27$$
$$0 = (x - 9)(x + 3)$$
$$x - 9 = 0 \quad \text{or} \quad x + 3 = 0$$
$$x = 9 \quad\quad \text{or} \quad x = -3$$

Discard the negative solution.
The width of the box is 9 units and
the length is 11 units.

71. Let x = amount invested at 8%
then $4000 - x$ = amount invested at 9%

$$(x)(8\%) + (4000 - x)(9\%) = \$350$$
$$.08x + .09(4000 - x) = 350$$
$$100\big[.08x + .09(4000 - x)\big] = 100(350)$$
$$8x + 9(4000 - x) = 35000$$
$$8x + 36000 - 9x = 35000$$
$$-x + 36000 = 35000$$
$$-x = -1000$$
$$x = 1000$$

The amount invested at 8% is \$1,000 and
the amount invested at 9% is \$3,000.

73. Let x = amount invested at 7%
then $2000 - x$ = amount invested at 8%

$$(2000 - x)(8\%) - (x)(7\%) = \$40$$
$$.08(2000 - x) - .07x = 40$$
$$100\big[.08(2000 - x) - .07x\big] = 100(40)$$
$$8(2000 - x) - 7x = 4000$$
$$16000 - 8x - 7x = 4000$$
$$-15x + 16000 = 4000$$
$$-15x = -12000$$
$$x = 800$$

The amount invested at 7% is \$800 and
the amount invested at 8% is \$1,200.

75. Let x = the number of shares of stock
Barry bought

$$\frac{600}{x} = \text{cost per share}$$

$$\frac{600}{x} + 3 = \text{value of a share}$$
$$\text{one month later}$$

$$\begin{array}{c}\text{Value of}\\\text{a share}\end{array} \times \begin{array}{c}\text{Shares}\\\text{sold}\end{array} = \text{Cost} + \text{Profit}$$

$$\left(\frac{600}{x} + 3\right)\left(x - 10\right) = 600$$

$$\left(\frac{600}{x} + 3\right)\left(x - 10\right) = 600$$

$$600 - \frac{6000}{x} + 3x - 30 = 600$$

$$-\frac{6000}{x} + 3x - 30 = 0$$

$$x\left(-\frac{6000}{x} + 3x - 30\right) = 0$$

$$-6000 + 3x^2 - 30x = 0$$

$$3x^2 - 30x - 6000 = 0$$

$$3\left(x^2 - 10x - 2000\right) = 0$$

$$x^2 - 10x - 2000 = 0$$

$$\left(x - 50\right)\left(x + 40\right) = 0$$

$$x - 50 = 0 \quad \text{or} \quad x + 40 = 0$$

$$x = 50 \quad\quad \text{or} \quad x = -40$$

Discard the negative solution.
Barry sold $50 - 10 = 40$ shares
at $\frac{600}{50} + 3 = \$15$ a share.

77. 1 inch represents 4 feet.

Let x = the value of 3.5 inches in feet
$$\frac{1}{4} = \frac{3.5}{x}$$
$$x = (4)(3.5)$$
$$x = 14$$

Let y = the value of 4.25 inches in feet
$$\frac{1}{4} = \frac{4.25}{y}$$
$$y = (4)(4.25)$$
$$y = 17$$
The dimensions of the room are
14 inches by 17 inches.

79. Compare miles to gallons.

Let x = distance traveled on 15 gallons of gas
$$\frac{200}{10} = \frac{x}{15}$$
$$(200)(15) = 10x$$
$$3000 = 10x$$
$$300 = x$$

The car will travel 300 miles.

81. Compare pounds of fertilizer
to square feet.
Let x = pounds of fertilizer needed
for 1750 square feet

$$\frac{20}{1400} = \frac{x}{1750}$$
$$(20)(1750) = 1400x$$
$$35000 = 1400x$$
$$25 = x$$

The amount of fertilizer is 25 pounds.

83. $1\frac{1}{2}$ inch represents 25 miles.

Let x = the value of $5\frac{1}{4}$ inches in miles

$$\frac{1\frac{1}{2}}{25} = \frac{5\frac{1}{4}}{x}$$
$$1\frac{1}{2}x = (25)\left(5\frac{1}{4}\right)$$
$$\frac{3}{2}x = 25\left(\frac{21}{4}\right)$$
$$\frac{3}{2}x = \frac{525}{4}$$
$$\frac{2}{3}\left(\frac{3}{2}x\right) = \frac{2}{3}\left(\frac{525}{4}\right)$$
$$x = \frac{525}{6} = 87\frac{1}{2}$$

The two cities are $87\frac{1}{2}$ miles apart.

85. Let x = number of females
$4400 - x$ = number of males

$$\frac{5}{3} = \frac{x}{4400 - x}$$
$$5(4400 - x) = 3x$$
$$22000 - 5x = 3x$$
$$22000 = 8x$$
$$2750 = x$$
The number of female students is 2750.
and the number of male students is 1650.

Appendix B

87. Let x = percent in decimal form

$$46 = x(40)$$
$$\frac{46}{40} = x$$
$$\frac{23}{20} = x$$
$$1.15 = x$$
$$115\% = x$$

Forty-six is 115% of 40.

89. Let x = the number

$$1.10x = 60.5$$
$$x = \frac{60.5}{1.10}$$
$$x = 55$$

One hundred ten percent of 55 is 60.5.

91. Let x = the original price

$$x - 0.30x = 36.40$$
$$0.70x = 36.40$$
$$x = \frac{36.40}{0.70} = 52$$

The original price is \$52.00.

93. $\$120 - \$72 = \$48$
discount $= \$48$

$$\left.\begin{matrix}\text{rate of discount} \\ \text{on selling price}\end{matrix}\right\} = \frac{48}{120} = 0.40 = 40\%$$

95. Selling price = cost + profit
Selling price = $\$5.00 + 0.60(\$5.00)$
Selling price = $\$5.00 + \$3.00$
Selling price = $\$8.00$

97. Let x = selling price

$$x = 22.00 + 0.20x$$
$$0.80x = 22.00$$
$$x = \frac{22.00}{0.80} = 27.50$$

The selling price is \$27.50.

99. Let r = speed of the first car
and $r + 7$ = speed of the second car

distance of 1st car + distance of 2nd car = 369
$$3r + 3(r + 7) = 369$$
$$3r + 3r + 21 = 369$$
$$6r + 21 = 369$$
$$6r = 348$$
$$r = 58$$

The speed of the first car is 58 mph and the speed of the second car is 65 mph.

101. Let t = time Macrina jogs
then $t - \dfrac{1}{2}$ = time Gordon jogs

$$4t = 6(t - \frac{1}{2})$$
$$4t = 6t - 3$$
$$-2t = -3$$
$$t = \frac{-3}{-2} = \frac{3}{2}$$

It takes Gordon $1.5 - .5 = 1$ hour to catch Macrina.

103. Let r = Kaitlin's speed
and $r + 3$ = Kent's speed

Kaitlin's time = Kent's time
$$\frac{27}{r} = \frac{36}{r + 3}$$
$$27(r + 3) = 36r$$
$$27r + 81 = 36r$$
$$81 = 9r$$
$$9 = r$$

Kaitlin's rate is 9 mph and Kent's rate is $9 + 3 = 12$ mph.

105. Let r = Debbie's rate on return trip
and $r + 4$ = Debbie's rate going out

362

$$\underset{\substack{\text{Debbie's}\\\text{time out}}}{\underbrace{}} - \tfrac{1}{2}\text{hour} = \underset{\substack{\text{Debbie's}\\\text{return trip}}}{\underbrace{}}$$

$$\frac{24}{r+4} - \frac{1}{2} = \frac{12}{r}$$

$$2r(r+4)\left[\frac{24}{r+4} - \frac{1}{2}\right] = 2r(r+4)\left(\frac{12}{r}\right)$$

$$48r - r(r+4) = 24(r+4)$$

$$48r - r^2 - 4r = 24r + 96$$

$$-r^2 + 44r = 24r + 96$$

$$-r^2 + 44r - 24r - 96 = 0$$

$$r^2 - 20r + 96 = 0$$

$$(r-8)(r-12) = 0$$

$$r - 8 = 0 \quad \text{or} \quad r - 12 = 0$$

$$r = 8 \qquad \text{or} \quad r = 12$$

Case 1: Return rate = 8 mph and
Rate going out = $8 + 4 = 12$ mph.

Case 2: Return rate = 12 mph and
Rate going out = $12 + 4 = 16$ mph.

107. Let x = the rate for the first part of the trip
then $x + 5$ = the rate for the last part

$$\underset{\substack{\text{time for the}\\\text{first part}}}{\underbrace{}} + \underset{\substack{\text{time for the}\\\text{second part}}}{\underbrace{}} = 10$$

$$\frac{330}{x} + \frac{240}{x+5} = 10$$

$$x(x+5)\left(\frac{330}{x} + \frac{240}{x+5}\right) = x(x+5)(10)$$

$$330(x+5) + 240x = 10x(x+5)$$

$$330x + 1650 + 240x = 10x^2 + 50x$$

$$570x + 1650 = 10x^2 + 50x$$

$$10x^2 - 520x - 1650 = 0$$

$$10(x^2 - 52x - 165) = 0$$

$$x^2 - 52x - 165 = 0$$

$$(x-55)(x+3) = 0$$

$$x - 55 = 0 \quad \text{or} \quad x + 3 = 0$$

$$x = 55 \qquad \text{or} \quad x = -3$$

Discard the negative solution.

The rate for the first part of the trip
is 55 mph. The rate for the second part
is $55 + 5 = 60$ mph.

109. Let r = Lorraine's rate
and $r + 5$ = Charlotte's rate

$$\underset{\substack{\text{Time for}\\\text{Charlotte}}}{\underbrace{}} - 1 = \underset{\substack{\text{Time for}\\\text{Lorraine}}}{\underbrace{}}$$

$$\frac{250}{r+5} - 1 = \frac{180}{r}$$

$$r(r+5)\left[\frac{250}{r+5} - 1\right] = r(r+5)\left[\frac{180}{r}\right]$$

$$250r - r(r+5) = 180(r+5)$$

$$250r - r^2 - 5r = 180r + 900$$

$$-r^2 + 65r - 900 = 0$$

$$r^2 - 65r + 900 = 0$$

$$(r-20)(r-45) = 0$$

$$r - 20 = 0 \quad \text{or} \quad r - 45 = 0$$

$$r = 20 \qquad \text{or} \quad r = 45$$

Case 1: Lorraine's rate = 20 mph and
Charlotte's rate = $20 + 5 = 25$ mph
Case 2: Lorraine's rate = 45 mph and
Charlotte's rate = $45 + 5 = 50$ mph

111. Let x = cups of grapefruit juice
to be added

$$\underset{\substack{\text{Pure}\\\text{juice}}}{\underbrace{}} + \underset{\substack{\text{Juice in}\\\text{5\% punch}}}{\underbrace{}} = \underset{\substack{\text{Juice in}\\\text{10\% punch}}}{\underbrace{}}$$

$$x + 0.05(40) = 0.10(x+40)$$

$$100\left[x + 0.05(40)\right] = 100\left[0.10(x+40)\right]$$

$$100x + 5(40) = 10(x+40)$$

$$100x + 200 = 10x + 400$$

$$100x - 200 = 10x$$

$$-200 = -90x$$

$$\frac{20}{9} = \frac{-200}{-90} = x$$

$2\frac{2}{9}$ cups of grapefruit juice should be added.

113. Let x = quarts of water removed

$$\underset{\substack{\text{Salt in 20\% solution}\\\text{before water removed}}}{\underbrace{}} = \underset{\substack{\text{Salt in}\\\text{30\% solution}}}{\underbrace{}}$$

Appendix B

$$.20(20) = 0.30(20 - x)$$
$$4 = 0.30(20 - x)$$
$$100(4) = 100[0.30(20 - x)]$$
$$400 = 30(20 - x)$$
$$400 = 600 - 30x$$
$$-200 = -30x$$
$$\frac{20}{3} = \frac{-200}{-30} = x$$

Six and $\frac{2}{3}$ quarts of water
should be added.

115. Let x = percent of the mixed solution

$$\underset{\substack{\text{Salt in} \\ \text{30\% solution}}}{} + \underset{\substack{\text{Salt in} \\ \text{50\% solution}}}{} = \underset{\substack{\text{Salt in} \\ \text{mixture}}}{}$$

$$0.30(10) + 0.50(20) = x(10 + 20)$$
$$100[0.30(10) + 0.50(20)] = 100[x(10 + 20)]$$
$$30(10) + 50(20) = 100x(30)$$
$$300 + 1000 = 3000x$$
$$1300 = 3000x$$
$$\frac{1300}{3000} = x$$
$$0.43\overline{3} = x$$
$$x \approx 43.3\%$$

The final mixture would be
approximately 43.3%.

117. Let x = quarts of of antifreeze to
be drained.

$$\underset{\substack{\text{Antifreeze in} \\ \text{50\% solution}}}{} + \underset{\substack{\text{Pure} \\ \text{antifreeze}}}{} = \underset{\substack{\text{Antifreeze in} \\ \text{60\% solution}}}{}$$

$$0.50(16 - x) + x = 0.60(16)$$
$$100[0.50(16 - x) + x] = 100[0.60(16)]$$
$$50(16 - x) + 100x = 60(16)$$
$$800 - 50x + 100x = 960$$
$$800 + 50x = 960$$
$$50x = 160$$
$$x = \frac{160}{50} = 3.2$$

The amount to be drained is 3.2 quarts.

119. Let x = time for Sean to do the job
working alone

$$\underset{\substack{\text{Ramon's} \\ \text{rate}}}{} + \underset{\substack{\text{Sean's} \\ \text{rate}}}{} = \underset{\substack{\text{Rate working} \\ \text{together}}}{}$$

$$\frac{1}{25} + \frac{1}{x} = \frac{1}{15}$$
$$75x\left(\frac{1}{25} + \frac{1}{x}\right) = 75x\left(\frac{1}{15}\right)$$
$$75x\left(\frac{1}{25}\right) + 75x\left(\frac{1}{x}\right) = 75x\left(\frac{1}{15}\right)$$
$$3x + 75 = 5x$$
$$75 = 2x$$
$$x = \frac{75}{2} = 37\frac{1}{2}$$

It would take Sean $37\frac{1}{2}$ minutes to do
the job working alone.

121. Let x = time for tank to overflow

$$\underset{\substack{\text{Rate} \\ \text{Inlet}}}{} - \underset{\substack{\text{Rate} \\ \text{Draining}}}{} = \underset{\substack{\text{Rate} \\ \text{Filling}}}{}$$

$$\frac{1}{10} - \frac{1}{12} = \frac{1}{x}$$
$$120x\left(\frac{1}{10} - \frac{1}{12}\right) = 120x\left(\frac{1}{x}\right)$$
$$120x\left(\frac{1}{10}\right) - 120x\left(\frac{1}{12}\right) = 120$$
$$12x - 10x = 120$$
$$2x = 120$$
$$x = 60$$

It would be 60 minutes before
the tank overflows.

123. Let x = time it takes to mow the lawn
with the push mower
$x - 40$ = time it takes to mow the lawn
with the power mower

$$\underset{\substack{\text{Time x Rate} \\ \text{for power mower}}}{} + \underset{\substack{\text{Time x Rate} \\ \text{for push mower}}}{} = \underset{\substack{\text{1job} \\ \text{done}}}{}$$

$$30\left(\frac{1}{x-40}\right)+(20)\left(\frac{1}{x}\right)=1$$

$$\frac{30}{x-40}+\frac{20}{x}=1$$

$$x(x-40)\left[\frac{30}{x-40}+\frac{20}{x}\right]=x(x\text{-}40)1$$

$$x(x-40)\left(\frac{30}{x-40}\right)+x(x-40)\left(\frac{20}{x}\right)=x^2-40x$$

$$30x+20(x-40)=x^2-40x$$

$$30x+20x-800=x^2-40x$$

$$50x-800=x^2-40x$$

$$-x^2+90x-800=0$$

$$x^2-90x+800=0$$

$$(x-80)(x-10)=0$$

$$x-80=0 \quad\text{or}\quad x-10=0$$

$$x=80 \quad\text{or}\quad x=10$$

Discard the solution $x=10$ because a timecould not be -30 minutes. So it takes Arlene $80-40=40$ minutes to mow with the power mower.

125. amount x value = total value

	amount	value	total
nickels	x	0.05	$.05x$
dimes	$30-x$	0.10	$.10(30-x)$

$$0.05x+0.10(30-x)=2.30$$

$$100\left[0.05x+0.10(30-x)\right]=100(2.30)$$

$$5x+10(30-x)=230$$

$$5x+300-10x=230$$

$$-5x+300=230$$

$$-5x=-70$$

$$x=14$$

Carson has 14 nickels and $30-14=16$ dimes.

127. amount x value = total value

	amount	value	total
nickels	$\frac{1}{4}x$	0.05	$.05\left(\frac{1}{4}x\right)$
dimes	$\frac{1}{2}x$	0.10	$.10\left(\frac{1}{2}x\right)$
quarters	x	0.25	$0.25x$

$$0.05\left(\frac{1}{4}x\right)+0.10\left(\frac{1}{2}x\right)+0.25x=12.50$$

$$100\left[0.05\left(\frac{1}{4}x\right)+0.10\left(\frac{1}{2}x\right)+0.25x\right]=100(12.50)$$

$$5\left(\frac{1}{4}x\right)+10\left(\frac{1}{2}x\right)+25x=1250$$

$$\frac{5}{4}x+5x+25x=1250$$

$$\frac{5}{4}x+30x=1250$$

$$\frac{5}{4}x+\frac{120}{4}x=1250$$

$$\frac{125}{4}x=1250$$

$$\frac{4}{125}\left(\frac{125}{4}x\right)=\frac{4}{125}(1250)$$

$$x=40$$

Amanda has 40 quarters, $\frac{1}{4}\left(40\right)=10$ nickels, and $\frac{1}{2}\left(40\right)=20$ dimes.

129. amount x value = total value

	amount	value	total
dimes	x	0.10	$0.10x$
quarters	$23-x$	0.25	$.25(23-x)$

$$0.10x+0.25(23-x)=4.55$$

$$100\left[0.10x+0.25(23-x)\right]=100(4.55)$$

$$10x+25(23-x)=455$$

$$10x+575-25x=455$$

$$-15x+575=455$$

$$-15x=-120$$

$$x=8$$

Mona has 8 dimes and $23-8=15$ quarters.

131. Let $x=$ earning per hour

$$7x=66.50$$

$$x=\frac{66.50}{7}=9.50$$

Cheryl earned $9.50 per hour.

Appendix B

133. Let x = price of labor per hour
Parts + 2(cost of labor per hour) = 75

$$15 + 2x = 75$$
$$2x = 60$$
$$x = 30$$
Labor costs $30 per hour.

135. Let p = price of the calculator
$3p + 5$ = price of the math book

$$p + 3p + 5 = 85$$
$$4p + 5 = 85$$
$$4p = 80$$
$$p = 20$$
The price of the mathematics book
is $3(20) + 5 = \$65$.

137. Let b = number of boys
$b + 40$ = number of girls

$$b + b + 40 = 680$$
$$2b + 40 = 680$$
$$2b = 640$$
$$b = 320$$
The number of girls is $320 + 40 = 360$.

139. Let h = normal hourly rate

$$40h + 3(2h) = 552$$
$$40h + 6h = 552$$
$$46h = 552$$
$$h = 12$$
Michael's normal hourly rate is $12.

141. Let n = number of rows
$3n - 2$ = number of trees per row

$$\frac{\text{number}}{\text{of rows}} + \frac{\text{trees}}{\text{per row}} = \frac{\text{number}}{\text{or trees}}$$

$$n(3n - 2) = 65$$
$$3n^2 - 2n = 65$$
$$3n^2 - 2n - 65 = 0$$
$$(3n + 13)(n - 5) = 0$$

$3n + 13 = 0$ or $n - 5 = 0$
$3n = -13$ or $n = 5$
$$n = -\frac{13}{3}$$
Discard the negative solution.
There are 5 rows and $3(5) - 2 = 13$ trees
per row.

143. Let x = price per pound of Gala apples
y = price per pound of Fuji apples

$$2x + 3y = 6.55$$
$$x + 4y = 5.75$$

Multiply the second equation by -2,
then add the equations.
$$2x + 3y = 6.55$$
$$\underline{-2x - 8y = -11.50}$$
$$-5y = -4.95$$
$$y = 0.99$$

Then substitute $y = 0.99$ into the second
equation and solve for x.
$$x + 4y = 5.75$$
$$x + 4(0.99) = 5.75$$
$$x + 3.96 = 5.75$$
$$x = 1.79$$

The Gala apples cost $1.79 per pound and
the Fuji apples cost $0.99 per pound.

145. Let x = cost of corn flakes
y = cost of wheat flakes

$$2x + 3y = 13.25$$
$$3x + 2y = 12.65$$

Multiply the first equation by -3 and
multiply the second equation by 2, and
then add the equations.
$$-6x - 9y = -39.75$$
$$\underline{6x + 4y = 25.30}$$
$$-5y = -14.45$$
$$y = 2.89$$

Substitute $y = 2.89$ into the first equation
and solve for x.

$$2x + 3y = 13.25$$
$$2x + 3(2.89) = 13.25$$
$$2x + 8.67 = 13.25$$
$$2x = 4.58$$
$$x = 2.29$$

The corn flakes cost $2.29 a box and the wheat flakes cost $2.89.

147. Let x = number of students involved in the party

$x - 5$ = original number of students involved

$\dfrac{100}{x}$ = cost for each student

$\dfrac{100}{x - 5}$ = original cost for each student

$$\underset{\text{cost}}{\text{Original}} \times \underset{\text{time}}{\text{Actual}} = 360$$

$$\frac{100}{x - 5} - 1 = \frac{100}{x}$$

$$x(x-5)\left[\frac{100}{x-5} - 1\right] = x(x-5)\left(\frac{100}{x}\right)$$

$$x(x-5)\left(\frac{100}{x-5}\right) - x(x-5)(1) = 100(x-5)$$

$$100x - x^2 + 5x = 100x - 500$$
$$105x - x^2 = 100x - 500$$
$$-x^2 - 5x + 500 = 0$$
$$x^2 + 5x - 500 = 0$$
$$(x + 20)(x - 25) = 0$$
$$x + 20 = 0 \quad \text{or} \quad x - 25 = 0$$
$$x = -20 \quad \text{or} \quad x = 25$$

Discard the negative solution.

There were 25 students and each paid $\dfrac{100}{25} = \$4$.

149. Let x = number of mugs bought

$x - 2$ = number of mugs sold

$\dfrac{48}{x}$ = original sellling price

$\dfrac{48}{x} + 3$ = new selling price

$$\underset{\text{sold}}{\text{Number}} \times \underset{\text{price}}{\text{Selling}} = \text{cost} + \text{profit}$$

$$(x - 2)\left(\frac{48}{x} + 3\right) = 48 + 22$$

$$48 + 3x - \frac{96}{x} - 6 = 70$$

$$x\left(42 + 3x - \frac{96}{x}\right) = x(70)$$

$$42x + 3x^2 - 96 = 70x$$
$$3x^2 - 28x - 96 = 0$$
$$(3x + 8)(x - 12) = 0$$
$$3x + 8 = 0 \quad \text{or} \quad x - 12 = 0$$
$$3x = -8 \quad \text{or} \quad x = 12$$
$$x = -\frac{8}{3}$$

Discard the negative solution.
There were 12 mugs bought and they were sold at $\dfrac{48}{12} + 3 = \$7$ each.